About Island Press

Since 1984, the nonprofit organization Island Press has been stimulating, shaping, and communicating ideas that are essential for solving environmental problems worldwide. With more than 1,000 titles in print and some 30 new releases each year, we are the nation's leading publisher on environmental issues. We identify innovative thinkers and emerging trends in the environmental field. We work with world-renowned experts and authors to develop cross-disciplinary solutions to environmental challenges.

Island Press designs and executes educational campaigns, in conjunction with our authors, to communicate their critical messages in print, in person, and online using the latest technologies, innovative programs, and the media. Our goal is to reach targeted audiences—scientists, policy makers, environmental advocates, urban planners, the media, and concerned citizens—with information that can be used to create the framework for long-term ecological health and human well-being.

Island Press gratefully acknowledges major support from The Bobolink Foundation, Caldera Foundation, The Curtis and Edith Munson Foundation, The Forrest C. and Frances H. Lattner Foundation, The JPB Foundation, The Kresge Foundation, The Summit Charitable Foundation, Inc., and many other generous organizations and individuals.

Generous support for this publication was provided by Deborah Wiley.

The opinions expressed in this book are those of the author(s) and do not necessarily reflect the views of our supporters.

Water Management

Water Management

Prioritizing Justice and Sustainability

Shimon C. Anisfeld

© 2024 Shimon C. Anisfeld

All rights reserved under International and Pan-American Copyright Conventions. No part of this book may be reproduced in any form or by any means without permission in writing from the publisher: Island Press, 2000 M Street, NW, Suite 480-B, Washington, DC 20036-3319.

Library of Congress Control Number: 2023945708

All Island Press books are printed on environmentally responsible materials.

Manufactured in the United States of America
10 9 8 7 6 5 4 3 2 1

Keywords: dams, drought, floods, global water crisis, groundwater, hydrographs, hydropower, irrigation, sanitation, stocks and flows, surface water depletion, wastewater, water allocation, water–energy nexus, water footprint, water hard path, water infrastructure, water quality and health, water supply, watershed, wetlands

Contents

Preface	xvii
Acknowledgments	xxi
Units and Conversions	xxiii

1 Introduction — 1
  1. The Great Acceleration and the Just Transition — 1
  2. The Centrality of Water — 4
    2.1. Water and the Human Spirit — 4
    2.2. Water as a Human Right — 4
    2.3. Water Use — 5
    2.4. Water as a Resource — 5
    2.5. Water and Global Development — 7
    2.6. The Water–Energy Nexus — 9
    2.7. Water and Ecosystems — 9
  3. The Water Hard Path and the Global Water Crisis — 10
  4. A Water Ethic: Justice and Sustainability — 12
    4.1. Justice — 13
    4.2. Sustainability — 13
  5. How We Get There: Infrastructure, Institutions, Incentives, and Information — 15

Part I. Water Availability and Use: Supply and Demand, Scarcity and Change — 19

2 Water Availability: Spatial and Temporal Variability — 21
  1. Global Stocks and Flows — 21
  2. Spatial Variation in Water Availability — 24

3. Temporal Variation in Water Availability 27
   - *3.1. Hydrographs* 27
   - *3.2. Flow Duration Curves* 28
   - *3.3. Flood Frequency Analysis* 29
   - *3.4. Rainfall Intensity–Duration–Frequency Curves* 30

4. Drought 32
   - *4.1. Defining Drought* 32
   - *4.2. Causes of Drought* 33
   - *4.3. The Great Drought and the Dust Bowl: The Role of Management* 36
   - *4.4. Paleoclimatology and Megadroughts* 38

3 Global and Local Change: The End of Stationarity 41

1. Climate Change 41
   - *1.1. Theory and Prediction* 42
   - *1.2. Observation* 43
   - *1.3. Indirect Effects* 45

2. Land-Use Change 46
   - *2.1. Deforestation: Watershed Effects* 46
   - *2.2. Deforestation: Impacts on Precipitation* 48
   - *2.3. Managing Forests for Water Supply* 50
   - *2.4. Desertification* 52
   - *2.5. Urbanization* 54

3. Managing Change 54

4 Water Use: From Ancient to Modern Times 58

1. Premodern Agriculture 58
   - *1.1. Large-River Irrigation* 59
   - *1.2. Small-River Irrigation* 60
   - *1.3. Runoff Farming* 61
   - *1.4. Groundwater Use* 62
   - *1.5. Wetlands and Springs* 62
   - *1.6. Water Storage* 63

2. Premodern Urban Water Supply 64

3. Modern Water Management 65
   - *3.1. The Hard Path* 65
   - *3.2. US Water Management: From Hard to Soft* 67
   - *3.3. Water in the Global Development Agenda* 70

4. Water Use in the Modern Era: A Quantitative Analysis 71
   - *4.1. Definitions* 71
   - *4.2. Water Footprints* 72
   - *4.3. Global and US Water Use* 74

5 Water Scarcity and Depletion: Are We Reaching the Limits of Our Supply? — 79

  1. Scarcity Indicators — 79
     - *1.1. Falkenmark and WTA Indicators* — *79*
     - *1.2. Incorporating Environmental Flows* — *80*
     - *1.3. Utility Shortage* — *81*
     - *1.4. Economic Water Scarcity* — *81*
  2. Assessing Water Scarcity — 82
  3. Surface Water Depletion — 85
     - *3.1. Shrinking Lakes* — *85*
     - *3.2. River Depletion and Altered Flow Regimes* — *86*
  4. Groundwater Use and Overuse — 88
     - *4.1. Groundwater Pumping* — *89*
     - *4.2. Extent of Groundwater Depletion* — *92*
     - *4.3. Groundwater Depletion in California's Central Valley* — *93*
  5. Managing Water Scarcity — 94

Part II. Instream Water Management: Rivers and Dams, Flooding and Hydropower — 99

6 Instream Uses: Navigation, Hydropower, Fishing, and Recreation — 101

  1. Navigation — 101
  2. Water Power — 106
     - *2.1. History of Water Power* — *107*
     - *2.2. Hydroelectric Power* — *107*
     - *2.3. Hydropower: Renewable Energy?* — *111*
     - *2.4. Sustainable Hydropower* — *112*
  3. Fishing — 113
  4. Recreation — 116

7 Flood Management: Learning to Live with "Too Much" Water — 120

  1. Types of Floods — 120
  2. Flooding Impacts and Trends — 123
  3. Flood Hazard — 124
  4. Flood Exposure — 126
  5. Flood Vulnerability — 127
     - *5.1. Structural Flood Control* — *127*
     - *5.2. The Levee Effect: The Illusion of Dryness* — *130*
     - *5.3. Benefit–Cost Analysis* — *131*
  6. The Way Forward — 133
     - *6.1. Building in Floodplains: The Problem* — *133*

6.2. The NFIP: The Solution? ... *135*
6.3. Buyouts and Managed Retreat ... *136*

8 Water Quality and the Clean Water Act ... 141

1. Water Quality and Health ... 142
2. The Clean Water Act ... 144
   2.1. Addressing Chemical Pressures ... *145*
   2.2. Defining and Meeting Water Quality Goals ... *148*
   2.3. Incorporating Nonchemical Stressors ... *151*
   2.4. Effectiveness of the Clean Water Act ... *153*
3. Other US Environmental Laws ... 155

9 Dams and Their Discontents: Are the Benefits Worth the Costs? ... 159

1. Dam Basics ... 159
   1.1. Types of Dams ... *159*
   1.2. Dam Components and Terminology ... *160*
   1.3. Managing Reservoirs for Multiple Purposes ... *161*
   1.4. Reservoir Storage–Yield Relationships ... *163*
2. The Dam Debate: The Changing Attitudes of the Past Century ... 164
   2.1. The Dam-Building Era(s) ... *164*
   2.2. From Zeal to Skepticism in the Twentieth Century ... *165*
   2.3. The World Commission on Dams and Its Aftermath ... *165*
3. Snapshot of US and Global Dams ... 166
   3.1. How Many Dams Are There? ... *167*
   3.2. What Are Dams Used For, and Who Manages Them? ... *167*
   3.3. How Has Storage Volume Changed over Time? ... *168*
4. Ecological Impacts of Dams ... 170
5. Social Impacts of Dams ... 171
   5.1. Population Displacement ... *171*
   5.2. Dam Safety ... *172*
6. Building the Right Dams and Managing Them the Right Way ... 174
   6.1. Building Better Dams ... *174*
   6.2. Managing Dams Better ... *176*
7. Dam Removal ... 176
8. Aqueducts ... 178

Part III. Water Governance: Allocation and Reallocation, Cooperation and Conflict ... 181

10 Water Allocation: Sharing the Common Pool ... 183

1. Water as a Common-Pool Resource ... 184
   1.1. Economic Efficiency in Water Allocation ... *184*

	1.2. CPRs and the Tragedy of the Commons	*186*
	1.3. Ostrom and Community Management	*187*
2.	Water Allocation in the United States	190
	2.1. The Riparian Doctrine	*190*
	2.2. Regulated Riparianism	*191*
	2.3. Prior Appropriation	*191*
	2.4. Groundwater Allocation	*194*
	2.5. Reserved Rights	*195*
	2.6. Interstate Water Allocation	*196*

11 Reallocation and Coordination for Improved Water Governance — 200

1. Reallocation for Cities, Tribes, and the Environment — 200
   - *1.1. Water Markets* — *201*
   - *1.2. Negotiating Ag-to-Urban Reallocation* — *202*
   - *1.3. Environmental Flow Requirements* — *203*
   - *1.4. Legal Tools for EFRs* — *205*
   - *1.5. WaterBack: Decolonizing Water Management* — *207*
2. Coordination and Planning — 210
   - *2.1. State Coordination and Planning* — *210*
   - *2.2. Interstate Coordination and Planning* — *213*

12 Transboundary Water Management: Conflict and Cooperation — 217

1. Introduction to Water Conflict — 218
2. International Basins and Aquifers — 219
   - *2.1. The Extent of Transboundary Waters* — *220*
   - *2.2. International Law of Transboundary Waters* — *222*
   - *2.3. Water Wars or Water Peace?* — *223*
   - *2.4. Drivers of Water Conflict* — *224*
   - *2.5. Building Institutional Resilience* — *226*
3. Nonstate Water Conflicts — 228

Part IV. Offstream Water Use: Cities and Farms, Mines and Factories — 233

13 Beyond Dams: Old and New Solutions for Water Supply — 235

1. Desalination — 236
2. Wastewater Reuse — 239
3. Water Harvesting — 243
4. Aquifer Storage — 246
5. Energy for Water Supply — 247
6. Summary — 249

14 Drinking Water, Sanitation, and Health: Water 3.0 — 253

1. Water and Infectious Disease — 254
2. History of Urban Water and Sanitation — 255
   - 2.1. *Ancient and Medieval Sanitation and Drainage* — 255
   - 2.2. *Epidemics and the Sanitary Revolution* — 256
   - 2.3. *The American Experience* — 258
3. Water 3.0: Drinking Water and Sanitation in the United States — 258
   - 3.1. *Defining Safe Drinking Water* — 258
   - 3.2. *Source Protection* — 260
   - 3.3. *Water Treatment* — 260
   - 3.4. *The Distribution System* — 261
   - 3.5. *Urban Sanitation* — 262
   - 3.6. *Combined Sewers* — 262
   - 3.7. *Rural Sanitation* — 264
4. Evaluating Water 3.0 — 265
   - 4.1. *Health Effects* — 265
   - 4.2. *Disparities in Access* — 265
   - 4.3. *Utility Structure* — 267
   - 4.4. *The Rise of Bottled Water* — 270
   - 4.5. *The Problem of Linear Flows* — 270

15 Urban Water Management: Water Conservation, Stormwater Management, and Beyond — 274

1. Defining and Achieving Reliability — 275
2. Conservation — 277
   - 2.1. *How Much Do We Use?* — 277
   - 2.2. *Tools for Conservation* — 279
3. Pricing — 280
   - 3.1. *Cost Recovery* — 280
   - 3.2. *Conservation* — 281
   - 3.3. *Affordability* — 283
   - 3.4. *Rate Structures* — 284
4. Reuse and Decentralization — 285
5. Urban Stormwater Management — 288
   - 5.1. *Stormwater and the Urban Stream Syndrome* — 288
   - 5.2. *Urban Stream Restoration* — 288
   - 5.3. *Legal and Financial Tools for Stormwater Management* — 291

16 Water, Sanitation, and Health in Low- and Middle-Income Countries: Leaving No One Behind — 295

1. Water and Sustainable Development — 295
   - 1.1. *The Burdens of Inadequate WASH* — 295
   - 1.2. *WASH Goals* — 297

2. Drinking Water 298
   2.1. From Goals to Indicators *298*
   2.2. Where We Stand *299*
   2.3. Affordability *301*

3. Sanitation 301
   3.1. Sanitation Solutions *302*
   3.2. From Goals to Indicators *303*
   3.3. Where We Stand *304*

4. Hygiene 307

5. The Way Forward 308

6. WASH Funding 310
   6.1. Tariffs *311*
   6.2. Taxes *311*
   6.3. Transfers *312*

17 Industrial Water Use: Our Invisible Water Footprints — 316

1. Water for Energy 316
   1.1. Water for Electricity *316*
   1.2. Water for Transportation *318*
   1.3. Pollution from Energy Production *319*

2. Water for Manufacturing 323
   2.1. Direct Water Use *323*
   2.2. Upstream and Downstream Water Use *324*
   2.3. Pollution from Manufacturing *326*

3. Water and Mining 330

4. Corporate Water Stewardship 332

18 Agricultural Water Use: Water's Central Role in Our Food Supply — 337

1. From the Green Revolution to Sustainable Intensification ... 337

2. Water and Agriculture: Overview 342
   2.1. Water Use in Global Agriculture *342*
   2.2. How Much Agricultural Water Do We Need? *343*

3. What We Eat 345
   3.1. Water Footprints of Various Foods *345*
   3.2. Animal Products: A Closer Look *347*
   3.3. Sustainability of Blue Water Use *349*

4. Where We Grow It 351

5. How We Use Water 352
   5.1. Irrigation *353*
   5.2. Rainfed Agriculture *355*

6. Pollution 356
   6.1. Salinization *356*

6.2. Nutrient Pollution from Croplands	*357*
6.3. Pollution from Animal Feeding Operations	*359*
Abbreviations	365
Glossary	369
About the Author	387
References	389
Index	413

*To Sharon,
for teaching me
about things that flow
and things that don't change.*

Preface

The summer of 2023 brought the global water crisis into the media spotlight once again, with headlines reflecting the range of devastating water problems being experienced around the world: Flooding in California. Drought and famine in the Horn of Africa. Massive fish kills in Texas and Australia. "Forever chemicals" in US drinking water. Escalating tensions between Egypt and Ethiopia over the latter's new dam.

Similar headlines are sure to dominate the news in the years ahead.

What is sometimes missing from the headlines, though, is an understanding that these diverse water problems are related, that these water "hotspots" are manifestations of serious underlying stresses on our interconnected social–physical water systems. These stresses require sustained attention from water managers, scientists, policymakers, and the public, even after the headlines have faded. That attention, in turn, requires a shared understanding of how water systems function, the stresses they are experiencing, and the tools available to increase their resilience.

This book aims to fill that need by providing the necessary knowledge base for understanding and managing water problems. It is geared primarily toward students in water management courses at the undergraduate and graduate levels but will also be a helpful resource for practicing water professionals who want to get new ideas or a broader view of water management. In addition, I hope that this book will prove useful for laypeople who want to understand this vital resource and their own role in protecting it. After all, water touches every aspect of our society, so every one of us is affected by water issues—and we each make decisions every day that affect our collective water future. The book is written primarily for a US audience, although case studies from around the world will make it useful for students from other countries as well.

Rather than focusing on one type of water problem (as many water books do), this book provides reasonably comprehensive coverage of the entire gamut of water issues, from dams to desalination, from flooding to famine, from prior appropriation to pumped storage, from sanitation to stormwater. And rather than teaching from one disciplinary perspective (as many water books do), this book looks at water problems through a variety of lenses: hydrology, climate science, ecology, and engineering, but also law, economics, history, and environmental justice.

Anyone writing a water management textbook (or teaching a water management course) faces two interconnected dilemmas. First, given the largely local nature of water problems—and the diversity of place-based solutions—how do I balance general principles with local applications? Second, given the rapidly changing nature of the water space, how do I provide an accurate, textured description of where we stand today, while also recognizing that things will look different tomorrow? I don't have a perfect solution to these dilemmas, of course, but I have tried to ensure that readers—whether they are in New Haven or New Delhi, whether they are reading in 2024 or 2034—will find in this book both principles and examples that they can apply to their own settings. In addition, the companion website (https://watermanagement.yale.edu) is regularly updated with the latest global data, along with the latest developments in the various case studies.

This book is designed to be accessible and engaging to readers from a broad range of backgrounds. Several features help make the book readable and student centered without losing rigor and comprehensiveness:

- Modularity: For readers who are interested in particular issues and don't have time to read the whole book, the book is designed to be modular. Once you have read the Introduction (Chapter 1) and Part I, each of the remaining parts—Part II (river and ecosystem management), Part III (water governance), and Part IV (urban, industrial, and agricultural uses)—can be read independently, depending on your interests. Similarly, within Part IV, readers who are interested in industrial or agricultural issues can turn directly to Chapters 17 or 18, respectively, while readers interested in drinking water issues in the United States can turn to Chapter 14. Fuller descriptions of reading suggestions for different interests, and syllabi for different types of courses, are available online at the companion website.
- Glossary: A detailed glossary is included, and every phrase in the glossary is presented in ***bold italics*** the first time it appears in each chapter. Thus, readers who haven't read (or have forgotten!) previous chapters have ready access to definitions of key concepts.
- Chapter highlights: Each chapter ends with bullet points that summarize the ideas introduced in the chapter, with critical concepts highlighted in ***bold italics*** (and appearing in the glossary).
- Case studies: Since water issues are largely local in nature, I use case studies (in both the text and the companion website) to illustrate, concretize, and enliven the concepts covered in the book.
- Water science primer: The book is focused on water *management* rather than water science and thus assumes some familiarity with the hydrologic cycle, water quality, and related scientific concepts. For those who lack this background, the companion website offers an "introduction to water science," which will quickly get you up to speed.

After an introductory chapter, the book is divided into four parts. Part I provides an overview of water as a resource for human use, addressing basic questions about *supply* (how much water is available, including how availability is affected by climate change), *demand* (how much water we are using), and *scarcity* (the gap between supply and demand). Part II looks at how rivers are managed for *instream uses* (navigation, hydropower, fishing, recreation, flood management, waste disposal, dams), the ecological and social consequences of those uses, and how these consequences could be mitigated. Part III examines water governance, including issues of water allocation, environmental flows, tribal water rights, planning and coordination, and transboundary conflict and cooperation. Part IV delves into offstream water uses—municipal, industrial, and agricultural—with the goal of describing how those uses could better serve the people who are reliant on them while minimizing impacts on other communities and ecosystems.

Acknowledgments

This book project evolved over several years and benefited significantly from many people's input along the way. I am grateful to them for their contributions to this book and for greatly enriching my own understanding of water issues.

Several people provided thoughtful feedback on various sections of the book: Mark Ashton, Chris Conover, Whitman Constantineau, Marissa Grenon, Dawn Henning, David Katz, Jacques Leslie, Jonathan Rak, Sara Schwartz, and Kevin Zak. Julia Sullivan provided helpful, detailed comments on the entire manuscript.

A number of people generously shared data and images: Mary Becker, Justin Brown, Chris Conover, Whitman Constantineau, Bethan Davies, Rick Dove, Ellen Hanak, Dawn Henning, Brenden Jongman, David Katz, Landon Marston, Caitlin Martin, Gokce Sencan, Lan Wang-Erlandsson, and Piotr Wolski. Space considerations prevented me from using all these materials in the book itself, but several of them are featured on the companion website. Anna Yue Yu made maps for this project, and Maureen Gately created several original images; their creativity and competence are impressive and much appreciated.

This book reflects what I have learned over the years from valued colleagues at Yale and elsewhere. In addition to those mentioned above, I am especially grateful for conversations with Gabe Benoit, Gidon Bromberg, Edgar Hertwich, John Hudak, Debbie Humphries, Arturo Llaxacóndor, Gary McKinney, Sheila Olmstead, Jordan Peccia, Jennifer Pitt, Zyg Plater, Ivan Rodriguez Cabanillas, Jim Saiers, Roy Schiff, Alon Tal, and Laura Wildman.

A first-century rabbi, Rabbi Chanina, is said to have remarked, "I have learned much from my teachers, and more from my colleagues, but most of all from my students." I can relate. Over the last two and a half decades, my students at the Yale School of the Environment have inspired me with their passion, challenged me with their thoughtful questions, and helped me learn from their diverse experiences and perspectives. I cherish the opportunity to teach water management to such an intelligent and committed group of people, and their influence is felt throughout the book.

Indy Burke, dean of the Yale School of the Environment, has supported me and this book wholeheartedly with helpful advice and timely encouragement and has nurtured a learning community that has helped me grow and contribute. I am grateful and proud to be part of an institution with such an important mission, distinguished history, and bright future.

Emily Turner of Island Press exhibited remarkable patience as I missed one deadline after another, and outstanding creativity, kindness, and good judgment as we worked to shape the final product. It has been a pleasure to work with her and the rest of the Island Press team.

One of the unexpected joys of this project has been connecting with my brother-in-law, Tim Lytton, around our shared interest in regulation and risk; Tim's feedback was invariably insightful, helpful, and kind.

My mother, Elizabeth Anisfeld, provided characteristically detailed and loving edits on each chapter—and gave me the confidence that I have something worth saying. My father, Moshe Anisfeld, did not live to see this book, but his presence as my role model for teaching and writing is felt on every page.

My family—Sharon, Daniel, and Tali—provided helpful feedback in multiple ways (including, in Daniel's case, reading every chapter!) and supported me through every day of work, play, and rest. They make life worth living, love worth sharing, and writing worth laboring over. They are my rock, the foundation from which I interact with the world. Or, better, they are my shipmates—my guides, companions, and crew—as we navigate together through sometimes turbulent waters toward a better world that is just over the horizon.

Units and Conversions

Abbreviations and Symbols

AF	acre-foot
BCM	billion cubic meters
cfs	cubic feet per second, ft^3/sec
Δ	difference
g	gram
ha	hectare
hr	hour
J	joule
L	liter
Lpcd	liters per capita per day
m	meter
MAF	million acre-feet
MCM	million cubic meters
MGD	million gallons per day
Σ	summation
sec	second
t	metric ton (1,000 kg)
TAF	thousand acre-feet
W	watt
Wh	watt-hour
yr	year

SI Prefixes

Note: These prefixes are used in front of units to indicate orders of magnitude. For example, nm = nanometer = 10^{-9} meters.

n	nano (10^{-9})
μ	micro (10^{-6})

m	milli (10^{-3})
c	centi (10^{-2})
k	kilo (10^{3})
M	mega (10^{6})
G	giga (10^{9})
T	tera (10^{12})
P	peta (10^{15})
E	exa (10^{18})

Useful Conversions

1 ha = 10,000 m^2 = 0.01 km^2 = 2.47 acres
1 yr = 3.15 × 10^7 sec
1 m^3 = 1,000 L = 264 gallons = 35.3 ft^3 = 1 metric ton (for freshwater)
10^3 m^3 = 10^6 L = 1 ML = 264,000 gallons = 0.811 AF
10^6 m^3 = 10^9 L = 1 GL = 1 MCM = 264 million gallons = 811 AF
10^9 m^3 = 1 km^3 = 1 BCM = 0.811 MAF
1 AF = 325,900 gallons = 1,233 m^3
1 TAF = 1,000 AF = 325.9 MG = 1.233 GL
1 MAF = 10^6 AF = 1.233 km^3
1 MGD = 1.55 cfs = 1.38 MCM/yr = 1.12 kAF/yr
1 Lpcd = 0.365 m^3/person/yr
1 Sv (Sverdrup) = 1 MCM/sec = 31.5 km^3/yr
1 US ton = 2,000 pounds = 907 kg = 0.907 metric tons

1 Introduction

This chapter sets the stage for the rest of the book by introducing some major themes:

- Water plays a central role in both ecosystems and society. All the most important environmental issues of our day—from climate change to the biodiversity crisis, from environmental justice to global development—have a strong water component.
- We are experiencing a multifaceted water crisis with serious impacts on human and ecological health; this crisis is playing out in location-specific ways but also has global aspects.
- The water crisis demands increased attention to justice and sustainability in water management.
- Better water management will require a mix of locally appropriate green and gray infrastructure, as well as increased attention to institutions, incentives, and information.

We start the chapter by placing the water crisis in the broader context of ongoing societal changes, sometimes called the Great Acceleration. We then turn to describing the central role that water plays in society and the way that the water "hard path" has contributed to the Great Acceleration. A brief summary of the global water crisis provides the pivot point of the chapter, leading us to a discussion of water ethics and tools for improved water management.

1. The Great Acceleration and the Just Transition

We are at a critical juncture in the history of humanity and of the planet we inhabit. The human footprint on Earth has grown so rapidly over the last few generations that we are now endangering the planetary systems that have allowed us to thrive. The decisions we make now will affect Earth and all its inhabitants for many millennia.

Let's take a moment to reflect on how we got here.

Many of us take for granted the basic parameters of life in a developed economy in the twenty-first century. Our lives are built around certain assumptions: that our homes will have running water and flush toilets, that antibiotics and surgical care are available when we need them, that we can pursue careers that don't necessarily involve growing food, and that we can quickly communicate and travel across great distances.

All these features of our lives were (to a greater or lesser extent) unavailable to people anywhere in the world a mere 200 years ago. And some of them are still unavailable today to many people around the world, including in the United States.

To provide a bit of historical perspective, Figure 1-1 shows some indicators of how human society has changed over the last 10,000 years. The most notable feature of these graphs is the ***Great Acceleration*** that began with the Industrial Revolution and continues to this day: the exponential growth in human population and impact on the natural world. We have truly entered the ***Anthropocene***: the geologic epoch characterized by the dominance of humanity as the primary force shaping the state of our planet.

With these graphs in mind, allow me to sketch four alternative narratives about where we have come from and where we stand now.

The first narrative celebrates our accomplishments as humans. Beginning with the invention of agriculture, continuing with the development of writing and complex civilizations, and culminating with the Industrial Revolution and the Digital Revolution, our story is one of increasing control over our own fate. Our lives are longer than ever before and are filled with possessions and technologies that previous generations couldn't dream of. We have figured out how to grow food so efficiently that most of us can specialize in a variety of other activities; this specialization has allowed us to improve our standard of living, create beautiful works of art, and explore the frontiers of science. Most fundamentally, our success as a species is told by our increasing numbers—a result of greater food production, improved health care, and lower mortality. This is the story told by the technological optimist, who believes human ingenuity will continue to increase our productivity and solve any problems that emerge.

The second narrative focuses on the uneven distribution of the benefits and costs of our technological age. This story brings to the fore the billions of people who lack access to adequate water, food, and other basic services, even as others spend extravagantly on luxuries. It reminds us that our consumer lifestyles depend on exploitation of people half a world away (or in our own neighborhoods), through invisible networks of mining, environmental degradation, dispossession, forced labor, and toxic waste. It highlights the central role of colonialism in creating and maintaining global inequalities. It questions whether emerging technologies will be used to level the playing field or to bring more power to the powerful. It demands that all our decisions take into account the needs and rights of the most vulnerable.

The third narrative sees an existential threat to humanity itself in our increasing numbers and our global impact. This story recognizes that a finite planet cannot sustain indefinite growth and that our boom must inevitably be followed by a bust—unless we can quickly stabilize, and then reduce, our population and our consumption rates. In this narrative, the unrelenting growth of the human footprint on the planet has endangered our own survival by undermining the life support systems of the planet, as exemplified by climate change, pollution, and habitat loss.

The fourth narrative has a longer time frame and a larger lens. In this story, we try to see the perspectives of plants, animals, and ecosystems. We acknowledge the long

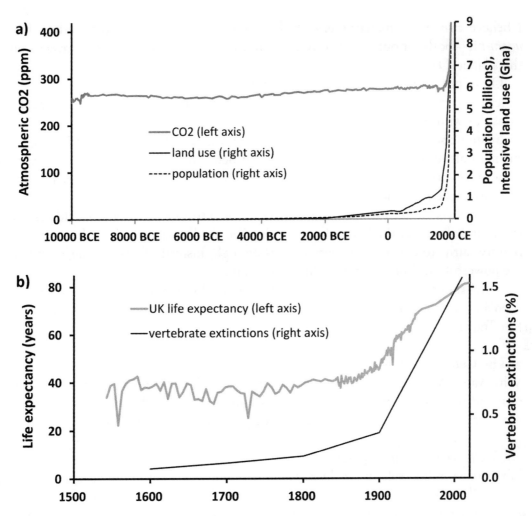

Figure 1-1. Some indicators of change during the Great Acceleration. (a) Atmospheric concentration of carbon dioxide (CO_2), the primary driver of climate change (https://ourworldindata.org/co2-and-other-greenhouse-gas-emissions), area of land in intensive human land use (Ellis et al. 2020), and global human population (https://ourworldindata.org/grapher/population). (b) Life expectancy in the United Kingdom, where good records go back to the sixteenth century (https://ourworldindata.org/life-expectancy), and the percentage of known vertebrate species that have gone extinct (Ceballos et al. 2015; conservative estimate).

history of the globe before *Homo sapiens* first stepped on the scene around 300,000 years ago. We bring to mind the two million species currently known to science, along with the many millions more not yet identified. We recognize that the unprecedented "success" of humanity has come at the expense of other species and at the expense of the planet's complexity, resilience, and beauty. We see ourselves, perhaps, as having forgotten our role as "the younger siblings of Creation,"[1] with much to learn from other species and with reciprocal relations of dependence and responsibility.

The approach I take in this book draws from all four narratives. I believe that humanity's manipulation of the water cycle has brought us many good things (narrative A) but has also contributed significantly to oppression and inequity (B). And

I believe that water management needs to start working *with* nature rather than *against* it—both for our own survival (C) and for the sake of our fellow travelers on this planet (D).

Or, to put it slightly differently, I believe that the central challenge of our age—in water management and beyond—is to bring sustainability and justice into our relationships with the natural world and each other.

Sustainability: We desperately need to reexamine and reclaim our relationship to the natural world and start living in ways that better respect the planet. We need to do what no other species has done: purposely limit ourselves to avoid exceeding the carrying capacity of the planet.

Justice: As we transition to sustainability, we also urgently need to address the power imbalances at the heart of our society. We need to learn how to share nature's bounty fairly, to ensure that everyone has enough, instead of continuing to enrich the powerful. In short, we need a Just Transition.

The issues of sustainability and justice go beyond water, of course. But water has been a central player in the Great Acceleration, and it must be equally central to the Just Transition.

2. The Centrality of Water

Why water? What makes water such an important part of both the Great Acceleration and the Just Transition? This section summarizes the underlying characteristics of water that make its impact on human society so far-reaching.

2.1. Water and the Human Spirit

Water plays a dual role in our lives. On one hand, it is a commonplace, as close as our bodies and as familiar as our kitchens. But it is also the animating force of the planet, the power of nature manifest in thunderstorms, mighty rivers, the vast ocean. In nature, water tends to inspire awe, ecstasy, contemplation. In our homes, water tends to evoke indifference—except when the taps go dry and we become suddenly, humbly aware of our supreme dependence on the thin thread that connects nature's water to ours.

Water is considered sacred in many religious traditions. Water is often viewed as the source of all life, and its ritual use tends to symbolize rebirth, cleansing, and purity. Many religions treat specific water bodies as particularly holy: the Lourdes Spring in Catholicism, the Zamzam Well in Islam, the Ganges River in Hinduism. In indigenous traditions, water is often seen as "an autonomous and primeval element to be encountered with humility, respect, joy and caution."[2]

2.2. Water as a Human Right

Water is, of course, indispensable for all living things on Earth. For people, water is a daily necessity, not just for drinking but also for preparing food and for cleaning ourselves. In addition, access to adequate sanitation is necessary for protecting human health, since inadequate disposal of human wastes leads to water contamination and disease transmission.

In 2010, the United Nations acknowledged these basic needs by passing resolution 64/292, which recognized "safe and clean drinking water and sanitation as a human right that is essential for the full enjoyment of life and all human rights." Still, millions of people around the world (including in the United States) struggle to access sufficient safe water and adequate sanitation solutions, and 1.5 million people die annually from preventable water-related diseases.

2.3. Water Use

Water is used in all aspects of our society and economy: growing food, generating power, making consumer products, transporting goods. Almost any daily activity you can think of has some water requirement, whether obvious or hidden.

In thinking about how we use water, we should note that water can take different forms, with differing availability for human use. Most obviously, water can be found as a solid (ice), liquid, and gas (water vapor); while human water use relies mostly on liquid water, the solid and gaseous forms are also of interest to water managers, whether as storage (seasonal snowpack) or as a potential new supply (water vapor capture). Less obvious but equally important is the distinction between **blue water**—water in the simple liquid or solid state—and **green water**: water in the soil that is adsorbed to soil particles. Green water can't be moved around and used to meet human water demand, so water management focuses primarily on blue water, but green water can be used by plants as a water source, so it is critical for supporting vegetation, including rainfed agriculture (in contrast to irrigated agriculture, which involves adding blue water to fields).

Uses of blue water can be grouped into two categories, which provide much of the structure for this book: **offstream** uses, where water is withdrawn from rivers or other water sources for use in households, farms, industry, or power plants; and **instream** uses, where water provides benefits to humans without being removed from the environment, as when we use a river for navigation, power generation, fishing, or recreation. These different uses are often competing for a limited supply of water, potentially leading to conflict at scales ranging from an individual irrigation ditch to international basins.

Many of the ways we use water—even when they involve offstream use—don't actually *use up* that water. The concept of **nonconsumptive** water use will be covered in more detail in Chapter 4, but for now just imagine the water you use to flush the toilet; it goes down the pipes, but then where does it go? Unless you live in a rural area or right along the coast, it probably goes to a wastewater treatment plant and then back into a river, where it is subsequently reused by another community downstream. This illustrates two fundamental facts about water: It can be reused multiple times, and upstream users may contaminate the water needed by downstream users.

2.4. Water as a Resource

The ubiquity of water in our lives means that water is a vital economic resource for individuals, companies, cities, and countries. As an economic resource, water has several unique characteristics that affect how it is managed.

- Water is both visible and invisible. Globally, most of the water we use is visible surface water drawn from lakes, rivers, or reservoirs. But a significant (and increasing) fraction of our water use comes from belowground water resources—ground-

water—found in formations known as ***aquifers***. Groundwater poses a serious management challenge because it is widely distributed, hard to monitor, and susceptible to overuse.
- Unlike oil or coal, water is a ***renewable resource***. The hydrologic cycle constantly replenishes the supply of water in rivers, and each year's supply is independent of whether last year's supply was used up.[3] On the other hand, there are also nonrenewable (or slowly renewable) stores of water, such as lakes and aquifers; when we use those stores more rapidly than they are replenished, the stock of water is depleted, so this water use is ultimately unsustainable.
- Unlike oil or coal, water is not substitutable in many of its most important uses. If we run out of oil, we can find other ways to generate energy. But there is no substitute for water for drinking, growing food, or providing habitat. At the same time, some uses of water *are* substitutable—we could use composting toilets instead of flushing with water, we could clean our sidewalks with a broom instead of a hose—so in times of scarcity, water should be reserved for the most essential uses.
- Water is considered a ***fugitive resource***, meaning that it moves on its own and can't be completely constrained by any one owner. For example, if I install a well on my property and pump water from a shared aquifer, groundwater will move from under my neighbor's property to my well. Rules for water allocation must acknowledge this physical fact about water and figure out how to deal fairly with these types of situations.
- Water availability (supply) is highly variable in both space and time. Some regions have lots of water, some have much less, and many face both droughts and floods. This variability has always been one of the central challenges facing water managers, but climate change and land-use change are increasing this variability.
- Water supply and demand are hard to monitor, and since "you can't manage what you don't measure," this makes water hard to manage. On the supply side, the difficulty stems from water's high variability, multiple forms and locations, and variable quality. On the demand side, some of the most important uses are inherently

---

Box 1-1. Local and Global Water Problems

The local versus global nature of water problems is an important theme that we will return to throughout the book. On one hand, water—unlike oil—is not a globally traded commodity, so each watershed must deal with its own scarcity, flooding, and pollution problems, thus giving water problems a fundamentally local character. On the other hand, as we will see, there are several aspects of water issues that do operate at larger scales:

- Complex atmospheric ***teleconnections*** mean that water consumption and land use in one region can affect water availability in other regions.
- Climate change is a global issue (since greenhouse gas emissions anywhere affect climate everywhere) with significant impacts on water availability around the world.
- Global trade in agricultural and industrial products results in global movement of ***virtual water***: the water that it took to make those products.
- Certain pollutants are transported around the world and thus are regulated by international treaties.

difficult to measure (e.g., groundwater irrigation), but we have also done a poor job investing in monitoring systems.
- Water is heavy, or, as an economist would put it, water trades have high ***transaction costs***. What this means is simply that significant energy is needed to move water from one place to another—especially uphill—in amounts that are large enough to matter. One implication of this is that—unlike oil, which is profitably transported around the world—water supply networks tend to be local and constrained by topography. Thus, water problems tend to be local, though with some important exceptions (Box 1-1).
- Despite being essential to life, water is generally cheap. Eighteenth-century economist Adam Smith wondered about this in his famous diamond–water paradox: Water is clearly more useful than diamonds, so why is it so much cheaper? The short answer is that water is much more abundant than diamonds, so the ***marginal benefit*** of an additional increment of water is low. Box 1-2 explores these ideas in more detail.

2.5. Water and Global Development

Given its strong links to both human health and economic development, water is featured prominently in the agendas of the World Bank, the World Health Organization (WHO), and other global development agencies. These agencies generally work on water issues within the framework of ***sustainable development***, a concept that, broadly speaking, means improving people's lives, especially in low-income countries (*development*), without destroying the ecological life-support systems needed by future generations (*sustainability*).

In order for ***low- and middle-income countries*** (LMICs) to achieve growth and lift their citizens from poverty, they must develop their water resources (although whether they need to do so following the same path as high-income countries is a question that we will wrestle with in later chapters). As two development economists put it in an influential 2007 article, "For those countries that have not achieved water security, this objective lies at the heart of their struggle for sustainable development, growth and poverty reduction."[4] They go on to define water security as "the availability of an acceptable quantity and quality of water for health, livelihoods, ecosystems and production, coupled with an acceptable level of water-related risks to people, environments and economies." This definition draws our attention to both the positive (the use of water to support our health, economies, and ecosystems) and the negative (the potential of water-related hazards, such as droughts and floods, to damage those same values).

Over the last several decades, international goals for sustainable development have included specific targets for water management. The ***Millennium Development Goals (MDGs)*** were a series of eight goals, meant to set the international development agenda for 2000–2015, ranging from "eradicate extreme poverty and hunger" to "develop a global partnership for development." These goals were translated into twenty-two targets, but water issues appeared in only one of these targets.

For the post-2015 period, the MDGs were replaced by ***the Sustainable Development Goals (SDGs)***, a more ambitious set of seventeen goals and 169 targets. Water issues are treated much more broadly in the SDGs, with Goal 6 ("ensure availability

Box 1-2. Marginal Benefits and the Diamond–Water Paradox

The solution to the diamond–water paradox lies in the concept of marginal benefits, coupled with the higher abundance of water relative to diamonds. If we think about the benefit provided to an individual by different levels of water use, it might look something like Figure 1-B1. The benefit of a small, initial increment of water is high (the curve rises quickly), but as more water is available, the water is being used for uses that are increasingly less valuable, and the benefit curve levels off. Thus, the marginal benefit—the benefit provided by an additional increment of water (i.e., the slope of the curve shown in Figure 1-B1)—decreases as more water is available (Figure 1-B2).

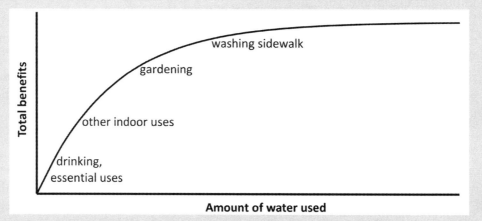

Figure 1-B1. Benefits provided by water use. As more water is available, it is put to increasingly less valuable uses. (Uses shown are illustrative only.)

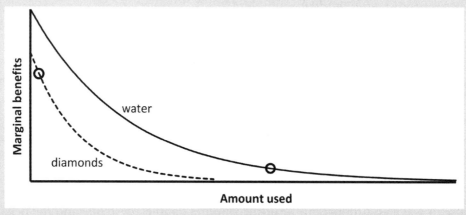

Figure 1-B2. Marginal benefits provided by water and diamond use. The curve shown for water corresponds to the slope of the curve in Figure 1-B1. Circles indicate typical amounts of water and diamonds.

Diamonds, too, have a declining marginal benefit curve, but since diamonds are much less abundant than water, our position on the diamond curve (indicated by the circle in Figure 1-B2) is much higher than our position on the water curve. In other words, given that water is often abundant—and diamonds are not—the marginal benefit of the next increment of water is lower than that of the next diamond. And it is the marginal benefit that determines the price. In fact, the marginal benefit curve is also known as the demand curve, since it expresses how much a person would be willing to pay for a given amount of the product.

Table 1-1. Summary of targets and indicators under Goal 6 of the SDGs. Several of these are discussed in this book; the remainder are covered on the companion website.

Target	Indicator
6.1. Drinking water	6.1.1. Drinking water access
6.2. Sanitation	6.2.1. Sanitation access
6.3. Water quality	6.3.1. Wastewater treatment
	6.3.2. Ambient water quality
6.4. Water scarcity	6.4.1. Water-use efficiency
	6.4.2. Water stress
6.5. Integrated water resource management (IWRM)	6.5.1. IWRM implementation
	6.5.2. Transboundary cooperation
6.6. Aquatic ecosystems	6.6.1. Aquatic ecosystems
6.a. International funding	6.a.1. International funding
6.b. Local participation	6.b.1. Local participation

and sustainable management of water and sanitation for all") being broken into eight targets and eleven indicators (Table 1-1).[5]

2.6. The Water–Energy Nexus

Energy, like water, is used in every part of our society, and there are close linkages between energy and water in both natural cycling and human use. There has been increasing interest in understanding and managing the linkages between the two sectors, a task that is made more difficult by the traditional siloing of energy and water into different academic fields and management agencies. Throughout the book, we will explore several of the linkages that make up the ***water–energy nexus***:

- Climate change, driven primarily by our use of ***fossil fuels*** for energy, is changing the spatial and temporal patterns of water availability (Chapter 3).
- The availability of cheap energy is one of the underlying drivers of modern water management (Chapter 4).
- The search for "clean energy" is driving massive hydroelectric dam projects (Chapter 6).
- Water–energy exchanges are potential tools for transboundary cooperative management (Chapter 12).
- Water and wastewater treatment use significant amounts of energy. Some newer technologies for water supply, such as desalination, wastewater reuse, and water vapor capture, are quite energy intensive (Chapter 13).
- Energy systems require large amounts of water and can also be significant sources of water pollution (Chapter 17).

2.7. Water and Ecosystems

Freshwater ecosystems—rivers, lakes, and wetlands—are hotspots of biodiversity, serving as home to about 10 percent of all species and 30 percent of all vertebrate species while covering less than 1 percent of the Earth's surface.[6] These ecosystems are particularly vulnerable to habitat degradation and anthropogenic changes in the quantity and quality of freshwater flows. The Living Planet Index—a measure of species abundance around the world—suggests that freshwater organisms are declining rapidly, with a 76 percent decrease in migratory freshwater fish and an 83 percent decrease in freshwater populations more broadly over the period 1970–2018 (compared with a 69 percent decrease for all species measured).[7]

In one of the saddest consequences of water management, anthropogenic impacts on freshwater ecosystems have led to several species extinctions. A recent example is the Chinese paddlefish, a large (up to 7 meters long) migratory fish that survived for tens of millions of years but could not withstand overfishing and damming of the Yangtze River.[8]

In response to this crisis, a group of scientists has called for an Emergency Recovery Plan for Freshwater Biodiversity,[9] focused on addressing the six primary threats to these organisms, which we will cover throughout the book:

- Changes in flow (Chapter 5)
- Water pollution (Chapter 8)
- Habitat degradation (Chapters 6 and 7)
- Overexploitation of fish (Chapter 6) and river sand (Chapter 17)
- Invasive species (Chapter 6)
- Habitat connectivity (Chapters 7 and 9).

Besides their value as habitat, aquatic ecosystems also provide great value to society, as expressed in the concept of **ecosystem services**, defined as "the benefits that people derive from functioning ecosystems" either directly or indirectly.[10] Ecosystem services are often categorized as provisioning services (e.g., water supply), regulating services (e.g., moderation of extreme events, wastewater treatment), habitat services (e.g., maintenance of genetic diversity), and cultural services (e.g., recreation, mental and spiritual health).[11] Freshwater ecosystems are estimated to have higher ecosystem service values (per unit area) than any other nonmarine biome.[12]

3. The Water Hard Path and the Global Water Crisis

Given that water is indispensable to our health and our economy, it is not surprising that the manipulation of water has played a central part in the Great Acceleration and the creation of the modern age. We will explore that history more fully in Chapter 4, but here we briefly introduce the **hard path** that has characterized modern water management.

The last two centuries have seen rapid increases in water use, increases that have improved our health and quality of life in many ways but have also driven widespread water pollution and the devastation of aquatic ecosystems. Water management during this period has been dominated by an approach that has been called the hard path, which can be characterized by its technologies and attitudes:

- Technologies: The hard path relies heavily on **gray infrastructure**—large-scale, centralized, highly engineered infrastructure requiring high inputs of materials and energy—especially aqueducts, levees, dams, wells, and water treatment plants. These technologies are typically implemented in a uniform manner, regardless of local conditions or community preferences.

- Attitudes: The hard path is grounded in the belief that nature is a set of resources for human exploitation and that the best way to advance human well-being is to dominate and control the natural world. In addition, the hard path is typically technocratic, with little room for the knowledge and values of local communities.

Despite the engineering successes associated with the hard path, we now face a moment of crisis.

The media and many water experts agree that we are facing a global water crisis, but there is less consensus on exactly what the crisis is. In fact, it often seems that there are multiple distinct water crises; this diversity of crises reflects both the multifaceted ways that water touches our lives and the local nature of water systems.

The flooding crisis: Unprecedented rainfall. Rising seas. Swollen rivers. Failing levees. Catastrophic mudslides. Homes and infrastructure destroyed. Survivors plucked from rooftops.

The scarcity crisis: Drought. Crop failures. Household taps running dry. Dropping water tables. Shrinking reservoirs. More straws desperately trying to suck from a dwindling supply.

The access crisis: Billions of people in poor countries (and millions in the US) still lacking safe, accessible water and sanitation. Aging treatment plants and leaky pipes in our cities. Skyrocketing water prices as we foot the bill for repairs.

The health crisis: Children (2,000 a day!) dying from contaminated water. Cholera resurgent. Carcinogens in our drinking water. Lead in our blood. Our beaches making us sick. Our cultural, social, and mental health frayed by lost connections to healthy waterscapes.

The displacement crisis: Millions of people displaced directly by dams, reservoirs, and levees—or indirectly by the ecological impacts of that infrastructure. Farmers migrating to cities as their crops suffer from drought or flooding.

The conflict crisis: Countries fighting over how to share a limited water supply. Insurgents using water as a weapon. Farmers and environmentalists squaring off over endangered fish. Endless litigation over the arcana of water law.

The ecological crisis: Rivers that don't reach the ocean. Dry lakebeds that send up clouds of toxic dust. Species that evolved over millions of years, gone in a generation. "Sacrifice zones" to support our consumptive lifestyles. The loss of food sovereignty for fish-dependent cultures.

Each of these is a genuine crisis. Yet these problems are—at least at the surface—different enough from each other that we are often tempted to focus on one or another as the *true* crisis, with the choice determined mostly by where we are looking and what lens we are looking through. Our challenge in this book is to pay serious attention to each of these crises and to understand how they are linked.

Is there a common thread, then, that unites these crises? Can we look underneath these symptoms of crisis and find deeper causes of crisis—and corresponding solutions?

Broadly speaking, the underlying causes of crisis lie in the stresses of the Great Acceleration and the shortcomings of hard path water management. The remainder of this chapter will explore causes and solutions in more detail, from two perspectives: values (Section 4) and management tools (Section 5).

4. A Water Ethic: Justice and Sustainability

To better understand the global water crisis, it is helpful to back up a step and think about what values we want water management to support. Many water decisions involve ethical choices, and seeing the underlying values more clearly may help us understand the strengths and weaknesses of current water management. In this section, we discuss the dominant ethic of the hard path—utilitarianism—and make the argument for emphasizing justice and sustainability in a modern water ethic.

Imagine a river with multiple demands on it: Farmers want water for irrigation; city residents want water for household use, including watering their lawns; an endangered freshwater mussel needs water in the river to survive. How do we divide up the limited supply? Or imagine a proposed hydroelectric dam project that will provide green energy for a developing country but will also inundate traditional fishing grounds: Should the dam be built? Or: Should we build levees that will protect a community from flooding but destroy wetlands? Competing values are at play in each of these scenarios.

Until recently, the (often implicit) ethic driving such decisions in modern water management was ***utilitarianism***, a philosophy encapsulated in the maxim "the greatest good for the greatest number." Under a utilitarian ethic, the right course of action is that which will maximize overall human utility (often equated to well-being or happiness). While some utilitarian thinkers do include the well-being of sentient animals in their moral calculus, utilitarianism in practice tends to focus on maximizing benefits to humans only. In addition, utilitarianism in practice tends to quantify utility solely in monetary terms, using tools such as ***benefit–cost analysis (BCA)***, which evaluates potential projects based on comparing costs and benefits through the common currency of dollar value.

This focus on quantifiable benefits and costs to humans means that utilitarianism has a bias toward actions that produce large short-term economic benefits, even if the long-term ecological or social costs—which are often hard to quantify—may be equally large. Thus, in the examples above, utilitarianism would generally favor cities and farms over mussels, hydroelectric power over small-scale fishing, and levees over wetlands.

Utilitarianism also tends to focus on aggregate human utility and sidestep the issue of the distribution of benefits across different populations. For example, BCA operates under the concept of *Kaldor–Hicks efficiency*, in which a project is beneficial if the gains to the winners are larger than the costs to the losers. The logic is that if the project produces aggregate gains, those gains could be redistributed so that everyone comes out ahead—but there is no requirement that this redistribution actually take place. This contrasts with the concept of ***Pareto efficiency***, in which a project is beneficial only if no one is made worse off (and there is benefit to at least one person). Since most public policies have negative impacts on some individuals, Pareto efficiency is probably too strict a standard; still, fairness seems to require that people who are giving up something for the public good should be compensated in some way.

In sum, utilitarianism tends to focus on a narrow slice of water's potential values—its ability to drive economic development—and ignore other social and ecological values. To incorporate those other values, we need to draw on concepts of justice and sustainability.

4.1. Justice
So far, we have used the term *justice* broadly to evoke the social values not included in utilitarianism. But we can be a bit more precise about what we mean, by identifying three aspects of justice that should be part of our water ethic:

- Participatory justice: Since water is central to life, water management decisions affect everyone. Communities and individuals thus have a right—even a responsibility—to participate in those decisions. In the past, a specialized water elite has made decisions that affect us all, through opaque and exclusionary processes. A contemporary water ethic must give communities—especially those who have historically been marginalized—a seat at the table.
- Distributive justice: The costs and benefits of water management need to be shared fairly. Ethicists have identified a variety of ways to define a "fair" distribution of goods:
  - Under *strict equality*, every person should receive an equal share of costs and benefits (or at least equal opportunity to access those benefits). Western societies have generally rejected strict equality for various reasons, most notably the desire to create incentives for entrepreneurial wealth generation.
  - Under the *sufficiency principle*, an unequal distribution of benefits is acceptable as long as everyone has "enough" (a term that admittedly can sometimes be hard to define).
  - Under the *difference principle* (proposed by philosopher John Rawls), social and economic inequalities are acceptable as long as they raise the absolute well-being of the least advantaged members of society.

  In this book, we will draw on both the sufficiency principle and the difference principle; that is, we will be thinking about ways to ensure that everyone has access to the basic water services they need and that water management is particularly focused on outcomes for the most vulnerable.
- Environmental justice: The environmental justice (EJ) movement is rooted in the insight that systemic racism and other deep-rooted societal forces have strongly influenced both the ability of different groups to participate in environmental decision making and the distribution of environmental harms and benefits. EJ initially focused on the location of hazardous waste sites, finding that they were disproportionately sited in neighborhoods of color, and has since expanded to include questions of water access and water quality, among other environmental issues. EJ recognizes that historical patterns of segregation, migration, racism, and underinvestment are reflected in today's environmental impacts and health outcomes, and it encourages us to use the lenses of race, gender, culture, and identity to understand power differences and how they manifest in water management.

4.2. Sustainability
We are using the term *sustainability* loosely to refer to the incorporation of ecological health as a water management goal. At the risk of oversimplifying, we can identify two distinct reasons that we might consider this an important goal, reasons that have their roots in the early-twentieth-century debate between **conservationism** and **preservationism** in the United States.

- Enlightened self-interest: The conservationist ethic—articulated by Gifford Pinchot (1865–1946), among others—saw nature as a resource to be exploited for the benefit of humanity but was concerned about the destruction of that resource by improper management, especially overharvesting, erosion, and pollution. Pinchot added a sustainability element to the utilitarian maxim, calling for "the greatest good for the greatest number *for the longest time*." Today's incarnation of conservationism might be best articulated as a form of enlightened self-interest, which sees the good of humanity as the primary goal but also understands that humanity cannot survive without intact, functioning natural ecosystems. The modern version of conservationism also recognizes that people need nature in ways that go beyond the physical, including spiritual, emotional, and communal needs for connection to healthy ecosystems.
- The community of nature: The preservationist ethic—articulated by John Muir (1838–1914), among others—posited a moral duty to protect nature for its own sake, not just for the benefits it provided to humans, and argued for the preservation of wilderness as a place for nature to thrive. We now understand that wilderness is not truly separate from people, both because of our global impact and because people have been actively managing "wild" places for many millennia. Still, many modern environmentalists share Muir's sense that the duty to protect ecosystems is not just about protecting their value to humans. There are many variants of this belief, each providing a different framework for reenvisioning our relationship to nonhuman species and ecosystems. Some focus on the beauty of nature as a moral good to be preserved, some speak of the moral (and potentially legal) standing of nonhuman beings and even inanimate objects, and some see people as members of a larger natural community, with reciprocal relationships that make a claim on us. Many of these perspectives draw from indigenous traditions, which often see features of the natural world—and water in particular—as sacred, animate, and in relationship with people.

Clearly these two different versions of sustainability might have somewhat different implications for water management decisions, but in practice the differences are often small. To use the examples we presented at the start of this section, both of these perspectives would give significant weight—much more than in utilitarianism—to the endangered mussel, the fishing ground, the wetlands.

Does articulating a water ethic shed light on the nature of the global water crisis? I believe it does. I believe that many of the symptoms we described above (e.g., scarcity, pollution, population displacement, flooding, lack of access) reflect the fact that we have not paid enough attention to justice and sustainability. We have not thought hard enough about how the fruits of the hard path are distributed or who has had to pay the price for our engineering accomplishments. We have not realized that efforts to dominate nature will not have long-term success if they undermine the base of natural productivity on which all life depends.

This realization has led many to call for a ***soft path***, with a corresponding shift in both technologies and attitudes:

- Technologies: The soft path relies on **_green infrastructure_** or nature-based solutions: infrastructure that uses or mimics natural ecosystems and cycles and requires low inputs of external materials and energy. Soft-path infrastructure often brings benefits to local communities as well as the larger society.
- Attitudes: The soft path calls for working with, rather than against, the power of nature; for making room for water rather than constraining it; for restoring aquatic ecosystems rather than destroying them; for working collaboratively with communities rather than imposing external solutions.

Our challenge, then, is to combine the best of both paths: to synthesize the most promising features of ancient, modern, and emerging technologies to produce solutions that equitably satisfy human needs while protecting and respecting nature.

5. How We Get There: Infrastructure, Institutions, Incentives, and Information

If the previous section focused on clarifying *what* we want water management to achieve, this section starts our conversation about *how* to achieve those goals. We discuss four topics that we will return to throughout the book; each has played a role in creating the current water crisis, and each must play a role in achieving a just and sustainable water future.

Infrastructure: The issue of appropriate infrastructure has already featured prominently in our description of the hard and soft paths, and throughout the book we will explore the strengths and weaknesses of various water technologies, from dams (classic gray infrastructure) to water harvesting (a suite of green infrastructure approaches).

The water crisis reflects, in part, infrastructural failings:

- Gray infrastructure requires significant inputs of energy and materials during construction, operation, and decommissioning, thus contributing to climate change and other environmental problems. In addition, gray infrastructure often does serious damage to local ecosystems and communities (as when dams displace people and destroy rivers).
- When gray infrastructure fails (e.g., a levee is overtopped), "natural" disasters such as floods can do *more* damage than they would have done in the absence of that infrastructure.
- While we have relied too heavily on certain types of gray infrastructure (e.g., dams and levees), we have also not built enough urban water infrastructure (pipes, treatment plants) to ensure safe water and sanitation access for all, especially in rapidly growing slums.
- In many cases, we have not kept up with infrastructure maintenance and are now footing the bill for much-needed repairs.

How do we find the right mix of hard and soft infrastructure and ensure that we pay attention to maintenance as well as construction? We will come back to these critical questions.

Institutions: Infrastructure is only part of the picture. Equally important is water governance: the institutions and rules that determine what infrastructure is built, how it is managed, and how its benefits and costs are distributed.

The water crisis reflects, in part, the weaknesses of our water governance systems:

- Our rules for water use—many of them legacies of previous centuries—are not fair, sustainable, effective, or adaptable.
- The very centrality of water to every aspect of life means that many different agencies are involved in water governance: agencies with conflicting agendas and overlapping spheres of influence, often pulling in different directions.
- We don't do well at integrating across various parts of the hydrologic cycle, so we treat different water problems in isolation.
- We lack holistic and visionary planning; our reactive management style solves each emergency by laying the groundwork for the next one.
- Water governance is traditionally expert driven and undemocratic.

Can we reform our water institutions to align them with our goals of justice and sustainability? Can water governance become more effective, democratic, and integrated? Can we solve multiple problems at once by seeing various water flows (rain, runoff, drinking water, wastewater) as part of One Water? Stay tuned.

Incentives: Since water touches every aspect of our economy, most "water managers" are the individuals or companies who use water—and whose choices about how much water to use are affected by price (as well as other incentives). If we want people to make choices that are compatible with justice and sustainability, we need to harness the power of the marketplace to ensure that those choices are incentivized. The water crisis reflects, in part, the perverse incentives imbedded in water prices:

- Prices are signals of value. The price paid by a water user should reflect the ***opportunity cost*** of that water: its value in other uses, such as supporting a healthy ecosystem. But a healthy ecosystem is a ***public good***—one that everyone can benefit from—so it is consistently undervalued by the market. Thus, the cost of water does not reflect the value of the ecosystem, and the destruction of the ecosystem is an ***externality***—a cost that is not borne by the water user and so is not considered in deciding how much water to use.
- Even when different human uses are competing for the same limited water supply—so scarcity should in theory drive up the price—water prices are kept artificially low for complex historical, economic, and social reasons. Thus, those who are lucky enough to have water rights often use that water wastefully because they are paying nowhere near water's true cost.
- At the same time, the simple economic prescription—raise water prices—runs up against the equally simple objection: Affordable water is a basic human right.

Can we raise the price of water to better reflect scarcity and the value of water to ecosystems, while ensuring that this basic necessity is affordable to everyone? What types of market and nonmarket mechanisms can we use to square these two goals? How can the value of water be felt and expressed in nonmonetary ways as well? To be discussed.

Information: Without discounting the importance of values, it is also true that some of the most important questions of water management are empirical, not normative: To what extent do reservoirs release greenhouse gases? What types of flows do various fish need to survive and reproduce? What levels of various pollutants are safe for people? How is climate change affecting water availability? These are the questions that science is good at answering, and we need our water management to be informed by the best available science. Indeed, this book delves deeply into the latest scientific findings, and we will frequently turn to data to answer important questions. But we also need to heed Aldo Leopold's admonition that "to keep every cog and wheel is the first precaution of intelligent tinkering"[13]; our understanding of complex ecological systems is inevitably incomplete, so we should err on the side of protection.

The water crisis reflects, in part, our limited understanding of the consequences of our actions:

- We failed to see that building dams would break rivers into disconnected fragments and destroy the biological and economic productivity of those rivers.
- We somehow didn't get that diverting water from rivers would lead to salinization and drying of wetlands, lakes, and estuaries, loss of fisheries, and cultural and health impacts on local communities.
- We didn't understand that building levees to protect riverside communities from flooding would increase flood damage by taking away natural flood valves and by encouraging development in flood-prone areas.
- We ignored the buildup of greenhouse gases in the atmosphere and failed to fully grasp how climate change would exacerbate both drought and flooding.

Can we learn humility from these failures without giving up on using science to inform management? Can we draw on theory and data to make water management decisions while also incorporating the ***precautionary principle***? Let's find out.

Chapter Highlights
1. The ***Great Acceleration*** that we have experienced in the modern era has brought much good to humanity, but its benefits have been unevenly distributed, and our growing impact on the natural world threatens our well-being and that of our fellow travelers on this planet. We urgently need a Just Transition to sustainability.
2. Water management has been a central player in the creation of modernity and must also be central to the Just Transition.
3. Water is simultaneously a human right and an economic resource, two characteristics that are sometimes in tension.
4. As a resource, water has unique properties that affect how we manage it: It is ***renewable***, ***fugitive***, largely nonsubstitutable, and highly variable in space and

time. We draw on it for both ***instream*** and ***offstream*** uses, in both ***consumptive*** and ***nonconsumptive*** ways.
5. Water is closely linked to other resources such as energy and food. Unlike those resources, water is not traded globally, so the focus of water management is largely local, although factors such as climate change and ***virtual water*** make water a global issue as well.
6. Water is central to ***sustainable development*** and features prominently in the ***Sustainable Development Goals***.
7. The global water crisis is really a set of interconnected crises: flooding, scarcity, inequitable access, health impacts, population displacement, water conflict, and ecosystem degradation.
8. Modern water management has been dominated by the ***hard path***, reflecting an underlying ethic of ***utilitarianism***. The global water crisis calls on us to incorporate justice and sustainability into our water ethic.
9. In the past, communities of color and low-income communities have often borne the impacts of water infrastructure without fully sharing in its benefits. Justice requires that we work toward a fairer distribution of costs and benefits, along with fuller participation of affected communities in water decision making.
10. The early-twentieth-century debate between ***conservationism*** and ***preservationism*** provides a template for two attitudes toward sustainability: enlightened self-interest and membership in the community of nature. In either case, utilitarianism must be supplemented with a concern for healthy ecosystems.
11. To achieve our goal of just and sustainable water management, we need to pay attention to four factors that have contributed to the current crisis but can be harnessed for a better future: infrastructure, institutions, incentives, and information.

Notes

1. Kimmerer (2013).
2. Greeley (2017).
3. This second feature makes water different from renewable resources such as fisheries and timber, whose supply in a given time period depends on the amount of reproducing stock in the previous time period.
4. Grey and Sadoff (2007).
5. https://www.un.org/sustainabledevelopment/water-and-sanitation/; https://www.sdg6data.org/en/node/1; and https://www.un.org/millenniumgoals/environ.shtml.
6. Tickner et al. (2020).
7. https://www.livingplanetindex.org/latest_results.
8. Zhang et al. (2020).
9. Tickner et al. (2020).
10. Costanza et al. (2017).
11. https://teebweb.org/.
12. Costanza et al. (2014).
13. Leopold (1949).

Part I
Water Availability and Use: Supply and Demand, Scarcity and Change

The four chapters in Part I provide an overview of water as a resource for human use, addressing basic questions about *supply* (how much water is available and how that varies over space and time), *demand* (how much water we are using and for what purposes), and *scarcity* (the gap between supply and demand). A central theme of these chapters is that water supply is not a fixed quantity but rather is affected by both natural variability and anthropogenic factors, including land-use change, climate change, and water use itself.

We start in Chapter 2 with an analysis of water stocks and flows, paying special attention to the spatial and temporal variability in flows and to the hydrologic tools—largely rooted in the concept of "stationarity"—that are used to quantify that variability. Chapter 2 also includes a discussion of the megadroughts that have been discovered in the paleoclimate record, a discussion that ultimately leads us to question the stationarity model. Chapter 3 further undermines the stationarity model by discussing the two contemporary large-scale drivers of changing water availability—climate change and land-use change—and their implications for water management and planning. Chapter 4 turns to the demand side, with a brief history of how people have used water over the last 10,000 years, a more detailed discussion of the last hundred years or so, and a quantitative look at how much water we use for what purposes; this involves delving into some important definitional issues, including the concept of water footprints. Chapter 5 defines scarcity, introduces various scarcity indicators, identifies where scarcity is (and isn't) a problem, and examines the phenomenon of *depletion*: rivers, lakes, and aquifers where human water use has changed the amount and timing of water availability, with significant impacts on both communities and ecosystems.

2 Water Availability: Spatial and Temporal Variability

- How much water is available for human use, and how does this vary across the world?
- How do floods and droughts affect water availability?
- What can we learn about drought from the paleoclimate record?

In this chapter, we define what we mean by water availability and analyze that availability globally and regionally. A critical consideration in this discussion is the uneven spatial and temporal distribution of water resources. We spend some time on tools for describing hydrologic variability and introduce the concept of stationarity. We then turn to understanding the causes and impacts of droughts, including past megadroughts, which leads us to question the utility of the stationarity model.

This chapter assumes some familiarity with hydrology. If you find that you need to strengthen or refresh your background, visit the companion website at https://watermanagement.yale.edu. Remember that terms in ***bold italics*** are defined in the glossary, and a guide to units and conversions is provided at the beginning of the book.

1. Global Stocks and Flows

Discussions of global water resources often begin with the volumes of water in various environmental compartments, referred to as water stocks. These numbers (Table 2-1) reveal that our "water planet" is, of course, mostly saltwater, which supports a great abundance of life but is unsuitable for human use and thus not considered part of the water "resource." Still, even excluding saltwater and frozen water, the earth has large volumes of freshwater, mostly in the form of lakes and ***groundwater***.

However, a focus on stocks is misleading, because water is a ***renewable resource*** that is continually replenished through the hydrologic cycle. The large stocks of water in ***aquifers*** and lakes may present tempting targets for water use, but if that water is not being replenished (or if we are using it more rapidly than it is being replenished), then we are depleting the stock and it will eventually run out. If we want to achieve

Table 2-1. Estimates of water volumes in various stocks (in thousands of km³). Data from Abbott et al. (2019), except for groundwater data from Ferguson et al. (2021).

Saltwater	
Oceans	1,300,000
Saline groundwater	28,000
Saline lakes	95
Freshwater	
Liquid	
Fresh groundwater[a]	15,900
Fresh lakes	110
Soil moisture (green water)[b]	54
Wetlands	14
Reservoirs	11
Rivers	2
Solid and Gas	
Glaciers	26,000
Permafrost	200
Atmosphere (water vapor)	13
Snowpack (annual maximum)	2.7

[a]The fraction of global groundwater that is fresh is poorly known; here I assume that groundwater in the top 1 km of Earth's crust is fresh, while groundwater below that level is saline. This probably overestimates the volume of recoverable fresh groundwater, both because the fresh/brackish transition is shallower than 1 km in many locations and because it is usually not economical to extract groundwater from deeper than a few hundred meters.

[b]The **green water** held in soil pores is available to plants, but—in contrast to the **blue water** in rivers, lakes, and aquifers—it can't be transported from its location for other uses, such as drinking.

sustainable water use, our emphasis should be on flows: the conversion of water vapor to precipitation and ultimately to streamflow. To put it slightly differently, rivers—despite their small volume—are of particular importance to water managers, because they are the primary mechanism for the renewable flow of liquid freshwater.

Figure 2-1 shows estimated water flows at a global scale. Evaporation from the ocean (450,000 km³/yr) is greater than precipitation over the ocean (404,000 km³/yr). The difference of 46,000 km³/yr is transported as atmospheric water vapor to the continents, where it contributes to an excess of precipitation over ***evapotranspiration***. This excess precipitation then flows as ***runoff*** back to the oceans, completing the cycle. It is this runoff—the flow back to the ocean of 46,000 km³ of freshwater—that constitutes the renewable water resource that is potentially available for human use year after year (although it should be clear that appropriating *all* this flow for human use would severely damage aquatic ecosystems). This dynamic is described by the basic water budget equation $P = ET + R$: precipitation (P) either returns to the atmosphere as evapotranspiration (ET) or becomes runoff (R).

This 46,000 km³/yr of runoff consists primarily of river flow but also includes about 2,300 km³/yr of ice discharge from Antarctica and a small amount (probably under 300 km³/yr)[1] of submarine groundwater discharge (i.e., groundwater that

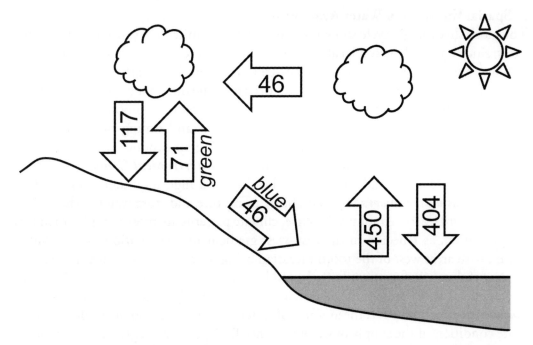

Figure 2-1. Simplified schematic of the global hydrologic cycle (units: thousands of km³/yr). Upward arrows indicate evapotranspiration and downward arrows indicate precipitation. Lateral arrows indicate water vapor transport from oceans to continents and water runoff from continents to oceans. Data from Rodell et al. (2015).

discharges directly to the oceans). In addition, much of the world's river flow has spent some time as groundwater before reaching the river. That is, we can think of **groundwater recharge** (~16,000 km³ per year)[2] as a component of renewable runoff, but since most of that recharge ends up discharging to rivers and becoming streamflow, we should be careful not to double-count it.

Several caveats apply to Table 2-1 and Figure 2-1:

- Our knowledge of the global hydrologic cycle is improving but is still incomplete, and there are substantial uncertainties in the values shown for both stocks and flows.
- Flows and stocks are changing over time, due to both natural and human causes; the data shown can be viewed as a snapshot in time representing roughly the first decades of the twenty-first century.
- Even within this time period, these data should be viewed as an average; the coefficient of variation of annual global runoff is about 3 percent[3] and is much higher for individual rivers or regions.

In addition to the 46,000 km³/yr of runoff (blue water), we should also pay attention to the precipitation that does not become runoff but instead returns to the atmosphere as ET (71,000 km³/yr). This green water flow supports the growth of terrestrial vegetation and is appropriated for human use when we grow crops using **rainfed agriculture** (as opposed to **irrigation**, which uses blue water).

2. Spatial Variation in Water Availability

The global hydrologic cycle shown in Figure 2-1 has limited relevance to water managers, since water is largely a local resource. Water resources—whether they are stocks or flows, blue or green—are not evenly distributed around the world. Complex atmospheric and oceanic circulation patterns—driven by the differential heating of Earth's surface and the Coriolis force resulting from Earth's rotation—produce a global mosaic of climates, with great variation in the amount and timing of precipitation. The wettest areas of the planet are those where rising air produces intense rainfall, including a zone near the equator (referred to as the ***Intertropical Convergence Zone [ITCZ]***), where air is made buoyant by heating, and high mountain ranges (often referred to as ***water towers*** for their ability to capture and store water), where air is forced upward by topography. Of course, climate patterns are more complex than this simplification, as reflected in the east–west gradient in precipitation in the United States, with areas west of the 100th meridian being much drier than areas to the east.

A useful metric for quantifying local blue water generation is the ***aridity index*** (AI), defined as the ratio of mean annual precipitation (P) to mean annual ***potential evapotranspiration*** (PET, the amount of water that would evaporate under local climate conditions if the supply of water were not limited). A ratio greater than 1 means that there is "excess" water that can become runoff, while a ratio lower than 1 suggests that rainfall will evaporate without generating any blue water. In practice, since AI is based on annual averages, even areas with AI less than 1 can produce some runoff during the wet season, when P can be temporarily greater than PET.

A global map of AI (Figure 2-2) can be used to identify ***drylands***: areas with AI less than 0.65 and therefore little or no runoff generation. Drylands occupy some 41 percent of the land surface and are home to 38 percent of the global population.[4] Subcategories of drylands are dry subhumid (AI = 0.5–0.65), semiarid (0.2–0.5), arid (0.05–0.2), and hyperarid (less than 0.05). While drylands include desert biomes, they also include grasslands and savannas (mixed grassland–woodland biomes).

As a picture of available blue water resources, Figure 2-2 is misleading in one important way: It shows only local AI (and by implication blue water generation) but ignores nonlocal inputs of blue water. In reality, runoff generated in wetter areas is channeled through river systems, which can transport blue water large distances from where it is generated; the ***watershed*** or ***river basin*** (Box 2-1) is thus a critical spatial unit for understanding and analyzing runoff. For example, Cairo, which experiences essentially no precipitation or runoff generation, sits on the banks of the Nile River, which delivers large amounts of runoff generated in the wetter portions of the Nile Basin.

Taking into account all these factors, the most useful metric for blue water availability is ***total renewable water resources (TRWR)***, expressed in flow units (volume/time) and usually calculated on an annual basis. At the basin scale, this corresponds to average annual runoff (streamflow). At other scales (e.g., a country), it includes internally generated runoff but also external inputs of blue water from upstream.

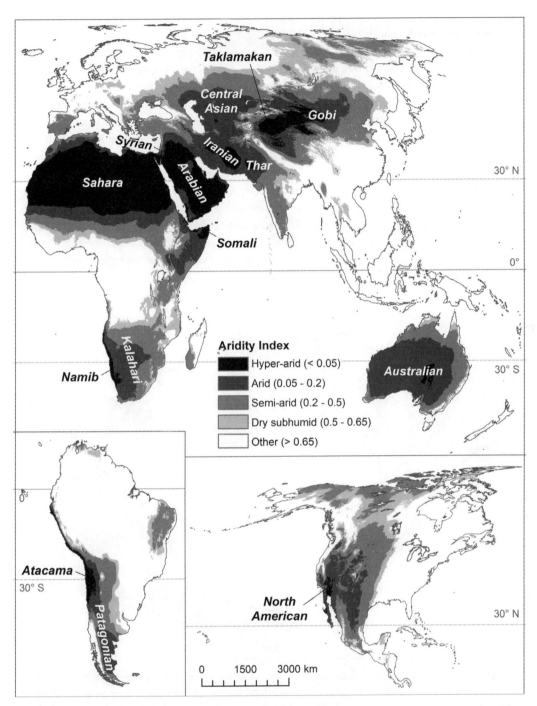

Figure 2-2. Global map of the aridity index, with major world deserts shown. Areas in white generate significant blue water runoff, while shaded areas do not. Map by Anna Yue Yu. Data from https://cgiarcsi.community/2019/01/24/global-aridity-index-and-potential-evapotranspiration-climate-database-v2/.

Box 2-1. Watershed Primer

A *watershed* can be defined as "the topographic area within which apparent surface water runoff drains to a specific point on a stream or to a water body such as a lake."[a] In other words, the watershed defined by a given point is the land area that contributes water to that point. Watersheds are delineated based on surface topography, with the understanding that water that falls as precipitation will move downhill to the stream (perhaps entering groundwater temporarily on the way there) and then flow downstream past our selected point. Of course, not all the precipitation that falls in the watershed will become streamflow (some will undergo ET), but all the water that leaves the watershed in liquid form must flow past our point.

A river's full watershed, often called a *river basin*, is defined by its outlet—the point where it enters the ocean or other water body—but we can also define nested subwatersheds at other points along the river's course (Figure 2-B1). To delineate the watershed for any point along a stream in the United States (and get flow estimates for that point), go to https://streamstats.usgs.gov/ss/. To see a visualization of the water flow pathway from any location in the United States, see https://river-runner.samlearner.com/.

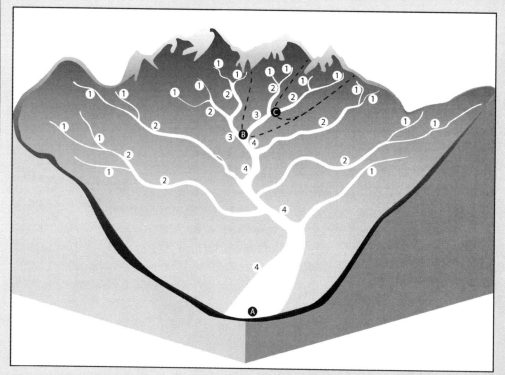

Figure 2-B1. The watershed defined by point A. In addition, two smaller subwatersheds, defined by points B and C, are also delineated. Watershed C is nested within watershed B, which is nested within watershed A. Numbers are Strahler stream orders, a measure of stream size. (To calculate Strahler stream order, follow these rules: The smallest perennial stream that is shown on the map is considered first order. Where two streams of equal order come together, the combined stream has an order that is 1 greater than each of its tributary streams. Where two streams of unequal order come together, the combined stream has an order equal to that of the higher-order tributary. In other words, in this version of math, 1 + 1 = 2 but 1 + 2 = 2.) Graphic by Maureen Gately.

> **Box 2-1** *continued*
>
> The watershed has long been recognized by hydrologists and aquatic chemists as a fundamental unit of analysis. Rivers have been described as the blood vessels of watersheds, and river hydrology and chemistry have been seen as a watershed's vital signs, critical for assessing its health. The underlying insight is that rivers are affected—in their flow and their chemistry—by how land is used within their watersheds, which means that water management must also involve land management.
>
> At the most basic level, the watershed is useful as a unit for water budgets. Since the only surface water flow across a watershed boundary consists of runoff at the outlet, the equation $P = R + ET$ can be applied at the watershed scale (but not at the scale of, say, a state). Two hydrologic parameters are often derived from this budget:
>
> - The **water yield** (typical units: mm/yr) is simply the runoff from the watershed (average annual streamflow) divided by the area of the watershed.
> - The **runoff ratio** (unitless) is the ratio of R to P and describes how efficiently precipitation is converted into blue water. Based on Figure 2-1, the global average runoff ratio is 46/117, or 39 percent.
>
> Beyond the scientific convenience of the watershed, there are benefits to the watershed as a unit of management. First, water users within a watershed are, in the most fundamental way, sharing the same resource. Water that is used upstream is not available for use downstream. Pollution that is discharged upstream affects the quality of water downstream.
>
> Second, the watershed is the natural unit within which anthropogenic movement of water has the lowest ecological and economic costs. When water is withdrawn from a stream and used within the same watershed, the return flow from that use (i.e., the portion of the water that is not consumed) remains within the watershed and ultimately flows back to the stream where it came from, thus reducing the ecological impacts of that withdrawal. In addition, it is often cheaper to move water within a watershed, since an **interbasin transfer** will require pumping water over the watershed divide (or tunneling through it).
>
> Still, water managers recognize that the watershed is not the only relevant unit. Given the ubiquity of interbasin transfers and the fact that water agencies are defined by state or national boundaries—which generally do not align with watershed boundaries—water managers must be prepared to deal with multiple overlapping scales simultaneously.
>
> **Note**
> [a]Omernik and Bailey (1997).

3. Temporal Variation in Water Availability

In addition to the spatial variability discussed above, water availability also fluctuates dramatically over time, posing a critical challenge for water managers. In this section, we introduce several tools for describing hydrologic variability, while the next section focuses specifically on drought as an often devastating manifestation of this variability.

3.1. Hydrographs

Perhaps the most obvious way to examine variability in streamflow is to look closely at a **hydrograph**, such as the one shown in Figure 2-3 for the Quinnipiac River (Connecticut) for water years 2014–2016. (A water year in the United States runs from October 1 to September 30.)

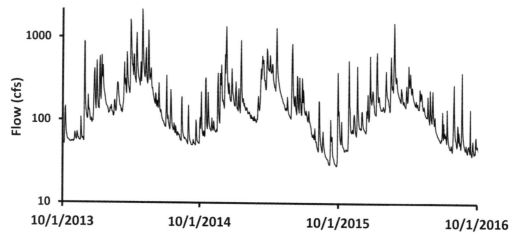

Figure 2-3. Hydrograph for the Quinnipiac River, Connecticut, for water years 2014 through 2016 (daily data). Baseflow can be envisioned as a smooth line running under the graph. Note the log scale on the vertical axis. Data from US Geological Survey (https://waterdata.usgs.gov/nwis/uv?01196500).

Figure 2-3 shows variation at three time scales:

- Storms: During storm events, flow rises rapidly in response to rainfall and then slowly declines. The time scale of the storm response is generally shorter for shorter rain events, smaller streams, and more urbanized watersheds. The flow of water in a stream in direct response to a storm event is referred to as stormflow, while flow between storm events is referred to as baseflow. The pathways by which water travels to the stream can be quite complex, but stormflow pathways are generally more rapid and closer to the surface relative to baseflow, which is fed by the slow discharge of groundwater.
- Seasonal: For the Quinnipiac River, precipitation is (on average) evenly distributed throughout the year, but flows are generally lower in the summer months because of higher ET. Other sites may show more extreme seasonal differences, either because of seasonality in precipitation (Box 2-2) or because precipitation accumulates as snow over the winter and then is converted to runoff in the form of a large spring snowmelt peak.
- Interannual: At longer time scales, flow varies in response to interannual variation in precipitation and temperature.

3.2. Flow Duration Curves

Despite the utility of hydrographs in identifying obvious patterns, we need statistical tools to quantitatively summarize long-term flow records. For example, if we want to use the Quinnipiac River to generate hydropower or to supply a town with a water source, it would be helpful to be able to say something like, "We can count on having at least X cfs (cubic feet per second) 95 percent of the time." That is exactly the logic of the **_flow duration curve_** (FDC), which plots how often different flow levels were equaled or exceeded (their **_exceedance probabilities_**) in the period of record

> **Box 2-2. Monsoon Systems**
>
> The word *monsoon*, as used in climate and water science, refers to a seasonal reversal of wind direction that drives a strong seasonal pattern in precipitation. Eight distinct monsoon systems can be identified: Indian, East Asian, Western North Pacific, Australian, North American, South American, North African, and South African. More than half the world's population lives in regions affected by these monsoons.
>
> Perhaps the best-understood of these is the Indian Monsoon, which is driven by the seasonal movement of the ITCZ relative to the Indian subcontinent. In January, the ITCZ is located just south of the equator in the Indian Ocean, and winds over India blow toward this low-pressure zone, resulting in dry continental winds from the northeast and little precipitation for most of the country. By July, the Asian land mass has heated up more than the waters of the Indian Ocean, and the ITCZ moves north over the Tibetan Plateau. This draws in moist air from the Indian Ocean, resulting in southwesterly winds over much of India that can produce torrential rainfall as the air is forced upward by the Himalayas.

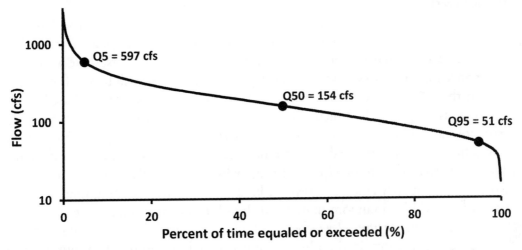

Figure 2-4. Flow duration curve, based on US Geological Survey daily flow data for the Quinnipiac River from 1931 to 2019.

(Figure 2-4). Monthly or seasonal FDCs can be used to understand typical patterns in different times of year.

Flow statistics can be directly read off FDCs; commonly used metrics include the following:

- Q_{95}: the flow that is equaled or exceeded 95 percent of the time (i.e., the fifth percentile flow), often used as a measure of low-flow conditions;
- Q_{50}: the flow that is equaled or exceeded 50 percent of the time (i.e., the median flow);
- Q_5/Q_{95}: the ratio of the 5 percent-exceedance flow to the 95 percent-exceedance flow; the higher this ratio, the greater the flow variability in the system.

3.3. Flood Frequency Analysis

What if we are especially interested in the risk of the highest flows? Since these flows are rare, an FDC is of limited utility, and we turn instead to an approach known as ***flood frequency analysis***. Like an FDC, flood frequency analysis uses past data to predict the likelihood of future flows, but unlike an FDC, it uses only the highest flow for each water year in the period of record. Using these peak flow data (and certain statistical assumptions), we can estimate the annual probability (p) that a river will exceed a certain flow level. For example (Figure 2-5), the Quinnipiac River has a 1% annual probability of reaching a flow of at least 6,663 cfs and a 50 percent annual probability of reaching a flow of at least 2,214 cfs.

The **recurrence interval** for a flood of a given magnitude is defined as $1/p$ (i.e., the inverse of the annual exceedance probability). Thus, a flood with an annual exceedance probability of 1 percent (0.01) is the hundred-year flood, a flood with a probability of 50 percent (0.5) is the two-year flood, and so on. Importantly, the phrase "hundred-year flood" is not meant to convey that a flood of this magnitude will occur like clockwork every hundred years. Rather, it indicates that a flood of this size (or larger) has a 1 percent chance of occurring in any given year, regardless of what has happened in previous years (since each year is assumed to be independent of the previous year).

The logic of flood frequency analysis can also be used to characterize low-flow events, most commonly through the 7Q10, defined as the seven-day average flow with a ten-year recurrence interval. For example, if the 7Q10 is 50 cfs, there is a 10 percent chance that the lowest seven-day average flow of the year will be 50 cfs or lower.

Both the flow duration curve and the flood frequency analysis are based on an important assumption: that the past is a good guide to the future. Put more formally, these tools (and others widely used by hydrologists) assume that a hydrologic variable such as flow is a random variable with a nonchanging **probability density function** (PDF) that can be determined from existing hydrologic data. For example, in the case of the flood frequency analysis shown in Figure 2-5, we are assuming that each year in the future will be a random draw from the PDF shown in Figure 2-6. Note that this assumption of **stationarity** recognizes hydrologic variability and even recognizes that there is uncertainty in the PDF, since it is based on limited data, but assumes that the underlying pattern of variability is not systematically changing over time. Is this assumption of stationarity a valid one? We will discuss this further at the end of this chapter and in Chapter 3.

One conservative alternative to flood frequency analysis is to estimate the probable maximum flood (PMF), defined as the maximum flood that is likely to occur in a given area under the most extreme conditions. This is derived from the probable maximum precipitation (PMP), estimated by meteorologists based on the maximum amount of moisture that the atmosphere can hold; the amount of the PMP that will run off to form the PMF is then estimated based on watershed soils, land use, and topography. The PMF is typically used in situations where a conservative estimate of possible flooding is needed, such as in dam safety design.

3.4. Rainfall Intensity–Duration–Frequency Curves

Temporal variability in precipitation can be treated similarly to the variability in streamflow discussed above, with two additional complications. First, unlike streamflow—which is relevant only in certain parts of the landscape (rivers)—precipitation falls everywhere, so we would ideally like to have point estimates of precipitation variability for every location of interest—but these estimates must be constructed from

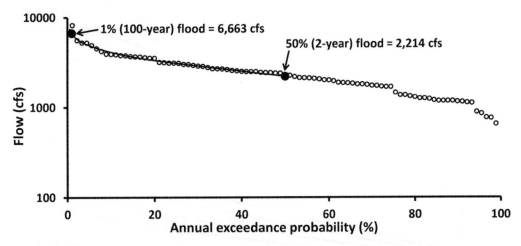

Figure 2-5. Flood frequency analysis for the Quinnipiac River. The circles are the peak annual flows for water years 1931–2019, and the line is the best fit to the data. (The best-fit curve uses the log Pearson type III relationship, which is traditionally used for flood frequency analysis.)

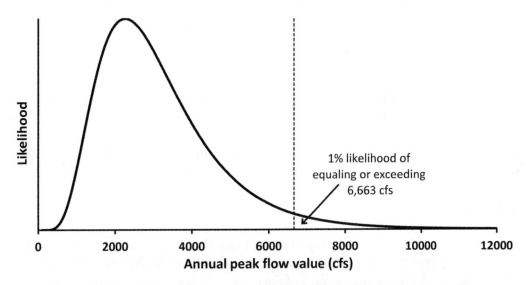

Figure 2-6. Probability density function for annual peak flows for the Quinnipiac River, based on the data shown in Figure 2-5. The area under the curve to the right of the dashed line represents the likelihood of equaling or exceeding 6,663 cfs (the hundred-year flood) and makes up 1 percent of the entire area under the curve.

a limited network of precipitation stations. Second, precipitation is episodic, so the time frame of analysis matters more than for streamflow; that is, we'd like to know how much rain we are likely to get in a day but also how much we are likely to get in an hour or a month, since each of those has a different type of relevance for water management. Precipitation estimates in the United States are produced regionally by the National Oceanic and Atmospheric Administration, in a process that is fundamentally based on the same logic as the flood frequency analysis described above (including the assumption that the past is a good guide to the future).

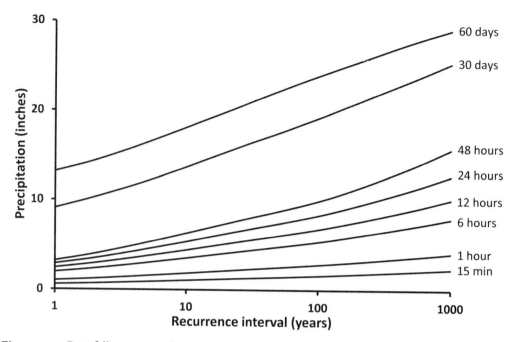

Figure 2-7. Rainfall intensity–duration–frequency curves for New Haven, Connecticut, from the National Oceanic and Atmospheric Administration's Atlas 14 (https://hdsc.nws.noaa.gov/hdsc/pfds/). PMP values (not shown) are approximately 26 inches for a six-hour storm and 37 inches for a forty-eight-hour storm (https://www.weather.gov/owp/hdsc_pmp).

Estimates of precipitation variability are usually presented as ***intensity–duration–frequency (IDF) curves***. For example, in New Haven, Connecticut, the hundred-year six-hour rainfall—the six-hour rainfall with a 1 percent annual exceedance probability—is 5.4 inches, while the hundred-year twenty-four-hour rainfall is 8.3 inches (Figure 2-7). These types of numbers are widely used in designing urban infrastructure.

4. Drought

> And it never failed that during the dry years the people forgot about the rich years, and during the wet years they lost all memory of the dry years. It was always that way. —John Steinbeck, *East of Eden*

Along with floods (Chapter 7), droughts are extreme manifestations of the temporal variability inherent in the hydrologic cycle. A good understanding of drought is essential to water management, and this section will cover the phenomenon in some detail. We start by defining the different aspects of drought, then turn to understanding the causes of drought, and finally look at some important historical droughts.

4.1. Defining Drought

Broadly speaking, drought is a short-term deviation from normal water availability; this is distinct from chronically low water availability (aridity). Drought is a complex phenomenon in which water shortages of varying spatial and temporal scales cascade through different parts of the hydrologic cycle. Droughts are sometimes broken down into four aspects:

- *Meteorological droughts* are characterized by a deficiency in precipitation relative to normal conditions and are generally the first manifestation of drought.
- In *agricultural droughts*, low soil moisture has negative impacts on crop growth. This is generally caused by meteorological drought but can also be amplified by poor land management that reduces the soil's water-holding capacity.
- *Hydrologic droughts* are characterized by diminished flows in rivers and lower water levels in lakes and aquifers. Recovery of rainfall does not necessarily resolve agricultural and hydrologic droughts, due to accumulated deficits in soil moisture, groundwater, and surface water, sometimes accompanied by long-term changes in the runoff ratio (the ratio of runoff to precipitation).[5]
- *Socioeconomic drought* refers to the wide-ranging ways in which drought can affect human society, including agricultural losses, famine, a decrease in hydropower production, wildfire, and strained water supplies for households and industry.

Droughts can strike any part of the world, although some regions are more susceptible than others. In general, drier locations experience greater interannual variability and are thus more drought-prone. Semiarid locations that receive most of their annual rainfall in just a few events are particularly vulnerable to drought, as are places near the boundary of a climate zone.

Unlike other natural hazards such as earthquakes or floods, drought usually occurs gradually, and it can be hard to define exactly when a region enters or exits drought conditions. The US Drought Monitor prepares drought maps with five categories of dryness ranging from Abnormally Dry to Exceptional Drought, using what has been called "a state-of-the-art blend of science and subjectivity."[6] The indicators used in this blend include the **Palmer Drought Severity Index (PDSI)**, which considers deviations in temperature (and thus ET) as well as precipitation; PDSI is normalized, so a given value is roughly comparable between locations.

4.2. Causes of Drought

There is still a great deal that we don't understand about the proximate and ultimate causes of droughts, but two concepts have emerged as central to the phenomenon: **climate variability** and **feedbacks**.

CLIMATE VARIABILITY

On one level, drought is simply a manifestation of random climate variability: A high-pressure system happens to settle over a region and blocks normal moisture patterns. Or the jet stream (the strong high-level winds that steer weather systems in mid-latitudes) shifts away from its normal course, bringing unusual rains to one location and unusual dryness to another.

Alongside this short-term variability, longer-term and larger-scale semiperiodic oscillations in key oceanic and atmospheric parameters—referred to as modes of climate variability—are dominant features of the earth's climate system. Each of these modes is centered in a certain region of the planet but can affect climate in distant areas through *teleconnections*. At least fourteen oscillations have been identified, each with a different time scale and a different degree of influence on precipitation and temperature in various parts of the world. The in-phase interaction of these modes, in which multiple modes are amplifying the same effect, can lead to extreme climatic events, while out-of-phase interactions tend to have muted effects; this makes it challenging to predict the effects of any one mode on its own.[7]

Probably the most important of the climate variability modes is **El Niño–Southern Oscillation (ENSO)**, a coupled ocean–atmosphere oscillation in the equatorial Pacific. In the neutral phase of this oscillation, easterly winds push warm water toward the western Pacific, leading to cool water dominating the eastern Pacific. In the cool phase (La Niña), the easterly winds are stronger than average, leading to cooler-than-average water in the eastern Pacific. In the warm phase (El Niño), the winds weaken and the warm pool of water stretches out toward the eastern Pacific. This oscillation affects precipitation around the world, including the southern United States, where La Niña often brings drought, and India, where El Niño tends to bring lower-than-average monsoon rains ("monsoon failure"). Since the El Niño phase brings drought to more of Earth's terrestrial surface, it has a significant negative effect on total global runoff.[8]

Feedbacks

(If you are not familiar with the concept of a **feedback loop**, read Box 2-3 before continuing.) Feedbacks between land and atmosphere are an important part of the phenomenon of drought. Changes on the land surface can affect weather patterns, including suppressing precipitation; when these land changes are themselves caused or exacerbated by drought, there is potential for a positive feedback loop of self-perpetuating drought. Three mechanisms are illustrated in Figure 2-8. First (Figure 2-8a), drought can cause loss of vegetation, which lowers ET, which decreases the supply of moisture for precipitation. Second (Figure 2-8b), loss of vegetation can lead to an increase in the reflectivity (**albedo**) of the land surface, which can diminish the heating of surface air that is necessary for convective precipitation. Third (Figure 2-8c), wind erosion of dry, unvegetated surfaces can lead to high levels of **aerosols** (solids or droplets suspended in air), which can block radiation from reaching the surface and suppress the upward air movement that generates precipitation.

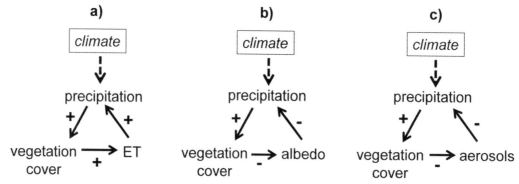

Figure 2-8. Three land–atmosphere positive feedback loops. As an example, in part (a), precipitation has a positive effect on vegetation cover, which has a positive effect on ET, which has a positive effect on precipitation. Thus, an externally driven increase (or decrease) in precipitation will tend to be amplified by the positive feedback loop, leading to a further increase (decrease) in precipitation.

Box 2-3. Feedback Loops

A *feedback loop* occurs when a perturbation caused by an external forcer leads to secondary processes that either amplify or diminish the initial effects of the forcer (positive and negative feedback loops, respectively). Two simple examples will help illustrate (Figure 2-B2).

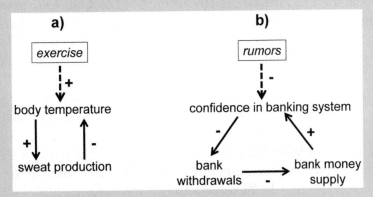

Figure 2-B2. Simple feedback loops: (a) negative and (b) positive. External forcers are shown with boxes and dashed lines; feedbacks are shown with solid lines.

Example 1. When you exercise and your body heats up, you sweat, which cools you down. This is a negative feedback loop, also known as a stabilizing feedback, since it tends to keep your body temperature stable within a certain range. This example is illustrated in Figure 2-B2a, with each arrow illustrating the causal relationship between two parameters: Higher body temperature leads to greater sweat production (+), and greater sweat production leads to lower body temperature (−). (Note that arrows between two parameters are designated as positive if an increase in the first parameter drives an increase in the second and negative if an increase in the first drives a decrease in the second.) Since the sweat–temperature loop has one positive and one negative arrow, the overall loop is negative (a positive multiplied by a negative gives a negative), and the feedback loop tends to counter the original perturbation.

Example 2. When rumors cause investors to lose confidence in a banking system, they will try to withdraw their money, which can cause the banking system to run out of money, further undermining confidence. A bank run is an example of a destabilizing, positive feedback, as illustrated in Figure 2-B2b. This loop is positive, since a positive multiplied by two negatives gives a positive, meaning that it amplifies the initial perturbation caused by the external forcer. Positive loops can lead to rapid changes from one state (banks with plenty of money and confidence) to another (banks with no money and shaken investor confidence). Note that positive feedback loops can operate in either direction depending on the initial perturbation. For example, good financial news can quell a bank run since the positive feedback loop will amplify the initial increase in confidence.

Positive feedback loops feature prominently in modern climate science and ecology, particularly in explaining how a small perturbation can lead a complex system to transition semipermanently from one state to another. For example, Earth's oscillations between glacial (cold) and interglacial (warm) conditions over the last 2 million years are driven by orbital forcing (changes in Earth's orientation relative to the sun), but climate feedbacks are essential in understanding why a small forcing leads to two distinct states that are so different in temperature and ice extent.

4.3. The Great Drought and the Dust Bowl: The Role of Management
In this section, we discuss two historical droughts: the global drought of the 1870s and the Dust Bowl of the 1930s. Both cases illustrate the role that water and land management play in determining the impacts of drought.

Probably the most significant drought of the last 200 years took place in 1876 to 1878, contributing to one of the most severe and widespread famines in recorded history—really a group of famines in India, China, Egypt, Brazil, and elsewhere that together killed at least 50 million people (3 percent of world population). The proximate cause of the famine was crop failure caused by extremely low rainfall throughout the tropics. Scientists have attributed the Great Drought to a very strong El Niño event in the Pacific, probably augmented by anomalies in sea surface temperatures elsewhere.

However, historian Mike Davis has pointed out that "the famines were not food shortages per se, but complex economic crises induced by the market impacts of drought and crop failure."[9] In British-controlled India, for example, where 12 to 29 million people died of hunger and disease, multiple factors were at play:

- The integration of rural India into the economy of the British Empire led to a shift from village-based subsistence farming to an export-oriented agricultural production system.
- The new political and economic order undermined traditional Indian approaches to weathering drought, such as maintaining local food reserves and shifting from crops to pasturing during dry spells.
- The British ignored traditional irrigation systems (often based on small wells) and implemented a system of canal-based irrigation that was more vulnerable to variations in surface water availability.
- During the drought, grain speculation and global demand drove food prices in India out of the reach of rural Indians. In adherence to the dictates of laissez-faire economics, the Raj continued to export large amounts of grain to British consumers (who could afford to pay for it), even as Indian peasants were dying by the millions.
- Among British officials in India, racism and **Malthusianism** combined to produce a stubborn resistance to implementing meaningful drought relief. The loss of life was, in some ways, seen as a "natural" solution to the "problem" of overpopulation.

The Great Drought illustrates an important feature of natural disasters, which we will return to throughout the book: The damage caused by a disaster is related not just to the magnitude of the natural event but also to how societies deal with that event. In other words (Equation 2-1), risk is the product of the likelihood of an event and the consequences of that event. Or, put slightly differently (Equation 2-2), the risk posed by a natural hazard such as drought is a product of three factors: the physical **hazard** (the likelihood of a drought of a given magnitude and duration), the societal **exposure** (how many people are affected by the drought), and the societal **vulnerability** (how the drought affects those people). We may not be able to reduce drought frequency, but we should be able to manage droughts better by reducing exposure and vulnerability.

Equation 2-1. Risk = Likelihood × Consequence

Equation 2-2. Risk = Hazard × Exposure × Vulnerability

Similarly, the Dust Bowl of the 1930s is considered one of the worst "natural disasters" in American history, but in reality it was far from natural, as a brief history of White settlement on the Great Plains will illustrate. In the 1870s, westward expansion brought cattle ranchers to the Great Plains, but overgrazing and harsh winters led the cattle boom to go bust. Ranchers were followed by farmers, who had a brief period of prosperity in the late 1880s, before drought in the 1890s busted that boom as well. In 1909, the Enlarged Homestead Act gave homesteaders 320 acres instead of 160 (although acreage limits were often ignored in any case), and—together with improved farming techniques—brought another wave of immigration to the southern Great Plains. The homesteaders broke up the native sod with their plows and started to grow wheat. Mechanization and the wartime demand for grain accelerated these trends. When wheat prices dropped, farmers responded by increasing production to meet the fixed costs of equipment, leading to a wheat glut.

The drought of the 1930s caught the farmers unprepared and the land unprotected. Without rain, crops couldn't grow. Without crops, the soil—stripped of its drought-tolerant native vegetation—succumbed to the windstorms and left the farms by the ton in massive dust clouds that reached all the way to the East Coast. Some 2.5 million people migrated out of the Dust Bowl region (centered on the panhandle of Oklahoma but including parts of Texas, New Mexico, Colorado, Kansas, and Nebraska), heading to California and the Pacific Northwest in the desperate hope of finding work amidst the Great Depression. The plight of these "Okies" was captured powerfully by Dorothea Lange's photographs, John Steinbeck's fiction, and Woody Guthrie's songs. A 2009 study showed that while the Dust Bowl drought was initiated by La Niña conditions, it was deepened by the feedback loops in Figure 2-8, especially loop (c): the suppression of precipitation by airborne dust.[10]

New Deal programs tried to heal the scars of the Dust Bowl through various soil and moisture conservation techniques and removing land from production. But another Great Plains wheat boom in the late 1940s was followed by the return of drought and wind erosion in the "filthy fifties." One of the tensions of the post–Dust Bowl era was the question of whether and how to reduce the footprint of farming in these marginal lands, a question that is reverberating today as semiarid areas around the world face choices on whether to use limited water supplies for farming or for cities.

If you visit or fly over the Dust Bowl region today, you will find a very different landscape than you would have seen in the 1930s or 1950s, one that is heavily dependent on groundwater irrigation from the Ogallala Aquifer rather than the vagaries of rainfall. While farmers in the southern Plains no longer worry about losing their livelihoods to drought, they do worry about how long the water in the Ogallala will last.

4.4. Paleoclimatology and Megadroughts

Given that droughts are, by definition, unusual events, the historical climate record may be too short to give us a good sense of drought frequency. **Paleoclimatology**, the study of past climates, can be helpful in supplementing the historical record. A variety of proxies have been used to reconstruct past precipitation and temperature, including tree rings, tree stumps, charcoal, lake and coastal sediment deposits, and speleothems (precipitated mineral formations in caves).

Megadroughts—long periods (generally more than a decade) of unusually dry conditions—appear to be recurrent features of the paleoclimatological record. In the US Southwest, PDSI reconstructions (Figure 2-9) show several megadroughts during the ninth to twelfth centuries, as well as a megadrought in the late sixteenth century that is also attested from records of the Spanish conquistadors. Several of these megadroughts lasted for over forty years and were clustered in time, to the point that some have identified them as multicentury megadroughts. Also noteworthy in Figure 2-9 are extended wet periods—megapluvials—when water levels in lakes, wetlands, and rivers must have been much higher than they are today. A new megapluvial today would flood much of our infrastructure and wreak almost as much havoc on our society as a return to megadrought.

Megadroughts are probably driven in part by the climate variability discussed in Section 4.2, but natural variability, amplified by feedbacks, is probably insufficient to explain the length and clustering of these events. Thus, most explanations of megadrought also invoke the fact that climate has changed systematically (not just stochastically) over time.

Over the 5 billion years of Earth history, our planet has experienced dramatic temperature changes, from the Paleocene–Eocene Thermal Maximum of 56 million years ago, when Earth was some 14°C warmer than today, to the Last Glacial Maximum of 22,000 years ago, when much of North America was covered by ice. Changes in regional precipitation have been equally dramatic; what is now the Sahara Desert was lush and green during the African Humid Period of 12,000–5,000 years ago.

In other words, internal climate variability is superimposed on long-term changes in climate, which are driven by external **climate forcers**: factors that change Earth's global or regional energy budget. The most important climate forcers include changes in the sun's output; changes in Earth's position relative to the sun (referred to as orbital forcing or Milankovitch cycles); inputs of light-blocking aerosols to the atmosphere, whether from volcanic eruptions or human emissions; and changes in the concentrations of **greenhouse gases** such as CO_2, whether due to natural or anthropogenic causes. Climate forcers—amplified by feedbacks—affect the basic drivers of global circulation and can thus dramatically change precipitation patterns around the globe, including changing the likelihood of droughts.

But if megadroughts reflect a changed climate, are they really droughts? Yes, one can conceptualize a drought as simply a sample from the lower end of a PDF. But it does not seem useful to conceptualize a fifty-year megadrought as fifty random draws that all happen to come from the lower end of the PDF. Rather, it seems clear that a fifty-year megadrought represents a shift (even if temporary) in the underlying PDF: a change in climate rather than fifty years of unusual weather. Thus, megadroughts pose a challenge to the stationarity model introduced in Section 3.

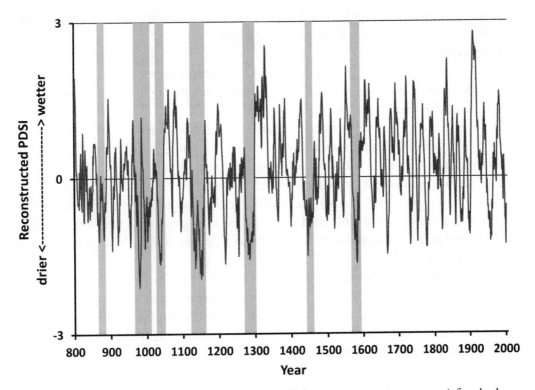

Figure 2-9. Reconstructed June–July–August PDSI (nine-year running average) for the last 1,200 years for the Southwest (32°–43°N, 117°–109°W). Shading indicates megadroughts, defined as time periods when the twenty-year average PDSI was lower than –0.9. Even during these droughts, there were wetter and drier years, but the climate was drier at the multidecadal scale. Data from the North American Drought Atlas (http://drought.memphis.edu/NADA/Default.aspx).

The potential return of megadroughts poses a unique challenge for water managers. The primary tools for drought management are usually focused on short-term efforts to get by until "normal" conditions return. By that measure, a fifty-year dry period is no longer a drought but a "new normal" necessitating permanent changes in water management and land-use patterns. The right term for an extended dry period is ***aridification*** rather than *drought*, since the latter implies that we can get by with emergency measures rather than fundamental changes in how we use water. As we will see in the next chapter, aridification driven by climate change is exactly what certain regions are now facing.

Chapter Highlights
1. ***Blue water*** for human use can come from appropriating renewable flows (rivers and groundwater recharge) or from drawing down nonrenewable stocks (lakes and aquifers).
2. Humanity also taps into the flow of ***green water*** (the ***transpiration*** of soil moisture by plants) by using it to grow ***rainfed*** crops.
3. Renewable flows generated by the global hydrologic cycle amount to about 46,000 km^3/yr of blue water (global ***runoff***) and 71,000 km^3/yr of green water (terrestrial ***ET***).

4. Water availability is highly variable spatially and is determined by climate patterns and the flow of runoff within **river basins**.
5. The temporal variability in runoff can be described using several hydrologic tools—largely rooted in assumptions of **stationarity**—including **hydrographs**, **flow duration curves**, **flood frequency analysis**, and rainfall **intensity–duration–frequency** curves.
6. Drought is a complex phenomenon that can devastate societies built around average conditions. **Climate variability**, land–atmosphere **feedbacks**, and **climate forcing** can each play a role in causing and perpetuating drought.
7. The impact of drought on society is determined by the physical **hazard** (magnitude and extent of drought) but also by societal **exposure** and **vulnerability**. Societies are most vulnerable to drought when they push land and water to their limits and in the process destroy natural resources (such as soils) that can temper drought's worst effects.
8. **Paleoclimatology** has revealed the existence of historical **megadroughts**, challenging assumptions of climate stationarity. **Aridification**, rather than drought, is probably a better term for megadrought, since megadroughts are driven by climate forcing rather than just climate variability and demand deep societal change rather than short-term measures.

Notes

1. Luijendijk et al. (2020).
2. Rockström et al. (2023).
3. Dai (2016).
4. Huang et al. (2017).
5. Peterson et al. (2021).
6. https://droughtmonitor.unl.edu/.
7. Guo et al. (2021).
8. Dai (2016).
9. Davis (2001).
10. Cook et al. (2009).

3 Global and Local Change: The End of Stationarity

- How is climate change affecting the hydrologic cycle?
- How does deforestation affect water supply, and how should we manage forests to protect water quality?
- Is desertification a threat to water availability in drylands?
- If "stationarity is dead," what should replace it?

This chapter picks up where the last left off: with the observation that historical **megadroughts** pose a challenge to the **stationarity** model, since they demonstrate that variability in water supply is not driven solely by random fluctuations around a static average. There are two additional reasons to question the usefulness of stationarity in the **Anthropocene**: Climate change and land-use change (LUC) are significantly affecting water availability, which means that one of the underlying premises of traditional water management—that the future will be like the past—is no longer valid.

1. Climate Change

Human activities—***fossil fuel*** combustion, deforestation, cement production, changes in the nitrogen cycle, and the use of certain synthetic chemicals—have led to increasing atmospheric concentrations of **greenhouse gases (GHGs)**, especially CO_2 but also CH_4, N_2O, O_3, and halogenated molecules. This has led to an average surface warming of 1.1°C between the period 1850–1900 and 2010–2019,[1] with further warming almost certain in the coming decades. Pollution-related ***aerosols*** have masked some of the warming that would have otherwise occurred, by reflecting sunlight and increasing Earth's ***albedo***. (As noted in Chapter 2, aerosols have also in some cases suppressed precipitation by reducing the surface heating that can drive upward air movement.)

Climate change has profound implications for water management. The fundamental parameters of the hydrologic cycle (precipitation, ***evapotranspiration***,

groundwater recharge, streamflow), which set the rules of the game for water managers, are the visible manifestation of the cycling of heat and water vapor through the land–ocean–atmosphere system. Climate change represents a fundamental change in these drivers and thus has far-reaching impacts on water availability.

1.1. Theory and Prediction

Based on climate models and the physical principles that underlie them, climate change is predicted to have multiple, complex effects on the water cycle, summarized below. While they are stated as predictions of the future, many of these changes are well under way, as we will discuss in Section 1.2.

- **Global increases in evapotranspiration (ET) and precipitation (P)**: Water evaporates more at higher temperatures. Thus, climate change will increase ***potential evapotranspiration (PET)*** and ET, leading (all other things being equal) to less water in soils, rivers, lakes, and reservoirs. However, higher ET from land and oceans will increase atmospheric moisture, which will, in turn, lead to higher global P, counteracting at least some of the effects of higher ET. The increase in both P and ET has been referred to as an intensification of the hydrologic cycle.

- **Regional increases and decreases in P**: Despite the global increase in precipitation, some areas will see decreases in P driven by changes in atmosphere and ocean circulation. Models suggest that, in many cases, wet areas in the tropics and high latitudes will get wetter, while dry areas in the subtropics will get drier. This is a consequence of the fact that the circulation patterns that currently move moisture from dry areas to wet ones will have greater capacity for water vapor transport.

- **Lower runoff in many areas**: In any particular location, the change in ***blue water*** availability (***runoff [R]***) will be determined by the balance between the changes in P and ET. (Recall that $R = P - ET$.) Areas where P is unchanged (or decreases) will see lower R. Put another way, many watersheds will see a decrease in ***runoff ratio*** (R/P) due to higher ET. Higher ET also implies more rapid drought development during dry periods.

- **Increased temporal variability**: In addition to changes in average water availability, climate change will bring greater temporal variability to most areas: longer dry spells along with heavier rain events. For areas with seasonal precipitation, the wet season is likely to get wetter (but perhaps shorter) and the dry season hotter and drier.

- **Less snow and ice**: Some two billion people depend on seasonal melting of snowpack and glaciers for a significant portion of their water supply and are likely to experience changes in the timing of water availability as higher temperatures drive earlier snowmelt and a shift in precipitation from snow to rain.[2] There is also potential for a positive ***feedback loop*** as the decrease in winter snowpack decreases albedo, leading to greater heat retention, increased ET,[3] and further loss of snow.[4] For regions such as California, where seasonal snowpack forms an important natural reservoir, earlier snowmelt will exacerbate the seasonal mismatch between supply and demand.

Although many of these effects are inevitable, their magnitude is dependent on the path humanity takes in the coming decades with respect to GHG emissions. Perhaps the most important thing we can do to improve our water future is to mitigate climate change through deep global cuts in GHG emissions. For example, in a business-as-usual scenario, the California snowpack is predicted to decrease by 64 percent by the end of this century, whereas a mitigation scenario will lead to "only" a 30 percent loss in snowpack.[5]

1.2. Observation

To what extent is climate change already influencing the hydrologic cycle? Remember that these anthropogenic changes are occurring in the context of a hydrologic system that is naturally highly variable, so that it may take some time for the climate change signal to fully emerge from the noise. Nonetheless, several studies have found that patterns in P, ET, and R over the last century or so show the increasing influence of climate change.

The Sixth Assessment Report (AR6) of the Intergovernmental Panel on Climate Change (IPCC) reviewed the evidence to date and concluded (with medium or high confidence) that anthropogenic climate change has already caused the following changes in the hydrologic cycle:

- An overall increase in terrestrial ET,
- An increase in precipitation and streamflow in northern high latitudes,
- An overall increase in precipitation intensity,
- Greater seasonal variability in precipitation in the tropics,
- Drier dry-season conditions in several locations, including western North America,
- Earlier spring snowmelt.

One of the more confident and consistent predictions associated with climate change is an increase in hydrologic variability, implying more frequent droughts even in regions that do not see a decrease in average water availability. The emerging field of **extreme event attribution** tries to determine how climate change affects the likelihood or intensity of a given drought (or other event), generally by comparing the actual event to a counterfactual no-climate-change scenario, using either models or a long historical record.[6]

Several drought attribution studies are worth highlighting:

- A study of the 2007–2010 Syrian drought (which some have blamed for helping initiate the Syrian civil war) found that climate change shifted the precipitation **probability density function (PDF)** to lower values, substantially increasing the likelihood of severe drought (Figure 3-1).[7]
- Studies of the 2011–2015 California drought suggest that the precipitation deficit was within natural variability, but warming caused by climate change exacerbated the drought by increasing ET and reducing snowpack.[8]
- A study of the 2000–2021 drought in the US Southwest found that anthropogenic climate change accounted for 42 percent of the decline in soil moisture, converting what would have otherwise been a minor event into a megadrought more intense than any other twenty-two-year period in the last 1,200 years.[9] Even

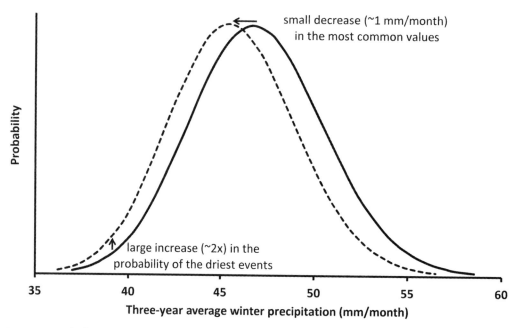

Figure 3-1. Shift in the precipitation PDF for Syria due to climate change. The solid line is the modeled PDF without climate change, and the dashed line is the modeled PDF with climate change. Redrawn from Kelley et al. (2015).

worse, climate change has vastly increased the probability that the drought will continue into the future.

An important theme of these studies is the importance of rising temperatures and ET in creating drought conditions. This is especially significant given that temperatures are almost certain to continue rising, while changes in precipitation are more uncertain. Thus, these results suggest that what we are really seeing in these places is ongoing *aridification* rather than episodic drought. We will still see wetter and drier periods in the Southwest, but average river flows are very likely to trend downwards. There won't be enough water in the wetter periods to pay off the debts accrued during the increasingly common and severe dry periods.

Another clear outcome of rising temperatures has been the recession of glaciers in the world's high-mountain regions (which are often called Earth's **water towers** because of their role in water storage for downstream communities). In the short term, meltwater from these glaciers increases downstream water availability, but over the long term, runoff will decline as glaciers disappear. The point of maximum meltwater flow—sometimes called peak water—has probably already been passed in regions with smaller or more marginal glaciers, including the tropical glaciers of the Andes and many of the glaciers of Europe and western Canada.[10] This phenomenon illustrates clearly the central challenge of water management in a time of change: how to best take advantage of the water that is available now while preparing for a different future.

Glacial melting can also raise water quality concerns through the release of pollutants accumulated over many decades, especially chemicals that tend to preferentially concentrate in colder areas. In addition, newly exposed rock from deglaciation can lead to high concentrations of toxic metals in the meltwater, in a process similar to **acid mine drainage**.

While scientific studies demonstrate the impacts of climate change on the hydrologic cycle, those most directly influenced are voting with their feet. The changing availability of water is already driving population movements both inside countries and across borders,[11] illustrating the way that climate impacts are felt mostly through water. In particular, changes in rainfall patterns are affecting farmers' ability to provide for their families, driving them to migrate to cities (where they are likely to end up living in rapidly growing urban slums with poor water access) or try to reach the United States and other high-income countries. The stresses of migration on international borders and urban centers (including their water systems) are likely to worsen in the decades ahead.

1.3. Indirect Effects

In addition to the direct effects of climate change on freshwater availability, two indirect effects are worth noting: saltwater intrusion and changes in transpiration by vegetation.

Saltwater intrusion: One of the most obvious effects of climate change over the last several decades has been **sea-level rise (SLR)**, driven primarily by the thermal expansion of seawater (i.e., the decrease in water density as the ocean warms) and the loss of ice from mountain glaciers and the ice sheets of Greenland and Antarctica.[12] SLR poses a flooding risk to coastal communities, but our primary concern here is its impact on the availability of groundwater. As sea level rises, the land-to-sea *hydraulic gradient* can be reversed, allowing saltwater to enter coastal aquifers, a phenomenon known as **saltwater intrusion** (Figure 3-2). Overpumping of groundwater can also initiate or exacerbate saltwater intrusion by lowering the water table relative to sea

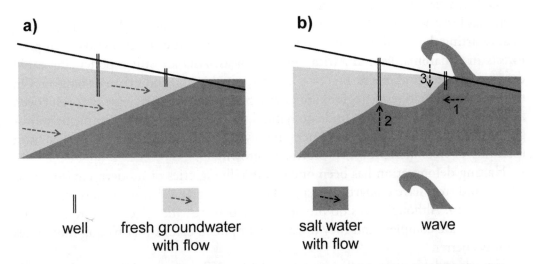

Figure 3-2. (a) Before disturbance, fresh groundwater flows to the ocean, keeping the salt front stable. (b) Three pathways of saltwater intrusion (arrows): (1) lateral intrusion due to sea-level rise and well pumping, (2) upward intrusion due to well pumping, and (3) downward intrusion due to waves.

level or by drawing in saltwater from deeper layers. In addition, storms and waves can introduce seawater at the land surface, contaminating aquifers from above (Figure 3-2b). A recent study suggests that many tropical atoll islands—which typically have average elevations less than 2 m above sea level—will become uninhabitable by midcentury due to **salinization** of the thin lens of fresh groundwater that is their sole source of drinking water.[13]

Changes in transpiration: In addition to its role in causing climate change, CO_2 has two contrasting impacts on the physiology of terrestrial plants. First, where other factors (water, nutrients, temperature) are not limiting, higher CO_2 concentrations lead to higher rates of photosynthesis; this CO_2 fertilization effect is the primary driver of a net global increase in vegetation cover, known as ***global greening***. These higher rates of photosynthesis mean more water lost to transpiration. However, higher CO_2 concentrations also cause plants to reduce their stomatal conductance (i.e., the rate of gas exchange through their leaves). This results in less water loss for a given amount of photosynthesis (i.e., an increase in ***plant water-use efficiency***, defined as the ratio of carbon gained to water lost). This physiological effect can reduce the amount of water lost to transpiration; the magnitude of this effect is poorly understood but may be large.[14]

2. Land-Use Change

People have been modifying Earth's surface for a long time, but over the last 300 years, seminatural landscapes have increasingly been replaced with more intensive land uses, such as grazing land, cropland, villages, and cities.[15] These land-use changes have far-reaching impacts on the hydrologic cycle at local, regional, and global scales. In this section, we start by examining the effects of deforestation on the hydrologic cycle and then look at ways to better manage forests for water supply. We then turn to the topic of desertification and close with a discussion of the impacts of urbanization.

2.1. Deforestation: Watershed Effects

Earth has lost about one third of its forest cover to anthropogenic deforestation, with losses starting thousands of years ago but accelerating in the last 200 years.[16] Tropical forests in Amazonia, central Africa, and Southeast Asia are still experiencing rapid deforestation for timber, grazing, crop cultivation, mining, and urban expansion. On the other hand, temperate regions are now experiencing a net increase in forest cover, as previously deforested areas undergo natural recovery and countries implement programs of reforestation (planting of trees in previously forested areas) and afforestation (planting of trees where few had previously grown).

Halting deforestation has been one of the rallying cries of modern environmentalism, and trees have enjoyed widespread appeal as "green" icons in modern culture and activism. Although trees do provide many benefits, the effects of forests on the water cycle are complex, and they resist oversimplified claims that having more trees is always better.

How does deforestation affect water availability? To answer this, we need to examine how forests affect different components of the hydrologic cycle, keeping in mind our basic water budget, $P = ET + R$. Table 3-1 summarizes the effects we will discuss in this section.

Table 3-1. Typical effects of deforestation on water availability, in roughly the order discussed in the text. Simplified from a complex reality.

Parameter	Increase	Decrease
Roughness		✓
Albedo	✓	
Evapotranspiration		✓
Annual runoff	✓	
Infiltration capacity		✓
Soil water-holding capacity		✓
Dry season flows[a]	✓	✓
Magnitude of floods	✓	
Downwind precipitation		✓
Erosion	✓	

[a]Effects on dry season flows vary from site to site.

Transpiration by trees is usually greater than transpiration by other plants, for several reasons:

- Tree canopies have high **roughness**, which leads to more efficient transfer of moisture to the atmosphere.
- Compared with croplands or pastures, forests tend to have lower albedo (reflectivity), so they capture more radiation, meaning that there is more energy available for transpiration.
- Trees can tap more water sources with their extensive root systems. This is especially true for **phreatophytes**, plants with deep roots that reach the water table and use groundwater (blue water) rather than soil water (**green water**).

In addition, **interception** by trees is high, especially when the canopy is dense, which leads to evaporation of rain before it can reach the soil.

Given that forests have high ET, we expect deforestation to lead to a decrease in ET (Table 3-1). Assuming that there are no changes in precipitation (an assumption that we will question below), deforestation will lead to an increase in streamflow. Conversely, reforestation is likely to increase ET, leading to a decrease in R. Multiple studies have confirmed that annual streamflow is usually—though not always—increased by deforestation and decreased by afforestation and reforestation.[17] Replacement of grasslands with pasture or cropland is likely to have similar effects to deforestation, as loss of deep-rooted, perennial natural vegetation leads to a decrease in ET and an increase in R.[18]

However, there is another factor to consider: the ability of the watershed to store water and how this affects the distribution of runoff over time. In intact forested watersheds, the **porosity** and **infiltration capacity** of the soil are quite high, and the soil can act as a sponge, absorbing water during precipitation events and releasing it slowly during dry periods. This is especially important in highly seasonal climates, where dry-season streamflow is completely dependent on the ability of watershed soils and aquifers to store water and release it over many months. When trees (or grasslands) are replaced with croplands or pasture, there is often an associated degradation of the soil, which may include erosion, compaction, loss of soil organic matter, and decreases in porosity, infiltration rate, and water-storage capacity. This means that more of the precipitation that falls is likely to run off in the immediate aftermath of a rain event, and less will be stored in the soil. This will tend to cause **_flashier_** hydrology: both higher peak flows and lower water availability during the dry season.

The net impact of deforestation on dry-season flows depends on the balance between the two competing effects discussed above: the overall increase in annual flow and the decreased interseason storage of water. Where the former dominates, dry-season flows increase; where the latter dominates, dry-season flows decrease.[19] In terms of dry-season flows, reforestation is not simply the opposite of deforestation, since the sponge effect of healthy soils may recover more slowly than tree cover. The impacts of deforestation and reforestation on dry-season flows in particular locations have often proven to be controversial and difficult to resolve.

The effects of deforestation on water *quality* are simpler than the effects on quantity: Watershed deforestation generally results in a decline in streamwater quality. Bare soil—unprotected by leaf litter and tree roots—is prone to erosion, resulting in high sediment loads to streams, which can increase water **_turbidity_** and bury gravel stream beds in sand and silt. In addition, faster cycling of water through deforested landscapes means less opportunity for nutrient uptake and pollutant removal.

Of course, the water quality impacts of deforestation are highly dependent on what replaces the forest and how the land is managed, as we will discuss in future chapters. Croplands that are left bare in the winter and receive high doses of agrichemicals are much more polluting than agricultural systems that maintain continuous ground cover (e.g., through mulching and cover crops) and use fertilizers and pesticides more judiciously. Well-managed pastures can maintain healthy soils, while overgrazed areas often experience very high erosion rates. More intensive land uses, such as industry and cities, tend to have high pollutant loadings, although wastewater treatment facilities can significantly reduce those impacts.

2.2. Deforestation: Impacts on Precipitation

Up to now, we have assumed that precipitation is a fixed external input that is unaffected by LUC, so that changes in runoff can be understood by consideration of changes in ET alone. In fact, as we noted in Chapter 2, there are strong land–atmosphere feedbacks through which LUC can affect precipitation.

One specific context in which land use is known to affect net precipitation is in tropical montane cloud forests. In these unique, biodiverse ecosystems, estimated to cover about 600,000 km² globally,[20] the forest canopy draws moisture from the air through the interaction of trees and associated flora with ambient **_fog_**. At some sites, this process may increase net precipitation by over 1,000 mm/yr, as water vapor condenses on trees and then drips to the forest floor (or gets absorbed directly by leaves), making these forests potentially significant local water sources. Deforestation in these systems may lead to a decrease in water capture and thus a decrease in R.[21] In temperate coastal areas with frequent fog, forested ecosystems—while not technically cloud forests—can exhibit similar water cycles. In northern California, redwood forests obtain about a third of their annual water requirement from fog,[22] and declines in fog frequency and extent pose threats to their survival.[23]

Beyond the specific case of cloud forests, deforestation can affect precipitation in ways that are complex and site-specific. Several mechanisms can be at play (see also Figure 2-8):

- Deforestation reduces ET, which can lead to lower atmospheric moisture content and decreased precipitation in downwind areas. This is especially important in areas that are highly dependent on ***terrestrial moisture recycling*** (Box 3-1).
- In some cases, especially in semiarid locations, loss of vegetation can lead to sandstorms that vastly increase the concentration of inorganic aerosols, to the point where they inhibit precipitation by reflecting sunlight and preventing surface heating (Chapter 2).[24]
- Deforestation can change the distribution of energy, air pressure, and winds across the landscape in complex ways, leading to scale-dependent threshold effects on precipitation.

Note that many of these effects on precipitation will be felt both locally and downwind from the site of deforestation. For example, deforestation in the West African Rainforest (well outside the Nile Basin) can affect flow in the Nile River, since about half of the precipitation in the Ethiopian Highlands comes from air masses that have passed over the rainforest.[25]

One study estimates that LUC globally has reduced terrestrial precipitation by 590 km^3/yr.[26] Combined with the decrease in terrestrial ET—estimated by the same authors at 1,250 km^3/yr—this suggests that LUC has caused a net increase in global runoff of somewhere around 660 km^3/yr.

Box 3-1. Terrestrial Moisture Recycling and Precipitationsheds

Depending on location, water that is transpired by terrestrial vegetation can condense into clouds and fall as precipitation on downwind land areas, a phenomenon called terrestrial moisture recycling (TMR). For small islands and windward coasts such as California, TMR is of negligible importance since almost all the precipitation originates as oceanic moisture. However, for locations downwind of continental land masses, such as much of China, a large fraction of annual precipitation originates from terrestrial ET, making these locations vulnerable to ET reductions in upwind source areas, such as those driven by deforestation.[a]

Better understanding of TMR has lent increased importance to the concept of the ***precipitationshed***,[b] defined as the source area whose ET contributes to precipitation in a given downwind location. Precipitationsheds are more poorly defined than watersheds, because different seasons or wind conditions can bring moisture from different directions. Still, changes in land (and water) use within precipitationsheds can have real effects (both positive and negative) on precipitation and water availability.[c]

Notes
[a] Keys et al. (2019).
[b] Keys et al. (2018).
[c] Lo and Famiglietti (2013).

2.3. Managing Forests for Water Supply

> Water is one of the most important natural resources flowing from forests.
> —US Forest Service[27]

Forests provide water to both ecosystems and people, and the way that forests are managed affects the quantity and quality of that water. Because they are found in relatively wet areas, forests often contribute disproportionately to water supply. In the conterminous United States, forested land occupies 29 percent of the land surface but provides 46 percent of the surface water flow.[28]

Deforestation and forest degradation can increase erosion and pollutant loading, affecting the quality of drinking-water sources and increasing the costs of water treatment.[29] A 2016 study found that over the period 1900–2005, the source watersheds of large cities around the world had large increases in population density and agricultural activity, resulting in increases in sediment, P, and N loads of 40 percent, 47 percent, and 119 percent, respectively.[30] The study estimates that this has increased treatment requirements (and thus costs) for about 30 percent of large cities.

Although some forest management agencies (e.g., US Forest Service) recognize their responsibility to manage for water, many forests are managed by private entities for whom water may not be high on the agenda. This is especially a problem when forested lands are located some distance upstream of communities that rely on water from those lands; the spatial disconnect means that upstream land use has a large effect on downstream water supply, but upstream land users have no incentive to take that into account. The concept of ***payment for ecosystem services (PES)*** has emerged as a way for cities to compensate upstream communities for conservation activities that prevent the degradation of water supplies, rather than spending money on water treatment. The Nature Conservancy has led the way in implementing PES with its Water Funds, described as "enabl[ing] downstream water users—like cities, businesses, and utilities—to invest in upstream land management to improve water quality and quantity and generate long-term benefits for people and nature."[31]

How should PES funds be spent? More broadly, what does it mean to manage forests for water (beyond preventing conversion to agricultural and urban uses)? Of course, forest managers will also be managing for other goals, such as timber, wildlife, and recreation, but from a water perspective, we can identify three goals of forest management: maintain water quality, avoid catastrophic forest change (e.g., wildfire), and increase water yields.

MAINTAIN WATER QUALITY
Protecting the water quality of forest runoff can be compatible with other uses, including recreation and limited timber harvesting, as long as soil compaction, erosion, and surface runoff are minimized. Three areas are of particular concern:

- Roads should be constructed and maintained for good water drainage to avoid concentrated flow and erosion. When roads must cross streams, adequate passage of water, sediment, and aquatic organisms under the road should be provided.

- Buffers along streams should be protected from harvesting and other activities; these buffers should be larger where slopes are steeper.
- Harvesting should use equipment that minimizes impact to the soil surface. Clear-cutting of large areas should be avoided.

AVOID CATASTROPHIC CHANGE

Recent record-breaking fire seasons in the western United States have raised awareness of the way that hot and dry conditions—caused in part by climate change—can fuel massive, unpredictable wildfires. Past forest management has also played an important role in driving these megafires, as a history of fire suppression has left forests overstocked and carrying high fuel loads, while pest outbreaks (including by invasive species) have left millions of dead trees ready to burn. In contrast, pre-settlement fires in the western United States—used as a land management tool by Native Americans—were probably smaller, more frequent, better controlled, and targeted at specific outcomes.

Mega-wildfires can have long-term impacts on drinking-water supplies. Rainfall on denuded, fire-scarred surfaces tends to run off quickly, carrying massive loads of sediment and ash, which can contaminate water supplies and take up valuable space in reservoirs. When fires burn through communities, damaged water pipes and other infrastructure can leach contaminants.[32] At large enough scale, aerosols from wildfires can even suppress precipitation, potentially creating drought that can fuel future wildfires.[33] In Colorado, several communities have conducted large-scale aerial mulching operations to reduce postwildfire erosion, but more attention is also needed to preventing these catastrophic fires in the first place.

Water managers in forested regions have increasingly directed their efforts to activities that reduce fire risk, such as thinning of overgrown forests—either mechanically or through ***prescribed burns***—and creating mixed-age, mixed-species forests that are less susceptible to disease outbreaks. In California, in an example of PES in action, the Yuba Project, designed to protect water quality (and other values) by increasing forest resilience to wildfires, was funded through a Forest Resilience Bond by private investors, who are being paid back by the Yuba Water Agency and the state of California.[34]

INCREASE WATER YIELD

In most cases, increasing ***water yield*** (runoff per unit area) should be a lower priority than protecting water quality and preventing catastrophic wildfire. But there are situations where thinning or burning can increase water yield, thus mitigating the impacts of climate change on water supply while also reducing wildfire risk.[35] For example, thinning projects in central Arizona, such as the Four Forest Restoration Initiative, could increase water supply in the Salt and Verde River systems by 8 percent (representing a net present value of over $100 million) while posing minimal risk of excessive sediment loading.[36] Similarly, ***cultural burns*** are being used by tribes in California to restore culturally important meadows and plants that need higher water tables.[37] In Cape Town, South Africa, which experienced a devastating water supply crisis in 2018, invasive trees transpire more water than native shrubs and grasses;

removing these trees could generate over 50 MCM/yr at a cost that is lower than that of alternative sources,[38] and the Greater Cape Town Water Fund is moving forward with this "nature-based solution."[39]

In regions with seasonal snowpacks, forests can be managed to maximize the accumulation and retention of snow, allowing greater water storage and higher dry-season flows. Depending on conditions, recommended forest management for snowpack may range from large gap creation to light thinning[40]; the orientation and size of patches may be important as well.

We close this section with a word on tree plantations. Around the world, natural forests are being replaced with industrial tree plantations designed to efficiently produce timber or other products such as palm oil. Plantations differ from natural forests in that they generally consist of evenly spaced, even-aged, short-rotation monocultures, often using species that are not native to the region. Activists have pointed to the damage that plantations can do to ecosystems and land/food sovereignty and have argued that plantations are not forests and should not qualify for carbon credits.[41]

The effects of plantations on water also differ from those of natural forests:

- Because plantations use rapidly growing, high-yielding species, they often transpire more water than native forests, thus depleting streamflow.
- In plantations, large areas are often harvested at one time, leaving the surface susceptible to erosion.
- Because plantations are monospecific, they are susceptible to pest outbreaks, which are then treated with pesticides that can contaminate local waterways.

Interestingly, South African law recognizes that plantations are net water users and requires commercial forestry operations to obtain a Streamflow Reduction Activity license for the difference in ET between the natural land cover (often grassland) and the plantation.[42]

2.4. Desertification

Desertification is defined by the 1994 UN Convention to Combat Desertification as "land degradation in arid, semi-arid and dry sub-humid areas resulting from various factors, including climatic variations and human activities." The latter part of the definition hints at a historical and ongoing debate over whether external climate drivers ("climatic variations") or internal dynamics ("human activities") are primarily responsible for this form of LUC.

Internal dynamics: The anti-desertification movement has focused primarily on local land degradation caused by population growth, overgrazing, and deforestation. It is influenced by historical desertification theory (Box 3-2)—the idea that deserts have low precipitation because of the absence of vegetation, not the other way around—together with modern scientific understanding of the ways that land degradation can exacerbate or even create drought through positive feedbacks. The UN Convention to Combat Desertification itself arose from concerns that the devastating Sahel drought of the 1960s–1990s was a manifestation of those feedbacks: a shift to a drier state because of overgrazing and loss of vegetation. This was consistent with contemporary

> **Box 3-2. Desertification Theory in the Nineteenth-Century United States**
>
> > Rain Follows the Plow. By the repeated processes of sowing and planting with diligence the desert line is driven back, not only in Africa and Arabia, but in all regions where man has been aggressive, so that in reality there is no desert anywhere except by man's permission or neglect.—Charles Dana Wilber[a]
>
> In this famous (and incorrect) declaration, Wilber, a booster of settlement in the western United States, argued that planting of trees and crops could increase rainfall in semiarid areas, driving a positive feedback loop of increased water availability and allowing successful rainfed agriculture in the semiarid Great Plains. His argument was bolstered by a string of wet years that fortuitously followed the westward migration initiated by the Homestead Act of 1862. However, John Wesley Powell—one-armed Civil War veteran, geology professor, leader of an 1869 expedition on the Colorado River, and second director of the US Geological Survey—was skeptical, arguing that if rainfall *really* had increased (he didn't entirely trust the limited data), it would probably decrease again as part of natural variability; this is exactly what happened in the 1880s. The disagreement over the science had larger implications; Wilber supported the westward expansion of small-homestead Jeffersonian dryland farming across the Great Plains and beyond, while Powell believed that the most appropriate land uses in most of the arid West were irrigated agriculture and livestock grazing.
>
> **Note**
> [a]Wilber (1881).

science, which noted that this region on the edge of the Sahara Desert was particularly susceptible to climate shifts driven by land–atmosphere feedbacks.

External drivers: In recent years, the increasing power of Earth system models has allowed scientists to demonstrate that the primary driver of the Sahel drought was not local land degradation but external forcing from sea surface temperatures. In particular, warming of the tropical oceans (due to climate change), together with cooling of the North Atlantic (due to aerosols from US air pollution), limited the northward reach of the ***Intertropical Convergence Zone*** and weakened the West African ***monsoon***; land–atmosphere feedbacks then amplified this change. The end of the drought in the 1990s was probably driven more by the implementation of air pollution controls in the United States and subsequent warming of the North Atlantic than by anything else.[43] This realization has led some to call for the "end of desertification" as a concept, noting that the institutions set up to combat desertification are grounded in a misunderstanding of the process and an undervaluing of desert ecosystems.[44] Yet the dual problems of drought and land degradation are still important ones, even if they are not linked bidirectionally as strongly as we once imagined.

One project that reflects this evolution in our understanding of desertification is the Great Green Wall of Africa. Originally conceived as a 15-km-wide band of dense tree cover running 7,000 km across the Sahel to "hold back" the Sahara Desert, the project has evolved into a more complex, site-specific effort to improve the health of landscapes, soils, and people. This reconception was driven in part by the failure of initial, top-down efforts to plant large numbers of trees; trees often did not survive, and when they did, they hindered local livelihoods. At the same time, farmers were experimenting with more sustainable farming practices that included allowing favored tree species to regrow naturally at low densities. The incorporation of trees into productive farming landscapes, referred to as ***agroforestry***, has provided multiple benefits in the Sahel, including soil protection, fodder provision, decreased reliance on artificial fertilizers, and improved household food security.[45]

The world's other Great Green Wall project—China's top-down efforts to afforest millions of square kilometers of dryland in its northern provinces to prevent sandstorms and achieve other benefits—has seen poor tree survival rates, along with depletion of water supplies due to higher ET.[46] Trees in themselves are not a panacea for environmental problems, especially in semiarid grasslands.

2.5. Urbanization

Urban areas make up a small fraction of global land use and are unlikely to affect the water balance at global or regional scales. However, they have strong effects on the local water cycle, especially the distribution of runoff over time.

Most of the hydrologic impacts of urbanization stem from the presence of ***impervious surfaces***: areas where soil has been covered over with a surface that cannot absorb water, such as a building or road. Rainfall that hits an impervious surface will run off quickly to the nearest stream—or to the nearest storm drain, from where it will get transported in a pipe to the stream. This contrasts with the behavior of rainfall on a vegetated surface, where most of the water from a typical rain event gets absorbed into the soil and either infiltrates to groundwater (sustaining baseflow) or travels to the stream as stormflow through subsurface flow in the soil.

Thus, impervious surfaces fundamentally change a watershed's response to rain events. Streams in urban areas are uniformly flashier than their undeveloped counterparts, with a quicker and higher stormflow peak and a more rapid return to baseflow. In many cases, urban streams also have lower baseflow than undeveloped streams, since the rapid runoff of precipitation leads to less recharge of the groundwater that sustains baseflow.

The water that flows rapidly from urban landscapes in response to storm events is called ***urban runoff*** or ***stormwater***, and it is often polluted as well as hydrologically altered. We will discuss imperviousness and urban water quality issues in more detail in Chapter 15.

There is emerging evidence that urbanization can also increase the maximum precipitation rates during storm events by increasing surface warming and moisture convergence (due to higher roughness), both of which favor precipitation. For example, one study found that the probability of the unprecedented rainfall that Hurricane Harvey dumped on the Houston area in August 2017 (more than 1,300 mm in six days) was increased twenty-one-fold by urbanization.[47]

3. Managing Change

Both climate change and land-use change are modifying the water cycle at global to local scales. This phenomenon—together with increasing recognition of natural climate variability from paleoclimate studies—led a group of prominent scientists to proclaim in 2008 that "stationarity is dead."[48]

How should water managers and planners deal with a rapidly changing hydrologic environment, in which assumptions of stationarity—which underlie most traditional water management—are no longer valid? How do we meet the needs of both people and ecosystems in a world in which many places are seeing long-term declines in water availability, more frequent and severe drought, and more intense flooding? These questions will occupy us throughout the book, but I offer some preliminary thoughts here.

First, managers must acknowledge four basic facts about climate change: (1) it is real, (2) it is caused by human activity, (3) it is causing changes in water availability that must be planned for, and (4) mitigating its worst effects requires concerted efforts to reduce GHG emissions.

Beyond those basics, things get more complicated. How exactly should we plan for changes in water availability, given the uncertainty attached to climate change projections, especially when it comes to local precipitation? Some have argued that our science is now good enough to forecast future conditions with reasonable accuracy. In the short term, though, I believe it would be an illusion to expect that climate models can simply give us new (and evolving) PDFs that better reflect our new reality and can be used for planning purposes. Instead, global change gives us an opportunity to decrease our reliance on the old deterministic models that were flawed even without climate change and to find new ways of living with uncertainty. The truth is that water has always been variable and unpredictable; climate change just makes that even truer and more obvious.

In short, we need a new water management based on resilience. We need water management that takes seriously the possibility of unusual events that individual water managers have not previously experienced. We need water management with a diverse array of supply, demand, and storage solutions. We need water management that builds in a margin of safety rather than allowing water demand to push up against the limits of water availability. We need water management that responds nimbly and creatively to changing conditions.

But the changes we need are not limited to water management. Changing water availability will require fundamental changes in how we live on the land. In some regions, such as the US Southwest, this may mean shrinking the size of desert cities and farms. In other regions, such as the Horn of Africa, this may mean shifting from rain-dependent pastoralism to irrigated farming. In still others, such as coastal Louisiana, this may mean **managed retreat** from low-lying areas. These changes will not be easy. But they will be easier if we plan for them. And they will be easier to plan for if all of us—the general public as well as water professionals—are aware of the limits that water places on us and of the folly of believing that we can ignore those limits.

Chapter Highlights

1. Climate change and land-use change are modifying global and local water availability, posing a threat to infrastructure and institutions built around past conditions.
2. The clearest impacts of climate change on the hydrologic cycle are those that stem directly from higher temperatures: an increase in both ET and P, more intense droughts and floods, earlier snowmelt, a decrease in snow and ice storage in **water tower** regions, and a decrease in runoff ratio in many locations. Climate change is also likely to affect the spatiotemporal distribution of precipitation in complex ways, leading to effects on local runoff that are harder to predict.

3. Climate change is making some extreme events, including droughts, more likely; ***extreme event attribution*** can now estimate the extent to which a given event was made more likely by climate change.
4. Other hydrologic effects of climate change and CO_2 include sea-level rise and ***saltwater intrusion*** into coastal aquifers, ***global greening*** (resulting in higher transpiration rates), and a decrease in stomatal conductance (resulting in higher ***plant water-use efficiency***).
5. With some exceptions (such as cloud forests), the effects of deforestation on water availability are driven by three factors: Deforestation reduces ET, thus increasing blue water availability; deforestation reduces soil water-storage capacity, producing a ***flashier*** hydrology; and deforestation can reduce precipitation in downwind regions, especially those that are heavily dependent on ***terrestrial moisture recycling***.
6. Forests are an important source of water supply and should be managed, at least in part, to protect water quality by minimizing erosion, pest outbreaks, and catastrophic wildfire. In some cases, forest thinning and ***prescribed*** or ***cultural burns*** can increase ***water yields*** while also increasing forest diversity and resilience.
7. Great Green Walls are being implemented in Africa and China to fight ***desertification***. Although dense planting of forests in arid regions is unlikely to be successful in increasing water availability, appropriate ***agroforestry*** practices can improve soil health and productivity in these areas.
8. In urban areas, the presence of ***impervious surfaces*** creates flashy ***stormwater*** that poses a significant problem in terms of flooding, water quality, and the health of urban streams.
9. "Stationarity is dead," but models are not yet good enough to provide actionable predictions, so we need our water systems to be resilient to uncertain future conditions.

Notes

1. Gulev et al. (2021).
2. Mankin et al. (2015).
3. Milly and Dunne (2020).
4. Air pollution also decreases the albedo of the snowpack by increasing the concentration of black carbon and other light-absorbing particles deposited on the snow. Remarkably, the decrease in economic activity associated with COVID-19 led to cleaner snow in the mountains of the Indus River Basin, affecting the melt timing of 6.6 km^3 of water; Bair et al. (2021).
5. Reich et al. (2018).
6. Swain et al. (2020); see also https://www.worldweatherattribution.org/analysis/drought/.
7. Kelley et al. (2015).
8. Seager et al. (2015), Berg and Hall (2017).
9. Williams et al. (2022).
10. Hock et al. (2019).
11. Zaveri et al. (2021), Borgomeo et al. (2021), https://www.nytimes.com/interactive/2020/07/23/magazine/climate-migration.html.
12. Water management has also had a small impact on sea level. In particular, groundwater depletion has reduced the amount of water stored on the continents and contributed to higher sea levels, and construction of dams has increased terrestrial water storage (in reservoirs and in the soil and rocks beneath them) and thus has lowered sea levels relative to a no-dam scenario.

13. Storlazzi et al. (2018).
14. Mankin et al. (2019), Sampaio et al. (2021).
15. Ellis et al. (2021).
16. https://ourworldindata.org/deforestation.
17. Acreman et al. (2021), Filoso et al. (2017).
18. Replacement of grassland with cropland may sometimes be an exception; the denser spacing and higher growth rates of crops may lead to an increase in ET; Destouni et al. (2013).
19. Peña-Arancibia et al. (2019) suggest that about 20 percent of tropical forested area falls into the latter category.
20. Karger et al. (2021).
21. Nonetheless, Bruijnzeel et al. (2011) argue that for most cloud forests, the net effect of conversion to pasture would be an increase in runoff because the reduced transpiration would be larger than the reduced fog capture.
22. Dawson (1998).
23. Johnstone and Dawson (2010); https://www.usgs.gov/centers/western-geographic-science-center/science/pacific-coastal-fog-project.
24. In other cases, deforestation can inhibit cloud formation because trees release bioactive aerosols that can serve as efficient nuclei for droplet formation.
25. Gebrehiwot et al. (2019).
26. Wang-Erlandsson et al. (2018).
27. https://www.fs.usda.gov/managing-land/national-forests-grasslands/water-facts.
28. Liu et al. (2022).
29. Lopes et al. (2019) show that, in Portugal, drinking water systems with greater forest cover have lower water-treatment costs, although the effect is small. Mapulanga and Naito (2019) show that, in Malawi, populations living in areas with more deforestation are less likely to have access to clean drinking water (after correction for other variables).
30. McDonald et al. (2016).
31. https://waterfundstoolbox.org/.
32. Proctor et al. (2020).
33. Tosca et al. (2010).
34. https://nff.maps.arcgis.com/apps/MapJournal/index.html?appid=5efcaae52465483d9d6848c1affied50.
35. Sun et al. (2015).
36. Simonit et al. (2015).
37. https://californiasciencecenter.org/funlab/ever-wonder/2022-02-02/if-fire-can-be-good-with-chairman-ron-goode.
38. Stafford et al. (2019).
39. https://www.nature.org/en-us/about-us/where-we-work/africa/stories-in-africa/nature-based-solutions-could-protect-cape-town-s-water-supply/.
40. Dickerson-Lange et al. (2021).
41. https://redd-monitor.org/2021/09/21/plantations-are-not-forests-2/.
42. Edwards and Roberts (2006).
43. Giannini and Kaplan (2019).
44. Behnke and Mortimore (2016).
45. Reij and Garrity (2016).
46. Cao et al. (2011), Tao et al. (2020).
47. Zhang et al. (2018).
48. Milly et al. (2008); see also Milly et al. (2015).

4 Water Use: From Ancient to Modern Times

- What role did water play in the development of civilization?
- What technologies and approaches have societies developed to deal with water's variability?
- Is water management in the modern era fundamentally different from that of earlier periods?
- How much water does humanity use, and how has that changed over time?

The previous two chapters introduced us to the *supply* of water; this chapter focuses on *demand*, exploring how people and societies have used that supply, including coping with the spatial and temporal variability that featured so prominently in our discussion of water availability.

Water is the lifeblood of society. We humans need water for survival, for economic development, and for our physical, mental, and cultural health. Because of this dependence, the development of civilization over the last 12,000 years has been closely intertwined with water management.

This chapter starts with a brief and selective history of water management, with a focus on two uses of water—agricultural and urban—that were foundational to the development of human society. (The history of other water uses, including power, navigation, sanitation, and flood management, will be covered in later chapters.) We then turn to an overview of hard-path water management in the modern era and close by looking at how much water we use for different purposes.

1. Premodern Agriculture

The development of agriculture (starting sometime around 8000 BCE) was, for good or ill, a critical step in the evolution of human society, a step that was tightly tied to the control of water. Agriculture led to the first significant human interventions in the water cycle and still today accounts for the majority of anthropogenic water use.

In areas with adequate rainfall, crops can be grown with **green water** (**rainfed agriculture**), which we will not discuss here because it leaves little archeological evidence and requires less innovation in water infrastructure and institutions. In arid and semiarid regions, crops often need some form of *irrigation*, which will be our focus in this section. (In some *drylands* with growing-season rainfall of at least 200 mm, farmers practice a form of rainfed agriculture called *dry farming*, in which drought-tolerant crops are grown without supplemental irrigation.)

Three themes will emerge from this brief overview. First, the water solutions that people have developed are a function of their specific geographic and cultural realities and thus have varied dramatically in different places and times, although there has also been a move toward some common techniques through technology sharing and convergent cultural evolution. Second, in many places water management has been successful in allowing societies to thrive for centuries, only to fail at some point due to environmental degradation, climatic changes, or sociopolitical upheaval. Third, some premodern water solutions appear to be more resilient and sustainable than current technologies; we would do well to recover some of these practices and add them to our water management toolkit.

1.1. Large-River Irrigation

Among the first societies to practice large-scale irrigation were the four large-river civilizations of the ancient world. The rivers in Table 4-1 share two important characteristics. First, they are *exotic*, meaning that the water they carry is generated in a humid area (for the Nile, the wet zones of East Africa) and then flows into an arid region (the desert climate of Egypt). Second, they are highly seasonal large rivers that flow through large, flat floodplains, where seasonal floodwaters can easily spread (naturally or with the aid of irrigation canals) over great areas of *alluvial soils*. In many respects, these sunny regions with productive soils and largely predictable floodwater flows are ideal for agriculture.

Irrigation seems to have developed gradually from earlier practices that tapped into the benefits of flooding, including flood-rise agriculture, in which rice and other flood-tolerant crops were planted in the *floodplain* as the river was rising, and flood-recession agriculture, in which crops were planted in moist floodplain soils as floods were receding. From there, it was a small step to manipulating this flooding by constructing diversion structures, canals, and dikes to direct the floodwaters into individual fields; this allowed greater control, as individual canals and dikes could be opened and closed, and floodwater could be stored in depressions for later use. Eventually, these irrigation systems expanded and grew more sophisticated, with networks of earthen ditches extending over large areas.

The development of irrigation allowed greatly increased food production and

Table 4-1. Ancient large-river civilizations. All these civilizations developed cities, writing, and large-scale irrigation in the millennia before the Common Era.

River System	Civilizations
Nile	Egyptian
Tigris and Euphrates (Mesopotamia)	Sumerian, Akkadian, Babylonian, Assyrian
Indus	Harappan
Yellow and Yangtze	Chinese

> **Box 4-1. Hydraulic Societies**
>
> German historian Karl Wittfogel (1896–1988), in his 1957 book *Oriental Despotism*, argued that irrigation-based **hydraulic societies** are naturally driven toward despotism by the need to build and maintain large irrigation systems. His thesis is now seen as simplistic and marred by racism, but his insights have been modified and refined by others. American historian Donald Worster, in his 1985 book *Rivers of Empire*, identifies three modes of organization for irrigation societies: the local subsistence mode (small-scale irrigation creating a mostly equal society), the agrarian state mode (large-scale irrigation leading to radical social stratification, as in Mesopotamia), and the capitalist state mode, which he sees at work in the contemporary American West and which he describes as a "coercive, monolithical, and hierarchical system, ruled by a power elite based on the ownership of capital and expertise."

greater concentrations of people than ever before. The relationship between irrigation and political structure is complex (Box 4-1), but in Mesopotamia at least, there appears to be a link between the various characteristics of the early city-states: the use of irrigation to create food surpluses that could support a ruling class, the need for forced labor to maintain irrigation systems, and the development of writing as a way of keeping track of grain production and tax payments.

Mesopotamia also saw periodic collapses of its early city-states. Although many factors, including ***megadroughts***, are thought to be responsible (Box 4-2), one factor was probably the difficulty of sustaining irrigation systems over long time periods, due to the twin problems of ***siltation*** and ***salinization***. The silt carried by the Tigris and Euphrates from the Turkish highlands was responsible for generating the fertile alluvial soils of Mesopotamia, but it also tended to clog irrigation canals, especially since the entire region is very flat. The salt problem was even more insidious. Irrigation water—even when it is of high quality—brings with it some dissolved solids. As the water is transpired by crops, the salts are left behind and can accumulate to the point where they affect plant growth.

1.2. Small-River Irrigation

Unlike Asia and Africa, the Americas don't have large exotic rivers with vast floodplains; instead, smaller, steeper exotic rivers were important in the development of irrigation-based civilizations.

The Norte Chico of northern coastal Peru, one of the earliest civilizations in the Americas, reached its peak during the Late Archaic Period (3000–1800 BCE), producing a network of cities with impressive monumental architecture (though no writing or pottery). The region consists of an arid narrow strip where the foothills of the Andes meet the Pacific Ocean. Most Norte Chico cities appear to have been situated along four small rivers whose flow originates in the Andes. The relative importance of irrigation and marine food sources in this culture has been disputed by scholars, but it seems likely that corn and other irrigated crops contributed significantly to the food supply and that irrigated cotton was used for fishing nets. At one site, small irrigation canals have been found that date to about 4000 BCE.[1] Unlike the large-river floodplain irrigation systems described above, the Norte Chico irrigation systems were probably small and decentralized, with networks of small canals diverting water into ***terraced*** fields arranged along the hillside. The steep terrain brought advantages

> **Box 4-2. Drought and Collapse: What Does History Have to Teach Us?**
>
> The collapse of the Akkadian Empire in Mesopotamia (about 2200 BCE) is part of a long list of well-documented cases of sudden societal declines. What caused these flourishing societies to suddenly disappear, and does the same fate await modern civilization? Jared Diamond, in his controversial 2005 book *Collapse: How Societies Choose to Fail or Succeed*, identifies five sets of factors that tend to drive collapse, with the first two being especially relevant to this book: anthropogenic environmental damage, such as deforestation, that erodes the society's resource base, and natural climate change, especially drought. Karl Butzer argues that a focus on climate change and environmental damage is an example of oversimplified monocausal environmental determinism and that "societal inputs and feedbacks are more common [in cases of collapse] than environmental variables."[a] However, Harvey Weiss and colleagues draw on increasingly precise paleoclimate proxies to show the uncanny correspondence between megadroughts and documented collapses, and they argue that "archaeological data shows that preindustrial human societies could not withstand megadrought, that circumstances rendered irrigation innovations physically impossible, and that they adapted through political collapse and migration to sustainable environments."[b]
>
> **Notes**
> [a]Butzer (2012).
> [b]Weiss (2017).

(the natural flushing of salt from soils) and disadvantages (the lack of flat ground for cultivation and the threat of soil erosion, both of which were partly addressed by terracing).

Another noteworthy irrigation-based civilization was established by the Hohokam in the US Southwest, drawing on the waters of two small exotic rivers, the Gila and the Salt, which meet near the modern city of Phoenix. The Hohokam irrigation system reached its peak between 1100 and 1450 CE, with some 1,500 km of carefully engineered canals distributing water to a vast network of farms. These canals were rediscovered in the twentieth century and used as the basis for the modern irrigation network. The area was largely depopulated during the second half of the fifteenth century, perhaps because of the difficulty of maintaining such a complex canal system in a fragile desert environment.

1.3. Runoff Farming

In dry regions without access to perennial rivers, early agriculture depended on taking advantage of infrequent rainstorms by capturing as much surface runoff as possible and directing it to fields, a practice known as ***runoff farming***. There is evidence that farmers in Jordan were using stone walls in ***wadis*** to divert stormwater runoff to fields as early as the seventh millennium BCE. The Mar'ib Dam in Yemen, whose construction began as early as the third millennium BCE, was designed to divert flood flows from Wadi Adhana to support food production for some 30,000–50,000 people; the ruins of its stone abutments still stand today, but the earthen dam itself was breached repeatedly during large floods, a feature that may have been intentionally designed to avoid damage to the diversion canals and fields.

In the Negev Desert in southern Israel, Copper Age residents developed sophisticated runoff farming systems in which runoff from hillslopes was directed to valley-bottom farms that were terraced to retain water. Gravel was apparently raked from the hillsides into mounds to expose the underlying soil, which would form a surface crust in response to rain, leading to reduced infiltration and increased run-off.[2] These practices, along with diversion channels extending from the larger wadis, allowed sufficient food production to sustain the population, despite the low and unpredictable rainfall.

1.4. Groundwater Use

In areas where surface water is limited and the water table is shallow, the practice of digging down to access groundwater is an ancient one. Among the oldest known wells is one in Cyprus that has been dated to 8000 BCE. Of course, the depth, construction materials, and sophistication of wells have varied over time and space.

One unique variant on a well is the **cenote**, a water-filled sinkhole common in the northern Yucatan Peninsula that forms naturally through dissolution of limestone in this **karst** landscape. Especially given the absence of surface water (another common feature of karst landscapes), the Maya used these as water sources and religious sites, often modifying them to improve access.

In the Indian subcontinent, **stepwells**, in which stone steps are used to access groundwater, were being built as early as 200 CE and reached their pinnacle with magnificent works of architecture such as the Chand Baori, with its 3,500 steps. Stepwells are suited to India's **monsoonal** climate, in which groundwater levels fluctuate dramatically through the year; during the wet season, water rises to near the surface of the well, but during the dry season, one can still reach the water by descending into the depths. Many stepwells (and their close cousins, step ponds) are also designed to capture surface runoff during the rainy season and store it for use during the dry season. Most of India's stepwells have fallen into disrepair and collected layers of trash and silt that prevent them from functioning properly, but there are increasing efforts to clean, desilt, and restore them for use as water sources.

One of the historical limits on the use of groundwater has been the need for significant energy to lift water from depth. One ingenious solution to the lifting problem is the **qanat** (Figure 4-1), a gently sloping tunnel that intercepts the water table and directs groundwater to the surface, where it can be used for irrigation or household uses. The engineering sophistication needed to build qanats (especially ones that run for tens of kilometers) is impressive. Some qanats are still in use today, but the local knowledge needed to construct and repair them has been lost in many places. Given our modern problem of groundwater overpumping, it is worth pointing out that qanats were passive technologies; the flow was determined by the level of the water table and couldn't be artificially increased.

1.5. Wetlands and Springs

Wetlands are highly productive ecosystems, and the earliest societies often depended heavily on fish, birds, and other wetland resources. In terms of farming, wetlands are well suited for growing rice, but other crops generally need drier conditions. The Maya developed large areas of raised-bed agriculture, in which they dug canals through wetlands and dumped the soil on the land surface to create a higher and drier microenvironment[3]; the canals themselves were used for navigation and fishing

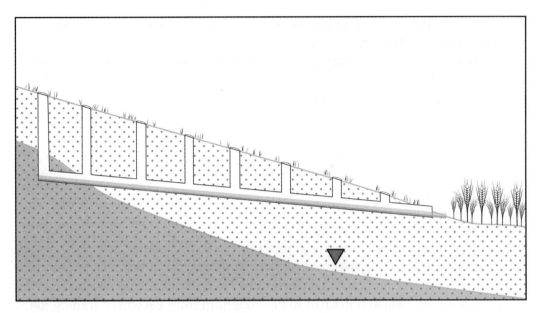

Figure 4-1. Schematic of a qanat. The tunnel intercepts the aquifer and directs water by gravity to the valley bottom. Note the vertical shafts for construction and maintenance. The qanat probably originated in Persia around 1000 BCE but quickly spread to other regions. Graphic by Maureen Gately.

and might have helped improve drainage and lower the water table. Similarly, in the shallow lakes of the Basin of Mexico, the Aztec used fences to trap debris and create small artificial islands called chinampas, which were used as agricultural plots.

Springs—places where groundwater discharges at the surface—have always been vital water sources for people and animals. Many of the largest and most important springs occur in karst formations, including the Baotu Spring in China, which has been used as a water source for thousands of years, and Nacimiento de Rio Frio in Mexico, which can flow at rates greater than 500 m^3/sec.

1.6. Water Storage
The seasonal and interannual variability in water availability has always posed a significant challenge, especially in places with extended dry seasons or frequent droughts. Storing water when it is available, for use when it is needed, is a central aspect of both ancient and modern water management. Different regions have developed a variety of water storage solutions for irrigation and other uses.

Cisterns (tanks for storing water, usually underground and often covered) have several advantages, including little water loss to evaporation and the ease of removing accumulated sediment as needed. A great variety of cistern types and sizes have been used throughout the world. Some of the oldest known cisterns (dating back to the Stone Age) have been unearthed on the island of Crete; these are typically cylindrical, lined with stones, and sealed with hydraulic plaster, and they have capacities up to about 100 m^3.

Surface reservoirs can store more water than cisterns, but they lose water to evaporation and can also lose capacity to sedimentation. The Bronze Age agropastoralist community of Jawa, Jordan, built a complex water collection and storage system, using diversion dams to direct episodic wadi flow to offstream reservoirs with a total storage capacity of about 70,000 m³.[4]

The Romans brought new techniques to dam and reservoir construction, including the use of "Roman concrete" and the development of the arch dam (which redistributes the pressure of the reservoir water toward the canyon sides, thus requiring a less massive structure). They constructed dams throughout the empire to impound rivers and create a water supply that could be drawn on during the dry season. Their largest dam—the Subiaco Dam, built in the first century CE to create a lake for Nero's villa—was an astonishing 40–50 m high and remained the highest in the world until its failure in the fourteenth century.

2. Premodern Urban Water Supply

Urban agglomerations are hotspots of water demand. Early cities, like early agriculture, often developed along rivers, where water supply was readily available and where waste could be diluted and carried downstream. However, cities also developed in less favorable locations, where water had to be brought in from elsewhere, leading to the development of the *aqueduct*, a general term for a water-carrying structure, whether aboveground or belowground, whether made of stone or metal or plastic.

Until the invention of the steam engine in the eighteenth century, a guiding principle in water conveyance was reliance on gravity, since it was impractical to move water any significant distance uphill. This meant that water sources needed to be higher in elevation than the cities they served. The elevation capital (the difference in elevation between the source and the destination) had to be carefully expended in traversing the distance to the city. If the slope of the aqueduct was too high, the water would erode the channel; if it was too low, the flow would be sluggish and the aqueduct would fill with sediment.

Minoan civilization (about 3000–1000 BCE) developed on the mountainous island of Crete, which lacks large rivers because of its karst geology. Aqueducts made of terra cotta pipes (both open and closed) delivered water from mountain springs to palaces up to 7 km away. Some of the closed pipes were tapered (larger pipes flowing into smaller pipes), implying an increase in water pressure that may have been used to create fountains; similar engineering has been found in the Mayan city of Palenque.

The city of Rome—which reached a population of about one million in the first century CE—was probably larger than any previous agglomeration of people, and the water delivery system that the city developed was correspondingly large and sophisticated. As the city grew, its water network—which began with its first aqueduct, the Appia, in 312 BCE—reached farther and farther into the surrounding hills. By the third century CE, 11 aqueducts with a total length of some 400 km were delivering vast quantities of water daily. The continuous expansion of the water supply—something familiar to modern water managers—was driven in part by population growth but also by a desire to display power.

In delivering water over these large distances, Roman engineers had to solve several technical problems. Roman aqueducts were made mostly of stone and included both aboveground and belowground sections. To prevent sedimentation of the aqueducts (and water quality problems associated with suspended sediment), *settling*

basins were built that could be cleaned out regularly. To get their aqueducts across valleys, the Romans used viaducts and inverted siphons.

Once water reached Rome, it was stored in water towers and distributed in lead pipes to various locations, including fountains, latrines, some private homes, artificial lakes (sometimes used to stage mock naval battles), and the ubiquitous baths that were so important to Roman culture. Roman water managers understood that not all uses required the same water quality, so the cleanest water was reserved for drinking. While wealthy citizens could pay to have water piped to their homes, water was also available for free at small public fountains in the streets. There were also the large ornamental fountains, which served to beautify the city and aggrandize the emperor, the gods, and those who had contributed to the water system. As David Sedlak has written, "Perhaps if our water utilities took a cue from the Romans and advertised their accomplishments with beautiful fountains, they would have an easier time convincing the public about the need to invest in the upkeep of the system."[5]

3. Modern Water Management

We now turn to an overview of water management in the modern period, introducing themes that will be explored more fully in later chapters. We start by describing the ***hard path*** and its core attitudes and technologies and then summarize the history of US water management and the transition to ***soft-path*** water management. We close out the section with a brief overview of recent global developments in water management.

3.1. The Hard Path

The Industrial Revolution kicked off a major increase in water use, as burgeoning industries and growing cities started tapping increasingly distant water sources, while population growth and globalization drove increased use of irrigation to grow food and fiber for global trade. These changes in water-use patterns were closely intertwined with changes in both technologies and attitudes, ultimately forming a water management approach called the hard path or the hydraulic mission.

At the risk of oversimplifying, we can identify four interlinked assumptions that characterize the water management hard path.

- Assumption 1. Nature consists of resources for human exploitation. We can use those resources without constraint, either because they are essentially unlimited or because technology will allow us to find substitutes when they are fully exploited. We have nothing to learn from cultures that see nature as a set of relationships rather than resources.
- Assumption 2. Where nature provides us with an unruly or unpredictable water supply, we can, with the help of modern science and engineering, improve on nature by imposing our will on it—and indeed we have an obligation to do so.
- Assumption 3. Water scarcity can best be solved by creating a larger supply of water rather than managing demand. Development of large cities and agricultural megaprojects is desirable even in arid environments, and water supplies must be increased to ensure that development is not constrained.

- Assumption 4. The main tool for water management should be massive, highly engineered, centrally managed *gray infrastructure*. This infrastructure should be applied in a uniform way around the world, without any need to consider local environments or cultures.

Five specific types of gray infrastructure have dominated hard-path water management. These water technologies—dams, aqueducts, wells, irrigation projects, and treatment plants—have been central to the creation of the modern world.

Dams: The twentieth century saw a massive increase in both the number and size of dams worldwide, a trend that continues in many parts of the world. These dams serve multiple functions: hydropower generation, water storage (for agricultural, urban, and industrial use), navigation, flood management, and recreation.

Aqueducts: The twentieth century saw a huge increase in long-distance movement of water from reservoirs and other sources to cities and farms that needed the water, using aqueducts similar in concept, but not in scale, to the Roman ones discussed above. In some cases, these water transfers made no economic sense for society but were driven by ideology, politics, or private profit making. These transfers set in motion a vicious cycle in which the new water supply allowed urban or agricultural expansion, which soon outgrew the supply, creating demand for still another new water source. The existence of large farms and urban centers in the desert is one expression of this dynamic.

Wells: Although shallow wells have been used for many millennia, the modern era gave us the ability to drill deeper wells and pump out vast quantities of groundwater, to the point where human use is now depleting global groundwater stocks. The availability of centrifugal pumps driven by gasoline engines, starting in the first decades of the twentieth century, was revolutionary in allowing ordinary farmers to tap groundwater cheaply, freeing them from reliance on rain or centrally managed surface irrigation works.

Irrigation: The combination of aqueducts, wells, and new irrigation technologies has allowed the rapid expansion of irrigation, with a twenty-one-fold increase in irrigated area worldwide over the nineteenth and twentieth centuries.

Treatment plants: The nineteenth and twentieth centuries saw the development of a new approach to urban water supply, which builds on the Roman example but adds drinking-water treatment plants to ensure safe water and sewage treatment plants to mitigate the downstream environmental and social impacts of human waste. These developments have led to the virtual elimination, in high-income countries, of cholera and similar *waterborne diseases* but have also led to large increases in water use and the treatment of valuable nutrients as wastes. At the same time, many people in *low- and middle-income countries (LMICs)* still do not have access to safe drinking water and sanitation, and water-related diseases still kill over a million people annually, mostly children.

How different are these modern attitudes and technologies from those of previous eras? Of course, both the ancient and modern worlds encompass a variety of water management approaches, so the question itself suffers from overgeneralization. Still, some see strong through-lines between ancient and modern water management approaches. For example, Steven Mithen writes, "The ambition within the ancient world to entirely redesign nature to meet the human desire for water is astonishing, as are the technical ingenuity and ability of past communities to do so."[6] At the same time, there are clearly important differences too, such as the greater uniformity of modern technology. At the very least, the scale and impact of our water infrastructure have grown tremendously; a few premodern water projects may have had substantial local impacts, but we now live in a world where almost every river, lake, and aquifer, no matter how large or remote, is affected by human manipulation of the water cycle.

One feature of modern water infrastructure that distinguishes it from previous eras is the extravagant use of two inputs that had previously been quite limited: energy and materials. Ancient water systems relied primarily on human and animal power to dig wells and move water short distances. In contrast, modern water systems use much more energy—to build dams, move water large distances, pump groundwater from great depths, treat water and wastewater—and therefore rely heavily on the availability of cheap energy, mostly in the form of fossil fuels.

Similarly, ancient water systems used locally available materials, such as earth and rock, for building small dams and canals. In contrast, modern water systems use a variety of industrial chemicals (chlorine for water treatment, chemical fertilizers and pesticides for agriculture, plastics for water pipes, cement for concrete structures such as dams) and huge volumes of **aggregate** for those concrete structures.

In many places, the hard path has supplanted traditional, smaller-scale water management approaches. But it has not completely erased them; those older, place-based practices still exist around the world. The rediscovery of these practices, coupled with selective adoption of modern materials and energy sources, is part of an exciting movement in water management toward a softer approach. Whether this softer approach can sustainably provide water services to an exploding global population is a question we will struggle with throughout the book.

3.2. US Water Management: From Hard to Soft

The hard path described in the previous section found particularly fertile ground in the American West, where manifest destiny and water technology ultimately combined to create a society organized around large-scale water projects—and, as recent years have started to show us, vulnerable to disruptions in water supply.

The first non-Indigenous irrigation projects in the western US were generally small and privately organized, including the Spanish **acequias** of the eighteenth century, Mormon irrigation in Utah (starting in 1847), the Union Colony in Greeley, Colorado (1870), and the San Joaquin and King's River Canal and Irrigation Company in California (1871). However, recognition of the difficulty of farming in this capricious landscape, along with the desire to settle the West with small farmers rather than wealthy cattlemen, led to increased federal involvement in irrigation. This took the form of the 1902 Reclamation Act, which created the Reclamation Service (later called the Bureau of Reclamation [BuRec]) and provided federal funding for construction of reservoirs and aqueducts to serve small settler-farmers, who would repay the funding with proceeds from their farms. The economic reality of this system has been far different, requiring continuing taxpayer subsidies to sustain irrigated agriculture, with those subsidies increasingly flowing into the pockets of corporate investors.

One of BuRec's proudest accomplishments was the construction of Hoover Dam

(1931–1936), a project unprecedented in ambition and scale and a symbol of the country's ability to pull itself out of the Great Depression through ingenuity, technology, and hard work. A flurry of BuRec dams and aqueducts followed throughout the West, while in the East, the US Army Corps of Engineers (USACE) was building dams for flood control, hydropower, and navigation.

The rapid development of water infrastructure in the United States in the early twentieth century brought to the fore the issue of integrated ***river-basin*** planning. As single-sector projects started to claim the same river for different uses—mining, irrigation, municipal supply, power, navigation, and flood-control projects, often managed by different agencies—Progressive Era policymakers came to realize that uncoordinated development was inefficient and would lead to competition and scarcity. President Theodore Roosevelt, who appointed two commissions to study the issue, believed that "Every stream should be used to its utmost. No stream can be so used unless such use is planned far in advance. When such plans are met, we shall find that, instead of interfering, one use can be made to assist another. Each river system, from its headwaters in the forest to its mouth on the coast, is a single unit and should be treated as such."[7]

The river-basin approach prompted basin surveys, known as "308 reports," in the 1920s–1940s but found perhaps its fullest expression in the 1933 creation of the Tennessee Valley Authority (TVA) to help the impoverished Tennessee Valley recover from the Great Depression by more fully using its abundant water (and coal) resources. Within its operating area of about 200,000 km^2, the TVA was given wide-ranging authority over economic development, navigation, and flood control, as well as electricity generation from both hydropower and fossil fuels. The TVA was viewed internationally as a model of integrated basin development.

The United States struggled to apply the river-basin approach more broadly, since such an approach was often perceived as a threat by vested federal, state, and private actors. For example, a TVA-like agency was proposed for the Missouri Valley in the 1940s, but opposition from existing federal agencies (USACE and BuRec) doomed the proposal.

Still, the realization that hydropower could pay for dams that would also serve agricultural and urban uses led to a sort of integration in the form of multipurpose dam projects. In addition, as agencies gained experience, they started to advance more complex projects involving multiple dams, aqueducts, and other structures, such as the Pick–Sloan Plan on the Missouri and the Colorado River Storage Project on the upper Colorado; these projects are similar in scale to the TVA but involve coordination between multiple federal and state agencies. To win approval for these projects, planners often needed to promise something for everyone, even if those promises were not realistic or didn't optimize resource use.

In 1950, a new presidential commission on water policy, headed by hydrologist Gilbert White, again called for integrated river-basin planning—but in the process also questioned some of the assumptions behind the hard path. White's commission emphasized several concepts. (1) "Water is limited in relation to the many and varied needs for its use," so its use must be coordinated to maximize economic and social benefits. (2) Given the effects of land use on the hydrologic cycle, land management must be a central part of water management. (3) Although the river basin is an appropriate unit for management, the federal government should also coordinate across basins. (4) In a democracy, water management requires "active cooperation of all the people, in a long-range program in which individuals, local, state and federal governments must jointly participate." (5) Government investment should support the "maximum sustained use of lakes, rivers, and their associated land and ground water resources, to support a continuing high level of prosperity throughout the country." (6) Project evaluations should include nonmonetary costs, such as "displacement of population, loss of land and minerals, loss of wildlife and loss of scenic or historic values."[8]

Inclusion in the Commission Report of these last points—the call for sustainability and the recognition of the social and ecological impacts of water infrastructure—marks the beginning of a shift in attitudes and practices, a shift that gained momentum in the 1970s and ultimately developed into the soft path.[9] Although some large projects continued to be built, focus in the last fifty years has shifted toward other activities (Table 4-2): better management of existing infrastructure, more efficient use of water, restoration of rivers and wetlands, incorporation of **green infrastructure** into urban water systems, and even dam removal.

The drivers behind these changes are complex. **Conservationist** values, such as those expressed in White's report, were certainly important. The declining marginal value of additional dams played a role, especially because the earliest dams generally occupied the most favorable sites. Since the 1970s, the modern environmental movement, with its focus on clean water and intact ecosystems, has been a major driver. A desire to rein in pork-barrel special interest spending has also been a factor, perhaps most notably in President Carter's "hit list" of boondoggle water projects that were economically unsound. Increasing awareness of past and present racial injustice has led a new generation of activists to raise issues of environmental justice and to demand a seat at the table for people of color and Indigenous communities.

The decline in water megaprojects seen in the United States has not been felt around the world. In Asia, Africa, and South America, large water infrastructure

Table 4-2. Differences between the hard-path and soft-path approaches to water management (simplified from a complex reality).

	Hard Path	*Soft Path*
Approach to large water projects	Building new projects	Managing existing projects
Type of new infrastructure	Gray	Green
Effects on aquatic ecosystems	Degradation	Protection and restoration
Approach to scarcity	Expanding supply	Managing demand
Stakeholder engagement	Top-down	Collaborative
Management locus	Centralized	Decentralized
Technology	Supply oriented	Efficiency oriented
Interest in water quality	Low	High
Metrics	Water supply	Water productivity
Meaning of *conservation*	Saving water from running off to the sea by fully tapping river flows	Using less water, often with the goal of letting more run off to the sea

projects are being built at an unrelenting pace and unimaginable scale. Some argue that LMICs have every right to follow the same hard path that the United States did, and they see hypocrisy in environmentalists from rich countries trying to block dam projects that are sorely needed by poor countries. Others argue that the environmental and social costs of these projects are borne locally, while the benefits often accrue to the Global North or regional elites; they see an opportunity for LMICs to leapfrog over the environmental and social degradation caused by the hard path and find more just and sustainable ways to meet their water, energy, and food needs. These issues will occupy us throughout the book.

3.3. Water in the Global Development Agenda

To round out our summary of water management in the modern era, we turn to the recent history of water governance in the international arena, focusing on two issues: the role of integrated water management and the tension between water as an economic good and as a human right.

Integrated Water Resource Management

Integrated water resource management (IWRM) has been a central focus of global water policy discourse over the last three decades. The Global Water Partnership defines IWRM as "a process which promotes the coordinated development and management of water, land and related resources, in order to maximize the resultant economic and social welfare in an equitable manner without compromising the sustainability of vital ecosystems." The 1992 Dublin Principles articulated four elements of IWRM:

1. "Fresh water is a finite and vulnerable resource, essential to sustain life, development, and the environment.
2. Water development and management should be based on a participatory approach, involving users, planners, and policy-makers at all levels.
3. Women play a central part in the provision, management, and safeguarding of water.
4. Water has an economic value in all its competing uses and should be recognized as an economic good."[10]

In addition, the river basin as a unit of management has been viewed as a key element of IWRM.

The echoes of Roosevelt and White are clear in several of these concepts: the importance of land use, the recognition of the many demands on a limited supply of water, the call for participation, the use of the basin as an integrating unit, and the goal of sustainability. However, the third and fourth principles introduce new concepts. The importance of gender in water management (#3) has been broadly accepted in water policy circles, although it has lagged far behind in implementation. The fourth principle, in contrast, has been the focus of much controversy, particularly around what it means to treat water as an economic good, as we will discuss in the next section.

IWRM has been broadly criticized from several different perspectives.[11] Perhaps most damning is the argument that it has become an end in itself—a series of checklists to justify whatever you wanted to do anyway—rather than a means to better management.

WATER ECONOMICS AND WATER JUSTICE
The last few decades have witnessed a growing debate over the fourth Dublin Principle: the identification of water as an economic good, as opposed to a public, social good.

On one hand, economists and policymakers have invoked increased water scarcity to argue for treatment of water as an economic good. This is reflected in several common prescriptions for improved water management: the use of markets to allocate water, the pricing of water at its true cost, and the involvement of the private sector in water management. Treating water as a commodity, the argument goes, will ensure that it is used efficiently, protect environmental values, and help avoid the type of government boondoggles that characterized the hard-path era.

On the other hand, a growing water justice movement sees these economic remedies as perpetuating power inequities, pricing low-income people out of a basic human right, allowing private actors to monetize a public good, and "shift[ing] accountability relations, from publicly-elected governments or local water user groups to non-democratic multilateral financial institutions."[12]

4. Water Use in the Modern Era: A Quantitative Analysis

Having described modern water management in qualitative terms, we now turn to a quantitative analysis of how water use has changed over the last century or so. We start with definitions, introduce the concept of water footprints, and look at global water use.

4.1. Definitions

Quantifying water use turns out to be trickier than it might seem at first. First, we need to distinguish between green water use (the appropriation of naturally occurring soil moisture to grow crops or other desired vegetation) and **blue water** use (use of liquid water, e.g., for household use or irrigation). Most water-use data deal with blue water, with water footprints (below) being a notable exception.

Within blue water use, we need to distinguish between **offstream water use**, in which water is removed from a river, lake, or aquifer for human use, and **instream water use**, in which water is used within a river or other water body (e.g., for navigation or hydropower); most water-use data deal with offstream use.

Within offstream uses, we need to distinguish between water **withdrawal** (gross water use) and water **consumption** (net water use). The difference between the two (Equation 4-1) is nonconsumptive water, also called **return flow**: water that is withdrawn from the environment, used for a certain purpose, and then returned to the environment, where it can be used again by other users.[13] In contrast, consumptive uses are those that make the water unavailable for further use in the short term. (In the long term, of course, no water is ever truly consumed, as water continues to move globally through the hydrologic cycle.)

Equation 4-1. Withdrawal = Consumption + Return flow

Consumptive uses include the following:

- ***Evapotranspiration (ET)***: Water that is transpired by crops or evaporated from an irrigation ditch is not available for further local use, since it ends up as water vapor in the atmosphere. This also means that evaporation from reservoirs (including reservoirs used for instream uses such as hydropower) is considered a consumptive use, even though there is no water withdrawal. Since reservoir evaporation is harder to measure than other consumptive uses, it is sometimes ignored, but as we will see, it can be quite large.
- Incorporation into products: Water that ends up in a product (e.g., agricultural crop, industrial product) is not available for further use, although this is generally a small portion of the water used in making the product.
- Interbasin transfers: Water that is transported across watershed boundaries for use in a different basin is considered consumed from the perspective of the originating watershed, since any return flow will end up in the receiving basin.
- Flow to the ocean: Return flow that is discharged directly to the ocean is no longer available for use as freshwater.
- Pollution: When return flow is so polluted that it can't be used by others, that could be considered a consumptive use. In practice, however, polluted return flow is often considered nonconsumptive, both because the pollutants can (at least in theory) be removed and because the level of pollution that is needed to make the water unusable depends on the nature of the downstream use.

In general, agricultural water use tends to have a high consumptive component, since transpiration of water (a consumptive use) is closely linked to crop growth. At the other extreme, most water used by power plants for cooling is discharged back into a river and can be withdrawn again by other users downstream.

4.2. Water Footprints

The concept of an ecological footprint is useful for shedding light on the often opaque linkages between our actions and the sometimes remote ecological impacts of those actions. Many readers will be familiar with the concept of a carbon footprint: the amount of greenhouse gases emitted due to consuming a given product (e.g., hamburger) or partaking in a given activity (e.g., a plane trip). The ***water footprint***, developed by Arjen Hoekstra and others at the Water Footprint Network (WFN),[14] falls into the broader category of ecological footprints but poses several unique complications.

Water-footprint accounting starts with an estimate of the water needed per unit of a given activity (e.g., the water needed to grow a kilogram of wheat) and then scales this up in various ways (Figure 4-2). For example (Figure 4-2, right side), the water footprint of a product includes all the water used to manufacture that product across its entire supply chain, which may span several continents. At the country scale, water footprints can be calculated on both a *production* basis (water used within that country to produce food and industrial products, regardless of where they are consumed (often referred to simply as water use) and a *consumption* basis (water used anywhere

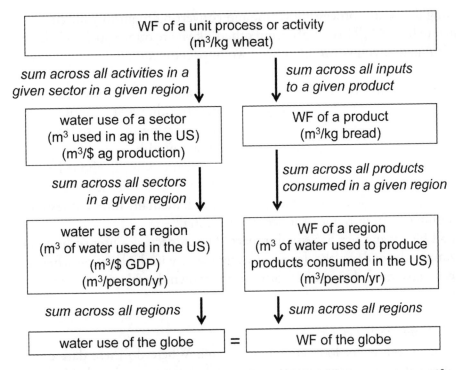

Figure 4-2. Comparison of production and consumption perspectives on water footprints (WFs), with typical units shown in parentheses. The WF from a production perspective is often referred to simply as water use, as I have done here. Modified from Hoekstra (2017).

in the world to produce the food and industrial products that are consumed by that country). The difference between the two represents trade in ***virtual water***.

The water footprint includes both blue and green water use (although these should be tracked separately). Note that the water footprint is defined based on water *consumption*, not water withdrawal.[15] This is both because consumption is considered a more useful metric (since it represents water that is unavailable for other users) and because consumption is more uniform across production systems and thus easier to model on a global level.

The WFN and others also sometimes calculate a country's (or a product's) "gray water" footprint, which is meant to account for the water needed to dilute pollution caused by manufacturing the products consumed by the country. However, this metric has not been widely accepted because of the difficulty of determining which pollutants to account for and what thresholds to use. From my perspective, the desire to incorporate water quantity and quality into one all-inclusive number is understandable but ultimately unworkable; the diversity of water-quality parameters and their differential impacts on different uses means that pollution is impossible to capture with one number.

Product water footprints can sometimes be misleading. For example, the average water footprint of a cup of coffee—all the water consumed through the supply chain of growing the beans, processing them, and converting them to your cup of joe—is estimated to be 128 L. This number and others like it are often shockingly high at first glance. But before you swear off coffee, two important caveats:

- This number includes both green and blue water. In fact, for coffee, all but 1 L of the average water footprint is green water. In other words, most of the water you are "consuming" when you drink coffee is rainwater or soil moisture that was used to grow coffee plants, not water that could be used for drinking or to sustain endangered fish. Arguably, green water use is really about land use (the choice to grow coffee rather than letting natural vegetation use the soil moisture), not about water. Green and blue water are fundamentally different resources, and combining them into one number is misleading.
- Not all blue water use has equal impact. The impact of a liter of blue water use very much depends on the degree of water scarcity in the location where that water was consumed. The water footprint on its own doesn't give you that information.

4.3. Global and US Water Use

One last issue before we look at water-use data: Where do these data come from, and how accurate are they? Unfortunately, the water sector lags behind energy in monitoring usage, in part because water is cheaper (often free) and in part because water can often be accessed more easily by individuals. Many agricultural water users (especially those using groundwater) still aren't required to meter or report their usage, and even for urban consumers, water metering is not universal (or may cover a whole apartment building rather than individual units). In this section, we will rely on country-level data compiled by the Food and Agriculture Organization (FAO) in their ***AQUASTAT*** database,[16] but we should be aware that data accuracy varies from country to country.

As an alternative source of water-use data, several global hydrologic models assess water availability and use for individual grid cells and then aggregate those data up to country or basin scales; the models are calibrated against actual water-use data, but the different models inevitably produce somewhat different results.[17] In addition, satellite data are starting to play a more important role in water-use monitoring. One exciting development is the OpenET platform,[18] which uses satellite data and an energy balance to estimate ET from agricultural fields (and other surfaces, such as reservoirs), a parameter that is both one of the hardest to measure and one of the largest components of water use.

Whenever you read an article on water use, make sure you understand what is meant: Models or measurements? Footprints or direct use? Blue or green? Withdrawal or consumption? With reservoir evaporation or without?

Figure 4-3 shows total global withdrawal and consumption of blue water from 1900 to 2020. Both withdrawal and consumption have increased about sevenfold over this period. Since much of this increase is due to population growth, it is helpful to also look at per capita water use (Figure 4-4), expressed in units of liters per capita per day (Lpcd).

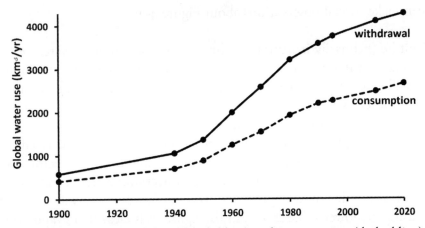

Figure 4-3. Global blue water withdrawal (solid line) and consumption (dashed line), 1900–2020. Data through 1995 are from Shiklomanov (2003), post-1995 withdrawal data are from AQUASTAT, post-1995 consumption data are estimated by applying 1995 consumptive fractions to AQUASTAT withdrawal data, and post-1995 reservoir evaporation data are from Shiklomanov's projections.

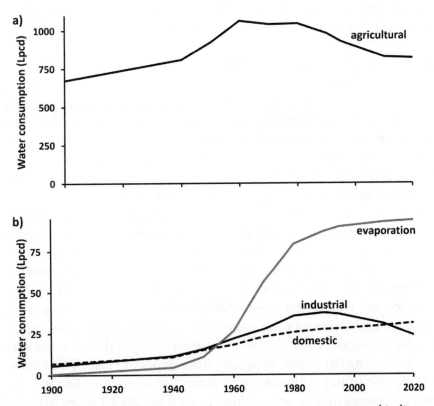

Figure 4-4. Global blue water consumption by sector, 1900–2020, expressed in liters per capita per day. Data sources as in Figure 4-3. Note the different scales in the two parts. "Agricultural" is mostly irrigation but also includes livestock watering and other uses (Chapter 18). "Evaporation" refers to evaporation from reservoirs. "Industrial" includes manufacturing, mining, and energy production (Chapter 17). "Domestic," also known as "municipal," includes rural self-supplied users (e.g., domestic wells) and all urban users who obtain their water from a public water supply (Chapter 15). For a discussion of uncertainty in these estimates, see the companion website.

We can make several observations about Figure 4-4:

- Agriculture (primarily irrigation) is the largest global user of water, representing 85 percent of water consumption. One implication of agriculture's dominance is that small changes in agricultural water use can free up significant amounts of water for other uses.
- The industrial and domestic sectors are roughly similar in their water consumption. Perhaps surprisingly, reservoir evaporation—estimated at 94 Lpcd in 2020—consumes more water globally than the industrial and domestic sectors combined. Estimates of reservoir evaporation are quite uncertain; a recent estimate suggests an even higher value of 140 Lpcd,[19] and another source uses a value of 497 Lpcd.[20] Even using the lower value shown in the figure, it is clear that evaporation represents a huge loss of water—and thus a huge opportunity for conservation.
- Per capita domestic water use has grown consistently over the past hundred years. In contrast, per capita water use for agriculture and industry peaked around 1960–1980 and have since declined, even while per capita agricultural and industrial production are larger than they have ever been. This is an encouraging theme that will recur throughout the book: We can produce more value per unit of water by being more efficient in our water use.

In the United States, the responsibility for collecting national water-use data falls to the US Geological Survey, which has produced reports every five years since 1950.[21] Unfortunately, the program has been scaled back in recent years, including the elimination of most consumptive-use estimates. The most recent data (Figure 4-5) show that in 2015, agriculture (mostly irrigation) was responsible for the largest water withdrawals in the United States, with cooling water for power plants (a largely nonconsumptive use) running a close second and domestic and industrial uses following.

An encouraging aspect of this figure is the decline in per capita water use in every category over the past few decades, even while per capita gross domestic product (GDP) has continued to rise; the amount of water needed to produce a dollar of GDP (the water footprint of GDP) has dropped from over 100 L in 1960 to about 22 L in 2015. This decoupling of economic growth from water use is due in part to greater ***water-use efficiency*** in each water-using sector, where efficiency (production per unit water, also referred to as ***water productivity***) is the inverse of the water footprint. However, structural changes in the economy (e.g., less water-intensive manufacturing) also play a role, as we will discuss in Chapter 17. Overall, US water withdrawal per capita has declined by almost 50 percent from its peak in 1980. This has more than offset the increase in population over the same time period, leading to a decline of more than 20 percent in total water withdrawals. This doesn't mean that the United States has no water scarcity problems, but it suggests that doing more with less water is part of the solution.

Chapter Highlights
1. Throughout human history, societies have met their water needs by adapting to—and modifying—their local water cycle. This has often involved the alteration of water's temporal distribution through storage and of its spatial distribution through water transfers.

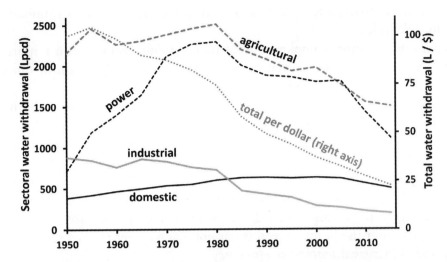

Figure 4-5. Per-capita freshwater withdrawal by sector over time in the United States (left axis); data from the US Geological Survey (USGS) (Dieter et al. 2018). Dotted gray line shows total freshwater withdrawal per unit gross domestic product (GDP) (right axis), using real GDP data (2012 dollars) from the Federal Reserve. (USGS water use reports have used different categories over time. For the purpose of this figure, "domestic" includes public supply and self-supplied domestic; "agriculture" includes irrigation, livestock, and aquaculture; and "industrial" includes self-supplied industrial and mining, as well as self-supplied commercial for the years when that category was included [1990, 1995]. Saltwater use was subtracted from USGS totals based on the assumption that all saline surface water was used in power plants and all saline groundwater was used in mining.)

2. The development of civilization was closely linked to an improved ability to harness water supplies for settlements and agriculture. Early civilizations developed a variety of water technologies, including **dry farming**, irrigation, **runoff farming**, **aqueducts**, wells, dams, cisterns, **cenotes**, **stepwells**, and **qanats**; the diversity of approaches reflects the range of local conditions encountered.
3. Droughts and other hydrologic changes posed challenges to early civilizations, sometimes contributing to collapse and migration.
4. The Romans developed a model of urban water supply that has had significant influence on modern water systems: the importation of water from the surrounding countryside in far-reaching aqueduct systems, coupled with the flushing of wastes from cities into waterways.
5. Water management in the modern era is characterized by an approach called the hard path, which has used gray infrastructure in a centralized, uniform way to dominate nature and change the distribution of water to match human demand. The hard path is different from earlier water management approaches in its global reach, the massive scale of its projects, and its profligate use of energy and materials.
6. Water management in the United States and elsewhere is transitioning to a softer path, with more of a focus on collaborative, decentralized, and nature-based solutions. Many LMICs continue to build massive dams and aqueducts.
7. Given the wide range of uses for water, water management responsibilities are often distributed across various agencies with different interests, so integration across agencies and scales is a central challenge. The river basin is a natural unit

for integrated management, but water managers must often work simultaneously at multiple scales.
8. "Water use" is a more complex concept than it might seem, so it is important to clarify what is meant in a given context: Blue or green? Instream or offstream? Withdrawal or consumption?
9. The ***water footprint*** is a tool for understanding water consumption—both direct and indirect—associated with products, companies, and countries. Water footprints have blue and green components, but these components are not directly comparable.
10. Globally, agriculture—especially irrigation—accounts for the majority of blue water withdrawal and consumption. Reservoir evaporation is a consumptive use that is large and growing but hard to quantify and easy to ignore.
11. Per capita water use in many sectors has gone down over the last several decades because of increased ***water-use efficiency***.

Notes

1. Dillehay et al. (2005).
2. Evenari et al. (1961).
3. Beach et al. (2019).
4. Helms (1981).
5. Sedlak (2014).
6. Mithen (2012).
7. Quoted in President's Water Resources Policy Commission (1951).
8. Ibid.
9. The term *soft path* in the water context is most associated with Peter Gleick, who articulated it powerfully in several forums, including Gleick (2003). My use draws from Gleick but also includes other elements.
10. https://www.gwp.org/en/gwp-SAS/ABOUT-GWP-SAS/WHY/About-IWRM/.
11. Biswas (2004), Saravanan et al. (2009), Giordano and Shah (2014), Benson et al. (2020).
12. Boelens et al. (2018).
13. Return flows to groundwater (e.g., percolation of irrigation water) are hard to measure and in practice often end up being counted as consumptive use. In theory, however, if these return flows replenish groundwater and allow other users to pump more water, they really should be considered nonconsumptive.
14. Sadly, Hoekstra died suddenly in 2019 at the age of fifty-two.
15. Note that we are (rather confusingly!) using the word *consumption* in two different ways: to contrast with production (e.g., my smartphone was produced in China but consumed by me in the United States) and to contrast with withdrawal (e.g., the water footprint of my smartphone includes only consumed water and does not include any cooling or process water that was returned to the environment).
16. http://www.fao.org/aquastat/en/.
17. Wada et al. (2016).
18. https://openetdata.org/.
19. Kohli and Frenken (2015).
20. Jaramillo and Destouni (2015).
21. https://www.usgs.gov/mission-areas/water-resources/science/water-use-united-states.

5 Water Scarcity and Depletion: Are We Reaching the Limits of Our Supply?

- Are we using too much water?
- Which parts of the world are reaching the limits of their water supply?
- How is human water use affecting the amount of water in rivers, lakes, and aquifers?
- What are river flow regimes, and how have people changed them?
- What are the impacts of using groundwater more rapidly than it is replenished?

Having explored both water supply (Chapters 2 and 3) and water use (Chapter 4), we now turn to one of the central issues of water management: What happens when use reaches the limits of supply? We examine different definitions of scarcity, look at where scarcity is—and isn't—a problem, and explore how scarcity is manifested in depletion of surface water and groundwater.

1. Scarcity Indicators

Broadly speaking, scarcity occurs when water demand exceeds supply. Defining scarcity more precisely turns out to be rather complex, and a large number of scarcity indicators have been developed, each meant to capture a different facet of the phenomenon.[1] In this section, we explore a few of these indicators.

1.1. Falkenmark and WTA Indicators

Table 5-1 introduces two of the most basic water scarcity indicators. The **Falkenmark indicator** is based on how much renewable blue water (**total renewable water resources [TRWR]**; Chapter 2) is available per person. The values shown in Table 5-1 are not meant to indicate how much water each person needs to use but rather to identify thresholds of availability below which there tend to be problems with obtaining sufficient water for all societal and ecosystem needs. One of the weaknesses of the Falkenmark indicator is that it does not take into account differences in water *needs* between different countries or regions; the thresholds in Table 5-1 apply both to countries where conditions are conducive to rainfed agriculture and to countries where all food must be grown using irrigation.

Table 5-1. Thresholds for two water scarcity indicators. Each indicator can be defined at multiple scales, ranging from a small watershed to the globe.

Scarcity Level	Falkenmark Indicator (m³/person/year)	WTA_1 Indicator (%)
None	>1,700	<10
Slight	1,000–1,700	10–20
Moderate	500–1,000	20–40
Severe	<500	>40

The ***withdrawal-to-availability (WTA) indicator*** is perhaps the most widely used indicator of water scarcity. WTA_1 is defined as the ratio of water withdrawals to renewable water availability (Equation 5-1), drawing our attention to the problems that arise when an area is using a significant fraction of its available water supply: At levels of 40 percent, we expect to see symptoms such as ecosystem degradation, water conflict, and difficulties in supplying adequate water to all users at all times. A variant of the WTA indicator uses consumption, rather than withdrawal, as the numerator (Equation 5-2).

Equation 5-1. $WTA_1 (\%) = \dfrac{\text{Withdrawal}}{\text{TRWR}} \times 100\%$

Equation 5-2. $WTA_2 (\%) = \dfrac{\text{Consumption}}{\text{TRWR}} \times 100\%$

Unlike the Falkenmark indicator, which focuses on natural water endowment, the WTA indicator focuses on how regions or countries are using that endowment. Thus, a country with apparently ample water resources per person by the Falkenmark indicator may be considered stressed by the WTA indicator if it is using a great deal of water per person, either because it is wasteful or because it is exporting water-intensive products.

1.2. Incorporating Environmental Flows
We have defined scarcity as a mismatch between water supply and demand, but so far we have included only *human* demand for water, not the water needs of rivers themselves (and other aquatic ecosystems). To be fair, ecosystem demands are implicitly built in to the thresholds in Table 5-1; that is one reason the WTA threshold for severe scarcity is 40 percent rather than 100 percent. But there have been several attempts to explicitly incorporate environmental needs into the definition of scarcity. One of these is the United Nations indicator for monitoring the scarcity aspect of Sustainable Development Goals (SDG) Target 6.4.

SDG Target 6.4. By 2030, substantially increase water-use efficiency across all sectors and ensure sustainable withdrawals and supply of freshwater to address water scarcity and substantially reduce the number of people suffering from water scarcity.

SDG Indicator 6.4.2. Level of water stress: freshwater withdrawal as a proportion of available freshwater resources.

Indicator 6.4.2 is defined in Equation 5-3, in which an estimate of the ***environmental flow requirement (EFR)*** is subtracted from the TRWR, thus making the point that this water should be considered off-limits and not part of the "available" water resource. Other studies account for EFR within a consumption-based WTA (Equation 5-4). Defining and implementing the EFR is a complex topic that we will discuss in Chapter 11.

Equation 5-3. $\text{WTA}_3(\%) = \dfrac{\text{Withdrawals}}{\text{TRWR} - \text{EFR}} \times 100\%$

Equation 5-4. $\text{WTA}_4(\%) = \dfrac{\text{Consumption}}{\text{TRWR} - \text{EFR}} \times 100\%$

Another approach to defining environmental limits to water use is the "planetary boundary" framework, which argues that for each global environmental stressor (e.g., climate change, ozone depletion, water use), we can identify thresholds which we must avoid exceeding in order to protect the life-support systems of the planet. For water use, the identified threshold is 4,000 km^3/yr of water consumption. Given that global consumptive use is only about 2,700 km^3/yr (Chapter 4), the planetary boundary assessment suggests that we are safely below the water planetary boundary as of now.[2] However, others have pointed out that if you take seriously the upper-end estimates of irrigation consumption, reservoir evaporation, and the effects of land-use change, consumptive use might be as high as 4,700 km^3/yr—well above the threshold.[3] Still others have argued that the whole notion of a global threshold for water use is unsupported and misleading given the inherently local impacts of excessive water use.[4]

Partly in response to these criticisms, a modified boundary framework has recently defined two water-related Earth system boundaries, which we will discuss in Section 4 (below) and Chapter 11 (Table 11-1).[5]

1.3. Utility Shortage

Most analyses of water scarcity focus on global, country-level, or regional scarcity, using the indicators above. However, in practice most water gets delivered to customers by a water utility (in the case of households) or an irrigation district (in the case of farmers). The amount of water that these entities can supply to their customers depends on regional water supply but also on physical infrastructure (reservoirs, pipes, pumping stations) and legal rights. We define utility-scale scarcity—often referred to as shortage—as occurring when customer demand for water exceeds the amount of water a utility (or irrigation district) can supply.

1.4. Economic Water Scarcity

Many people around the world don't have access to the water they need for their household use or livelihoods. In most cases, this is not because they live in water-scarce regions but because of poverty, lack of infrastructure, and inequities in how water and infrastructure are developed and managed. This phenomenon is sometimes called economic water scarcity or water poverty, terms that have received various definitions but generally center around the lack of funding to address water supply infrastructure needs.

Others have identified inequality in water access as the core issue and have objected to the emphasis on scarcity in the water management discourse. This perspective sees the quantification of scarcity indicators as an example of *making technical*, the process by which technocratic elites convert ethical issues into scientific and technological ones. For example, the Santa Cruz Declaration on the Global Water Crisis argues that "the global water crisis is not, as some suggest, primarily driven by water scarcity. Although limited water supply and inadequate institutions are indeed part of the problem, we assert that the global water crisis is fundamentally one of injustice and inequality."[6]

My own perspective is that scarcity and injustice are both very real—but different—aspects of the global water crisis. Scarcity is about not having enough water supply to meet demand, while injustice is about who has access to that water. Conflating the two can obscure both problems, as the needed solutions are fundamentally different. Examples of scarcity are Cape Town's Day Zero, in which the city's taps came close to running dry; conflicts between California and Arizona over dwindling Colorado River water; and the death of Australian livestock during the Millennium Drought. In contrast, a family in Detroit facing a water shutoff, an unhoused person in Los Angeles who has to scrounge to access a toilet, and a slum-dweller in Dhaka who waits on line for poor-quality water are all facing injustice, not scarcity; there is enough water to meet their needs if infrastructure and resources were allocated fairly. This chapter focuses on physical scarcity, and issues of access and water allocation will be addressed in later chapters.

2. Assessing Water Scarcity

In this section, we use the WTA indicator to assess water scarcity at different spatial and temporal scales.

Using our estimates of global water withdrawals (Chapter 4) and global runoff (Chapter 2), we can calculate that the global value of the WTA_1 indicator is 4,300 km^3/yr ÷ 46,000 km^3/yr = 9.3 percent, which is in the "none" category in Table 5-1. In other words, water scarcity does not appear to be a global problem (but see Box 5-1).

An analysis of the WTA indicator at the country scale (Figure 5-1) shows that seventy-three countries have some level of water stress, with thirty-one countries in the severe scarcity category (withdrawals greater than 40 percent of availability). Not surprisingly, most of the countries with severe scarcity are in arid parts of the world. Fifteen countries (mostly in the Middle East but also Barbados, Turkmenistan, and Uzbekistan) have WTA_1 values greater than 100 percent (i.e., they are withdrawing more than 100 percent of their TRWR). Think for a moment about how it might be possible to use more water than you have.

There are several ways that a country can use more than 100 percent of its available water:

- A given parcel of water can be withdrawn multiple times before it is consumed. In Turkmenistan and Uzbekistan, for example, agricultural return flow is often reused, allowing those countries to withdraw more than 100 percent of their supply.

> **Box 5-1. Global or Local?**
>
> To what extent is water scarcity a global issue? The low values of the WTA indicator for the globe and for most countries (Figure 5-1) suggest that the world as a whole is not water scarce; there is plenty of water, but it is in the "wrong" places. Underlying this perspective are two important facts: (1) Unlike climate change, where carbon emissions *anywhere* affect climate *everywhere*, water use in one watershed doesn't affect water availability in other watersheds around the world. (2) Given how heavy it is, water can't readily be moved from water-rich areas to water-scarce areas, so each region needs to deal with its own water problems.
>
> But it's more complicated than that, since the global trade in food and industrial products is also a trade in **virtual water** (the water needed to produce the traded items). In fact, some of the most water-scarce countries survive by importing water-intensive goods, thus transferring much of their water burden to other countries, and some countries that have adequate water supplies for their own needs have become water scarce because of global trade patterns. In that sense, water scarcity is very much a global issue.
>
> One tool for highlighting the flow of virtual water around the world is the water footprint (Chapter 4), but standard water footprints don't reflect the degree of scarcity in source regions. In other words, I shouldn't really care what the volumetric water footprint of my T-shirt is but rather whether the cotton was grown in water-scarce Uzbekistan (where it contributes to the death of the Aral Sea) or in water-abundant Mississippi. A scarcity-weighted water footprint has been developed[a] but has not been widely accepted in the water management community.
>
> **Note**
> [a] Boulay et al. (2018).

- Desalination allows a country to use more than its TRWR by creating an additional water supply that is not traditionally included in the TRWR since it is not considered renewable. Desalination is a dominant factor for most of the countries that are using more than 100 percent of their TRWR.
- TRWR includes only renewable annual flows of surface water and groundwater. However, as we will discuss below, many countries are depleting stocks (lakes and aquifers), allowing them to use more than 100 percent of TRWR, at least in the short term.

In one important way, Figure 5-1 understates the true extent of scarcity: It is calculated based on average annual data (for both withdrawal and availability). Given the strong seasonal and interannual variability in water supply in many places—and the fact that water demand tends to go up precisely when less water is available—water scarcity is certainly higher during dry seasons and droughts.[7]

Analyses of scarcity can also be carried out at smaller spatial scales, using either disaggregated data or hydrologic models. This is especially important for large, climatically diverse countries such as the United States, where a low country-level WTA value (14 percent) hides the fact that some hydrologic units, especially in the West, are experiencing severe scarcity.[8]

Urban areas are particularly vulnerable to scarcity, given their concentrated water demand and rapid growth. Long-distance water transfers have allowed cities to grow well beyond what local water sources could support, but they have also put cities in

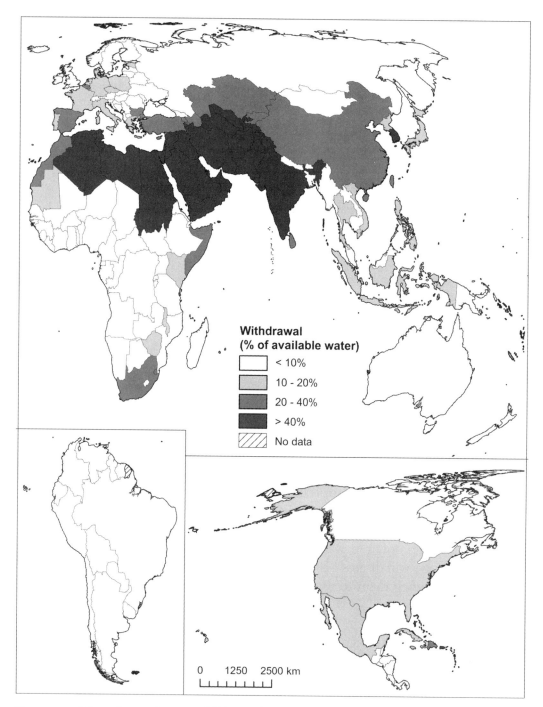

Figure 5-1. Map of annual average WTA$_1$ indicator. Map by Anna Yue Yu. Data from AQUASTAT.

competition with agriculture for rural water supplies. The scale of rural-to-urban water transfer is impressive; source watersheds for large cities make up 41 percent of Earth's surface (while the cities themselves occupy only 1 percent).[9] In the absence of these transfers, some 39 percent of the residents of large cities would be living in severe water scarcity (WTA$_1$ greater than 40 percent); even with these transfers, 25 percent of this population is suffering severe scarcity.[10]

With urban water demand projected to increase by 80 percent by 2050[11] and climate change likely to increase drought frequency, the next few decades may see a significant increase in urban water scarcity. Already, several major cities have experienced drought-induced water supply crises, including Barcelona in 2008, São Paulo in 2014, and Cape Town in 2018. One study of 482 cities projects that urban water scarcity will increase dramatically by 2050 and that cities will increasingly look to agricultural water for solutions.[12]

Leaving aside the question of whether water scarcity is a global problem (Box 5-1), it is clearly a problem for substantial portions of the world, at least during dry seasons or years. What happens in these places when there is not enough water supply to meet demand? First, some water demand may go unmet; this may take various forms, from households cutting back on water use to farmers fallowing fields or power plants shutting down, with potentially serious impacts on food security and economic development. Second, supplies are overused, by overtapping renewable flows or drawing down stocks of water such as lakes and aquifers; this overuse is the subject of the remainder of this chapter.

3. Surface Water Depletion

Perhaps the most visible aspect of the global water crisis has been the decline in surface water, especially lakes and rivers, as the human appetite for water increasingly runs into the limits of water supply.

3.1. Shrinking Lakes

One obvious manifestation of human overuse of water is a decline in the area and volume of lakes and reservoirs.[13] The most vulnerable lakes are **terminal lakes** (lakes with no surface outflow), which are particularly sensitive to changes in inflow or evaporation, whether driven by climatic fluctuations or human use of water from tributary rivers. Terminal lakes tend to be saline because of evaporation-induced concentration of salts, and decreases in water inflow will increase lake salinity. It can be difficult to fully disentangle the causes of these changes, and local water managers may prefer to blame global climate change rather than water use. Two examples will illustrate the myriad consequences of these declines.

The Aral Sea—a terminal lake in Central Asia fed by the Syr and Amu Rivers—is a prominent example of the ecological and human costs of twentieth-century water management. Starting in the 1960s, the diversion of river water for cotton irrigation decreased the flows of freshwater into the Aral, causing the lake to shrink dramatically and separate into two smaller water bodies. The northern Small Aral has recently stabilized in size and salinity, in large part because of a dam built to raise water levels and prevent outflow to the remainder of the former lake. In contrast, the southern Large Aral has continued to shrink, and its salinity has increased from about 10 parts per thousand to over 100 parts per thousand[14] (compared to 35 parts per thousand for seawater) because of the absence of freshwater inflow and the influence of evaporation. Fish can't survive under these conditions, and the once-thriving fisheries of the Aral have collapsed. The recession of the lake shoreline has left former port cities far from the water and has exposed sediments rich in salt and agricultural chemicals, which are blown around the region by windstorms. A number of chronic diseases have increased in the region, and life expectancy has dropped.[15]

In Great Salt Lake in Utah, consumptive water use has also lowered lake levels, though not as dramatically as at the Aral Sea. The recession of the lake has exposed large areas of ***playa*** (dry lakebed) and left microbialites (underwater microbial reefs) stranded above the water level. AccuWeather and the *Salt Lake Tribune* announced in 2021 that they would replace the familiar image of the lake with one that more accurately reflects its shrunken condition.[16] The proposed Bear River Development project would accelerate this shrinking trend by diverting an additional 0.25 km³/yr from one of the lake's tributaries. Likely effects of further water-level declines in Great Salt Lake include loss of the open water and wetlands that are vital habitat for millions of migratory birds; a decline in the brine shrimp fishery, the mineral extraction industry, and recreational use; air quality problems associated with dust from lakebed exposure; and a decline in the lake-effect snow that sustains the skiing industry. Even leaving aside some hard-to-monetize effects (such as migratory birds), these changes could have economic costs of about $2 billion per year.[17]

3.2. River Depletion and Altered Flow Regimes

Given the global increase in consumptive water use, it is probably not surprising that many rivers, especially in water-scarce areas, have less water flowing in them than they did in previous generations. Most dramatic, perhaps, are the large rivers that are so overtapped that they dry up before reaching the ocean, such as the Colorado River (Box 5-2). River flow depletion has impacts that are both ecological (e.g., habitat loss, fish kills) and human (e.g., loss of cultural connection, inadequate supply to meet all uses, conflict over limited flows).

Besides the impacts on the rivers themselves, flow depletion also has serious effects on some of the most productive and diverse ecosystems on Earth: the estuaries and deltas where freshwater and saltwater mix. Without freshwater flows to sustain them, these ecosystems turn dry and saline and lose much of their habitat value. Even in rivers that still reach the ocean, a decline in flow, coupled with global ***sea-level rise***, can lead to upstream incursion of saltwater, sometimes endangering urban water supplies as well as freshwater ecosystems.

In addition to the obvious declines in streamflow caused by consumptive water use, water management activities often have other, subtler effects on rivers. Some rivers have higher-than-natural flows from being on the receiving end of ***interbasin transfers***. Some rivers see huge short-term fluctuations in streamflow, as hydropower dams respond to peak daily electricity demand (Figure 5-2a). Many rivers see a flattening of natural seasonal variability (Figure 5-2b), as dams capture high flows and release them slowly during the dry season, serving their desired function of tempering natural variability, but at a high cost to the river ecosystem. The loss of high flows cuts rivers off from their ***floodplains*** and destroys the rich wetland habitats of the ***riparian*** zone (Box 5-3).

Thus, keeping a river healthy means not just making sure that it doesn't dry up but also maintaining the characteristic flow patterns that make up its unique natural ***flow regime***: the range of different flows it experiences, including the magnitude, timing, duration, frequency, rate of change, and predictability of those flows.[18] Underlying

Box 5-2. The Overworked Colorado River

> The river was nowhere and everywhere, for he could not decide which of a hundred green lagoons offered the most pleasant and least speedy path to the Gulf.—Aldo Leopold, *Sketches Here and There*[a]

By most standards the Colorado is a small river (5 percent of the flow of the Columbia, despite having a slightly larger watershed), but it has outsized importance as the only perennial water source in a large area of the arid Southwest. The river and its tributaries rise in the mountains of Colorado, Wyoming, New Mexico, and Utah, where they gain water from rain and snowmelt, making for a highly seasonal natural flow pattern (see Figure 5-2b). The Colorado then flows through the arid states of Arizona, Nevada, and California, cutting deep canyons (including the Grand Canyon) and carrying the vast sediment load that gives the river its name before flowing into Mexico and ultimately discharging into the Gulf of California. Along the way, the waters of "The Hardest Working River in the West"[b] are tapped to provide drinking water to some 40 million people and to irrigate some 5 million acres (both inside and outside its natural watershed), to the point that the river no longer reaches the sea, and its delta—which enchanted Leopold with its verdure when he visited it in 1922—is now dry and salinized.

The increasing mismatch of supply and demand in the Colorado is shown in Figure 5-B1, which illustrates that consumptive use rose steadily through the twentieth century, while water availability has been highly variable but has generally been decreasing due to drought and climate change (Chapter 3). Large reservoirs in this system—most notably Lakes Mead and Powell—can store up to four years' worth of average flow, buffering the system from short-term variability. But it is becoming clear that there is a structural deficit in this river; more water is promised than the river can deliver, and reservoir water levels are declining, leaving the basin states, federal government, and Mexico scrambling to figure out how to share the shortfall. At the same time, endangered species such as the Colorado pikeminnow are struggling to survive the altered flow regime, along with other stressors. The complex and evolving rules for water allocation in the Colorado are based on the principles discussed in chapters 10–12 and are covered in detail on the companion website.

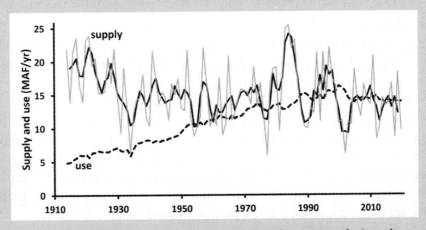

Figure 5-B1. Consumptive use (dashed line, annual) and water supply (gray line, annual; solid line, three-year running average) in the Colorado River Basin. MAF: millions of acre-feet. Data from the Bureau of Reclamation, obtained from Brian Richter.

Notes

[a]Leopold (1949).
[b]https://storymaps.arcgis.com/stories/2efeafc8613440dba5b56cb83cd790ba.

> **Box 5-3. Effects of Flow Reduction on a Nigerian Wetland**
>
> The Hadejia-Nguru floodplain wetlands in semiarid northern Nigeria provide a good example of the negative effects of hydrologic alteration. The wetlands experience seasonal flooding (July–October) due to summer rains in the upstream (southern) portion of the watershed, combined with the flat landscape and braiding river channels of the floodplain area itself. This flooding supports vital habitat for waterbirds and a diversity of human uses,[a] including several types of agriculture over the course of the seasonal cycle: rainfed cultivation of sorghum and millet in upland areas; wet cultivation of rice in flooded areas; irrigation of wheat and vegetables; flood recession agriculture; and grazing of livestock on the floodplains in the dry season, when the residual moisture from the floods allows natural vegetation growth. In addition, the seasonal flood pulse allows fish to grow and breed rapidly throughout the flooded area, and different fishing activities have developed around this seasonal cycle. Flooding is also important for recharging groundwater, which can then be used during the dry season.
>
> The Tiga Dam (constructed in 1974) and Chowalla Gorge Dam (1992) have increased upstream diversion of water for irrigation and urban use, reducing the amount of water flowing downstream to the Hadejia-Nguru floodplain. The area inundated at the height of the flood season dropped from 2,000–3,000 km^2 in the 1960s to generally less than 1,000 km^2 in the 1980s and 1990s, leading to a decline in fisheries and agricultural productivity. An economic study[b] calculates that, when all floodplain uses are accounted for, the net value of water is much higher in the floodplain than in upstream irrigation, so the diversion of water is economically unsound as well as ecologically destructive. This illustrates one of the blind spots of hard-path water management; it tends to favor the uniform, engineered productivity of a dam diversion irrigation scheme over the complex, site-specific, evolving indigenous productivity of working with the river's natural rhythms.
>
> **Notes**
> [a]Thompson and Polet (2000).
> [b]Barbier (2003).

the flow regime paradigm is the notion that streamflow serves as a master variable controlling the physical, chemical, and biological character of a given river. The biotic community in a particular river is adapted to that river's flow regime and to the conditions—water temperature, turbidity, riverbed substrate—created by that flow regime. Changes to flow will alter both the abiotic conditions and the biotic community, presumably to ones that are less desirable or at least more homogeneous across systems. Thus, the consequences of flow alteration can include changes to water temperature and sediment transport, accumulation of sediment in the channel, disruption of fish migration and spawning, encroachment of upland vegetation into the floodplain and the channel margins, impoverishment of the invertebrate community, invasion of nonnative species, wetland loss, and a decline in biodiversity.

4. Groundwater Use and Overuse

Groundwater depletion is less visible than surface water depletion but poses an equally significant threat to the sustainability of our current water systems. Groundwater accounts for roughly one third of global water use,[19] and some 130 million people in the United States rely on groundwater as their primary source of drinking water.[20] Groundwater use has some significant advantages over surface water use. It doesn't require the damming of rivers and the building of canals. It is decentralized and locally

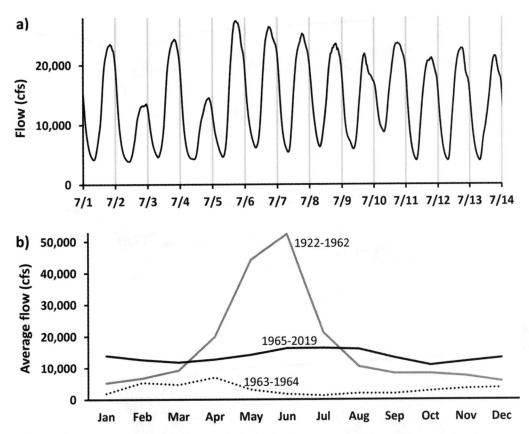

Figure 5-2. Hydrologic alteration in the Colorado River at Lees Ferry (US Geological Survey station 09380000), downstream of Glen Canyon Dam. (a) Thirty-minute flow data over two weeks in July 1989, showing massive subdaily fluctuations resulting from hydropower production to meet peak demand. (b) Average monthly flows for three periods: 1922–1962 (no dam), 1963–1964 (filling of the reservoir), and 1965–2019 (dam in operation). The natural seasonal snowmelt peak has been eliminated, as has the natural low-flow period (September–February).

controlled. It allows farmers to achieve higher yields by having water available when they need it. It doesn't have evaporation losses associated with storage, as surface-water reservoirs do. It is drought-proof, or, to put it more precisely, groundwater responds more slowly than river flow to climatic changes, so managers tend to turn to it when surface water is depleted by drought. But groundwater overuse threatens the sustainability of this resource.

4.1. Groundwater Pumping

To understand groundwater use, we need a better picture of the changes that happen when we start pumping water from an aquifer (for simplicity, an **unconfined aquifer**). We can assume that before pumping began, the aquifer was in balance (over annual or longer time scales), with water gains from **recharge** (percolation of rainwater) balanced by water losses to **discharge** (outflow to streams or springs). Once pumping begins, the removal of water creates a cone of depression centered at the well, which changes the local **hydraulic gradient** and draws water toward the well (Figure 5-3).

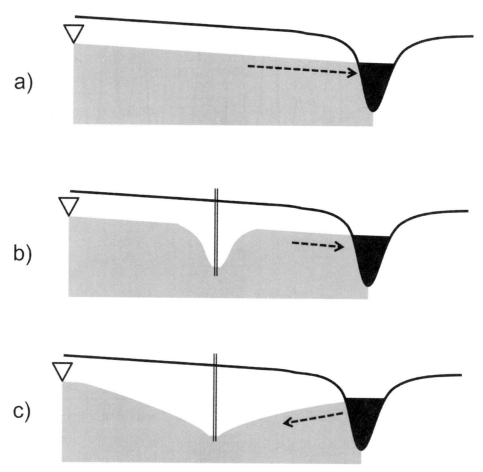

Figure 5-3. Change in the hydraulic gradient with pumping. (a) Before pumping, groundwater discharges to the stream. (b) After pumping starts, discharge to the stream is reduced. (c) At a higher pumping rate, the cone of depression reaches the stream and starts to draw surface water toward the well, reversing the groundwater flow path.

The size of the cone of depression depends on the pumping rate and the hydrogeology of the aquifer, and generally it can be fully understood only with a computer model coupled with pumping tests.

To simplify what can be a very complex situation, it is fair to say that the water that is pumped from a well comes from some combination of three sources:

- Decline in storage: Most obviously, if water is being removed from the aquifer, this will result in less water being stored in the aquifer. In the short term, the decline in storage is the main source of pumped water.
- Decrease in discharge: As the hydraulic gradient shifts, the outward discharge of the aquifer will decrease (Figure 5-3b); thus, groundwater pumping may lead to declines in river flows, drying of springs, or a reduction in plant transpiration.
- Increase in recharge: The change in hydraulic gradient often means that more water flows into the aquifer as recharge than before pumping began. This water can come from surface water (Figure 5-3c), rainfall that would otherwise have flowed

elsewhere, lateral flow from an adjacent aquifer, or percolation of irrigation water back into the aquifer from which it was pumped.

The last two categories (decrease in discharge and increase in recharge) are collectively called ***capture***.

To simplify still further, we can conclude that groundwater pumping ultimately operates in one of two modes (or, more realistically, some combination of the two), depending on the physics of the aquifer, the extent of pumping, and the time since pumping began: Pumping either reduces surface water flows or reduces the volume of groundwater. The former is more important for shallow, surface-connected aquifers, and the latter tends to dominate for deeper, ***confined aquifers***, especially those with minimal active recharge (often called ***fossil aquifers***).

The two modes of groundwater pumping have different implications and must be managed differently. In situations where groundwater pumping reduces surface water flows, we are using renewable water. Of course, the reduction in surface flows may have serious ecological consequences for rivers and surface-water habitats, especially during the dry season, when baseflow is dependent on groundwater discharge. In any case, for this mode of pumping, groundwater should be managed together with surface water, with the goal of balancing human and ecosystem needs for that renewable flow.

On the other hand, where groundwater pumping pulls primarily from storage, it is drawing down a nonrenewable resource. This phenomenon, known as ***groundwater mining*** or ***groundwater depletion***, is manifested in a dropping water table, decreasing saturated thickness of the aquifer, and a reduced volume of water in the aquifer.

Groundwater depletion can cause a number of problems:

- When water tables fall below the depth of a well, that well can no longer provide water. A study of the western United States estimates that 3.3 percent of wells have dried up because of dropping water tables; because domestic wells tend to be shallower than agricultural wells, more domestic wells have gone dry.[21] One common response to declining water tables is to dig deeper wells, but this may not be economically feasible, both because of the high cost of digging deeper wells and because of the increased energy needed to pump deeper water. In addition, depending on the aquifer, deeper wells may be less productive or may tap brackish water. For these reasons, deeper wells have been called "an unsustainable stopgap to groundwater depletion."[22]
- There can be a strong equity component to these problems. Those wealthy enough to do so can drill deeper wells to extract more water, drying up the wells of neighbors who don't have the resources to deepen their own wells. There is also an intergenerational equity problem; when groundwater runs out, future generations will bear the consequences of drastically reduced water availability. For example, groundwater depletion is expected to decrease cropping intensity by 20 percent for India as a whole and by 68 percent for the hardest-hit regions of the country, posing a serious threat to the country's food security.[23]
- When fine-grained materials are dewatered by groundwater pumping, they may undergo compaction, which leads to irreversible loss of groundwater storage

capacity and ***land subsidence***. The extent of subsidence depends on the presence of compactible clays in the formation, so it can be highly variable spatially, leading to changes in surface topography, which can cause serious problems for surface and subsurface infrastructure. Because compaction is driven by the slow dewatering of low-conductivity materials, there can be a time lag between groundwater extraction and subsidence.[24]

- Water that is no longer stored as groundwater ends up in the ocean and contributes to sea level rise. In addition, groundwater depletion results in greenhouse gas emissions, since groundwater is often high in dissolved CO_2.[25]
- Groundwater depletion can change hydraulic gradients and affect groundwater quality. In coastal areas, a drop in the water table can reverse the land-to-sea hydraulic gradient and draw in seawater, which can contaminate the aquifer through saltwater intrusion (Chapter 3). The likelihood of saltwater intrusion depends on local hydrogeology and pumping rates but appears to be more common along the relatively flat US Atlantic and Gulf Coasts, compared to the Pacific Coast with its steeper land–sea margin.[26] Sea-level rise and land subsidence both exacerbate the threat of saltwater intrusion.

Because of the problems associated with groundwater depletion, regulators often require that groundwater use be limited to the ***safe yield*** of the aquifer, defined as the rate of recharge (natural plus artificial). Likewise, the safe Earth system boundary is defined as "annual [groundwater use that] does not exceed average annual recharge." However, in practice, groundwater depletion is common; about half of the global land area is estimated to violate this boundary.[27]

4.2. Extent of Groundwater Depletion

Data on groundwater availability and use are hard to come by. Unlike river flow, groundwater recharge is often spatially diffuse and hard to quantify. Unlike surface water abstraction, groundwater use is widely distributed across millions of wells, whose pumping rate is poorly regulated and often measured inaccurately, if at all. Even the total volume of water in an aquifer is hard to estimate, given that aquifers are often spatially heterogeneous in their physical properties.

Nonetheless, four types of information can be used to assess the extent of groundwater depletion in an aquifer.[28] First, local measurements over time in multiple wells across an aquifer can provide data on changes in water-table elevations (which can be converted to water volumes using aquifer properties); these data are good in some locations but practically absent in others. Second, estimation of the local water budget (precipitation, evapotranspiration [ET], surface runoff, storage) can allow estimation of groundwater changes by difference. Third, global hydrologic models can be used to estimate both groundwater recharge and groundwater use. Fourth, the paired-satellite system known as ***Gravity Recovery and Climate Experiment (GRACE)*** measured changes in the distribution of mass on Earth from 2002 to 2017; a second pair of satellites, known as GRACE-FO (GRACE Follow-On), was launched in 2018. GRACE data can be used to "see" places where large amounts of water have been gained or lost and calculate changes in aquifer water volumes, as well as other changes in terrestrial water storage.

Table 5-2. Estimates of global groundwater depletion.

Model Used	Time Period	Groundwater Depletion (km³/yr)	Groundwater Withdrawal (km³/yr)	Depletion/ Withdrawal (%)
WaterGAP[a]	1960–2000	56	470	12
	2000–2009	113	660	17
	2002–2016	112		
PCR-GLOBWB[b]	2000	195	1,000	20
	2010	241	1,200	20

[a]Döll et al. (2014), An et al. (2021).
[b]Wada et al. (2014), Dalin et al. (2017).

The best estimates of groundwater depletion come from combining multiple data sources, but even then estimates can vary widely depending on the models used and how the GRACE data are processed. Two estimates of global groundwater depletion are shown in Table 5-2. Despite their different numbers, the estimates agree on where the majority of global groundwater depletion is taking place: India, Iran, Pakistan, China, the United States, and Saudi Arabia. Also shown in Table 5-2 are estimates of what percentage of groundwater use comes from depletion, a metric that you can think of as analogous to the fraction of your spending that comes from depletion of your savings rather than from your income.

4.3. Groundwater Depletion in California's Central Valley

To make the concept of groundwater depletion a little more concrete, let's take a brief look at one of the most important aquifers in the United States: the unconfined aquifer underlying California's Central Valley. The Central Valley—which includes the Sacramento Valley north of the Sacramento–San Joaquin Delta and the San Joaquin Valley south of it—is among the most productive agricultural regions in the country, growing some 250 different crops and producing about 25 percent of the nation's food supply. Water for farming in the Central Valley comes from both surface water and groundwater. On average, groundwater provides about 40 percent of water use, but this rises to 70 percent during droughts, when less surface water is available.[29]

Before people began using large volumes of groundwater, the Central Valley aquifer was approximately in balance, with recharge from precipitation and **losing streams** being balanced by ET (mostly from wetlands) and discharge to **gaining streams** (Figure 5-4). Pumping and other changes have increased groundwater discharge by a factor of 10 (Figure 5-4). Although recharge has also increased—mostly because of excess irrigation water percolating down to the water table—the aquifer is now losing water from storage at an average rate of about 2 km³/yr.

Of course, the depletion rate varies around this average (in both time and space) with variation in recharge and discharge rates. Since the 1970s, Central Valley water tables have tended to drop during droughts (when less surface water is available) and then recover during wetter periods. More recently, though, depletion periods have gotten longer and steeper, and recovery periods are recovering only a small fraction of the water lost.[30]

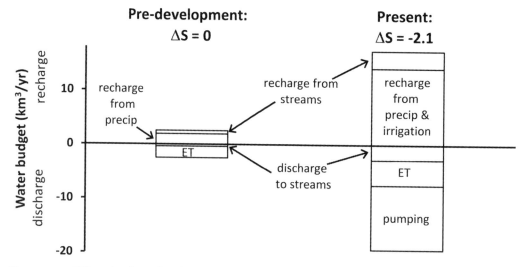

Figure 5-4. Water budgets for the Central Valley aquifer for predevelopment (steady state) conditions and recent conditions. Data from Scanlon et al. (2012). ΔS represents the change in storage (in km³/yr).

Is the Central Valley aquifer "running out of water"? Total depletion from the 1920s to 2013 is about 240 km³. The aquifer is thought to have a total (predevelopment) water storage of about 1,000 km³, suggesting a lifetime of 200–300 years at current depletion rates. However, depletion is not evenly distributed through the Central Valley, and areas experiencing the highest depletion (e.g., the Tulare Basin in the southern part of the San Joaquin Valley) will have the shortest lifetimes.

Even if the aquifer has lots of water left, there have already been negative effects of groundwater depletion in the Central Valley:

- Land subsidence started in the 1920s and has continued into the present, especially in certain parts of the southern San Joaquin Valley.[31] Uneven land subsidence wreaks havoc on water delivery and flood control infrastructure. Ironically, the capacity of the Friant–Kern Canal—an aqueduct delivering surface water to farmers—has been reduced by subsidence, causing greater reliance on groundwater. Restoring the Friant–Kern's capacity may cost up to $1 billion.
- The compaction that causes land subsidence also results in permanent loss of storage capacity in the aquifer. During the 2012–2015 drought, 0.4–3.2 percent of aquifer storage capacity was lost.[32]
- Groundwater depletion affects water quality. For example, groundwater nitrate concentrations are higher when water levels are lower, presumably because higher pumping rates draw in near-surface contaminants.[33]

5. Managing Water Scarcity

If we want to avoid the undesirable outcomes outlined in the previous sections, we need to bring supply and demand into balance. Solutions to water scarcity problems generally fall into four categories, which will be explored in future chapters (Table 5-3):

Table 5-3. Comparison of twentieth- and twenty-first-century solutions to water scarcity.

Category	Twentieth-Century Approaches	Twenty-First-Century Approaches
Increase supply	Aqueducts (Chapter 9)	• Desalination (Chapter 13) • Wastewater reuse (Chapter 13)
Increase storage	Dams and reservoirs (Chapter 9)	• Improved management of existing storage (Chapter 9) • Water harvesting (Chapter 13) • Aquifer storage (Chapter 13)
Decrease demand	Not a significant focus	• Domestic (Chapter 15) • Industrial (Chapter 17) • Agricultural (Chapter 18)
Reallocate	Not a significant focus	• Agricultural-to-urban relocation (Chapter 11)

- Increase supply: This is the primary twentieth-century response to water scarcity: build new infrastructure, especially aqueducts, to deliver more water to areas facing scarcity. In the coming chapters, we will explore some of the shortcomings of this approach and explore alternative ways to increase water supply.
- Increase storage: Given the high (and increasing) temporal variability in water supply, storage plays a critical role in alleviating scarcity. In the coming chapters, we will explore dams and other forms of storage.
- Decrease demand: Conservation and efficiency measures have great potential to solve water scarcity problems by decreasing the need for water. Conservation tools and how much water they can actually save will be discussed by sector (domestic, agricultural, industrial) in Part IV.
- Reallocate water: Who gets the water is an important factor in determining whether we are getting the maximum utility from a limited supply, and reallocation can help solve scarcity issues. We discuss water allocation and reallocation in Part III.

Chapter Highlights

1. Water scarcity—broadly, not enough supply to meet demand—can be quantified in many different ways, of which the **Falkenmark** and **WTA indicators** are the most widely used.
2. The world as a whole is not water scarce, but many countries, regions, and cities are struggling with water scarcity. Although water scarcity is largely a local issue, the trade of virtual water around the world adds a global dimension to the problem.
3. Human consumptive use of water has led to declines in reservoirs and lakes, especially *terminal lakes*, where these declines are often accompanied by an increase in salinity and the loss of fisheries and other ecosystem services. When large areas of *playa* are exposed, the resulting toxic dust storms can have significant impacts on air quality.
4. Consumptive use has also decreased the flow of many rivers around the world, with serious impacts on the rivers themselves and on river-dependent wetlands, floodplains, deltas, and estuaries. In addition to declines in flow, rivers have also

experienced broad changes to their *flow regimes*, changes that often include a dampening of natural seasonal variability. These hydrologic changes can affect sediment transport, water temperature, habitat connectivity, and biodiversity.

5. As a water source, groundwater has some advantages over surface water, but overuse of groundwater is leading to **groundwater depletion**, which has economic, social, and ecological impacts, including **land subsidence**, **saltwater intrusion**, increased water supply costs, and inequitable water access.

6. Groundwater pumping has complex, site-specific effects on hydraulic gradients and flow paths. When pumping draws primarily from *capture*, it is using renewable water, but it can affect river flow and should be co-managed with surface water. When pumping draws primarily from storage (as in a deep *fossil aquifer*), it is mining a nonrenewable resource.

7. Groundwater depletion can be measured at large scales with the **GRACE** satellites or at smaller scales with local data. Globally, about 15–20 percent of groundwater use consists of depletion.

8. In systems with access to both surface water and groundwater, such as California's Central Valley, groundwater can serve as a reservoir that is drawn on during dry periods and replenished during wet ones, but overuse and more frequent droughts have converted it into a time-limited, nonrenewable resource.

9. There are four basic approaches to mitigating scarcity: increase supply, increase storage, reduce demand, and reallocate water.

Notes

1. Gleick and Cooley (2021).
2. Steffen et al. (2015). This paper also identified local boundaries for consumptive water use.
3. Jaramillo and Destouni (2015).
4. Heistermann (2017).
5. Rockström et al. (2023).
6. https://www2.ucsc.edu/scdeclaration/.
7. Brauman et al. (2016).
8. Averyt et al. (2013).
9. McDonald et al. (2014).
10. Ibid.
11. Flörke et al. (2018).
12. Ibid.
13. Yao et al. (2023).
14. Izhitskiy et al. (2016).
15. Waehler and Dietrichs (2017).
16. https://www.sltrib.com/news/environment/2021/11/22/great-salt-lake-is-dying/.
17. ECONorthwest and Martin & Nicholson Environmental Consultants (2019).
18. Poff et al. (1997).
19. Wada et al. (2014).
20. DeSimone et al. (2014).
21. Perrone and Jasechko (2017).
22. Perrone and Jasechko (2019).
23. Jain et al. (2021).
24. Ojha et al. (2019).
25. Wood and Hyndman (2017).

26. Saltwater intrusion is also a natural phenomenon, but it is exacerbated by groundwater pumping; Jasechko et al. (2020).
27. Rockström et al. (2023).
28. Alam et al. (2021).
29. Faunt et al. (2016).
30. Alam et al. (2021).
31. Faunt et al. (2016).
32. Ojha et al. (2019).
33. Levy et al. (2021).

Part II
Instream Water Management: Rivers and Dams, Flooding and Hydropower

The four chapters in Part II focus on managing rivers for instream uses, including navigation, hydropower, fishing, recreation, flood management, and waste disposal. Our goal is to understand how people modify rivers for various purposes, what the ecological and social consequences of those modifications are, and how to better use rivers to satisfy economic, social, and ecological goals. Dams figure prominently in Part II, since they are a central component of many instream uses but can also be highly damaging to rivers and communities.

Chapter 6 introduces us to navigation, hydropower, fishing, and recreation, examining the historical development of these uses and analyzing the pros and cons of each. Chapter 7 covers flood management from ancient times to the present, with a focus on the shortcomings of structural flood control and the benefits of alternative ways to reduce flood risk. Chapter 8 addresses the use of rivers for waste disposal and how we can protect water quality by using the Clean Water Act. Chapter 9 looks in some detail at dams (and aqueducts): How many have we built? How are they managed? What are their social and ecological impacts? Should we build more? Can we successfully restore rivers by removing dams?

6 Instream Uses: Navigation, Hydropower, Fishing, and Recreation

- How do societies use rivers for transportation, energy, food, and recreation?
- Are those uses compatible with maintaining healthy river ecosystems?
- Is water-based transportation still important to the US economy?
- Will increasing reliance on hydropower help mitigate climate change or make us more vulnerable to it?
- Why are salmon and other iconic fish doing so badly?

In this chapter, we turn to the four most significant instream uses of water—navigation, hydropower, fishing, and recreation—trying to understand both the historical arc of these uses and their current importance and impacts.

1. Navigation

In the absence of motorized transport and good roads, the movement of people and goods by water is much more efficient than by land, so water-based navigation has historically been a major use of rivers and lakes and continues to be significant today. This section traces this history and assesses the current state of water-based transportation, with a focus on the United States.

Most early civilizations developed along rivers or coastlines, which served as routes for trade and migration. Over time, as trade volume increased and trading networks grew more extensive, societies began to bump up against two limitations of using natural rivers for navigation.

First, many rivers are not particularly suited to the movement of larger vessels; they are too shallow, too steep, too meandering, too dynamic, or too filled with obstacles such as boulders and snags (dead trees and other coarse woody debris). These problems led societies to modify rivers to improve navigability through measures such as riverbank stabilization and removal of channel obstructions.

The second "limitation" of a natural river is that its position on the landscape is determined by geomorphic forces, not human convenience. The desire to put waterways where they would best serve trading routes led societies to construct navigation canals. One of the earliest such canals was constructed in China in the fifth or sixth century BCE and ultimately grew into what is now called the Grand Canal, which runs more than 1,000 miles and connects China's two great river systems: the Yangtze in the south and the Yellow in the north. The strategic importance of the Grand Canal to various Chinese dynasties lay in its ability to transport grain and troops between the two centers. The collective effort involved in constructing and maintaining such a massive canal rivaled that associated with the Great Wall.

One of the central problems facing canal builders is how to deal with elevation differences along the route. Finding the flattest possible route for the canal can minimize this problem but can also dramatically increase the length of the canal as it meanders around hills and valleys. Alternatively, bridges or embankments can be used to carry canals directly across valleys.

The primary technology for dealing with elevation differences along a navigation route (whether canal or river) is the **lock**—either the flash lock (developed in China in the first century BCE and now rarely used) or the pound lock (developed in China in the tenth century CE and illustrated in Figure 6-1). Every **lockage** (passage through a lock) requires the release of a lockful of water, so canals need a water source at their highest elevation and can be in competition for water with offstream uses.

As engineering skills improved during the Renaissance, canals were increasingly used in Europe to promote economic development. The Canal du Midi (1681), which was used to transport wheat, wine, and textiles between interior France and the Mediterranean coast, represented a major step forward in its length, its elevation change, and its impoundment of water to feed the canal. During the Industrial Revolution in England, canals such as the Bridgewater Canal (1776) played a vital role in lowering the costs of delivering raw materials to industrial centers and allowing manufactured goods to be delivered efficiently to consumers.

In contrast to England's compact network of canals, the early United States faced a problem of a different scale: how to construct a navigation system that would unite the large and growing country into a coherent economic and political entity. The completion of the Erie Canal in 1825 was a major step in doing so and an early taste of the unprecedented scale of water projects that would emerge in the United States.

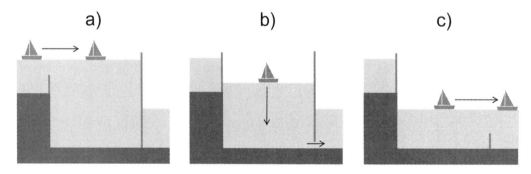

Figure 6-1. Pound lock operation for a vessel moving downstream. (a) The upper gate is opened, and the vessel enters the lock chamber. (b) The upper gate is closed, water is released from the chamber, and the water level in the chamber falls. (c) When the water level in the chamber equals the downstream water level, the water valve is closed, the lower gate is opened, and the vessel moves out of the chamber.

The Erie Canal was the largest public works project in the early history of the United States, reaching about 370 miles from Albany on the Hudson River westward across New York State to Buffalo on Lake Erie. The canal dramatically changed communication and transport in a country where, just forty-five years earlier, it had cost as much to move cargo 30 miles inland as it did to transport it across the Atlantic Ocean.[1] It opened upstate New York and the Great Lakes region to tourists and settlers (Box 6-1), enshrined New York City as a center of commerce, birthed several cultural and religious movements, and helped usher in the era of government-financed hydraulic megaprojects.

Although the Erie Canal flourished for decades and saw several expansions, it was ultimately supplanted by railroad, highway, and pipeline. The canal has not seen significant commercial traffic in several decades and has essentially become a linear park for boaters and bikers. The New York State Canal Corporation now takes in far less in revenue than it spends on operations and maintenance.[2] The same technological innovation that the canal once represented and fostered has now left the canal behind.

The success of the Erie Canal (financed by New York State) helped prompt the federal government to dip its toes into navigation as well. In the US federalist system,

Box 6-1. The Erie Canal and the Haudenosaunee

At the time of European contact, the area through which the Erie Canal would eventually run was home to the Five Nations of the Haudenosaunee Confederacy: from east to west, the Mohawk, Oneida, Onondaga, Cayuga, and Seneca. (Another Iroquoian-speaking nation, the Tuscarora, joined them in the eighteenth century to form the Six Nations.) By 1850, more than 99 percent of Haudenosaunee land had been taken through force, subterfuge, treaty, and illegal land purchases. Although the dispossession and genocide campaign against the Haudenosaunee predated the Erie Canal, the canal sped up this process in multiple ways, including raising the value of land along the canal route and vastly increasing the penetration of White settlers into the area.

In addition, the canal represented a violation of the Haudenosaunee's sacred practices and beliefs. In the words of Jake Haiwhagai'i Edwards, a contemporary Onondaga elder, "We should always share gratitude with what keeps us healthy: the water, the water that carries you for nine months before you take a breath of air. . . . In no way do you change its course for any reason; you walk around and you change *your* path to follow and respect the sacred journey of water. The destruction of a people's belief: [our land] was cut right in half and [they] built a water dam right through the heart of our home, called the Erie Canal."[a]

In their triumphal journey from Buffalo to New York City inaugurating the canal, New York governor DeWitt Clinton and other dignitaries, traveling in a boat called the *Seneca Chief*, were followed by the *Noah's Ark* carrying "specimens of all manners of living things," including "a bear, two eagles, two fawns, several fish and two Indian boys."[b] The symbolism was clear: The Erie Canal represented the end of wildness and the end of the Haudenosaunee. Yet, although true wildness may indeed be gone from upstate New York, the Haudenosaunee are still here, with several reservations and a vibrant cultural center.[c] Water managers have much to learn from them about gratitude and respect for water.

Notes
[a] https://www.youtube.com/watch?app=desktop&v=e4yystxtIoM.
[b] https://www.lockportjournal.com/news/lifestyles/niagara-discoveries-noahs-ark-on-the-erie-canal/article_79ffa9ea-9e3e-11eb-8acb-3364fb4bfaa0.html.
[c] https://www.skanonhcenter.org/.

navigation provides the clearest constitutional basis for the involvement of the federal government in rivers. An important theme in the history of US water policy is the expansion of federal authority (and funding) from navigation to flood control and ultimately to irrigation and hydropower (and, later, environmental protection). This expansion in federal involvement tended to increase the scale and uniformity of the projects, leading to an emphasis on **hard-path** river engineering.

Congress took its first steps into funding navigation improvements in 1824 by passing two laws: the General Survey Act, which allowed for surveys of potential road and canal routes; and An Act to Improve the Navigation of the Ohio and Mississippi Rivers, which provided $75,000 to clear sandbars and snags that were interfering with steamboat traffic.[3] Both of these laws were to be carried out by the US Army Corps of Engineers (USACE), which had been founded in 1802 and soon became a dominant force in river management. Since those initial forays into funding river improvements, Congress has consistently expanded both the scope of activities authorized and the amount of funding allocated ($79 billion for the USACE Civil Works Program in FY 2023).

Much of the early federal financing for navigation was spent on river ***channelization***, sometimes called "river improvement," a suite of practices designed to convert a natural, dynamic river into a stable transportation channel (Table 6-1). Much of this work initially took place on the Mississippi River and its tributaries. Mark Twain, who piloted a steamboat on the Mississippi from 1857 to 1861, described the difficulty of learning to navigate the river when it was still somewhat natural: "There's only one way to be a pilot, and that is to get this entire river by heart. You have to know it just like A B C. That was a dismal revelation to me; for my memory was never loaded with anything but blank cartridges."[4] Through its river channelization work, USACE aimed to eliminate the need for this kind of detailed, place-based knowledge by regularizing and homogenizing the river to a standard width and depth.

Table 6-1. Types of channelization practices used to "improve" rivers for navigation purposes. Many of these same practices are also used for flood control.

Practice	*Purpose*
Removal of snags and debris	Prevent collisions
Dredging of sandbars or riverbeds	Increase ease of navigation; increase water depth to accommodate larger vessels
Narrowing of channels; elimination of braided channels and backwaters	Increase ease of navigation; concentrate flow into one navigation channel
Cutting off of meanders	Shorten navigation distance; increase the slope of the channel, leading to bed scour and channel deepening
Revetments (armored banks)	Prevent the channel from moving laterally (especially important for straightened rivers that want to re-form their meanders)
Wing dikes (obstructions extending from both banks, designed to concentrate flow into the center of the channel)	Narrow the river; create a deep navigation channel

Along with channelization, dams are also widely used to assist navigation, in two distinct ways. First, upstream storage dams are used to store water during the wet season and release it when needed for downstream navigation. Second, where rivers are too shallow for navigation, dams are constructed to raise water levels and create deeper navigable channels.

A prime example of the use of navigation dams is the "Stairway of Water" on the upper Mississippi River between St. Louis and Minneapolis, where the steep and shallow character of the river had long posed a challenge to navigation. Over time, USACE used channelization techniques to achieve a 4.5-foot-deep channel (mandated by Congress in 1878) and a 6-foot channel (1907). But when the use of larger vessels led Congress to mandate a 9-foot channel (1930), USACE set about building twenty-nine low dams, each with a set of locks. A boat trip along the upper Mississippi is now a series of lake cruises separated by lock passages. The Mississippi saw over 100,000 lockages in 2018 (18 percent of the total nationwide), but the system is aging, resulting in increased delays.[5]

How do channelization and dams affect rivers? **Fluvial** geomorphology is the study of the shape of river channels and their floodplains and how they evolve over time. A key insight of this field is that a healthy river has a balance between sediment sources and the water velocities capable of moving those sediments. This balance results in a dynamic, slowly evolving channel and floodplain. Disruption to either sediment sources or water velocities can upset this equilibrium and lead to a channel that rapidly erodes or accumulates sediment.

Geomorphically, one of the main effects of channelization is to shorten a river's path (by about 10 percent in the case of the lower Missouri[6]), which increases its slope. The higher slope, together with the focusing of flow into a narrower channel, increases the river's ability to carry sediment. At the same time, the supply of sediment to the river is reduced by revetments and by sediment trapping behind dams and wing dikes. This combination creates a river that is erosive (**hungry water**) and tends to scour or downcut its bed (by several meters in the case of the lower Missouri).

In addition, channelization locks a river into one form rather than letting it evolve over time. Despite the human perception of rivers as static features of the landscape that can be drawn as lines on a map, rivers are highly dynamic. A management decision to keep a river locked in place implies a commitment to a long-term fight against the river's natural tendencies.

Besides the geomorphic effects, channelization also does damage to river ecology by replacing a complex mosaic of habitats—pools, runs, and riffles; woody debris and small eddies; shallow backwaters; narrow side channels; vegetated wetlands—with a straight trapezoidal channel of uniform width and depth. This habitat simplification, together with the removal of benthic and streamside vegetation, and the covering of the streambank with riprap, concrete, or other material, leads to impoverished biotic communities and the disconnection of the river from its riparian zone.

Is water-based navigation still important to the US economy, or has it, like the Erie Canal, been displaced by newer transportation modes? Table 6-2 provides a snapshot of the current status of inland waterway navigation in the United States and how it compares to truck, rail, and pipeline transport.

Table 6-2. Summary of US inland waterway navigation system.[a]

Parameter	Value
System size	>36,000 miles, 241 locks
Most-used waterways	Ohio, Mississippi, Tennessee-Tombigbee, St. Lawrence, Columbia
Main goods transported	Coal, petroleum, chemicals, agricultural products
Rank, in ton-miles carried	4th (after truck, rail, and pipeline)
Fraction of total ton-miles carried	6–7%
Fraction of total value carried	1–2%
Cost, energy use, and air pollution generated, compared with other modes	Lower when there is a direct water route (especially for travel downstream), higher otherwise
Susceptibility to climate change, compared with other modes	Higher; both droughts and floods can disrupt waterborne commerce[b]; increased precipitation variability makes multipurpose reservoir management more challenging
Federal share of cost, compared with other modes	Much higher (about 90%, compared with about 25% for highways and close to 0% for rail and pipeline)

[a]Based on https://www.bts.gov/content/water-transport-profile, https://www.bts.gov/us-ton-miles-freight, and National Academies (2015).
[b]For example, in 2021, drought caused low water levels in Argentina's Parana River, forcing shipping companies to reduce grain loads by 18–25%; https://www.reuters.com/business/environment/once-100-years-drought-seen-affecting-argentine-grains-exports-into-next-year-2021-08-12/.

The summary in Table 6-2 raises an obvious question: Are the costs of waterway navigation—particularly the associated environmental damage—really worth the small contribution the system makes to national transportation, especially given that (a) waterway navigation is economical only because of government subsidies, and (b) much of the material transported helps feed our societal addiction to ***fossil fuels***?

Although the question is a real one, there are several reasons that water-based navigation is here to stay for the foreseeable future. First, much of the system is multipurpose, serving for flood control and water supply as well as navigation. Second, eliminating waterborne navigation would severely hurt farmers in certain regions, who don't have other low-cost options for getting their products to market. Third, diverting current water-based transportation to other modes would increase congestion, air pollution, and energy use. Still, there are places where the benefits of maintaining navigation probably outweigh the costs; see the companion website for some examples.

2. Water Power

The downstream movement of water in a river involves the conversion of potential energy (the elevation of the water) into kinetic energy (the movement of water and sediment). This energy ultimately derives from the solar-powered hydrologic cycle, which evaporates ocean water and delivers it to mountaintops. This section traces the history of water power, from ancient mechanical power to modern hydroelectricity, and discusses the pros and cons of hydropower, including the complex question of whether it should be considered a sustainable energy source.

2.1. History of Water Power

Compared with other uses of water, hydropower technology was somewhat slower to develop. Both the Roman Empire and the Chinese Han Dynasty developed early versions of gristmills (flour-grinding operations) using low-efficiency horizontal water wheels. (The efficiency of a water power system is the amount of useful work obtained relative to the amount of water power theoretically available; the latter is a function of the flow rate and the ***head***, i.e., the vertical distance over which the water is descending.)

The Middle Ages in Europe saw incremental improvements in hydropower technology, along with a large increase in the amount of hydropower used and the diversity of industries that it served. By the late eleventh century, England had over 5,000 water mills (roughly one for every fifty households), most used for grinding flour. By the sixteenth century, water mills were used in Europe to power at least forty different industrial processes, including olive oil production, silk spinning, metal cutting, and wool fulling.[7]

Water played a critical role in the Industrial Revolution by providing the ***hydromechanical*** energy to power England's famous textile mills, as well as those in New England, including in Lowell, Massachusetts, where James Francis developed the more efficient Francis turbine, still in widespread use today. Dams played a critical role in hydropower facilities by increasing the head, simplifying the diversion of water to turbines, and storing water so power could be generated when it was needed. River locations with a large drop in elevation were particularly favorable for hydropower, and industry clustered around these sites, developing better ways to harness and share the available energy, including running power canals through factory sites so each user could take their share of the flow.

In the nineteenth century, competition from the steam engine led to a decline in the use of mechanical hydropower, although there was a considerable period of overlap, during which hydropower was cheaper than steam where it was available. Ultimately, the victory of the steam engine liberated industry from the need to locate along suitable rivers while enslaving it to a supply of wood or fossil fuels.

2.2. Hydroelectric Power

Hydroelectric power taps into the same energy source as the hydromechanical power discussed above but converts it into electricity via an electrical generator. The invention of hydroelectric power in the late nineteenth century ranks as one of the most significant advances in the history of water technology, and it links the energy and water sectors, to the point where I need to add an energy primer (Box 6-2) to this water book! The coupling of electrical generators to larger and larger Francis turbines and the development of alternating current allowed the energy of rivers to be transmitted long distances and used for a huge variety of purposes. It catalyzed the rapid industrial development of coal-poor, water-rich countries and contributed to a massive spree of dam building that continues to this day. The amount of hydropower used worldwide has risen steadily since the early twentieth century, while hydropower's share of global primary energy has stabilized at about 6 percent (Figure 6-2).

The United States, an early leader in hydroelectricity, has built a total of about 2,500 hydroelectric dams, of which the majority were built in the early to mid-twentieth

> **Box 6-2. Energy and Power**
>
> Energy is the ability to do work. One common unit for energy is the British thermal unit (BTU), defined as the amount of energy needed to raise the temperature of a pound of water by 1 degree Fahrenheit. The amount of energy used in the United States in a year is around 100 quads, where a quad is a quadrillion (10^{15}) BTUs. The metric unit for energy is the joule (J); there are 1,055 J in a BTU. Power is the *rate* of energy generation or use; the most common unit is the watt (W), which is a joule per second. A watt-hour (Wh) is the amount of energy generated by a 1-W power source over the course of 1 hour and is equal to 3,600 J.
>
> Hydropower plants are described by their ***installed capacity***, which is the amount of power they can generate at full operation. For example, Hoover Dam has an installed capacity of 2,080 MW, meaning that it can generate 2,080,000,000 J per second. If Hoover ran at full capacity for a year (8,760 hours), it would generate total energy of about 18 trillion Wh, or 18 TWh (equation 6-B1).
>
> **Equation 6-B1.** $2{,}080{,}000{,}000\,W \times 8{,}760\,h = 1.8 \times 10^{13}\,Wh = 18\,TWh$
>
> However, the power plant can run at full capacity only if there is sufficient water to move through the turbines. For Hoover, average annual generation is about 4 TWh (representing roughly 0.1 percent of total electricity generation in the United States). The ***capacity factor***—the average fraction of full-capacity generation—is thus 4 TWh/18 TWh, or about 22 percent.
>
> Drought and water shortage in the Colorado River Basin have affected Hoover's energy generation, both because less water is moving through the turbines and because the head is lower, meaning that each unit of water generates less energy. The average annual generation has dropped from 4.5 TWh in the 1990s to 4.1 TWh in the 2000s and 3.6 TWh in the 2010s. A lower head can also cause cavitation problems, in which bubbles form in the water as it moves through the turbines, causing mechanical damage. Some of the turbines at Hoover Dam were recently replaced with a different design to prevent cavitation.
>
> When discussing the contribution of hydropower to our energy supply, we need to distinguish between electricity and total energy use. Because hydropower is used only for electricity and not for other energy uses (e.g., transportation fuels or heating), its contribution to electricity generation (currently about 16 percent globally) is much larger than its share of total energy use (about 6 percent).

century, including some that arguably were critical in allowing the Allies to win World War II. By 1980, the United States had mostly stopped building large hydroelectric dams, in part because the best sites had been taken and in part because of growing awareness of the environmental and social costs of these projects. Thus, hydroelectricity output for the United States (and Canada) has been relatively stable for the last several decades (Figure 6-3). In contrast, later-developing countries with large rivers—especially Brazil and China—have been rapidly building hydropower dams in recent years, including the world's five largest by installed capacity: Three Gorges (22,500 MW, China, completed 2008), Baihetan (16,000 MW, China, 2022), Itaipu (14,000 MW, Brazil and Paraguay, 1984), Xiluodu (13,900 MW, China, 2014), and Belo Monte (11,200 MW, Brazil, 2016).

Globally, there is still a great deal of unexploited hydropower potential. A global compilation identifies 1,174 large hydropower dams at various stages of planning and construction (mostly in ***low- and middle-income countries*** [LMICs]), a significant

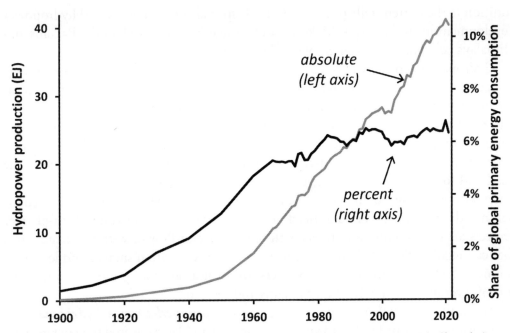

Figure 6-2. Global hydropower over time. Gray line: hydropower use, in EJ (10^{18} joules); dark line: hydropower as a percentage of global primary energy use. Data for 1900–1960 from Smil (2010), 1965–2021 from BP (2022).

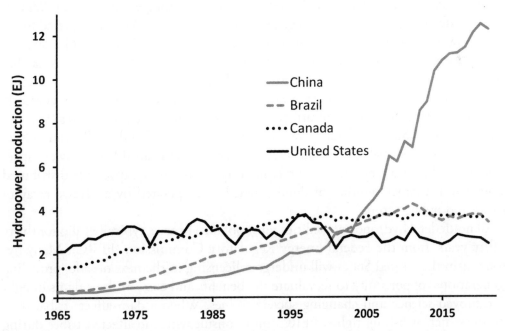

Figure 6-3. Hydropower use over time for the four countries with the largest hydropower production in 2021. Data from BP (2022). Notice the variability in US hydropower production over the last several decades, driven primarily by drought and fluctuations in water availability, despite a fairly constant installed capacity.

addition to the current tally of 2,573.[8] There is special concern over rapid hydropower development in the Amazon, Mekong, and Congo basins, as well as the Balkans and Himalaya, all considered vulnerable ecological hotspots.

Hydropower projects can be categorized as either ***storage*** or ***run-of-river***. Storage projects use dams to store water, allowing it to be released for electricity generation (or other uses) when needed; these are often large dams impounding large multi-purpose reservoirs with long residence times. In contrast, run-of-river projects are designed to generate electricity from whatever flow is moving through the system, without storing water for later use. Many run-of-river projects use dams to raise the head and to direct water through the turbine, but some run-of-river projects divert water without using a dam; the tradeoff is that these may need a long off-river channel in order to generate enough head, leaving a dewatered river reach. Run-of-river projects are most common in rivers with fairly constant flow or with upstream dams that can regulate flow. Although run-of-river projects are often smaller and less environmentally harmful than storage projects, this is not necessarily the case, and some run-of-river projects involve massive dams and reservoirs.[9]

An additional kind of hydropower project is ***pumped-storage hydropower***, in which water is pumped uphill to an upper reservoir and later released to generate electricity when needed. The basic technology has been used in the United States since 1929 as a way of storing energy when it is abundant (and cheap) for recovery when it is scarce (and expensive). There are currently forty-three pumped-storage projects in the United States, with a total storage capacity of 553 GWh,[10] including a facility in Bath County, Virginia, that is the largest in the world (storage capacity 24 GWh; elevation difference 380 m; maximum generation capacity 3,003 MW; completed 1985).

As power grids move toward greater emphasis on variable renewables such as wind and solar, there is increasing need for energy storage to smooth out mismatches between supply and demand. Compared with other solutions, such as batteries, pumped-storage hydropower offers much greater capacity at lower cost and currently accounts for the vast majority of utility energy storage in the United States.[11]

In the United States, federal agencies—especially USACE and the Bureau of Reclamation—are responsible for about half of the nation's installed hydropower capacity, generated at 160 large dams. The remaining hydropower capacity is distributed across a much larger number of dams, owned and operated by a diverse array of private and public actors.

All nonfederal hydropower facilities must obtain a license, typically valid for thirty to fifty years, from the Federal Energy Regulatory Commission (FERC). Many hydrodams in the United States will undergo relicensing in the next decade, providing a generational opportunity to reevaluate the benefits and costs of these dams in light of improved science and changing societal values. When a dam affects tribal trust resources such as fishing rights, FERC must consult with the affected tribes during licensing. In addition, since 1986 FERC has been required to give several social and environmental factors, including habitat and recreation, equal consideration with energy production when evaluating a relicensing application. This gives environmental groups a stronger hand in pushing for denial of relicensing or changes in operations to minimize environmental and social impact.

2.3. Hydropower: Renewable Energy?
The hydropower boom in LMICs has been driven in part by concern over climate change and the need to shift to carbon-free energy sources. In contrast to fossil fuels such as coal, oil, or natural gas, hydropower is considered renewable, because the flow of water that drives the turbines is continuously renewed by the hydrologic cycle. In fact, hydropower currently contributes more energy globally than all other renewables—solar, wind, geothermal, and biomass—combined.

Although hydropower is renewable in a narrow sense, we should be cautious in allowing climate change to push us toward building more hydrodams. First, hydropower reservoirs, like other reservoirs, lose a great deal of water to evaporation. Second, hydropower dams, like other dams, do tremendous damage to rivers and the communities (human and nonhuman) that depend on them, a topic we will discuss in detail in Chapter 9. Weighing this damage against the benefits of "clean energy" is not simple.

But is hydropower really clean energy? A hydrodam certainly *looks* cleaner than a smoky power plant, and for many pollutants—soot or sulfur dioxide, for example—emissions are in fact very low. But in terms of **greenhouse gas (GHG)** emissions—the driver of climate change—the story is more complicated.

Reservoirs—whether used for hydropower or other purposes—release both CO_2 and the more potent GHG **methane (CH_4)**. These compounds originate from the decomposition of organic matter in the reservoir, especially in bottom sediments, where anoxic (oxygen-depleted) conditions often lead to CH_4 production. Sources of organic matter for decomposition include particulate matter washed in from upstream, algae produced in the reservoir itself, and vegetation that occupied the area before flooding. Estimates suggest that reservoirs (which make up a tiny fraction of the earth's surface) are hotspots of methane production, accounting for 3–4 percent of global methane emissions,[12] and the United States has recently started including this source in its GHG inventory.

Perhaps the most relevant question for evaluating the climate impact of hydropower is this: How do GHG emissions from a hydropower reservoir compare to those from a fossil fuel power plant generating the same amount of energy? The comparison is complicated, in part because of the difficulty of comparing different GHGs (methane from hydropower vs. CO_2 from fossil fuels) and in part because power plant emissions are continuous through the lifetime of the plant, whereas hydropower GHG emissions often peak shortly after reservoir construction (when the decomposition of the flooded reservoir vegetation tends to produce a pulse of CH_4).

Still, we can draw some general conclusions, using such metrics as GHG intensity, defined as the amount of GHG released per unit of energy produced (units: kg of CO_2eq per MWh of energy); note that the use of CO_2eq (**CO_2 equivalent**) is a way to combine different GHGs based on their climate impact. Coal-fired power plants have a median GHG intensity of 1,010 kg CO_2eq/MWh (range 900–1,500), and natural gas plants have a median of 510 (range 410–980).[13] The GHG intensity of hydropower is tremendously variable, ranging from about 0.001 to 100,000 kg CO_2eq/MWh (twenty-year time scale).[14]

Thus, although it is likely that most hydropower facilities are lower GHG emitters than fossil fuel power plants, the high variability between dam sites means that *some hydropower installations have higher GHG emissions than natural gas plants or even coal plants.* One study suggests that, of about 1,500 hydropower facilities for which data exist, some 13 percent have higher emissions than comparable coal plants over a twenty-five-year time horizon.[15]

Can we identify which dam sites are likely to have high GHG emissions—preferably before we build a hydropower dam at that site? Several factors tend to drive high GHG emissions:

- Reservoir area: Because methane emissions are a function of the reservoir surface area, reservoirs with low power production per unit area have higher GHG intensities. The best hydropower sites are in steep mountain canyons, where a tall dam can impound a reservoir with high head and low surface area, resulting in high power density (power per unit area) and low emissions.
- Temperature: Tropical reservoirs are likely to have higher methane production than temperate or boreal ones, because of higher standing biomass and higher rates of decomposition.
- Productivity: Regions with high biological productivity (a function of temperature but also nutrient availability) have more organic matter available and thus higher methane production. Increases in anthropogenic inputs of phosphorus into reservoirs could significantly increase methane emissions.[16]

Based on these criteria, tropical reservoirs with low power densities, including most of the lowland dams proposed for the Brazilian Amazon, are not justifiable from a climate perspective, even if we ignore the other impacts on the river system. Higher-elevation Amazon dams, on the steeper slopes of Ecuador, Peru, and Bolivia, make more sense from this perspective.[17]

2.4. Sustainable Hydropower

Can we find ways to reap the benefits of renewable hydropower while minimizing the environmental and social costs mentioned above? Is there such a thing as "sustainable hydropower"?

One arena where these questions are playing out in the United States is in renewable portfolio standards, state-level mandates for how much renewable energy must be used by utilities. States vary in which types of hydropower can count as renewable, with many states requiring hydropower installations to be small, or run-of-river, or use an existing dam. The Low Impact Hydropower Institute (LIHI) has a certification program based on eight criteria, including safe fish passage; several states require LIHI certification for a project to be considered renewable. There are also encouraging technological developments, including fish-friendly turbines, instream "zero head" turbines that don't require dams, and turbines designed to run on existing water flows in irrigation channels and water or wastewater pipes.

Globally, there has been a rapid proliferation of "small hydropower" over the last two decades, driven by the perception that smaller size equates to lower social and environmental impact. The number of small hydropower dams (including those under construction) is estimated to be over 82,000, much higher than the number of large hydropower dams.[18] Despite a relative paucity of research, there is increasing evidence that these installations can have significant ecological and social impacts.

Given the problems associated with new dams, it makes sense to squeeze as much power as we can out of existing dams, including both those that are currently generating hydropower and "nonpowered dams," which account for most US dams (although most of these will be uneconomical to retrofit for power production). As of 2019, the United States has 209 such projects in the pipeline, which will add 1.4 GW of installed capacity to the 80 GW of hydropower capacity currently in place.[19]

One conundrum in thinking about sustainable hydropower is that hydropower is, arguably, the power source that is most vulnerable to climate change. If climate change decreases runoff or increases hydrologic variability, hydropower generation at existing sites will go down, presumably leading to an increase in fossil fuel combustion and ultimately to greater climate change and thus further increased hydrologic variability. Countries such as Brazil that are heavily dependent on hydropower already suffer doubly when drought threatens both water supply and electricity generation.[20]

3. Fishing

We now turn to a third important category of instream water use: harvesting of fish and other aquatic resources. Fishing highlights two aspects of rivers that we have not adequately considered to this point. First, rivers are not merely conduits for water, sources of power, or even geomorphic forces; they are complex and productive ecosystems. Second, the various demands that people place on rivers are not all mutually compatible; in fact, as we will see below, fish and fish-dependent societies have been the clearest losers in the race to fully use rivers for hydropower, navigation, waste dilution, irrigation, and industry.

Humans have been catching fish and other aquatic organisms for tens of thousands of years. From Polynesia to Egypt to the Norte Chico, many early societies thrived on the natural productivity of oceans, estuaries, rivers, and wetlands, developing increasingly sophisticated harvesting techniques.

Around the world, low dams known as fish weirs have long been used to improve fishing success. In what is now Boston, Native Americans harvested fish in tidal waters using wooden weirs made of interwoven branches and brush that have been dated to 2300 BCE to 800 BCE, suggesting some 1,500 years of nearly continuous use. In the Ashuelot River in New Hampshire, the Swanzey Fish Dam, a V-shaped arrangement of rocks (with the point of the V in the downstream direction), was in use from about 1800 BCE until European contact. This type of V-shaped weir was common, because it allowed harvesting of fish migrating in both directions: Downstream-moving fish were caught at the point of the V, and upstream migrators were caught along the banks. These dams were made of stone, wood, or both and needed regular maintenance. Fish weirs did not impound water, but their ability to efficiently harvest migrating fish could create tensions between upstream and downstream tribes.[21]

The Industrial Revolution brought new technologies and attitudes into the world of fishing. The ability to preserve large quantities of fish through canning and refrigeration, together with the transportation revolution, have turned fish into a global commodity and led to overexploitation of many freshwater and marine fish stocks. Other stressors, including dams, water withdrawals, and climate change, have exacerbated this decline. Aquaculture—the rearing of aquatic animals and plants in controlled conditions—has stepped in to fill the demand for seafood and has grown rapidly even while capture fisheries have stagnated.

The story of salmon in the Pacific Northwest illustrates the tragic consequences of the transformation of fishing in the modern age. Pacific salmonids—five species of salmon along with two species of trout, all from the genus Oncorhynchus—are ***anadromous fish***; they are born and spend their early lives in freshwater streams, then migrate downstream to the ocean, where they grow and live as adults before migrating back to their natal stream to spawn and (except for steelhead trout) die, bringing oceanic nutrients hundreds of miles inland. Because of salmon's fidelity to specific streams, the Pacific Northwest is home to hundreds of distinct salmon populations, each adapted to a particular path and timing of migration among the complex habitats of the river–ocean system. Because of their great abundance and ecological significance, salmon have long been considered **keystone species** of Pacific Northwest aquatic ecosystems.

For thousands of years, the tribes of the Pacific Northwest have harvested salmon sustainably and seen salmon as central to their identity. The traditional story of Salmon Woman reflects this, telling how the Mother of Salmon first came to live with the people and share her Salmon Children with them. When the people acted disrespectfully, Salmon Woman moved back to the ocean with her children, ultimately agreeing to continue sending her children back each year to feed the people, but only if the people respected that gift. Many tribes still mark the seasonal return of the salmon with the First Salmon Ceremony, in which salmon and water are consumed with gratitude and celebration. The First Salmon Ceremony—and the attitude that it reflects—helped prevent overharvest by ensuring that many of the returning spawners were allowed to continue upstream to produce the next generation of salmon.

Since the mid-nineteenth century, the incursion of settlers and industry into the Pacific Northwest has decimated wild salmon runs. Although the drivers of salmon decline are complex, we can identify several stressors along the entire stream-to-ocean life history of the salmon. The cumulative and synergistic impacts of all these factors are reflected in the fact that 28 populations of Pacific salmonids are listed as threatened or endangered under the Endangered Species Act.[22]

Habitat: Throughout the nineteenth century, one wave after another of settler resource exploitation hit the Pacific Northwest, with each wave leaving its marks on the streams and rivers that salmon depend on. Gold mining denuded hillsides and washed them into rivers, covering gravel spawning grounds with silt. Timber harvesting increased erosion rates and caused physical damage to rivers used for log transport; sawdust settled on riverbeds and decomposed, depriving streams of oxygen. Cattle grazing led to further erosion and soil compaction, while water withdrawals for irrigation killed migrating fish and depleted river flows. In the twentieth century, urbanization and industrialization further transformed the pathways of water flows and introduced toxic pollutants into streams.

Harvest: Overexploitation of salmon began in the nineteenth century and continues in a different form today. After the first salmon cannery opened on a barge in the Sacramento River in 1864, canneries quickly spread throughout California, Oregon, and Washington, until they were canning millions of pounds of salmon from all the region's important runs. Today, commercial fisheries still target salmon and have been joined by a growing group of recreational fishers. Although treaty rights in theory entitle tribes to half of the harvest, the decimation of salmon has hit tribal fisheries hard, affecting food security, sovereignty, and identity.

Dams: Fish, especially ones with migratory life histories such as salmon, can be highly affected by the presence of dams. Dams of almost any size create a barrier to upstream fish movement. In addition, large reservoirs and dams create problems for downstream fish movement as well, because of the risks of passing through a turbine and the longer time and greater energy needed to swim through a reservoir compared with being carried downstream by the current. For smaller streams, culverts used for road crossings can also impede fish movement.

The damming of rivers throughout the Pacific Northwest has blocked fish migration and shrunk the area available for salmon spawning. Various options have been used to try to reduce the impacts of dams on salmon migration, but they have had limited success.

Along the Columbia River, dams have inundated traditional fishing sites such as the famed Celilo Falls, also known as Wy-Am, where Native Americans from several different tribes would gather annually to spear fish from wooden platforms. When construction of the Dalles Dam inundated the Wy-Am waterfalls and surrounding communities in 1957, the tribes were promised new homes for their communities and new "in-lieu" fishing sites. But the former have yet to materialize (despite rapid relocation of the White community of North Bonneville) and the latter are a poor substitute for the traditional sites, especially with so few fish returning.[23]

Hatcheries: Although some forms of aquaculture have existed since ancient times, the nineteenth century saw increased interest in artificial propagation of salmon for release back into the wild. This ultimately grew into a vast network of salmon hatcheries across the Pacific Northwest (and elsewhere), funded mostly with taxpayer dollars,[24] that try to compensate for the vanishing wild salmon populations by producing and releasing tens of millions of juvenile salmon each year. The hatchery program has proven ineffective at stopping the decline in wild salmon and is probably accelerating their decline by homogenizing the tremendous genetic diversity and adaptability that constitute the wild salmon's greatest strength. As Jim Lichatowich points out, the ecological placelessness of hatchery salmon parallels the placelessness of the industrial mindset that favors fish factories over wild rivers.[25]

Climate change: Salmon in the Pacific Northwest are vulnerable to climatically driven changes in both freshwater and saltwater, including ocean acidification, higher stream and sea-surface temperatures, and changes in ocean circulation.[26] Some of these changes, such as warmer rivers, are exacerbated by other factors such as lower water volumes and longer residence times in reservoirs. The stresses of warmer, less oxygenated water can make salmon more vulnerable to parasites and disease.

Invasive species: *Nonnative invasive species* are those that have been intentionally or accidentally transported by people from one region to another and are spreading rapidly in their new home, doing damage to native species and ecosystems. Salmon in the Pacific Northwest must run a gauntlet of hundreds of nonnative invasive plants, invertebrates, and fish, including some that prey directly on juvenile or adult salmon (e.g., channel catfish, smallmouth bass), others that compete with salmon for food (e.g., American shad), and still others that outcompete important salmon food sources (e.g., New Zealand mud snail).[27] The best solution to biological invasions is to prevent them before they happen by targeting invasion pathways (e.g., canals, ballast water, aquarium trade, sports fish introductions), because it is very hard to eradicate an invasive species once it has become established in a new habitat.

In many cases, the flourishing of nonnative species at the expense of native ones is due in large part to climate change or other anthropogenic changes that alter the fundamental parameters (flow, **turbidity**, temperature) to which native flora and fauna have adapted over many centuries. Anthropogenic changes can even drive native species to become invasive (i.e., to spread rapidly and crowd out other species), as when sea lions decimate salmon who are attempting to climb fish ladders or when dredging creates island habitat for native terns, increasing their population to the point where they negatively affect salmon survival.[28]

4. Recreation

A fourth important instream use is water-based recreation, including both flatwater recreation (e.g., reservoir fishing, boating) and river recreation (e.g., fly fishing, whitewater kayaking). In areas that attract large numbers of tourists, recreation can be among the most valuable uses of water, at least in monetary value. For example, the recreational value of the Colorado River Basin is estimated at $26 billion per year,[29] and the tourism and recreation industry accounts for over 11 percent of jobs in the basin states.[30] Across the United States, recreational fishing has roughly the same economic and job-creation impact as commercial fishing.[31] One could even claim that outdoor recreation contributes more to the US economy than farming ($374 billion, 1.8 percent of GDP[32] vs. $135 billion, 0.6 percent of GDP[33]), although this is a little misleading because it includes knock-on effects of outdoor recreation (e.g., construction, travel) but not of agriculture (e.g., food manufacturing, food retail, restaurants).

Towns built around water-based recreation can experience devastating impacts from droughts and floods; for example, Page, Arizona, is suffering as water levels drop in nearby Lake Powell and houseboats can no longer access boat ramps.[34] Recreation interests often support ecological protection, although many recreationists prefer their nature somewhat tamed. In fact, many kayaking and rafting runs are dependent on dam releases, which are often scheduled far in advance, often for weekends and holidays.[35] Those running the river may not know it, but they are actually riding a short-term pulse of water that is released for their enjoyment. Similarly, some of the best trout fishing spots are just downstream of dams releasing deeper, cold water.

A fascinating example of balancing recreation and hydropower is provided by Niagara Falls, on the border of Ontario, Canada, and the state of New York. The falls' elevation drop of over 50 meters makes it a tempting target for hydropower, and in the early twentieth century, this site was the first to demonstrate the potential of large-scale hydroelectricity generation and transmission. However, water diverted for hydropower doesn't flow over the iconic waterfalls, creating an inherent tension between power generation and tourism. This tension has resulted in a compromise in which less flow is diverted to hydropower during peak tourism season (summer daytimes), although even in those cases the falls carry only about half of their natural volume. To compensate for the reduced flow, managers have made far-reaching modifications to the rock structure of the falls, with the goal of creating the "impression of volume," resulting in what Daniel Macfarlane calls an "envirotechnical system" that creates a simulacrum of nature to replace the real thing.[36] The 30 million tourists who visit the falls each year seem content with the trade, if also unaware of it.

The increasing popularity of water-based recreation raises important and difficult questions about the relationships of tourists with local landscapes and people. At its worst, water-based recreation can bring in outside wealth that displaces long-time residents and damages local ecosystems in the interests of an idealized experience of "nature." At its best, recreation can deepen mutual understanding between urban and rural communities, driving collaboration to protect these places for the benefit of both human and nonhuman residents.

Chapter Highlights

1. Water-based navigation, including the groundbreaking Erie Canal, played a central role in the economic and political development of the United States in the nineteenth century. The use of water for moving goods is still important along certain waterways, but for the United States as a whole it is a distant fourth to trucks, rails, and pipelines.
2. To facilitate navigation, many rivers have been extensively modified through ***channelization*** and damming, with significant impacts on river geomorphology and ecology.
3. ***Hydromechanical*** power has been used for millennia and played an important role in the Industrial Revolution. The development of ***hydroelectric*** power at the beginning of the twentieth century revolutionized water and energy management.
4. There are three kinds of modern hydropower projects: ***run-of-river***, ***storage***, and ***pumped storage***. Run-of-river projects are sometimes—but not always—less harmful to rivers than storage projects. Pumped storage accounts for the majority of power storage in the United States and is becoming even more important as the electric grid transitions to more variable renewables.
5. There is currently a boom in construction of large hydropower dams in LMICs, often in places where they will significantly affect biodiversity and Indigenous groups.
6. Hydropower is often considered a renewable energy source that can play an important role in the transition to a carbon-free energy system. However, hydropower has major impacts on rivers and communities, and at some locations it can produce emissions of GHGs (especially ***methane***) that are even greater than fossil fuel alternatives. There is increasing interest in finding ways to make hydropower more sustainable.
7. Fishing has long been a part of the human relationship to water, but native fish populations have plummeted in many locations because of a host of insults: dams, habitat destruction, overharvesting, ***nonnative invasive species***, and climate change. Attempts to recreate native fish runs by using ***hatcheries*** have been largely unsuccessful.

8. Water-based recreation can be an important source of revenue to rural areas but can also create social and ecological problems.

Notes

1. National Academies (2015).
2. Until the transfer of the canal from the Thruway Authority to the Power Authority in 2016, this shortfall was made up by highway tolls, prompting a lawsuit.
3. The steamboat was invented in the late eighteenth century, but the first commercial steamboat in the United States was Robert Fulton's *Clermont*, which provided service along the Hudson River between New York and Albany starting in 1807. In 1816, Henry Shreve developed a steamboat more suited for the shallow waters of the Mississippi, and by 1827, there were more than 100 steamboats in service on the Mississippi, including Shreve's *Post Boy*, which reduced the time for mail delivery from New Orleans to Louisville from a month to a week (Klein and Zellmer 2014).
4. Twain (1883).
5. https://datahub.transportation.gov/stories/s/vvk5-xjjp/#lock-characteristics-and-delays-on-rivers-with-10000-or-more-lockages.
6. Alexander et al. (2012).
7. Reynolds (1984).
8. https://globalenergymonitor.org/projects/global-hydropower-tracker/tracker-map/.
9. For example, the 56-meter-high John Jay Dam on the Columbia impounds Lake Umatilla (177 km long, volume of 3.1 km^3), but it is considered run-of-river because reservoir water levels are not actively controlled to provide storage.
10. Uría-Martínez et al. (2021).
11. Ibid.
12. Deemer et al. (2016), Yan et al. (2021). Reservoirs can also emit the greenhouse gas nitrous oxide (N_2O).
13. https://www.iea.org/data-and-statistics/charts/full-lifecycle-emissions-intensity-of-global-coal-and-gas-supply-for-power-generation-2018.
14. Calculated from Hertwich (2013). Since methane has a shorter atmospheric lifetime than CO_2, its climate impact is relatively larger when measured over shorter time scales.
15. Ocko and Hamburg (2019).
16. Yan et al. (2021).
17. Almeida et al. (2019).
18. Couto and Olden (2018).
19. There are twelve projects to add capacity at existing hydropower plants, eighty-eight projects to generate power at nonpowered dams, and 109 projects to generate power from existing conduits (e.g., aqueducts); Uría-Martínez et al. (2021).
20. https://www.reuters.com/world/americas/brazil-minister-warns-deeper-energy-crisis-amid-worsening-drought-2021-08-31/.
21. Ritchie and Angelbeck (2020).
22. https://www.fisheries.noaa.gov/species-directory/threatened-endangered.
23. For more on Celilo Falls, see https://www.critfc.org/salmon-culture/tribal-salmon-culture/celilo-falls/ and https://www.seattletimes.com/pacific-nw-magazine/covid-and-squalor-threaten-tribal-members-living-in-once-abundant-indian-fishing-sites-along-the-columbia-river/.
24. Each Columbia River Basin hatchery salmon that successfully returns to the river requires an investment of somewhere between $250 and $650 to produce; https://www.circleofblue.org/2022/world/the-u-s-has-spent-more-than-2-billion-on-a-plan-to-save-salmon-the-fish-are-vanishing-anyway/.
25. Lichatowich (2013).

26. Crozier et al. (2019).
27. Sanderson et al. (2009).
28. Carey et al. (2012).
29. James et al. (2014).
30. Taber (2012).
31. Hughes (2015).
32. https://www.bea.gov/data/special-topics/outdoor-recreation.
33. https://www.ers.usda.gov/data-products/chart-gallery/gallery/chart-detail/?chartId=58270.
34. https://www.hcn.org/articles/climate-change-sinks-lake-powell-local-rec-industry.
35. See, for example, https://amcbostonpaddlers.org/documents/river-releases/.
36. Macfarlane (2020).

7 Flood Management: Learning to Live with "Too Much" Water

- What are the different kinds of floods?
- Are all floods bad?
- Have floods been getting worse because of climate change?
- What can we do to reduce flood risk?
- Does flood insurance help or hurt in the effort to better manage floods?

Perhaps no other weather event makes us feel Nature's power and humanity's vulnerability like a hurricane or large flood. Even with all the tools of modern science and technology, large storm events can still devastate cities and shape history. Smaller, more frequent floods also shape society, though in more continuous and less obvious ways.

Flooding—a particularly obvious manifestation of the variability that is so central to the nature of water—has long been both a boon and a challenge to human populations, especially those living in river valleys and coastal areas. Recent changes in settlement patterns, along with increasing hydrologic variability driven by climate change, have made flood management more important than ever.

We start this chapter by introducing the different types of flood events and then address the question of whether floods are getting "worse." This leads us to distinguish and analyze the three factors contributing to flood damage: hazard, exposure, and vulnerability. We close with a discussion of ways to improve flood management.

1. Types of Floods

The National Weather Service defines a flood as "an overflow of water onto normally dry land."[1] This simple definition presupposes a sharp distinction between areas that are supposed to be wet and those that are supposed to be dry, and thus it underemphasizes the complex interplay of water and land in areas such as wetlands, where the periodic presence of water on land is normal and life sustaining. In order to better live with flooding, we have to stop thinking of it as aberrant and realize that in certain places flooding is normal, even vital.

One factor that makes this awareness difficult is that different areas experience flooding with different frequencies, ranging from coastal wetlands that experience tidal flooding twice a day to higher parts of river *floodplains* that are inundated only in rare floods. The most problematic areas in terms of flood damages are often those with less frequent flooding, where people—perceiving the area as "normally dry land"—are more likely to settle and be affected by floods that are infrequent on human time scales but frequent on geologic ones. This dynamic—in which people move into flood zones only to be surprised when flooding happens—will occupy much of our attention in this chapter.

Even from a purely physical perspective, flooding is a complex, multifarious phenomenon. Flood events can be classified into several sometimes-overlapping types.

Fluvial (riverine) flooding can take two forms:

- **Inundation flooding**: When the flow of water in a river is greater than the capacity of the channel to convey it, the river will overflow its banks and inundate its *floodplain*. Precipitation is obviously a key factor in these flood events (Box 7-1), but antecedent conditions in the watershed are also important; soils that are already saturated will generate more runoff, and rain-on-snow events can increase the amount of runoff by adding large volumes of melted snow. The duration of a river flood depends on the size of the watershed. For the smallest streams, several hours of high rainfall can lead to short-term flooding, while in large rivers, floods can last for a month or more and are driven by sustained rainfall over an extended period.
- **Erosion flooding**: Erosion flooding occurs when rivers erode their banks, either gradually, as part of the river's natural tendency to move over time, or abruptly during flood events. Erosion can lead to *avulsion*, in which the river abruptly changes course in response to high flows. Avulsion often occurs in settings with low slopes and large deposits of sediment or woody debris, where the river can find an easier route downstream by cutting off a meander or cutting through unconsolidated sediment. Flooding and erosion in avulsion-prone rivers can be unpredictable and can affect areas outside the river's floodplain.

Pluvial floods are caused directly by heavy rainfall and can occur far from river channels. These floods can manifest as either standing water (on flat land) or flowing water (on steeper slopes). In many cases, pluvial floods are driven by intense rainfall in urbanized areas and thus qualify as both flash floods and urban floods.

Flash floods are caused by intense, short-duration rainfall, especially in settings where the land can't absorb much water, such as deserts or urban areas (pluvial flash floods), or by the sudden failure of natural or anthropogenic obstructions to streamflow, such as the melting of an ice jam or the failure of a levee (fluvial flash floods).

Urban flooding is characterized by the failure of the human-built drainage system to accommodate the necessary volume of stormwater runoff (pluvial urban floods) or by the overflow of small, flashy urban streams (fluvial urban floods). (Remember that many urban drainage systems must deal with increasing flows due to *impervious surfaces* and climate change.) Because of the complexities of urban drainage systems, areas prone to urban flooding can be patchy and hard to map, and they may fall outside of recognized flood-prone areas such as the 100-year floodplain.

> **Box 7-1. Atmospheric Rivers**
>
> In some coastal regions, the most devastating floods are those associated with ***atmospheric rivers***, weather patterns involving intense horizontal transport of water vapor in a long, narrow band (a few hundred kilometers wide by a few thousand kilometers long). They typically form over the ocean in association with wintertime extratropical storms and can lead to remarkably intense precipitation when they encounter land, especially when they are forced upward by topographic barriers. Like hurricanes, atmospheric rivers are assigned a category (from 1 to 5) based on their duration and their water vapor transport. Studies have estimated that atmospheric rivers generate 22 percent of global runoff,[a] with values over 50 percent common in many coastal areas. In the eleven western US states, atmospheric rivers accounted for 84 percent of total flood damages over the period 1978–2017, with one category-4 event causing $3.7 billion in damage.[b] To date, the effects of climate change on atmospheric rivers have been mostly counterbalanced by the effects of aerosols, but future climate change is likely to make atmospheric rivers more intense.[c]
>
> **Notes**
> [a]Paltan et al. (2017).
> [b]Corringham et al. (2019).
> [c]Espinoza et al. (2018), Baek and Lora (2021).

Coastal flooding occurs when seawater reaches elevations higher than people expect it to:

- Climate change has led to ***sea-level rise*** (Chapter 3), and the rising sea is starting to reach coastal cities in the form of sunny-day flooding, in which high tides flood low-lying streets even in the absence of storm events.
- Hurricanes and other coastal storms can bring large ***storm surges***, in which water is pushed ashore by high winds. A storm surge can add several feet of water on top of the regular high tide, creating a storm tide (high tide plus storm surge) that reaches areas well above the normal tides. In addition to seawater flooding, hurricanes can cause wind damage and freshwater flooding from intense rainfall; distinguishing between these sources of damage can be difficult.
- A tsunami occurs when an undersea earthquake triggers a wave with a very long wavelength and high velocity. These waves are barely perceptible in the open ocean but rise dramatically in height when they come ashore and can do tremendous damage with little warning.

Lacustrine (lake) flooding, which can damage shoreline properties and infrastructure, occurs when more water (from precipitation and rivers) flows into a lake than flows out, resulting in a rise in lake levels.

Toxic flooding—which can fall into any of the categories above—refers to situations where floodwaters pose a chemical, as well as a physical, risk to people and ecosystems. Minor toxicity can arise simply from the chemicals associated with flooding of homes and vehicles. More serious toxicity occurs when highly contaminated industrial sites are flooded, when dams holding mining wastes fail, or when manure-laden animal feeding operations experience flooding. More than half of hazardous waste sites are susceptible to flooding.[2] In addition, river floods can remobilize previously buried contaminated sediments, leading to increased exposure for people and wildlife.[3]

Debris flows are a type of flood-related landslide in which a slurry of water, soil, and rock moves rapidly down a steep slope. They are particularly common in semi-arid regions with sparse vegetation and locations where natural vegetation cover has been removed.

2. Flooding Impacts and Trends

Not all floods are damaging. Regular flooding is part of the natural pulse of many ecosystems and can bring great benefits to these ecosystems and to the human communities that rely on them. Flood-dependent ecosystems include coastal salt marshes, which are defined by tidal flooding (and suffer when it is impeded); cypress swamps, where flooding creates a unique environment rich in animal life; ***riparian*** floodplains, where periodic flooding delivers nutrient-rich silt, allows fish to feed on terrestrial resources, and creates diverse habitats; and even ephemeral desert ponds, which briefly fill with water—and life—after rare rain events.

However, flooding can also cause loss of life and extensive property damage. From 1911 to 2022, inland (noncoastal) floods are estimated to have killed 4.5 million people worldwide, of whom 4.3 million were in China.[4] In the United States, the National Oceanic and Atmospheric Administration (NOAA) database of billion-dollar disasters (Table 7-1) suggests that flooding (especially coastal flooding) causes more monetary damage than all other weather disasters combined. Besides the obvious direct impacts, flooding can also lead to disease and hunger for weeks or months as large numbers of people struggle to find clean water, sanitation, food, and medical care, especially if the water supply infrastructure was damaged or contaminated with sewage or toxic chemicals.

Have floods gotten "worse" over the last fifty to one hundred years? One complication in trying to answer this question is the improving quality of global data over time; an apparent increase in the number of floods or the amount of flood damage may merely reflect better information on more recent events, especially for more remote parts of the world. To avoid this problem, we focus on the United States only and look at large coastal storms, for which there is a more complete database (Figure 7-1). Despite the noise introduced by individual events, it seems clear that the trend over time is toward higher damages.

Table 7-1. Impact of inland and coastal flooding in the United States, 1980–2022, based on NOAA's database of weather disasters with damage of at least $1 billion.[a] "All other disasters" includes drought, heat waves, freeze events, severe storms, wildfires, and winter storms. Damages are in real 2022 dollars (i.e., adjusted for inflation).

Event Type	*Number of Events*	*Fatalities*	*Damage (billion dollars)*
Inland flooding	40	701	184
Coastal storms[b]	60	6,890	1,348
All other disasters	248	8,267	990

[a]https://www.ncei.noaa.gov/access/billions/.
[b]Includes wind impacts as well as coastal flooding.

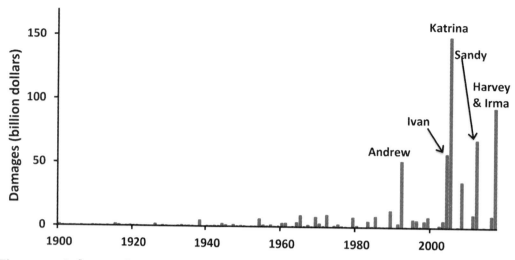

Figure 7-1. Inflation-adjusted economic damages by year for US coastal storms. Data are from Weinkle et al. (2018) and include all damages from hurricanes (i.e., wind damage as well as flooding) in the continental United States.

If flood damages are indeed getting worse, as Figure 7-1 suggests, why is this happening? Should we blame climate change? Can you think of other factors that might explain this pattern? If you can't, you might want to turn back to Chapter 2, Section 4.3, before reading on.

The damage associated with flooding is a function not just of the physical event itself but also of the ways humans live relative to that event. If a river floods but there are no people or structures in the floodplain, there is no economic damage. If a river floods homes but they are designed to withstand flooding, there may be little negative impact. This is a direct application of the concept we introduced in Chapter 2 (in the context of drought); risk is a function not just of hazard but also of exposure and vulnerability. Indeed, this concept will be our guide for thinking systematically about the factors affecting flood risk and understanding the patterns shown in Figure 7-1. The next three sections deal, in turn, with hazard, exposure, and vulnerability.

3. Flood Hazard

Hazard describes the probability of floods of different magnitudes. Remember from Chapter 2 that flood hazard is traditionally described under the assumption of stationarity and the language of flood frequency analysis (e.g., "The hundred-year flood is 10,000 cfs"). Recall also that in many cases this assumption of stationarity is no longer valid; the future will look different from the past because of climate change and land-use change. These changes in flooding can be expressed as a change in the magnitude of floods of a certain frequency (e.g., "The hundred-year flood has increased from 10,000 cfs to 20,000 cfs") or as a change in the frequency of floods of a given magnitude (e.g., "10,000 cfs is now the 50-year flood rather than the 100-year flood").

What do we know about how the flood hazard is changing and will change in the coming decades? For flooding from rain events, we need to think about three factors:

- ***Climate variability***, of the kind discussed in Chapter 2, plays a significant role in driving patterns of precipitation. For example, flooding in northern Europe is associated with a positive phase of the North Atlantic Oscillation, which tends to channel moisture from the Atlantic into the region.[5]
- As we noted in Chapter 3, climate change is likely to drive more intense storms, because a warmer atmosphere can hold more moisture. Indeed, the 2021 Intergovernmental Panel on Climate Change report noted with high confidence that global warming has already driven overall increases in precipitation intensity. Numerous extreme event attribution studies have found that climate change has made some floods more likely or more intense than they would otherwise have been.[6]
- As suggested in Chapter 3, land-use change tends to increase flooding, especially in urban areas, by increasing the fraction of precipitation that becomes stormflow and by increasing maximum precipitation rates. Irrigation may also increase the magnitude of storm events by increasing ET and thus atmospheric moisture supply.

A recent study[7] confirms that both precipitation extremes and flooding events have already increased over most of the globe, with flooding increasing at a rate even larger than precipitation extremes, consistent with the effect of land-use change.

The same three factors also affect coastal flooding:

- Climate variability affects the frequency and magnitude of hurricanes and other ***tropical cyclones***. For example, when the Atlantic Multidecadal Oscillation is in its warm phase, the North Atlantic tends to experience a period of higher-than-average hurricane activity.
- There has been intense scientific debate about whether climate change will lead to more frequent and intense hurricanes. On one hand, hurricanes draw their energy from warm water, so higher sea-surface temperatures will mean the potential for stronger hurricanes, with higher storm surge and more rainfall. On the other hand, many other factors affect the likelihood of hurricane formation, some of which may change in ways that decrease the number of hurricanes.[8] Storm tracks may also change in complex ways, affecting which parts of the coast are more likely to see landfall and possibly lengthening the amount of time that a hurricane sits over a given location. In addition, sea-level rise raises the baseline for storm tides, and a small increase in sea level can translate into a large increase in the frequency of flooding at a given elevation.
- Land-use change—especially the loss and degradation of coastal ecosystems that protect shoreline communities by dampening wave energy and attenuating storm surges—can increase the hazard of coastal storms. For example, the loss of Louisiana's coastal wetlands has left New Orleans less protected from hurricanes.

To sum up, then, we expect that the probability density function of flood events—both inland and coastal—will continue to widen at the upper end (i.e., extreme events will get larger) and the center of the distribution will continue to shift toward larger events.

4. Flood Exposure

Exposure is measured as the number of people (and the value of assets) that are in the path of flooding. Given the massive population growth and economic development that have taken place over the past century, it seems likely that at least some of the trend of increasing flood damages that we saw in Figure 7-1 can be attributed to increased exposure. One way to illustrate this phenomenon is to reconstruct how much damage historical storms would have done if settlement patterns had been similar to today. The perhaps-surprising results from one such analysis (for coastal storms only) are shown in Figure 7-2, which suggests that historic storms, such as the Great Miami Hurricane of 1926, would have done even more damage than recent storms if human development patterns had been similar. In other words, the trend of increasing damage in Figure 7-1 is attributable not just (or even primarily) to physical changes in the magnitude of flooding but also to changes in societal exposure.

Thus, if we are worried about increases in flood damage, we must think seriously about the increase in people and property located in flood-prone areas. Humans have always been drawn to floodplains, riverbanks, and coasts, but for much of human history people have used flood-prone areas in ways that were compatible with—or even benefited from—periodic flooding (e.g., agriculture or fishing). Recently, however, there has been a dramatic increase in permanent human settlement in flood-prone areas. One study found that from 1970 to 2010, the global population living in the hundred-year flood zone almost doubled, and the value of assets in the flood zone increased nineteenfold.[9] A more recent study produced equally troubling results: From 2000 to 2018, the population living in areas that experienced flooding at least once increased at a rate 80 percent higher than the overall rate of population growth.[10] Importantly, both studies held flood *hazard* constant (i.e., the area exposed to flooding was assumed to be constant over time).

Flood exposure is not evenly distributed between populations; those most exposed to floods tend to be disproportionately poor and members of historically disadvantaged groups. An analysis of forty-eight countries found that twenty-two of them had significant overrepresentation of the bottom income quintile in urban flood zones (only eight had significant underrepresentation); patterns in rural areas were weaker.[11] These patterns can form a vicious cycle of poverty, flood damage, and reduced ability to recover from floods. Studies in rural Bangladesh have found that lower-income groups tended to experience higher inundation levels, suffer more flood damage as a share of household income, and suffer more flood-related diarrhea due to contamination of drinking water sources; these groups are also less equipped to prepare for flooding before it happens.[12]

In the United States, an analysis of flood risk at the scale of individual properties found that risk is highest in census tracts that are more impoverished and (perhaps surprisingly) more White.[13] The latter finding is explained largely by the geography of flood risk, which is very high in Appalachia and rural areas of the Northeast and West.

Differential flood exposure within cities reflects the racist history of ***redlining*** in the United States and the continued underinvestment in infrastructure and services in communities of color. An analysis of thirty-eight US metropolitan areas by real estate company Redfin found that homes in areas that had been redlined or yellowlined were 22 percent more likely to be at risk of flooding than homes in greenlined

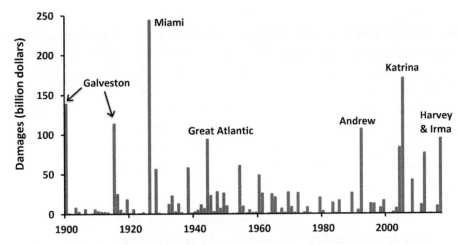

Figure 7-2. Estimated inflation-adjusted economic damages by year from US coastal storms, if those storms had hit the current (2018) landscape in terms of the locations and values of assets. Compare to Figure 7-1. Data from Weinkle et al. (2018).

or bluelined areas.[14] In some cities, such as Sacramento, New York, and Detroit, the difference in risk was twofold or more.[15]

Urban flooding in impoverished neighborhoods sometimes manifests in sewer backups, in which sewer capacity is overwhelmed by runoff from impervious areas, leading raw sewage to spew out of toilets and drains into basements. The resulting damage is difficult and expensive to clean up and may carry a veneer of shame, leading to underreporting of the problem. Underreporting is also driven by language barriers and a sense (often justified) that utilities and municipal governments are not listening to the voices of poor communities of color. Community activists around the country are leading the fight against this manifestation of environmental injustice, including convening communities to openly share their experiences of flooding and organize to demand change.

5. Flood Vulnerability

Once a community is exposed to a hazard (flood) of a given magnitude, its vulnerability determines how much impact occurs. We can decrease flood vulnerability (or, equivalently, increase resilience) by acting thoughtfully at all stages of the flood cycle: planning before floods happen, responding to the immediate flood event, and recovering and adapting after the flood has passed.

Tools available for reducing flood vulnerability include both structural and nonstructural approaches, but the hard path approach of the past century or so has emphasized the former. This section will introduce structural flood control and its drawbacks; we also take the opportunity to introduce benefit–cost analysis as a tool for evaluating flood control projects. Nonstructural tools will be discussed in Section 6.

5.1. Structural Flood Control

One way to reduce flood vulnerability is to prevent flooding from reaching areas of human settlement, an approach often called flood control. Several flood-control structures are part of the engineer's arsenal, but each has its drawbacks.

LEVEES

Levees—embankments or walls along a riverbank designed to keep a river from overflowing—are among the oldest flood-control measures, but they have been implemented at massive scale in the modern era. In the United States, there are over 49,000 km of levees listed in the US Army Corps of Engineers (USACE) National Levee Database[16] and perhaps another 16,000 km that haven't yet been included in the database because "their location and condition are unknown due to complex and varying local ownership."[17] Of the levees in the database, 97 percent are earthen embankments, and 3 percent are concrete, steel, or rock floodwalls.

Several issues typically arise from levee use:

- Increase in flooding: Because they concentrate water in the channel and prevent excess flow from spreading out over the floodplain, levees typically raise downstream water levels. In addition, levees often raise upstream water levels because they constrict the river and act as a bottleneck to water flow.[18] Thus, one of our primary structural tools for reducing vulnerability actually increases hazard throughout the river system!
- ***Levee arms race***: Levees built on one bank of the river will encourage the river to overflow the other bank; levees built on one reach will encourage the river to overflow in other reaches. Thus, once a levee is in place to protect one community, other communities will be compelled to build their own levees, until the river is completely leveed in with no place to overflow. This in turn increases water levels (as noted above), so levees must be built higher and higher. At each point in this levee arms race, the river tends to overtop or breach the weakest or lowest levees, often those protecting poor communities or communities of color. In many large rivers, levees alone will never be able to fully contain the power of the river at flood. Every levee has a point at which it will be overtopped or fail, so there is always a residual risk of flooding in areas protected by levees.
- Floodplain disconnection: Levee use is premised on "reclaiming" floodplains for productive use, such as growing crops on rich alluvial soils or establishing communities along the riverfront. However, as the levees get larger, they undermine the same attractions that brought people there in the first place; alluvial soils are no longer replenished by regular flooding, and riverfront communities are blocked off from the river. In addition, the valuable ecological functions of the floodplain—including sustaining fish and rare species—are lost.
- Safety and maintenance: As Hurricane Katrina demonstrated to devastating effect, levees can fail when they are most needed. To prevent failure, levees must be inspected and maintained regularly. A 2021 American Society of Civil Engineers (ASCE) report gives US levees a grade of D and notes that $21 billion is needed to improve and maintain moderate- and high-risk levees.[19,20] Historically, levee construction and maintenance have often been carried out with the conscripted labor of slaves, convicts, and minorities. During the Great Flood of 1927, for example, African American sharecroppers in Greenville, Mississippi, were forced to work on reinforcing the levee; many perished when the levee failed.[21]
- Governance: Levees in the United States fall under a patchwork of governance

structures. Many smaller levees are built and operated by local government agencies, but the country's largest and most important levees are part of the USACE portfolio, which consists of three categories: levees built and operated by USACE (about 7,000 km), levees built by USACE but maintained and operated by local agencies (mostly levee districts or water management districts; about 13,000 km), and levees built and operated by local agencies but inspected by USACE under its Rehabilitation and Inspection Program (about 3,000 km).[22] Levees in the first two categories are known as federal levees. Federal–local conflicts over responsibility and funding can jeopardize proper levee management.

River Channelization

Various physical changes are often made to river channels themselves (e.g., dredging deeper channels, straightening meanders) to increase the efficiency with which the river can carry water; many of these are the same changes made in the service of navigation (Chapter 6). In extreme cases, rivers are replaced with concrete channels or put underground into tunnels. Increased channel efficiency moves water downstream more quickly, which reduces upstream flooding but can exacerbate downstream flooding, especially if tributaries now deliver their water at the same time instead of in a staggered way. Channelization creates a river that is out of equilibrium and will need ongoing management (e.g., lining with revetments) to prevent it from reassuming its natural form.

In addition, channelization does significant damage to a river's habitat value. Natural streams are complex systems, with a variety of microhabitats created by the presence of various water velocities, depths, and substrates. Hard engineering of river channels tends to flatten out these differences and replace them with a homogenized, sterile system. A concrete channel—with no sand beds for fish spawning, no woody debris to hide behind, no particulate organic matter to feed on, no trees for shade, and no pools to rest in—is not a river. Furthermore, if this concrete channel is designed to be large enough to carry extreme floods, it will be too wide for normal flow conditions, leading to shallow, warm conditions that very few fish can survive.

Floodways

Historically, when major flooding threatened to overtop or destroy river levees, managers sometimes chose to create intentional crevasses (breaches in the levees) to relieve pressure on the system and protect higher-value property. This practice has evolved into the use of ***floodways***: essentially artificial floodplains, designated to receive water during major floods when needed to protect levees. Water is typically shunted into floodways by opening floodgates or fuse-plug levees: levees lower than the surrounding levee system that are designed to be overtopped and washed away (or blown up) when water reaches a certain level. Floodways can be quite effective in minimizing flood damage and can be designed to reap other benefits, such as wildlife habitat and agriculture; the Yolo Bypass near Sacramento is a good example. However, floodways can also raise complicated questions about how to value different areas and populations: Who gets flooded? And who decides?

DAMS

Flood-control dams, which store water during flood events and then release it afterwards, can be an effective tool for controlling floods, but they can also lengthen the duration of flooding and come with serious environmental and social problems.

5.2. The Levee Effect: The Illusion of Dryness

How effective has structural flood control been at reducing vulnerability to flood damage? One way to answer this question at the global scale is to look at how flood risk has changed over time relative to the population and assets exposed to flooding, as expressed in Equation 7-1 (a rearrangement of Equation 2-2).

Equation 7-1. $$\text{Vulnerability} = \frac{\text{Risk}}{\text{Hazard} \times \text{Exposure}}$$

In this approach, flood risk is generally quantified as economic damages or fatalities. At least two studies have used this approach; both found that vulnerability, whether expressed as economic damages per dollar value exposed or deaths per person exposed, has generally declined globally over the last several decades, suggesting that flood-protection measures have worked to some extent.[23]

Does that mean that we should double down on levees, dams, and other structures? There are two reasons that this logic is incorrect. The first is that these analyses do not take into account the lost environmental and economic value of the rivers and floodplains that are destroyed by structural flood control. Second, flood control itself has been a major contributor to increased exposure and thus high flood damages. In a phenomenon that geographer Gilbert F. White called "the levee effect," levees and other flood control measures tend to encourage land development by providing the illusion that an area is safe from flooding. Although damages per person exposed may have gone down, that doesn't mean much if levees are driving up the number of exposed people. The levee effect also means that when levees fail or are overtopped, people are unprepared, and damages are much higher than they would have been in the absence of levees.[24]

The vicious cycle of the levee effect is illustrated in Figure 7-3. The central loop (larger lettering) illustrates that the presence of flood control decreases people's awareness that an area is a flood zone, which increases exposure by encouraging settlement in these areas that—aside from the flood risk—are generally quite desirable. The greater population in turn leads to political and economic pressure to protect those assets by investing in more extensive flood control (e.g., raising levees).

In addition (bottom of diagram), the decimation of Indigenous peoples and the associated loss of sense of place means that people don't know the flood history of a site and see only the benefits of living near the river or coast, not the built-in risks. That lack of awareness is compounded by the short time scales of human perception, in which an area that hasn't flooded in several years is viewed as safe from flooding, and by climate change, which makes flood awareness a moving target.

Another factor at play in driving flood control measures that are ultimately harmful is a predisposition toward structural flood control solutions (top of diagram), due

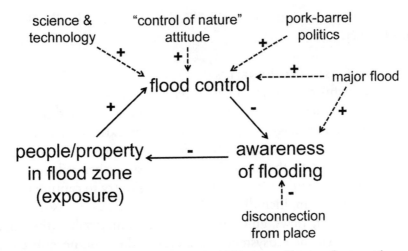

Figure 7-3. The vulnerability–exposure trap. See Chapter 2, Box 2-3 for a guide to reading feedback loops.

to the cultural dominance of technological optimism, which gives us the illusion that we can control the river; the attitude that nature needs to be tamed for human benefit; and the financial and political benefits that accrue—mostly to special interests—from investments in large infrastructure, especially when those investments are funded by taxpayers (who are often unaware of the costs). Even when there is a requirement that flood-control projects be evaluated to make sure they make economic sense, special interests and the absence of data on environmental damage can distort the analysis, as we will see in the next section.

5.3. Benefit–Cost Analysis

Structural flood-control projects are often expensive. Are they worth the money? How do we decide which projects to pursue? How do we avoid succumbing to the pressures shown in Figure 7-3 and building projects that don't make economic sense?

Benefit–cost analysis (BCA) is the primary economic tool used to answer these questions. The idea is simple: Estimate all the societal costs and benefits associated with a proposed flood control project. If the benefits (the flood damages avoided) exceed the costs (i.e., the **benefit/cost ratio** [BCR] is greater than 1) then the project is worth doing. If several alternatives are available, the one with the highest BCR is the best. Because some benefits (or costs) may come many years in the future, BCA involves **discounting** future benefits and costs, so that benefits and costs can be compared on a **net present value (NPV)** basis.

Although BCA was used by USACE as early as 1902, it was first mandated by Congress in the Flood Control Act of 1936, which stated that the federal government should build flood control projects only "if the benefits to whomsoever they may accrue are in excess of the estimated costs." Since then, BCA has become an integral part of project planning and decision making well beyond flood control, although the specific federal guidelines on how to implement BCA have changed several times over the years. BCAs can also be carried out ex post (after the fact) to examine the value of a project in hindsight.

Three thorny issues complicate the idealized picture of objective BCAs guiding flood-control policy.

Unquantifiable Costs and Benefits

Some costs and benefits of a project may be very difficult or impossible to quantify. How large is the cost when we submerge a river rapid that is sacred to a native tribe? What is the value of an endangered species that may go extinct if a river is dammed? These nonmarket goods—unlike, say, the concrete that goes into building a dam—are not traded in markets, so the normal ways of assessing their value don't exist.

Economists have come up with various approaches to try to capture the monetary value of a healthy ecosystem, including both use value (the value associated with recreation, fishing, clean air, views, and other uses) and nonuse value (the value people derive from the fact that the ecosystem exists, even if they will never use it). Stated-preference methods, such as contingent valuation, ask people how much they would be willing to pay to preserve a fish species or a river. Revealed-preference methods use indirect approaches to get at the same question. For example, hedonic valuation looks at how housing prices are affected by environmental amenities (e.g., an intact floodplain ecosystem), and the travel cost method looks at how much people spend to visit various natural settings. Each of these methods has its strengths and weaknesses, and data on nonmarket values are necessarily incomplete.

To many environmentalists and Indigenous leaders, the idea of putting a dollar value on Nature is not just practically difficult but morally wrong—a clash of categories that makes BCA a fundamentally flawed tool for environmental decision making. Simply put, what gives our species the right to turn a river into a commodity? Or, if a river is a living entity with whom I have a personal and communal relationship, should the loss of that relationship really be monetized?

Political Pressure

In theory, the goal of BCA is to provide an objective evaluation of a project to help decision makers resist the pressure to move forward with projects that don't make sense—whether that pressure originates from local communities who want flood protection or from politicians who want a shiny new dam with their name on it. However, those pressures are all too real, and the practice of BCA has proven highly susceptible to manipulation, given the data gaps discussed above.

Political scientists have identified one common pattern of influence, known as the ***iron triangle***. It is named for its three main players: executive-branch agencies such as the USACE and Bureau of Reclamation, who design and build projects; congressional committees, who allocate funding for the projects; and special interest groups, who will implement or benefit from a project (e.g., construction contractors, or developers who want to build in flood-prone areas). All three have an interest in moving forward with projects, because these projects will typically provide them with increased power or wealth.

Distributive Problems

A project with positive net benefits may still have losers: people who are worse off than they were before, such as those whose land is designated as a floodway. Although it may make sense to move ahead with such projects, it is important to be aware of the equity issues involved. Ideally, a BCA can be helpful in identifying which groups are net losers, and it should be possible to compensate those groups. But the BCA itself is not a tool for doing that. In fact, the BCA process can perpetuate preexisting inequalities, because property values in richer neighborhoods are higher than in poor ones, so an "objective" BCA would suggest that the former deserve more protection.

6. The Way Forward

Given the shortcomings of structural flood control, how should we move forward to better manage floods?

6.1. Building in Floodplains: The Problem

Nonstructural approaches to flood management largely fall into two categories:

- **Slow the flow**: Slowing the pathways by which water moves through the landscape allows water to be retained in soils and groundwater rather than contributing to surface flooding. **Green infrastructure** (Chapter 15), wetland restoration, and even beaver dams (Box 7-2) can be used to decrease the effective imperviousness of the land surface and maximize its ability to absorb and store water.
- **Room for the river**: Reconnecting rivers with their floodplains can reduce flood damage while also allowing for the natural productivity that comes with flooding. However, this reconnection is unlikely to happen if there are people living in the floodplain. Thus, one of the most important measures we can take to improve flood management is to minimize the number of people and structures in

---

Box 7-2. Beaver: Nature's Flood Managers

American beaver (*Castor canadensis*) were once abundant in stream and riparian habitats throughout North America, with a population that might have been as high as 400 million individuals.[a] Trapping and hunting for fur, along with changes in land use and hydrology, have decimated beaver populations and extirpated them from many stream systems. Scientists are just starting to understand the resulting impacts on stream geomorphology, ecology, hydrology, and chemistry—and to explore what might be gained from beaver reintroduction.

Beaver dams on small streams are small, porous, short-lived structures that have fundamentally different effects than modern anthropogenic large-river dams. Beaver dams—especially multiple dams along a stream reach—create a beaver-meadow wetland complex comprising a mosaic of habitats with high biodiversity, strong carbon sequestration potential, and good drought resilience. Geomorphically, beaver dams generate braided streams with strong lateral connectivity to their floodplains. Hydrologically, beaver dams slow the flow in small streams and thus contribute to downstream flood mitigation in large rivers. Reintroduction of beaver to streams in England has led to significant reductions in flashiness and peak flows,[b] and there is growing interest in the United States in "rewilding" with beaver for flood mitigation, climate resilience, and other benefits.[c]

Notes
[a] Scamardo et al. (2022).
[b] Puttock et al. (2021).
[c] Ripple et al. (2022), Jordan and Fairfax (2022).

flood-prone areas, which would simultaneously reduce hazard and exposure. The basic idea is the one that Gilbert F. White articulated many decades ago: We can minimize our flood risk by not building in floodplains.

Yet we are not following this commonsense advice. For example, an analysis by Zillow and Climate Central found that in many coastal states, the rates of new home construction over the period 2010–2017 were actually higher in high-flood-risk areas than in safer areas.[25] In the top ten states for risky construction, almost 16,000 homes, with a total value of over $13 billion, were built—in just eight years—in areas that by the end of a thirty-year mortgage will be flooding regularly. This does not seem like rational behavior!

Why do we continue to build in risky areas, and what can we do about it? We can identify three interrelated factors that drive these patterns.

Mismatch in Scale

Decisions on where to live and work are made by individuals, within the constraints imposed by local conditions. Neither individuals nor local communities necessarily have the information or incentives to evaluate and control flood risk. At the same time, government agencies charged with flood management have no say over local land-use decisions, at least in a society like the United States that values local control so highly. This mismatch in scale plays out in the realms of information and incentives, posing serious problems for flood risk governance.

Information

To make the right decisions about where to build, communities need fine-grained data on flood risk, data that are often not readily available. Information is expensive, especially when the risk is nonstationary and when we are dealing with infrequent events such as the hundred-year flood. For example, Atlas 14—NOAA's best estimate of ***rainfall intensity–duration–frequency curves*** (Figure 2-7)—has not been updated in at least fifteen years for about half the states[26]; NOAA's probable maximum precipitation (PMP) values are even older.[27] Relying on stationary flooding estimates in a time of climate change means that infrastructure designed for the hundred-year flood may actually only be adequate for the forty-year flood.[28] Even when accurate information is available, it may not be communicated effectively to prospective homebuyers, who may learn nothing about flood risk until late in the process.[29]

Incentives

Even when they do have good flood risk information, communities may allow development in floodplains because the costs of doing so are externalized. An ***externality*** is a cost or benefit that is not borne by the relevant actor but by other parties; the presence of externalities can skew decision making. In this case, the benefits of development accrue to the community (and the real estate developer), while two types of costs are externalized. First, the cost of additional flood control to protect the new development probably accrues to the USACE or another agency (that is, to US taxpayers as a whole). Second, the ecological costs of floodplain destruction are borne primarily by the ecosystem itself and secondarily by the public at large, not specifically by those who benefit from floodplain development.

A related problem of skewed decision making takes place at the scale of the individual deciding whether to live in a floodplain. Specifically, the existence of federal disaster relief (a fixture of the federal response to floods and other disasters since the late 1940s) and flood insurance (available since 1968, discussed in more detail below) creates a **moral hazard**, a situation where decision makers don't suffer the consequences of their risky decisions. Why should I avoid buying a home in a floodplain when I know that if a flood occurs, I will be bailed out by the government?

6.2. The NFIP: The Solution?

The 1968 creation of the National Flood Insurance Program (NFIP), administered since 1979 by the Federal Emergency Management Agency (FEMA), was an attempt to deal with all three of the problems outlined above: the inability of the federal government to regulate local floodplain development, the need for accurate flood-risk information, and the moral hazard created by disaster relief. The core of the program is the provision of federal flood insurance for homeowners—under the condition that the local municipality participates in the NFIP by developing floodplain regulations that meet certain minimum criteria. The goals of the program are to reduce exposure by working (through local communities) to limit development in floodplains while also reducing vulnerability by providing a safety net to people who suffer flood damage. This approach would also eliminate the moral hazard—if premiums accurately reflected flood risk, so that people who chose to live in risky places would bear the costs of those choices.

Yet the NFIP has not succeeded in reducing flooding exposure, as indicated by the continued increase in population in flood-prone areas and by the prevalence of **repetitive loss properties** (properties that have collected NFIP payouts more than once). The reasons come down, once again, to problems of scale, information, and incentives.

SCALE

While the federal government provides the insurance, the regulations to control floodplain development are written and enforced by the 22,000 communities that participate in the NFIP, with highly variable results. FEMA tries to provide incentives for more effective floodplain management through its Community Rating System, which can lower premiums in highly rated communities, but this incentive is not enough to compete against deep-pocketed developers promising increased property taxes. Yet given the central role of the federal government in flood recovery, FEMA does have some leverage that it could use more effectively (e.g., by conditioning recovery funds on not rebuilding in flood zones).

INFORMATION

A central challenge for FEMA has been to create accurate flood maps for the entire country, both for determining insurance requirements and rates and for guiding communities in floodplain management. Despite an expenditure of $10.6 billion (2019 dollars) over several decades, many areas of the country still have outdated maps (or no maps). One estimate suggests that adequate mapping would require an additional investment of $3.2–$11.8 billion, followed by an annual expenditure of $107–$480 million to keep the maps current.[30]

An alternative to FEMA maps has recently emerged, in the form of the FloodFactor model, a hydrodynamic model designed by a collaborative team of researchers and practitioners. This model is superior to FEMA's in several ways: It covers the entire United States, it includes all flood types (pluvial in addition to FEMA's focus on coastal and fluvial), it operates at the scale of individual buildings rather than large zones, it analyzes flood risk as a continuous variable rather than just identifying whether an area is subject to the hundred-year flood, it estimates the damage done by flooding as well as the physical hazard, it can account for future changes in flood risk, and it is user-friendly, allowing potential home buyers to look up any property at https://riskfactor.com/. In 2020, Realtor.com became the first major real estate company to show flood risk (both FEMA and FloodFactor) for listings on its website.[31] Unfortunately, a comparable model does not exist for other countries, although satellite imagery is increasingly being used to provide flood risk information around the world.[32]

INCENTIVES

How well has the NFIP done at providing incentives to homeowners to avoid risky areas? From its inception, the program has struggled with the tension between premiums that reflect flood risks and subsidies meant to share the burden. That tension has mounted in recent years, as flood damages have increased and it has become obvious that premiums are not covering payouts. In 2012, Congress attempted to fix the problems with the NFIP by passing the Biggert-Waters Flood Insurance Reform Act, which phased out subsidies and moved toward having insurance premiums reflect the actual risk of flooding. Public outcry over the increases in premiums led Congress in 2014 to reinstate many subsidies and to moderate the rate of premium increases. Still, many homeowners are facing rising premiums that will quickly become a significant burden as FEMA implements a new pricing structure called Risk Rating 2.0.

6.3. Buyouts and Managed Retreat

It is clear that we must stop building in flood-prone areas. But when it comes to existing floodplain development, there is no simple solution. On one hand, we should not be subsidizing people to live in high-risk places. On the other hand, one can't help but sympathize with a homeowner who can't afford to pay rapidly rising premiums on a home she has lived in for decades—a home that has been newly mapped into a flood zone and is suddenly impossible to sell even if she wanted to.

One solution to this conundrum is to provide funding to buy out flood-prone properties as part of a strategy of reducing exposure through ***managed retreat***. FEMA's Hazard Mitigation Grant Program (HMGP) is designed to do just that, with a focus on postflood buyouts. Once homes are damaged, the HMGP pays homeowners to move (by compensating them for the preflood value of the home) instead of paying them to rebuild. A recent analysis of HMGP found that 43,633 damaged properties were bought out over the period 1989–2017, including 8,000 in 1993 alone (a year of major flooding in the Midwest).[33] Buyouts generally have BCRs of 3 or more, meaning that every dollar spent on a buyout provides at least $3 in avoided flood damages and other benefits.[34] Unfortunately, buyouts are labor intensive, especially given the multiple layers of government involved and the small number of properties bought at one time; the median project size for HMGP is three properties. Flooding is often a communal issue, so managed retreat from flood-prone areas really needs to happen at a larger scale. In addition to HMGP, a number of smaller federal, state, and local programs exist, some of which allow proactive buyouts of homes in flood zones.[35]

Questions of justice and equity must be central to any managed retreat program. We need programs in which people in flood-prone areas are offered fair compensation for the preflood value of their house and given an opportunity to build a new life on higher ground and to move as a community where possible and desired. We do not want a program in which low-income residents of color are forcibly cleared from flood zones while their White and higher-income counterparts have levees built to protect them.[36] And, as was evident throughout this section, we need to carefully balance individual autonomy with wise public planning.

Buyouts also pose questions about what happens on the bought-out properties. There are two scenarios we clearly want to avoid: the one where the property is resold to new owners, who build a new structure on the same plot; and the one where the property becomes a hotspot for crime, driving neighborhood disintegration. The best uses for the bought-out lands involve some form of return to nature, preferably in a way that creates a green amenity for the neighborhood that can be enjoyed when the site is not flooded. For example, the town of Ottawa, Illinois, bought out sixty-three low-lying structures between 1997 and 2013 and created what is now known as Fox River Park. The buyouts cost $4.5 million but as of 2017 had already saved $8.5 million in avoided flood losses.[37]

One promising approach to managed retreat combines buyouts with **setback levees**, in which levees are repositioned farther away from the river. This allows levees to continue to protect the most valuable property while giving the river more room to flood and creating riparian habitat.

Given the high cost of land in urban areas, managed retreat from agricultural land may be more cost-effective. Although farmland is not eligible for buyouts under the HMGP and most other programs, several USDA programs can finance the retirement of frequently flooded farmland, including the Emergency Watershed Protection Program,[38] the Wetland Reserve Easement Program,[39] and the broader Conservation Reserve Easement Program.[40] Unfortunately, these programs are underfunded relative to farmer demand.[41]

The NFIP does not insure agricultural land (although it does insure farm structures), but subsidized crop insurance can allow farmers to make money from frequently flooded lands, providing a perverse incentive to keep these lands in use. From 1995 to 2020, farmers across the United States received $39 billion in insurance payments for crop failures due to excess moisture, second only to payments for drought ($49 billion).[42] Climate change is leading to increases over time in crop insurance payouts.[43] We need to change these patterns by taking vulnerable agricultural lands out of production and allowing them to revert to natural areas, where periodic flooding will be beneficial, not harmful.

Chapter Highlights

1. There are many different types of floods. Greatest attention is often paid to fluvial and coastal flooding, but ***pluvial***, ***lacustrine***, flash, toxic, and urban flooding are also important, as are ***debris flows***.
2. Flooding is often ecologically beneficial and can provide significant ecosystem services to society, but it can also cause great damage.
3. The physical flood ***hazard*** is increasing in many places due to climate change and land-use change, but the primary driver of increased flood damage over the last several decades is increased ***exposure*** (more people living in flood zones).
4. Flood exposure is not evenly distributed; poorer communities and communities of color are more likely to suffer from flooding.
5. Many rivers have been extensively modified for flood control with channelization, dams, and ***levees***. Levees have several major drawbacks: They cause ecological and social damage by cutting off rivers from their floodplains, they increase water levels in upstream and downstream reaches (leading to a ***levee arms race***), and they encourage development in flood-prone areas by providing the illusion of dryness (the ***levee effect***).
6. The hard-path approach to flood management—massive engineering projects designed to keep the river in its channel—is slowly being replaced with an approach that focuses more on slowing water down on its way to the river, providing room for the river to overflow, and managing risk by reducing exposure.
7. ***Benefit–cost analysis*** is meant to provide an objective measure of whether a given flood-control (or other) project makes sense, but it often ends up undervaluing ecosystems and favoring special interests over the public good.
8. Because of problems of information, incentives, and scales, we are continuing to build in flood-prone areas, increasing our exposure to future floods. The NFIP, which was meant to reduce exposure, has instead increased it by subsidizing (re)building in flood zones.
9. New models provide better data on flood risk, which can help individuals and government agencies make better decisions.
10. Efforts at ***managed retreat*** from flood-prone areas (sometimes using ***setback levees***) face financial, logistical, and equity problems, but they are having some success in reducing exposure, providing space for floodwaters, and reconnecting rivers with their floodplains.

Notes

1. https://www.weather.gov/mrx/flood_and_flash.
2. GAO (2019).
3. Crawford et al. (2022).
4. Data from the Emergency Events Database (https://www.emdat.be/).
5. Kundzewicz et al. (2019).
6. https://www.nature.com/articles/d41586-018-05849-9.
7. Yin et al. (2018).
8. For example, wind shear—the vertical difference in wind direction and intensity—may increase in a warmer world, thus destabilizing nascent hurricanes.

9. Jongman et al. (2012); the same study projects that by 2050, the flood-prone population will have increased (relative to 1970) 2.5-fold and assets will have risen sixty-four-fold.

10. Tellman et al. (2021).

11. Winsemius et al. (2018).

12. Brouwer et al. (2007), Hashizume et al. (2008).

13. Wing et al. (2022).

14. Maps created in the 1930s by the federal Home Owners' Loan Corporation for identifying favorable areas for home lending used four categories: A (green or "best"), B (blue or "still desirable"), C (yellow or "definitely declining"), and D (red or "hazardous"). Communities of color were assigned to categories C and D. https://dsl.richmond.edu/panorama/redlining/.

15. https://www.redfin.com/news/redlining-flood-risk/. See also Qiang (2019), who finds, using coarse spatial data, that people living in flood-prone coastal areas tend to be wealthier than average, and people living in flood-prone inland areas tend to be poorer than average.

16. https://levees.sec.usace.army.mil/#/.

17. ASCE (2021c).

18. Heine and Pinter (2012).

19. ASCE (2021c).

20. The USACE levee risk assessment program defines risk based on a combination of likelihood and consequence of failure.

21. Klein and Zellmer (2014).

22. https://www.mvp.usace.army.mil/Missions/Emergency-Management/Rehabilitation-Inspection/.

23. Jongman et al. (2015), Tanoue et al. (2016).

24. Logan et al. (2018).

25. Climate Central (2019); https://rzh.climatecentral.org/#12/40.7298/-74.0070?show=zillow_k17&projections=1&level=5&unit=feet&pois=hide.

26. https://www.npr.org/2022/02/09/1078261183/an-unexpected-item-is-blocking-cities-climate-change-prep-obsolete-rainfall-reco.

27. NOAA says this about the PMP data: "NOAA's National Weather Service has provided PMP guidance and studies since the late 1940s at the request of various federal agencies and with funding provided by those agencies. In recent years that funding has diminished and gradually ceased. As a result we are unable to continue our PMP activities. However, we will continue to provide copies of related documents on this site. We recognize that many of the current documents need updating. The Federal Advisory Committee on Water Information's Subcommittee on Hydrology is examining this issue."

28. Wright et al. (2019). In the upper Midwest, the rainfall event that TP-40 (the antecedent to Atlas 14) identified as the hundred-year event had become the forty-year event by the year 2017.

29. Ironically, the Department of Housing and Urban Development is particularly guilty of selling flood-prone homes without disclosure. https://www.npr.org/2021/09/13/1033993846/the-federal-government-sells-flood-prone-homes-to-often-unsuspecting-buyers-npr-.

30. Association of State Floodplain Managers (2020).

31. https://www.npr.org/2020/08/26/905551631/major-real-estate-website-now-shows-flood-risk-should-they-all.

32. https://floodmapping.inweh.unu.edu/.

33. Mach et al. (2019).

34. Nelson and Camp (2020).

35. Peterson et al. (2020).

36. Understanding of the influence of racial bias on buyouts is complicated by the paucity of demographic data on individual homeowners, but inferences from census tract data suggest a complex, multilevel process in which race certainly plays a part; Mach et al. (2019), Elliott et al. (2020).

37. https://www.illinoisfloods.org/content/documents/4c_estimating_the_economic_impact_of_buyouts.pdf.
38. https://www.nrcs.usda.gov/programs-initiatives/emergency-watershed-protection-ewp-program-buyouts.
39. https://www.nrcs.usda.gov/wps/portal/nrcs/main/national/programs/easements/acep/.
40. https://www.fsa.usda.gov/programs-and-services/conservation-programs/conservation-reserve-enhancement/crep_for_producers/index.
41. https://www.ewg.org/research/mississippi-river-region-billions-dollars-spent-crop-insurance-payouts-could-have-been.
42. https://farm.ewg.org/cropinsurance.php.
43. Diffenbaugh et al. (2021).

8 Water Quality and the Clean Water Act

- Which pollutants affect the health of aquatic ecosystems and the people who use them?
- How does the Clean Water Act work to protect and restore the ecological and human values of rivers, lakes, and wetlands? How successful has it been in doing so?
- What other laws are relevant to protecting aquatic ecosystems?

In the last three chapters, we discussed the various ways that rivers have been hydrologically, physically, and ecologically altered through anthropogenic water use, navigation, hydropower, fishing, recreation, and flood control. In this chapter, we explore the water-quality dimension of the human relationship with rivers by focusing on the use of rivers for waste disposal—and the ways that this use undermines the ecological and human value that healthy rivers can provide. We also explore the Clean Water Act as a tool for protecting aquatic ecosystems from pollution and other abuses.

People have been using rivers to flush away wastes for a long time, but population growth and the Industrial Revolution brought massive increases in the types and quantities of pollutants, with little concern over the impacts on rivers and other water bodies. In the nineteenth and twentieth centuries, recognition of the impacts of water pollution on human health, quality of life, and fish and wildlife habitat gradually led to greater pollution controls in many nations.

Pollution can come from a variety of sources, as we will discuss in later chapters: sewage (Chapter 14), urban stormwater runoff (Chapter 15), industry (Chapter 17), and agriculture (Chapter 18). The goal of this chapter is to provide an overview of water quality management in three parts: a brief background section on how water pollution affects the health of people and aquatic organisms; an exploration of the Clean Water Act, the primary tool in US law for water quality protection; and a brief summary of two other relevant laws.

1. Water Quality and Health

Water pollution is an ongoing problem for human and ecosystem health around the world. In the United States, recent headlines highlight a plethora of issues: "forever chemicals" in Michigan groundwater, harmful algal blooms off the coast of Florida, increasing turbidity in the blue waters of Lake Tahoe, beach closures due to bacteria in southern California, and arsenic in drinking water in the Southwest. Even in this short list, we can already see an important aspect of water pollution: Unlike air pollution, where a small number of parameters (particulate matter and ozone) dominate risk profiles, water pollution is quite diverse, with different types of problems in different locations; water quality monitoring is correspondingly complex and expensive.

Table 8-1 summarizes twelve important water-quality parameters, ranging from general descriptions of water characteristics (e.g., temperature) to specific chemical constituents (e.g., arsenic) and from easily monitored field parameters (e.g., salinity) to chemicals that require expensive lab analysis (e.g., organic contaminants). The pollutants listed in Table 8-1 can be introduced into water bodies from both ***point sources*** (discrete facilities that discharge through a pipe, such as a factory or a sewage treatment facility) and ***nonpoint sources*** (activities across the landscape that lead to diffuse flows of pollutants, such as ***stormwater*** runoff that picks up pollutants and delivers them to water bodies).

The last category in Table 8-1—health-related parameters—requires a bit more explanation. This category covers the two primary ways that polluted water can harm human health directly: infectious disease and chemical toxicity. Managing these health risks requires understanding what the disease agent is, how the agent enters the water, how people get exposed, what kind of disease they get, and how likely they are to get sick if they are exposed. The last of those factors is commonly thought of in terms of a ***dose–response curve***; the likelihood that a person will get (or die from) a given disease is a function of the dose they receive.

Infectious disease: Human and animal excreta can contain disease-causing organisms (pathogens), which we will discuss in more detail in Chapter 14. To identify the possible presence of pathogens in water, water quality assessments typically measure indicator bacteria: bacterial groups that are not in themselves harmful but that are easy to measure and that tend to co-occur with fecal pathogens. When concentrations of these indicators go above established thresholds, the water is considered potentially contaminated and unsafe for various uses. The most important indicator groups are as follows:

- Total coliform: a large group of bacteria, used to indicate possible water quality problems in drinking water and to warrant more specific testing.
- Fecal coliform: a subset of total coliform that is a bit more specific to human and animal waste, used in some countries for testing of both drinking water and bathing beaches.
- *Escherichia coli* (*E. coli*): a subset of fecal coliform that is used in the United States as the primary indicator for freshwater bathing beaches.
- Enterococci: a group of fecal bacteria that is used in the United States as the primary indicator for saltwater bathing beaches.

Table 8-1. Summary of some important water quality parameters. For background on each parameter, see the companion website.

Parameter	Ecological Impact	Human Impact	Typical Sources
Basic Descriptive Parameters			
Trash, plastic, *floatables*	Wildlife entanglement, potential toxicity of microplastics	Aesthetics, potential toxicity of microplastics	Street litter, urban stormwater, sewage
Temperature	Fundamental[a]	High temperature limits suitability for industrial use.	Urban stormwater, power plant discharges
pH (acid/base balance)	Fundamental[a]	Toxicity, corrosiveness	Urban stormwater, industrial discharges
Dissolved solids, salinity	Fundamental[a]	Taste, blood pressure	Urban and agricultural stormwater, sewage
Turbidity, total suspended solids (TSS)	Photosynthesis, fish behavior and reproduction	Sediment can carry microbes and toxics, and interferes with water treatment.	Urban and agricultural stormwater, sewage
Oxygen-Related Parameters			
Dissolved oxygen (DO)	*Hypoxia* (low DO) causes fish kills.	Low-DO water may have a sulfidic odor or high metal concentrations.	DO problems are typically caused by discharges of BOD or nutrients.
Biochemical oxygen demand (BOD)	High inputs of degradable organic matter (i.e., BOD) can fuel high rates of bacterial respiration and lead to hypoxia.	Organic matter can form toxic disinfection byproducts during drinking water treatment.	Urban and agricultural stormwater, sewage
Nutrients	High inputs of nutrients (especially nitrogen and phosphorus) can lead to algal blooms.	NO_3^- is toxic at high levels.	Urban and agricultural stormwater, sewage
Health-Related Parameters			
Indicator bacteria	No impact	Disease transmission	Urban and agricultural stormwater, sewage
Radioactive elements	Toxicity	Toxicity	Industry, oil and gas, mining
Trace metals	Toxicity	Toxicity	Urban and agricultural stormwater, sewage
Organic contaminants	Toxicity	Toxicity	Urban and agricultural stormwater, sewage, industry, oil and gas, mining

[a]This parameter describes the fundamental physical and chemical nature of the water body and determines what type of organisms it can support.

Chemical toxicants: For certain chemicals and radioactive elements, high-dose exposure for a short time (acute exposure) or low-dose exposure for a longer time (chronic exposure) can cause cancer or other toxic effects. Cancer is a diverse set of diseases that have in common the uncontrolled growth of cells as a result of genetic changes, which can occur naturally but can be sped up by exposure to certain chemicals, referred to as ***carcinogens***. Noncancer toxic effects vary widely and can affect the functioning of all of our bodily systems, including the immune system, nervous system, digestive system (including the liver), urinary system (including the kidneys), reproductive system, respiratory system, and cardiovascular system.

Carcinogens and noncarcinogens are regulated somewhat differently. Noncarcinogens, often called ***threshold toxicants***, are understood to have a "safe dose" below which there is no risk of disease, whereas carcinogens are assumed to have a linear dose–response curve at low doses, so any increase in dose leads to a corresponding increase in cancer risk (although scientists are increasingly questioning the validity of this assumption).

Every step of the chemical risk assessment process, from identifying the type of toxicity to determining the potency of the toxicant, is data-intensive and fraught with uncertainty, especially because it often involves extrapolating from high-dose animal testing to low-dose human exposure. Complicating matters further, a group of chemicals known as ***endocrine-disrupting chemicals*** appears to frequently show nonmonotonic dose–response curves: U shapes or inverted-U shapes or other complex patterns where risk does not uniformly increase with dose.[1] These chemicals—which are structurally similar to human hormones such as estrogen—operate by interfering with the body's hormone signals, but the effects of these disruptions appear to be capable of causing a variety of disease types, from immune problems to abnormal sexual development to cardiovascular disease. The implications of nonmonotonic dose–response curves are startling; these chemicals may cause harm at very low exposures, and our usual system for evaluating risk is not suited to the task.

In addition to affecting human health, chemical pollution can cause death and disease (including cancer) in aquatic organisms. Even levels of pollution that do not kill individual organisms can decimate populations if they interfere with reproduction.

2. The Clean Water Act

With this background on water quality, we can now turn to the 1972 Federal Water Pollution Control Act (amended in 1977, 1987, and 2000), commonly referred to as the Clean Water Act (CWA). As noted in previous chapters, the pressures on freshwater ecosystems extend well beyond the water quality focus implied by the term "clean water" to include numerous hydrologic and physical stressors. Despite its name, the stated goal of the Clean Water Act is fittingly broad: to "restore and maintain the *physical, chemical, and biological* integrity of the Nation's waters" (italics added). However, many of the CWA's provisions are focused primarily on pollution control, a legacy of the highly polluted waterways of the 1960s and a reflection of that period's limited scientific understanding of physical and hydrologic impacts.

Figure 8-1 summarizes the most relevant provisions of the CWA and outlines the structure of this section. The CWA works from two directions: reducing the pressures

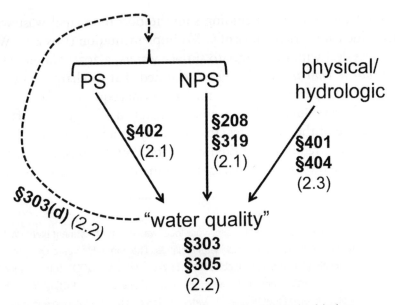

Figure 8-1. Overview of the structure of the CWA, highlighting the most relevant sections of the act and (in parentheses) the sections of this chapter where they will be discussed. Chemical and nonchemical stressors (top) affect water quality (bottom). The TMDL provision (§303(d), dashed arrow in the figure) is meant to allow the results of water quality assessment to feed back to the control of pollution sources. NPS, nonpoint source pollution; PS, point source pollution.

on aquatic ecosystems, in particular by controlling pollution (and to a lesser extent addressing physical and hydrologic stressors); and evaluating the health of aquatic ecosystems and having those results feed back to require additional pollution reduction as necessary. We will start our discussion in Section 2.1 by addressing provisions for reducing point and nonpoint source pollution, then turn in Section 2.2 to water quality standards and assessment, and finally, in Section 2.3, discuss nonchemical impacts.

2.1. Addressing Chemical Pressures

One of the reasons that the CWA has been more successful than previous water pollution control laws in the United States is its focus on **technology-based effluent limits** for point source discharges. In other words, the CWA requires factories and cities to treat their wastewater to a level determined by what is technologically feasible, regardless of the condition of the receiving waters. This has vastly simplified both the regulatory apparatus and the engineering of treatment plants by creating uniform categories of point sources that require the same type of treatment.

The regulatory mechanism for implementing these effluent limits is a system of pollution permits for point sources, known as the **National Pollutant Discharge Elimination System (NPDES)**. NPDES permits (which are required by §402 of the CWA) specify acceptable levels of pollutants in wastewater effluent, as well as monitoring and reporting requirements. There are more than 65,000 NPDES-permitted facilities around the country,[2] including some 16,000 municipal wastewater treatment plants, which are cornerstones of CWA implementation (Box 8-1). While most NPDES permits deal with discharges to surface waters, a 2020 Supreme Court case (*County of Maui v. Hawaii Wildlife Fund*) clarified that discharges to groundwater also require a NPDES permit if they are the "functional equivalent" of a surface water discharge (i.e., if the polluted groundwater ultimately discharges to surface water).

For all but three of the fifty states, the authority to administer the NPDES program has been transferred from the Environmental Protection Agency (EPA) to state government, under a provision known as state primary enforcement responsibility, or

---

Box 8-1. Municipal Wastewater Treatment

Unless you live in a rural area, you probably "use" your local wastewater treatment plant (WWTP) every day, but most people are barely aware of these vital facilities. The domestic wastewater that flows into WWTPs carries high levels of pathogens, biochemical oxygen demand (BOD), total suspended solids (TSS), nutrients, metals, some organic pollutants, and fat, oil, and grease (FOG).[a] In addition, urban sewer systems may also receive industrial wastewater, which can be quite variable in its composition; because WWTPs are not designed to treat toxics, industrial dischargers may be required to pretreat their waste before discharging to sewers.[b] WWTPs sometimes receive weirder items as well, such as bowling balls, opossums, prosthetic legs, and tires.[c]

NPDES permits for WWTPs require maximum effluent BOD and TSS concentrations of 30 mg/L, as well as 85 percent removal of these pollutants from the incoming wastewater. To achieve these permit requirements, WWTPs generally use the treatment sequence shown in Figure 8-B1. ***Primary treatment*** is aimed at removing solids through screening, followed by settling in a clarifier (a tank designed to let solids sink to the bottom and scum float to the top for removal). ***Secondary treatment*** goes farther and breaks down the organic matter or BOD that is present at very high concentrations even after primary treatment. The heart of secondary treatment is a living microbial community (often in the form of activated sludge) that decomposes the organic matter and converts it into CO_2 and its own biomass. This microbial community must be protected from toxins and from rapid changes in flow and water quality. In addition, because the microbes are rapidly consuming O_2, the activated sludge chamber must be continuously aerated to provide sufficient O_2 for ***aerobic respiration***.

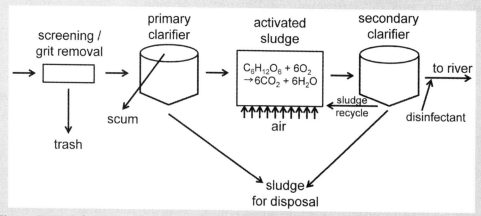

Figure 8-B1. Schematic of a wastewater treatment plant. The chemical reaction shown in the activated sludge portion of the treatment is aerobic respiration.

After traveling through the aeration chamber, the mixture of microbes and treated water travels into a secondary clarifier, where the microbes settle to the bottom and are partly recycled back to the start of the aeration chamber. The clarified effluent is then discharged to a river or other water body,

Box 8-1. *continued*

sometimes preceded by disinfection.[d] (Secondary treatment without disinfection typically results in 90–99.9 percent removal of *E. coli*.[e]) Material from the bottom of the clarifiers, consisting of microbes along with other solids, is called sludge or biosolids and must be disposed of.

Secondary treatment uses significant energy for water pumping, aeration, and other uses, with energy costs typically representing 25–40 percent of WWTP operating budgets. US WWTPs collectively consume more than 30 TWh/yr of electricity, but energy-efficiency retrofits can reduce energy usage by an average of 30 percent.[f]

WWTPs can be modified to provide additional (**tertiary**) treatment, which most commonly involves removal of nutrients (N or P), which are not effectively removed by secondary treatment and can cause eutrophication problems. However, adding tertiary treatment increases cost and energy use and generates additional sludge.

US sludge production amounts to 23 kg dry weight per person per year.[g] Because raw sludge (before dewatering or other treatment) is only about 1–2 percent dry matter, this translates into roughly a ton or two of thick brown gunk as the annual contribution from each of us. The primary pathways for sludge disposal are incineration, landfilling, and application to agricultural land as a fertilizer or conditioner, but each method has its problems, especially because the sludge may have high concentrations of toxic metals, organic contaminants, and pathogens, including newly emerging viruses.[h] Anaerobic digestion is an alternative that can capture some of the energy content of the sludge.

Most US WWTPs use some form of the technology shown in Figure 8-B1, but there is also growing interest in membrane bioreactors, which combine biological treatment with **membrane filtration** (use of semipermeable membranes to separate solids and sludge from liquid). Membrane bioreactors have a smaller footprint than traditional WWTPs and are often used for small-scale applications, including modular, prefabricated package plants for small communities. There are also low-tech, greener alternatives to WWTPs, including constructed wetlands, oxidation ponds, and algae-based systems.

Notes

[a]FOG can congeal inside sewers, along with wet wipes and other debris, to form "fatbergs" that can severely restrict sewer capacity and are difficult to remove.
[b]In 2022, the Conservation Law Foundation sued the Massachusetts Water Resources Authority for failing to enforce adequate pretreatment requirements on industrial dischargers, thus allowing toxics to flow into Boston Harbor.
[c]https://mwrd.org/water-reclamation-plants.
[d]WWTPs may be required to dechlorinate between disinfection and discharge to the environment to mitigate the impacts of chlorine on aquatic organisms. Alternatively, some WWTPs use nonchlorine disinfectants that don't persist in the effluent, such as ultraviolet radiation.
[e]EPA (2012).
[f]https://www.energy.gov/scep/slsc/wastewater-infrastructure.
[g]Peccia and Westerhoff (2015).
[h]Bibby and Peccia (2013).

---

state primacy. The 1987 CWA amendments allowed federally recognized tribes to apply for the authority to administer NPDES programs and other elements of the CWA on tribal territory; however, no tribes have received this treatment-as-state status for the NPDES program.

Nonpoint sources are more challenging to regulate than point sources. Unlike sewage or industrial effluent—where end-of-pipe monitoring can easily assess the concentrations and fluxes of pollutants—nonpoint sources reach waterways in complex ways that are often episodic (e.g., during rain events) and hidden from view (e.g., through groundwater), meaning that even assessing the pollutant contribution from nonpoint sources is very difficult. Approaches to reducing nonpoint source pollutant contributions generally involve implementing best management practices that are likely to have some impact on pollutant loading, but that impact is uncertain, site specific, and hard to quantify. Even requiring best management practices can be politically and legally fraught, because it is often seen as an invasion of federal or state government into private land use.

For these reasons, the CWA framework has struggled to address nonpoint sources such as agricultural runoff. Section 208 of the 1972 CWA and §319 of the 1987 amendments provide a framework and funding for states to design and implement management plans for nonpoint sources, but these provisions remain largely voluntary.

2.2. Defining and Meeting Water Quality Goals

In addition to reducing the pressures on aquatic ecosystems, the CWA also attempts to evaluate whether these ecosystems are in fact achieving the desired "physical, chemical, and biological integrity." In particular, §303 of the CWA calls on states and tribes[3] to establish ambient water quality standards (WQS), with three main components:

- **Designated uses**: States and tribes must produce a list of possible uses and assign a subset of those uses to each water body. Uses must conform to the basic "fishable/swimmable" goals of the CWA and generally fall into the following categories: habitat (referred to as ***aquatic life use support [ALUS]***), recreation, finfish and shellfish harvesting, aesthetics, municipal water supply, industrial water supply, agricultural water supply, and navigation.
- **Water quality criteria (WQC)**: WQC describe the levels of various water quality parameters that are considered sufficient to protect each designated use.
- Antidegradation policy: This policy is designed to protect high-quality water bodies from degrading to conditions that just barely meet WQC.

For toxic pollutants (metals and organics), separate WQC are calculated for protecting humans and aquatic organisms. The human criteria vary depending on the designated use (e.g., drinking vs. fishing). The aquatic life criteria include both acute criteria (levels that could be toxic even with just a one-hour exposure) and chronic criteria (somewhat lower levels that can be safely exceeded for a short time but should not be exceeded for a four-day period).

The numerical values of the toxics criteria are based on toxicity testing and are generally published by EPA and implemented similarly across the states. EPA updated the human health water quality criteria in 2015 "to reflect the latest scientific information, including updated exposure factors (body weight, drinking-water consumption rates, fish consumption rate), bioaccumulation factors, and toxicity factors (reference dose, cancer slope factor)." A comparison of the new and old criteria (Figure 8-2) highlights the continuing scientific uncertainty of these toxicity values; of the eighty-eight chemicals evaluated, thirty-five saw changes of at least an order of magnitude in their estimated human health criteria, and most of these changes were in the direction of increasing toxicity (lower criteria).

For indicator bacteria (e.g., *E. coli*), criteria are based on EPA guidance under the

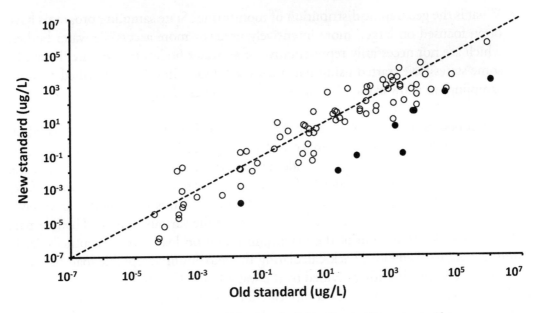

Figure 8-2. Comparison of old and new human health water quality criteria (for consumption of organisms only) for eighty-eight chemicals. Note log scale on both axes. Filled symbols highlight the ten chemicals whose new standards are less than 1/100 of the old standards (i.e., they are now considered 100 times as toxic as they were previously). Data from EPA (https://www.epa.gov/wqc/human-health-water-quality-criteria-and-methods-toxics#2015).

BEACH Act, a 2000 amendment to the CWA. The guidance allows states to choose between two maximum levels of gastrointestinal illness for swimmers—thirty-two or thirty-six illnesses per 1,000 swimming events—and set levels of indicator bacteria based on achieving those illness levels.[4] These arbitrary-seeming values are based on historical regulations, and their high values reflect the fact that swimming is a choice. I am skeptical, though, that the average swimmer is aware that a beach is considered safe when there is a 3–4 percent chance of getting sick; is that really "swimmable"?

Once WQC are established, §305(b) of the CWA requires states to assess their water bodies to determine whether the relevant WQC for each water body are being met. Importantly, §305(b) data from different states and time periods are not directly comparable, because states differ in their answers to three important questions:

- Which parameters are included in the WQC and in the assessment? All states include the basic conventional and toxic pollutants in their WQC and §305(b) reports, but there is much less consistency when it comes to parameters that are not purely chemical, such as the level of flow, the degree of channel alteration, or the health of biota.
- How frequently is a given water body monitored? Ecosystem health—especially water quality—can be quite variable over time, especially in rivers, where changes can occur over several different time scales (diurnal, storm, seasonal, long term). Yet resource constraints necessitate limited sampling—often only four to eight times per year.

- What is the geographic distribution of monitoring? State sampling programs have often focused on larger, more intensively used, or more accessible water bodies, which are not necessarily representative of all water bodies in the state. Recently, some states have started using statistically valid sampling (e.g., stratified random sampling) to allow extrapolation to all water bodies of a given type.

Water bodies that are not meeting the WQC for their designated uses—based on the sampling described above—are considered *impaired*; those in danger of becoming impaired in the future are considered threatened. In many of these cases, the state (or EPA) is required, under CWA §303(d), to prepare a ***total maximum daily load (TMDL)***, a determination of how much pollution this water body can receive, along with an allocation of that pollution between the various sources. The first part of this—the determination of the maximum pollution load—is conceptually fairly simple, because it involves back-calculating the maximum acceptable load based on the WQC and the dilution provided by the water body. The second part—the allocation of loads—can be politically complicated, often evoking finger-pointing between various point and nonpoint sources. Part of the complication is the difficulty of actually measuring the diffuse flux of pollutants from nonpoint sources or implementing measures that will reliably reduce that flux. Despite a slow start, more than 50,000 TMDLs have been approved.[5]

The TMDL approach is often held up as part of a holistic, watershed-based approach to ecosystem restoration. Indeed, the strength of this approach lies in two important features: its use of ecosystem health (as expressed by WQC) as an endpoint and its call to evaluate and control all sources of pollution. However, the TMDL does not in itself provide significant additional tools for controlling nonpoint sources, so it can sometimes devolve into an exercise in checking regulatory boxes rather than an implementable plan for actually fixing impairments. For example, of twenty-five long-established TMDLs reviewed by the US Government Accountability Office, seventeen "did not show that addressing identified stressors would help attain WQS; 12 contained vague or no information on actions that need to be taken, or by whom, for implementation; and 15 did not contain features to help ensure that TMDLs are revised if need be."[6]

The information required under §305(b) (water quality assessment) and §303(d) (TMDL status) is reported by each state biannually, either separately ("assessed waters" and "impaired waters" reports) or in the form of an Integrated Water Quality Report. EPA compiles state assessment data in a database called ATTAINS (Assessment, TMDL Tracking, and Implementation System)[7] and provides searchable information on local water bodies through the "How's My Waterway?" website.[8]

What do state water quality reports reveal about the health of US water bodies? A recent analysis of state data by the Environmental Integrity Project,[9] while suffering from the interstate comparability problems identified above, suggests that about half of assessed river miles in the United States are impaired for at least one use, and more than half of river miles have not been assessed. Similarly, about half of lake area and one quarter of estuary area are considered impaired.

The causes of river impairment vary by state and designated use but commonly include the following:

- Indicator bacteria: Violation of bacterial WQC is the most common cause of recreational impairment in rivers, lakes, and estuaries.
- Toxics: Many water bodies are impaired for fish consumption because of the presence of bioaccumulative toxics, especially mercury and polychlorinated biphenyls (PCBs).
- Phosphorus: As the limiting nutrient in freshwater, P—delivered in excess amounts from sewage and nonpoint sources—is leading to eutrophication of rivers and lakes across the country. Determining and implementing WQC for nutrients is more complex and controversial than for toxics.
- "Unknown": The fact that "unknown" is a commonly listed cause of ALUS impairment suggests that habitat problems in rivers are often caused by a complex mixture of stressors, including nonchemical ones, rather than a particular water quality problem.
- Hydrologic impairment: Depending on the state, flow problems may be frequently identified as impairing aquatic life. EPA defines hydromodification as "alteration of the hydrologic characteristics of coastal and non-coastal waters, which in turn could cause degradation of water resources." Despite this definition—which makes clear that this is not a water quality problem per se—EPA treats hydromodification as a nonpoint source of pollution,[10] which reflects the limited tools available under the CWA to address nonchemical stressors.

We turn now to an important issue that has emerged from this brief analysis of water quality assessment data: How does the CWA deal with nonchemical stressors?

2.3. Incorporating Nonchemical Stressors

Many aquatic ecosystems worldwide suffer from hydrologic and physical degradation rather than (or in addition to) impaired water quality. Having clean water does not make a river healthy if that river is channelized, fragmented by dams, or robbed of its water by excessive withdrawals. Although water-use law is generally the domain of individual states (as we will discuss in Chapter 10), the federal CWA does have some tools available to deal with these issues.

First, §404 of the CWA requires a permit from the US Army Corps of Engineers (USACE)[11] for discharging fill or dredge material into water bodies; this includes activities such as dam and levee construction, as well as wetland filling or drainage. The §404 permitting process is meant to minimize physical impacts to streams, lakes, and wetlands while ensuring that impacts that do occur are mitigated through the restoration, creation, enhancement, or preservation of similar habitat elsewhere (compensatory mitigation). However, the extent to which wetlands are covered by the CWA is a question that has been the subject of intense legal battles over the last forty years (Box 8-2).

A second tool for managing hydrologic and physical stressors is provided by §401 of the CWA, which requires state or tribal "water quality certification" for any project that produces a discharge and requires a federal permit, including USACE §404 permitting and Federal Energy Regulatory Commission licensing of hydroelectric dams. This gives states and tribes the power to defend instream flows by denying certification or imposing conditions on harmful projects, although this power has been used

> **Box 8-2. Wetlands and the Clean Water Act**
>
> Wetlands are ecological hotspots on the landscape, with exceptional importance for habitat, water storage, flood mitigation, water quality improvement, and other ecosystem services. Wetlands around the world have suffered extensive losses due to filling for urban development, flooding for reservoirs, discharges of sewage and other wastes, and conversion to open-water aquaculture systems.
>
> Wetlands are also routinely drained for conversion to agricultural use. Drainage activities can vary in sophistication from shallow ditches to modern tile drainage, in which a network of perforated plastic pipes is buried in the soil; in any case, the goal is to remove excess water and lower the water table to allow nonwetland crops to grow. Besides loss of wetland habitat, three additional problems can arise from this practice. First, aeration of the organic soil can lead to subsidence due to rapid decomposition. Second, in soils with iron sulfide minerals, aeration can result in acidic conditions due to sulfide oxidation. Third, the water drained from the soil is often salty and contaminated with agrichemicals and toxic metals.
>
> The Ramsar Convention on Wetlands, signed in 1971, was meant to reverse wetland loss by protecting wetlands of international importance. The Ramsar List includes over 2,400 sites in over 170 countries,[a] but enforcement has been inconsistent.
>
> Although their ecological and hydrologic importance suggests that wetlands should be subject to the jurisdiction of the CWA (including the permit requirements of §404, which provide protection against physical disturbance, including drainage), a four-decade legal battle has been waging to define which wetlands are included in the phrase "waters of the United States" (WOTUS), used in the CWA. This struggle has landed squarely in the middle of larger societal debates over federalism, property rights, and land-use regulation. In 2023 (*Sackett v. EPA*), a five-to-four Supreme Court majority dealt a blow to wetland protections by interpreting the CWA narrowly to include only wetlands that are "indistinguishable" from surface water bodies due to a "continuous surface connection," a ruling that is inconsistent with decades of practice and will lead to loss of federal protection for large swaths of the country's wetlands. In a blistering opinion dissenting from this new test, Justice Kagan gave a lesson in both water management and law:
>
>> Vital to the Clean Water Act's project is the protection of wetlands—both those contiguous to covered waters and others nearby. As this Court (again, formerly) recognized, wetlands "serve to filter and purify water draining into adjacent bodies of water, and to slow the flow of surface runoff." . . . Wetlands thus "function as integral parts of the aquatic environment." . . . At the same time, wetlands play a crucial part in flood control. . . .
>>
>> Today's majority, though, believes Congress went too far . . . [and believes] it must rescue property owners from Congress's too-ambitious program of pollution control. . . .
>>
>> The Court substitutes its own ideas about policymaking for Congress's. The Court will not allow the [CWA] to work as Congress instructed.
>
> **Note**
> [a] https://www.ramsar.org/.

infrequently. In a particularly significant example, the State of Washington in 1986 exercised its §401 certification power to require minimum streamflow releases from a proposed hydroelectric dam on the Dosewallips River, with the goal of protecting salmon runs, a designated use in this pristine river. The case reached the US Supreme Court (*PUD No. 1 of Jefferson County v. Washington Department of Ecology*), which sided with the state in 1994, noting, "Petitioners' assertion that the [Clean Water] Act is only concerned with water quality, not quantity, makes an artificial distinction, since a sufficient lowering of quantity could destroy all of a river's designated uses, and since the Act recognizes that reduced stream flow can constitute water pollution." One only wishes that this clear recognition of streamflow as a CWA issue would be consistently upheld throughout the act's implementation.

In addition, states often use WQC to set broad goals for the health of aquatic communities such as invertebrates or fish, which implicitly addresses any stressors that would prevent those goals from being met, including insufficient flows and unstable channels. For example, Connecticut states in its Water Quality Standards that "sustainable, diverse biological communities of indigenous taxa shall be present."

Establishing these ecological goals requires states to assess whether those goals are being met, using some type of **biomonitoring**, in which the health of ecological communities is measured directly. The most commonly used bioindicators are periphyton (riverbed algae), benthic macroinvertebrates (insects living on the river bottom), and fish. For each group, the numbers and types of organisms reveal something important about the health of the ecosystem as a whole.

Biomonitoring has several advantages over traditional water quality monitoring:

- The health of biota reflects conditions in the water body over a long time period, so the issues of short-term variability in water quality can be sidestepped.
- Biomonitoring is quick and cheap to perform, so more sites can be monitored.
- Biomonitoring tends to be more understandable and attractive to laypeople, generating both greater public interest in the results and greater volunteer participation in monitoring.

On the other hand, when biomonitoring reveals a degraded community, it does not tell you which of many possible stressors is causing that degradation, often leading to the "unknown cause of impairment" categorization discussed above. Unlike exceedances of traditional WQC such as DO or cadmium levels, which can be addressed through a combination of NPDES programs and TMDLs, it can be challenging to know what steps to take when biomonitoring reveals a degraded biological community but doesn't reveal why.

2.4. Effectiveness of the Clean Water Act

How effective has the CWA been at improving the health of US water bodies? Answering this question is challenging, in part because of the difficulties inherent in water monitoring:

- Ecological health is affected by many potential pollutants, as well as nonpollutant issues such as substrate and flow; monitoring all of these can be cost prohibitive.
- State water quality assessment programs are not designed to provide statistically representative data or to allow comparisons over time and space.
- Temporal variability in water quality can be quite high, but most water quality parameters (with the exception of temperature, dissolved oxygen, turbidity, and conductivity) can't be easily measured with automated loggers that would record conditions over time.

An additional challenge in assessing the effectiveness of water pollution control efforts is that these efforts were under way well before the implementation of the CWA—and well before the systematic collection of monitoring data. Looking at images from the nineteenth and early twentieth century, it is clear that we have come a long way, even if we don't have statistical data from that time period. The fact that many US water bodies are still considered impaired is, in some ways, a testament to our more rigorous WQC, not a lack of progress—although this observation doesn't contradict the fact that we still have work to do.

This general conclusion—water quality has improved over time but still has a long way to go—is confirmed by statistical studies of water quality, in particular Keiser and Shapiro's compilation of millions of pollution readings from river sites in the United States.[12] As shown in Figure 8-3, they found that water quality improved substantially over the period 1962–2001, with about half the improvement taking place before the CWA in 1972, due to the ongoing pollution control efforts mentioned above. Still, they found that CWA grants for construction of treatment plants did have measurable effects on water quality.

What about nonconventional pollutants? Keiser and Shapiro found significant declines over time in lead, mercury, phenols (a type of organic pollutant), nitrogen, and phosphorus. On the other hand, they did not find data to evaluate emerging contaminants such as per- and poly-fluorinated alkyl substances (PFAS), which are probably increasing over time.

Beyond assessing changes in water quality over time, economists such as Keiser and Shapiro have also tried to determine the cost-effectiveness of the CWA. Has the national expenditure on water quality of over a trillion dollars since 1972—much more than the sums spent on other environmental protection measures—produced equally large results? The analyses to date have generally shown low benefit/cost ratios (<1) for the CWA, but this probably reflects flaws in the benefit–cost analysis (BCA)

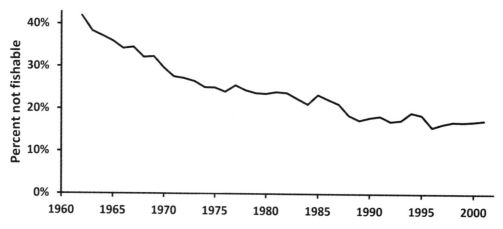

Figure 8-3. Change over time in the share of US rivers classified as not fishable. To be considered fishable, a site must meet all of the following criteria: biochemical oxygen demand <2.4 mg/L, dissolved oxygen >64 percent saturation, fecal coliform <1,000/100 mL, and total suspended solids <50 mg/L. Data from Keiser and Shapiro (2019), based on about 11,000,000 observations at about 180,000 monitoring sites, corrected for site, seasonal, and diurnal effects.

> **Box 8-3. Environmental Justice and the CWA**
>
> Environmental justice practitioners working on water issues have, understandably, focused their attention primarily on differential access to safe drinking water and sanitation and the impacts of industry and agriculture, rather than on the health of aquatic ecosystems in affected communities. Still, there are clear indications that ambient water quality is affected by racial and economic variables. In Rhode Island, Black and Hispanic populations live farther from coastal sites with no history of water quality impairment.[a] In California, stream restoration efforts are more common in areas with a highly educated, wealthy, White population.[b]
>
> Enforcement of NPDES permits (which ultimately affects ambient water quality) is also affected by the makeup of the local community. From 2010 to 2015, inspections of wastewater treatment plants were significantly less common—and permit violations were more common—when those facilities were located in communities of color, especially when they were tribally owned facilities.[c] In addition, when wastewater treatment plants are located in poor or Latinx communities, permitting authorities do a worse job of prioritizing noncompliant facilities for inspection over compliant ones.[d]
>
> **Notes**
> [a]Twichell et al. (2022).
> [b]Stanford et al. (2018).
> [c]Teodoro et al. (2018).
> [d]Konisky et al. (2021).

process more than in the CWA. In particular, BCAs of the CWA generally exclude human health outcomes, even though protecting ambient water quality is an essential part of providing clean drinking water. In addition, BCAs of the CWA often exclude existence values, the impacts of toxics, and coastal waters, all of which provide significant benefits. Analyses of the CWA's effectiveness and cost-effectiveness have rarely examined issues of environmental justice (Box 8-3).

3. Other US Environmental Laws

The CWA is the primary US law aimed at protecting aquatic ecosystems, but two broader environmental laws are highly relevant as well.

The National Environmental Policy Act (NEPA): NEPA, passed in 1969, aims to make federal agencies consider the environmental impact of their actions by establishing formal procedures for evaluating those impacts. Any proposed federal action—including the funding or permitting of state or private actions—that has the potential to cause significant environmental impacts requires the lead federal agency to prepare an Environmental Impact Statement (EIS). The EIS, which is available for public comment in draft form, lays out the potential impacts of the action and discusses alternative actions that would reduce those impacts. After finalizing the EIS, the agency must choose which action it will undertake and justify that choice in a Record of Decision. Although there is no requirement for the agency to choose the least environmentally damaging alternative, the procedural requirements of NEPA do bring transparency and science into the decision-making arena.

The Endangered Species Act (ESA): The ESA, passed in 1973, has played a significant role in protecting aquatic ecosystems in the United States. The goal of the ESA is to protect species that are in danger of extinction, specifically those that are

> **Box 8-4. Freshwater Mussels: Biodiversity and Extinction**
>
> Our grandfathers were less well-housed, well-fed, well-clothed than we are. The strivings by which they bettered their lot are also those which deprived us of [passenger pigeons, which were hunted to extinction]. Perhaps we now grieve because we are not sure, in our hearts, that we have gained by the exchange. . . . We know now what was unknown to all the preceding caravan of generations: that men are only fellow-voyagers with other creatures in the odyssey of evolution. This new knowledge should have given us, by this time, a sense of kinship with fellow-creatures; a wish to live and let live; a sense of wonder over the magnitude and duration of the biotic enterprise. —Aldo Leopold[a]
>
> Freshwater mussels—over 800 described species in the order Unionida—are **benthic filter feeders** whose larvae must parasitize fish in order to survive; some mussel species are dependent on one fish species for this purpose, while others can use up to fifty-three different fish hosts.[b] This unique life cycle has led some species of mussels to develop amazingly lifelike prey-mimicking "lures" that entice fish to strike, causing the larvae to be released and latch onto the fish.
>
> Freshwater mussels are among the most endangered groups on the planet. In the United States, a hotspot of mussel biodiversity, some 10 percent of species are already extinct and another 65 percent are imperiled.[c] The threats to these species include habitat modification and fragmentation (due to damming, **channelization**, and **aggregate** mining), changes in **flow regimes**, declines in fish hosts, poor water quality (including high sediment and nutrient loads), overharvesting (mostly for mother-of-pearl), and nonnative invasives such as the zebra mussel. The eight species of freshwater mussels that are now being delisted from the ESA due to extinction are all from the Southeast (Tennessee and Mobile River basins), where construction of the Tennessee–Tombigbee Waterway (a navigation route completed in 1984), siltation, and other impacts have contributed to their extinction.
>
> While these departed mussels may lack the numbers and grandeur of the passenger pigeons that once clouded North American skies, we should recognize that they were an equally wondrous part of the biotic enterprise, and we should, with Leopold, grieve their loss, both for what that loss means for them and for what it says about us.
>
> **Notes**
> [a] Leopold (1949).
> [b] Modesto et al. (2018).
> [c] Haag and Williams (2014).

formally *listed* by the Fish and Wildlife Service or National Marine Fisheries Service (also known as National Oceanic and Atmospheric Administration Fisheries) as endangered or threatened. Section 9 of the ESA prohibits anyone from harming listed species or their habitat, although Section 10 allows incidental take of listed species under certain circumstances, including implementation of a Habitat Conservation Plan. Section 7 prohibits all federal agencies from funding or authorizing actions that will jeopardize listed species or their habitat.

Since almost any river management activity involves federal permitting or funding, the ESA provides a strong tool for protecting river ecosystems as habitat for listed species. As a result, the ESA has been used to prescribe flow releases, prevent water withdrawals, improve water quality, require dam removal, and prevent development in floodplains. The use of the ESA for these purposes has proven controversial. Some argue that those bringing ESA lawsuits don't really care about the individual species per se but rather are using the ESA as a tool to rebalance water management toward broader environmental goals. In response, many environmentalists are happy to concede the point that their goals are broader than simply protection of one species and point out that using endangered species as a "canary in the coal mine"—a warning of systemic environmental degradation—is consistent with the logic of the ESA.

More than fifty ESA-listed species have recovered to the point where they have been delisted. Sadly, however, twenty-three listed species are proposed for delisting due to extinction, including eight species of freshwater mussels (Box 8-4).

Chapter Highlights

1. Rivers and other water bodies should be clean enough to protect human and ecological health. However, quantifying "safe levels" of chemicals is a complicated task that relies heavily on extrapolating from limited toxicity data, and monitoring of ambient water quality is challenging and expensive because of high temporal variability and the large number of water quality parameters that must be measured.
2. The Clean Water Act (CWA) aims to protect and restore aquatic ecosystems by restricting pollutant discharges through the ***National Pollutant Discharge Elimination System (NPDES)*** program, establishing ***water quality criteria*** to protect humans and aquatic organisms, assessing the health of water bodies relative to those criteria, and bringing ***impaired*** waters back to health through ***total maximum daily loads (TMDLs)*** and other tools.
3. The CWA has had success in addressing ***point sources*** of pollution through permits and ***wastewater treatment plants***, but it has had more trouble dealing with ***nonpoint sources*** and hydrologic, physical, and biotic impacts. ***Biomonitoring*** is increasingly being used to help identify and address these impacts.
4. The CWA has improved water quality in many ways, but many water bodies are still considered impaired for habitat, recreation, and fishing, primarily because of impacts from agriculture, hydrologic modification, and urban runoff.
5. Wetlands provide valuable ecosystem services, including habitat and flood mitigation, but the use of the CWA to protect wetlands has been stymied by legal battles over which wetlands are subject to federal jurisdiction.
6. The National Environmental Policy Act (NEPA) and the Endangered Species Act (ESA) provide additional tools for the protection of aquatic ecosystems.

Notes

1. Lagarde et al. (2015).
2. CRS (2016).
3. Eighty-four tribes have received authority to establish their own water quality standards and monitoring programs through the treatment-as-state provision; https://www.epa.gov/tribal/tribes-approved-treatment-state-tas.
4. EPA (2012a).
5. This statistic counts a single document that addresses two impairments (e.g., ALUS and recreation) in three river segments as six TMDLs.
6. GAO (2013).

7. https://www.epa.gov/waterdata/attains.
8. https://mywaterway.epa.gov/.
9. Environmental Integrity Project (2022).
10. EPA (2007).
11. Three states have received authorization to administer their own §404 programs (under US Army Corps of Engineers supervision).
12. Keiser and Shapiro (2019).

9 Dams and Their Discontents: Are the Benefits Worth the Costs?

- How has the discourse around dams changed over the last century?
- What are the impacts of dams on rivers and communities?
- Can we maximize the benefits of dams while minimizing their social and ecological impacts?
- What are the benefits and possible pitfalls of removing dams?
- What are the pros and cons of long-distance water transport through aqueducts?

Dams—with their capacity to store water, control flooding, generate power, improve navigation, and provide recreational opportunities—are among the most useful tools in the water manager's toolbox. At the same time, dams—with their displacement of communities and their disruptions to the natural movement of water, sediment, nutrients, and fish—can devastate rivers and the people who depend on them.

The chapter starts with an introduction to dams, followed by an overview of the changing societal attitudes toward large dams and a statistical snapshot of dams and their uses. We then turn to some of the negative impacts of dams, including ecological impacts, population displacement, and dam safety issues. We examine ways to build better dams and better manage existing dams and discuss the emerging field of dam removal. We close the chapter with a discussion of aqueducts, which make up significant parts of many dam projects and are a companion hard-path technology.

1. Dam Basics
In this section, we introduce the different types of dams and the various components of dam projects. Since reservoir storage is a central feature of many dams, we also discuss the allocation of this storage between various uses.

1.1. Types of Dams
The term "dam" covers a large variety of structures, which differ from each other in their benefits and their social and environmental impacts. There are several ways to categorize dams, although each dam project is unique and can cut across categories.

INSTREAM VERSUS OFFSTREAM
Most dams are placed across river channels, but dams can also be installed at the outlets of natural lakes to control water levels. ***Offstream dams***, on the other hand, create impoundments away from natural water bodies; these may create ***tailings ponds*** for storing mining waste or reservoirs for storing water diverted from a river. Offstream dams have lesser ongoing impacts than river dams, although failures of tailings dams are surprisingly common and often devastating.

STORAGE VERSUS RUN-OF-RIVER
Run-of-river dams raise the elevation of the water in a river but do not create a storage reservoir (though see Chapter 6); these dams are typically used to aid navigation, facilitate recreation or fishing, divert water from the river into a canal (diversion dams), or generate run-of-river hydropower. In contrast, storage dams are used to create reservoirs that can store large volumes of water for use.

SIZE
Dams can range in height from a foot-high ***weir*** to a skyscraper-sized structure, and they can vary similarly in dam length and reservoir size. The International Commission on Large Dams (ICOLD) defines large dams as those with heights of at least 15 m, plus those with heights between 5 and 15 m that have reservoirs of at least 3 MCM.[1]

MATERIALS
Dams can be made of various materials and designs:

- Embankment dams are made of compacted material, usually earth or rock. Timber crib dams, mostly constructed before the twentieth century, are embankment dams in which a framework of wooden planks is filled with earth or rock.
- Gravity dams are made of concrete or masonry and rely on the effects of gravity to keep the massive structure in place and resist the force of water.
- Arch dams—usually found in steep rock canyons—are also made of concrete or masonry but require much less material than gravity dams, since the pressure of water against the face of the dam is partly distributed into the rock on either side of the dam by the arch shape.
- Arch–gravity dams use features of both gravity and arch dams to maintain stability.

1.2. Dam Components and Terminology
An image of Oroville Dam in California (Figure 9-1) provides a useful guide to the components of a dam project. The dam face is the downstream side of the dam itself, and the abutment is the hillside that the dam rests against. (Left and right abutments are named looking downstream.) Since Oroville is a hydroelectric dam, most of the water released by the dam travels through a penstock to the power plant and is then released into the tailrace. However, when water levels get too high, managers can open the spillway headgates and release excess water through the spillway. If the spillway is inadequate to handle the water volume, the emergency spillway weir would be overtopped and water would flow down the hillside. (During high water in 2017,

Figure 9-1. Oroville Dam (235 m high, 2,109 m long), an embankment dam on the Feather River in California, showing different components of the dam project. Photo credit: California Department of Water Resources (https://commons.wikimedia.org/wiki/File:OrovilleDam.jpg).

both the spillway and the emergency spillway of the Oroville Dam were damaged, leading to the evacuation of almost 200,000 people downstream.)

1.3. Managing Reservoirs for Multiple Purposes

As noted previously, dams are used for many different purposes:

- Instream uses
    - Hydropower: Hydropower installations often use dams to increase the **head** and to store water for generating electricity when it is needed.
    - Navigation: Dams facilitate navigation by increasing water depths and providing flat water for easier travel. Dams can also be used to divert water into navigation canals or to release water into a river to maintain the necessary depths in downstream navigational reaches.

- Recreation: Reservoirs are attractive sites for boating, fishing, and swimming. Dam releases can provide whitewater for river-based recreation.
- Flood control: Dams can prevent downstream flooding by holding back water during flood events and releasing it later.
- Offstream uses: Dams can facilitate diversions for offstream uses by raising water elevations. In addition, since ancient times dams have been a central tool to manage water's variability by storing water during wet periods for use during dry ones. Given that many regions are already experiencing scarcity and that climate change is magnifying water's temporal variability, dam storage may become even more important in coming decades.

Many dams serve more than one purpose, and careful management may be needed to simultaneously achieve those different purposes. One of the trickiest balancing acts for dam managers involves the inherent tension between flood control and water supply. From a flood control perspective, the best reservoir is an empty reservoir, with plenty of room to hold floodwater, but from a water supply perspective, the best reservoir is a full reservoir, with plenty of water to get you through the dry season. In the United States, federally operated reservoirs have operation manuals that specify how much reservoir storage volume is allocated to each of the dam's congressionally authorized purposes. In Lake Mead, for example (Figure 9-2), the upper 1.5 MAF (million acre-feet) of reservoir capacity is devoted exclusively to flood control, while the rest of the active reservoir capacity is considered a *joint conservation pool*, available for municipal and industrial water supply, irrigation, and power production, as well as flood control. Below an elevation of 950 ft, the turbines can't generate hydropower, but there is still about 2 MAF of inactive pool that could theoretically be released for other uses.[2] Below 895 ft, the dam can't release any water, and the remaining water is considered dead pool.

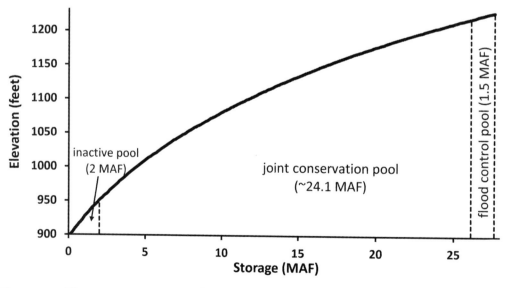

Figure 9-2. Elevation–storage curve for Lake Mead, showing the division of reservoir volume into the exclusive flood control pool, the joint conservation pool, and the inactive pool.

Balancing uses *within* the joint conservation pool can also be complicated. Water used for hydropower can also be supplied to offstream users downstream, although the desired timing of water releases may be in conflict. More importantly, irrigation and municipal water supply are directly competing for the same water; their allocations are determined in part by water rights that we will discuss in Chapter 10 but also by the availability of reservoir space for each purpose. Reservoir space can be reallocated in response to changing physical and societal conditions, although the process is often lengthy.[3]

1.4. Reservoir Storage–Yield Relationships

Given the crucial role that reservoir storage plays in providing water for offstream uses, several methods have been developed for estimating how large a reservoir is needed for a given water demand, known as the storage–yield relationship.[4] These methods simulate the filling and emptying of the reservoir over time in response to the variability in supply (streamflow) and demand (water use), trying to ensure that the reservoir can store enough water to get the system through a drought.

Figure 9-3 illustrates that the volume of storage needed depends on two variables:

- How does water demand compare to availability? At low levels of demand (near 0 on the *x*-axis), not much storage is needed, since there will be enough supply even in dry years. As demand increases and approaches average availability, storage needs increase dramatically.
- How variable is water availability? In Figure 9-3, this is expressed as the coefficient of variation (CV) of flows. Not surprisingly, higher variability requires higher storage, suggesting that climate change will increase the need for storage.

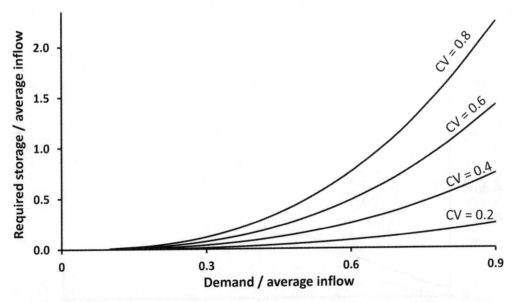

Figure 9-3. Reservoir storage–yield relationships, based on McMahon et al. (2007). (The curves shown assume a skewness of 0.4 and a desired reliability of 95 percent.) The required storage (*y*-axis) is a function of demand (*x*-axis) and the coefficient of variability (CV) of flows. Both demand and storage are relative to average inflow.

A third variable—reliability—will be discussed in Chapter 15.

Figure 9-3 also illustrates a basic point about dams: They provide water storage, not new water, so they can help water systems weather variability, but they can't help water systems that are chronically short on water. Once demand reaches average inflow (either because demand is growing or because climate change is reducing supply), you no longer have a storage problem, you have a supply problem. For example, under a future warm–dry climate, there are negligible benefits to building more storage in southern California; we won't have any water to put in the new storage.[5]

2. The Dam Debate: The Changing Attitudes of the Past Century
This section provides a brief overview of the changing attitudes toward dams in the modern era.

2.1. The Dam-Building Era(s)
Dams are perhaps the preeminent symbol of modern, hard-path water management. Although simple dams have been built since ancient times, technological developments in the twentieth century have allowed the construction of immense structures that claim to match the power of the world's large rivers. The middle decades of the century saw an unprecedented dam-building boom in the United States and other Western countries (Figure 9-4), which came to an end in the late twentieth century, in part because all the best sites had been taken and in part because of growing opposition on environmental grounds.

The end of dam construction in the United States, apparent in Figure 9-4, stands in stark contrast to the pattern in much of the world, where dam building continues. China has built about one large dam per day since the 1950s, amounting to over 23,000 large dams, about half the global total, and shows no sign of letting up. India is in third place globally (behind China and the United States) with over 5,000 large

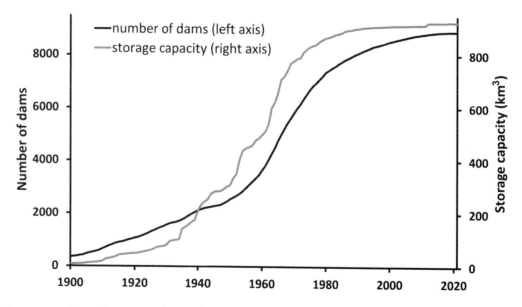

Figure 9-4. Cumulative number and storage capacity of dams in the United States over time. Data from National Inventory of Dams.

dams. As noted in Chapter 6, Africa, Latin America, and Asia are at the center of a new hydropower dam-building boom, with megaprojects that dwarf Hoover Dam in size and cost.

2.2. From Zeal to Skepticism in the Twentieth Century

Through much of the twentieth century, dams were viewed as the quintessential symbols of progress, icons of humanity's ability to use science and technology to harness nature for our benefit. American water managers, such as Floyd Dominy, commissioner of the Bureau of Reclamation from 1959 to 1969, had an almost religious zeal for large dams. Low-income countries too—encouraged and funded by the World Bank and other development banks, along with the United States and other rich countries—saw large dams as central to their visions of modernity and economic progress. Leaders around the world extolled both the tangible benefits and the powerful symbolism of these huge structures.

But slowly, as the true costs of dam projects became clear, opposition started to develop and strengthen. Historical highlights of the antidam movement include the following:

- 1901–1913: John Muir leads the Sierra Club in an unsuccessful campaign to stop the damming of Hetch Hetchy Valley to provide a water supply for San Francisco. Muir writes in his 1912 book *The Yosemite*, "Dam Hetch Hetchy! As well dam for water-tanks the people's cathedrals and churches, for no holier temple has ever been consecrated by the heart of man."
- 1955–1956: Environmentalists succeed in scuttling a dam proposed for the Yampa River (a tributary of the Colorado) in Dinosaur National Monument, in exchange for not opposing the Glen Canyon Dam on the Colorado River.
- 1966: The Sierra Club runs its famous "Sistine Chapel" ad, citing both utilitarian and moral arguments to successfully oppose dams that would have flooded part of the Grand Canyon.[6]
- 1985: Narmada Bachao Andolan (Save the Narmada Movement) is founded in India in opposition to the planned Sardar Sarovar Dam on the Narmada River and starts receiving worldwide attention for its dramatic methods and passionate leadership.
- 1993: In response to pressure, the World Bank pulls out of funding the Sardar Sarovar Dam and generally reduces its funding of large dam projects.
- 1996: Patrick McCully publishes *Silenced Rivers: The Ecology and Politics of Large Dams*, a deeply researched global catalog of the dangers of dams.

2.3. The World Commission on Dams and Its Aftermath

By 1998, the global antidam movement had gained enough momentum that the World Bank and the International Union for Conservation of Nature jointly established the World Commission on Dams (WCD) in an effort to reach some compromise in the battle between dam advocates and dam opponents. The twelve commissioners—leaders from both sides of the debate—published a final report in 2000.[7]

At the heart of the WCD report was the idea that dam building should continue—but only with much stronger processes in place to ensure equity and sustainability.

The report identified seven "strategic priorities" and twenty-six guidelines to support this vision. Underlying the WCD's guidelines is the insight that not all dams are the same. The social and environmental impacts of the worst dams are much larger than those of the best dams, and the benefits provided by the best dams are much greater than those provided by the worst. The WCD approach suggests that we should build the best dams but not the worst, and we should do so using best practices. In other words, if a dam project really is the best option to solve a particular problem—after full assessment of all economic, environmental, and social benefits, costs, and risks, and after full, voluntary agreement of all stakeholders, and with a real plan to mitigate social and environmental impacts—then, and only then, the WCD posits, we should go ahead and build it. But we shouldn't treat a large dam as a default option that should be built just because it can be.

Some environmentalists think that no dam is worth building, in part because we simply can't understand and quantify all the ecological impacts it will have. Still, most environmental groups responded fairly positively to the WCD report, recognizing that implementation of the WCD guidelines would lead to vastly improved (and fewer) new dams.

Many dam builders and funders, on the other hand—including the World Bank, India, China, and ICOLD—refused to commit to the WCD guidelines. Many low- and middle-income countries still see large dams, especially hydropower dams, as a path toward economic prosperity and global power. They see hypocrisy in the notion of the West telling the rest of the world not to build large dams, when much of the economic might of the West was built on those very same dams. As John Briscoe puts it in a slightly broader context, "The rise of the environmental movement in rich countries, however, was accompanied by a rise in activism against the sorts of investments that had made rich countries rich. This was due in part to legitimate and important concerns with social and environmental impacts of large infrastructure projects. But it was also fueled by an ahistorical paradigm that scorned the same types of investments that had been necessary to bring about the privilege those critics enjoyed."[8]

After a lull during the WCD process, the World Bank has returned to funding large dams, especially hydropower projects, although activist pressure has continued pushing it to withdraw funding from the most controversial projects and shift funding toward other, nonhydropower renewable energy projects. Other funding sources have stepped into the gap, especially China, which sees dam funding as part of its geopolitical and economic strategy under the Belt and Road Initiative. In addition, Chinese companies such as Sinohydro often provide the expertise to build and operate the newer generation of large dams. The waning influence of the traditional international funders means that environmental and social safeguards are not being enforced with any consistency. Still, antidam activists, both local and global, are continuing to fight against the most damaging projects.

3. Snapshot of US and Global Dams

In this section, we use global and US databases to quantify the number, purposes, and storage capacity of the world's existing dams, leading us to the question: Do we need to build more dams?

Table 9-1. Dam databases.

Database	Types of Structures Included	Number of Structures
ICOLD[a]	Height ≥15 m, or height ≥5 m and reservoir ≥3 MCM	58,713
US NID[b]	US dams with: (a) Height ≥25 ft and reservoir >15 AF; or (b) Height >6 ft and reservoir ≥50 AF; or (c) High or significant hazard classification	91,759 (of which 10,211 meet ICOLD criteria)
Connecticut Dam Safety[c]	"Any barrier of any kind whatsoever which is capable of impounding or controlling the flow of water"	3,906 (of which 834 are in the NID)
European Barrier Atlas[d]	All river barriers in Europe, including dams, weirs, culverts, fords, sluices, and ramps	>1,000,000
GOODD[e]	Dams with reservoirs visible in satellite imagery	38,667
GRanD version 1.3[f]	Dams with reservoirs >100 MCM, though some smaller dams are also included	7,320 (including 1,920 from the United States)

[a]ICOLD, International Commission on Large Dams. https://www.icold-cigb.org/GB/world_register/general_synthesis.asp; updated April 2020.
[b]NID, National Inventory of Dams. https://nid.sec.usace.army.mil/; updated June 2023.
[c]https://portal.ct.gov/DEEP/Water/Dams/Dams-Safety; updated January 2023.
[d]Belletti et al. (2020); https://amber.international/european-barrier-atlas/.
[e]GOODD, Global Georeferenced Database of Dams. Mulligan et al. (2020).
[f]GRanD, Global Reservoir and Dam Database. https://www.globaldamwatch.org/grand.

3.1. How Many Dams Are There?

The simple answers to this question are "Nobody knows exactly" and "It depends what you mean by a dam" and "More than you think." The diversity of dam types, the rapid proliferation of these structures, and the lack of transparency on the part of many countries—all these factors combine to make a complete count impossible, but there are several dam databases that can help provide a statistical snapshot (Table 9-1).

It is clear from this table that there are many more small dams than large ones, so databases that focus on large dams provide a serious undercount of total dams. For example, the National Inventory of Dams (NID) database has roughly ten times as many dams that don't qualify as large (by ICOLD criteria) as those that do. And even the NID database is far from comprehensive, as indicated by the much larger number of Connecticut dams that are known to the state compared with those that are included in the NID. Likewise, the European Barrier Atlas, with its estimate of over a million structures, illustrates the ubiquity of barriers in rivers; although many of these are probably small, they can still pose a barrier to the movement of fish, sediment, and water.

3.2. What Are Dams Used For, and Who Manages Them?

Figure 9-5 compares the dam purposes recorded in three different databases. Larger dams (GRanD and NID large) tend to be used mostly for hydropower, irrigation, municipal water supply, and flood control. Recreation is also an important purpose for large dams, although it is mostly a secondary purpose; once the reservoir is there,

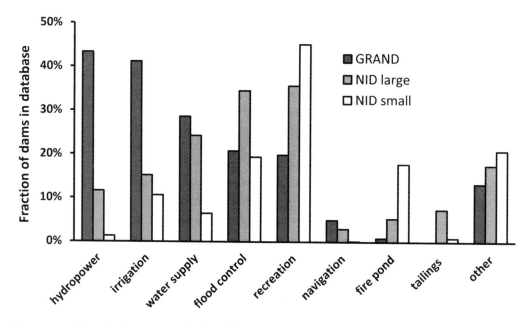

Figure 9-5. Recorded purposes of global large dams (Global Reservoir and Dam Database [GRanD] version 1.3, n = 5,758), US large dams (National Inventory of Dams [NID] dams meeting large-dam criteria, n = 9,486), and US small dams (NID dams not meeting large-dam criteria, n = 74,402). Dams with no listed purpose are not included. Both primary and secondary purposes are included, so the total for each group of dams is more than 100 percent. "Fire pond" includes small ponds used for firefighting, livestock watering, and fish stocking. "Water supply" includes municipal and industrial but excludes irrigation.

it makes sense to use it for boating and fishing. Less than 5 percent of dams in each database are used for navigation, although—as we discussed in Chapter 6—these dams are quite important for some navigation routes. Interestingly, many of the very largest dams in the United States are offstream tailings dams.

Smaller reservoirs (lightest bars in Figure 9-5) are often used for somewhat different purposes than larger ones. They are more likely to have recreation as the primary purpose and often serve as a local water supply for firefighting or livestock watering. Many of these small dams were built as mill dams in the eighteenth or nineteenth centuries and have long outlived their initial purpose; often these dams have been around for so long that they are perceived as part of the natural landscape.

With the exception of tailings dams, most of the largest dams in the United States are publicly owned, mostly by the US Army Corps of Engineers, the Bureau of Reclamation, and the Tennessee Valley Authority. In contrast, smaller dams in the United States are mostly privately owned, often by individuals or small businesses.

3.3. How Has Storage Volume Changed over Time?
Many of the benefits provided by dams are linked to the amount of water they can store, so examining the evolution of global storage capacity can help us understand the changing dam landscape. Three different dam-building phases over the last century can be identified (Figure 9-6, thick line):

- 1920–1950: the beginning of the large-dam era, with about 13 km³ of storage added per year
- 1950–1985: rapid growth, at about 140 km³/yr
- 1985–2020: continued global growth at about 40 km³/yr, despite the near end of dam building in the global North.

However, two factors complicate this picture. First, reservoirs—still-water settings in an otherwise moving-water system—tend to trap riverine sediments, which settle out and accumulate at the bottom of the reservoir (***reservoir sedimentation***). This has several consequences, but the one of concern to us right now is the loss of water storage capacity that occurs when sediment occupies reservoir space. For example, a 2022 elevation–storage curve for Lake Powell shows that "full pool" elevation (elevation 3,700 feet above sea level) translates into 6.8 percent lower water volume (and 1.3 percent lower surface area) than it did in 1963, because of sedimentation along the margins and bottom of the reservoir.[9] The rate of sedimentation is highly variable from reservoir to reservoir, because of variation in both trapping efficiency (the fraction of incoming sediment that is captured, which is usually a function of reservoir size relative to inflow) and river sediment load (which varies widely due to both natural and anthropogenic influences). In Figure 9-6, I use a conservative assumption that

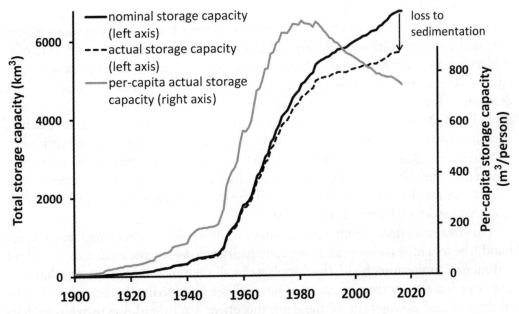

Figure 9-6. Global storage capacity (left axis) and per capita storage capacity (right axis). Nominal storage capacity (solid line) refers to the full volume of reservoirs, while actual storage capacity (dashed line) subtracts the storage space lost to reservoir sedimentation, assuming a loss rate of 0.35 percent/yr (Wisser et al. 2013). Based on Global Reservoir and Dam Database (GRanD) data for global storage over time. (Although GRanD is missing many large dams, its focus on the largest dams means it covers most of the storage capacity. For example, GRanD storage capacity for the 1,920 US dams it lists is 740 km³, which is 75 percent of the storage capacity for all 91,227 dams listed in the NID.)

0.35 percent of reservoir capacity is lost each year[10]; under this assumption (Figure 9-6, dashed line), about 16 percent of storage capacity has been lost to sedimentation,[11] and global storage capacity has leveled off in the last few years.

In addition, we should account for population growth by calculating per capita storage capacity. When we correct for both reservoir sedimentation and population growth (Figure 9-6, gray line), we find that per capita actual storage capacity peaked in 1981 and has fallen by about 25 percent since that peak.

Is the decline in per capita storage capacity a problem? Some certainly think so. For example, dam expert George Annandale comments, "This situation is obviously untenable," and he calls for increases in storage capacity through construction of new dams as well as preserving and restoring storage capacity in existing dams.[12] On the other hand, we need to take seriously the social and environmental impacts of large dams.

The better questions, then, might be these: How much storage do we really need? Are there ways to manage water that need less storage? Can we provide the needed storage (and other functions of dams) in ways other than large dams, ideally ways that don't lose their storage to sedimentation? Those questions will occupy us throughout the book, but for now, we turn to the dark side of dams.

4. Ecological Impacts of Dams

The ecological impacts of dams can be categorized into reservoir effects, downstream effects, and system-wide effects.

Reservoirs are fundamentally different ecosystems from the rivers they replace. The still-water conditions and long residence time created by large reservoirs often lead to sediment deposition, algal blooms, stratification, deep-water ***hypoxia***, warmer surface waters, increased evaporation, and the dominance of ***lentic*** fish species over ***lotic*** ones. Reservoirs can create barriers for fish migration in both the downstream direction (fish must expend more energy than normal to traverse the same distance) and the upstream direction (fish may have trouble orienting themselves without the flow of water). Reservoirs also have upstream geomorphic effects stemming from the rise in the river's base level (the elevation at which it discharges into the reservoir); the decrease in the river's gradient upstream of the reservoir leads to sediment accumulation that can extend some distance upstream.

Downstream effects result from changes in flow and sediment regimes. On one hand, the loss of seasonal peak flows can result in downstream accumulation of fine sediment. On the other hand, the trapping of sediment by reservoirs means that dams tend to release water that is clear and "hungry" for more sediment, leading to erosion of the riverbed downstream of the dam; this effect can extend downstream to deltas and other coastal areas, where the combined impacts of ***sea-level rise*** and sediment deprivation can lead to significant land loss. Downstream water quality impacts of dams can include abnormally warm water (for dams that release reservoir surface waters), abnormally cold or deoxygenated water (for dams that release deeper waters), and altered nutrient ratios.

Perhaps the most far-reaching ecological impact of dams is fragmentation: the alteration from a free-flowing, longitudinally connected ecosystem to a series of disconnected reaches separated by barriers. This underlies the phenomena previously described (reservoir evaporation, sediment trapping, fish migration disruption) and is part of a broader loss of connectivity within river systems, which also includes the loss of lateral connectivity due to levees and floodplain destruction.

5. Social Impacts of Dams

In this section, we discuss population displacement and other impacts associated with dam construction and then look at the issue of dam safety.

5.1. Population Displacement

> I have not seen a single case in which people have been compensated fairly for the disruptions to their lives caused by dams. If governments are arguing that these projects aren't viable without underpaying compensation, then maybe these projects aren't right for the country. —Ian Baird[13]

Arguably the greatest negative impacts of large dams come when communities are flooded by the rising waters of a new reservoir. In 2000, the WCD estimated that 40–80 million people worldwide had been directly displaced from their homes by the reservoirs of large dams; that number has no doubt increased since then. Many of the people displaced by dams do not leave voluntarily but are forced to move by governments or dam builders, often using violence or the threat of violence.

There are certain situations where the greater good dictates that people must give up their homes (e.g., the use of eminent domain), but fairness requires that the number of affected people should be minimized; consent of affected people should be obtained whenever possible; Indigenous and tribal communities should never be subject to involuntary displacement, as recognized by the "free, prior, and informed consent" provision of the UN Declaration on the Rights of Indigenous Peoples; and benefits of the project must be shared with those negatively affected. All these provisions have been routinely ignored (Box 9-1).

Although resettlement programs are supposed to provide new homes and livelihoods to dam refugees, most of these programs have been woefully inadequate. Anthropologist Thayer Scudder has documented the devastating effects of displacement on individuals and communities. He notes that of fifty large dams surveyed, only two achieved the goal of improved living standards for the majority of resettled people, and thirty-one resulted in significantly worsened conditions for the majority.[14] Those displaced by the dam often lose their livelihoods and cultural ties, in addition to their homes, and rarely share fully in the benefits that the dam brings. No wonder, then, that local communities around the world have resisted these projects.

Besides resettlement, other social impacts of dams may include the following:

- Submergence of sites of archaeological, cultural, or religious significance;
- Submergence of infrastructure that can release toxic chemicals (e.g., oil);
- Influxes of construction workers from other regions, leading to social conflict and strains on local infrastructure; and
- Increases in certain diseases, such as malaria (due to increased mosquito breeding grounds) and mercury poisoning (due to mercury methylation in reservoir sediments).

> **Box 9-1. Pick–Sloan's Impact on Native Americans**
>
> The Pick–Sloan Plan was, without doubt, the single most destructive act ever perpetrated on any tribe by the United States. —Vine Deloria, Jr., Standing Rock Sioux[a]
>
> The Pick–Sloan Program—a joint US Army Corps of Engineers/Bureau of Reclamation infrastructure project on the Missouri River—had devastating impacts on several Sioux-speaking tribes in North and South Dakota. Between 1946 and 1966, five massive dams were constructed on the mainstem of the Missouri in the Dakotas, flooding over 1,100 km of the river valley. In violation of the Treaty of Fort Laramie, land for the dams and reservoirs was taken from tribes by eminent domain without their consent, with eight reservations—Fort Berthold, Standing Rock, Cheyenne River, Lower Brule, Crow Creek, Rosebud, Yankton, and Santee (upstream-to-downstream)—losing over 1,400 km² of ecologically productive bottomlands and more than 3,500 people displaced. The selective flooding of Native territory over White communities reflected a planning process that saw Native land as empty and unused, with tribes' active management of, and dependence on, land and water resources all but invisible to planners.[b] Likewise, compensation to the tribes was, at best, based on market values and often ignored the loss of game, wood, medicinal and edible plants, water, and cultural connection. The remaining reservation lands—situated high above their former heartlands in the river valley—were often dry, unproductive, and fragmented by the massive reservoirs.
>
> Michael Lawson, in his thorough history of Pick–Sloan's impact on the Oceti Sakowin (Sioux),[c] analyzes how the purported benefits of the project flowed—or didn't flow—to the tribes most affected by its costs. He finds that flood control and navigation mostly benefited nontribal economies in downstream states (Nebraska and Missouri), while hydropower and water supply aspects of the project provided electricity and drinking water that were too expensive for the tribes to afford. The recreation dollars that have flowed into the area have mostly bypassed the tribes, and tribal irrigation has been mostly limited to the Lower Brule Reservation.
>
> **Notes**
> [a]Lawson (2009).
> [b]Schneiders (1997).
> [c]Lawson (2009).

In addition to those directly displaced by dams, other groups of people may lose their livelihoods due to the ecological changes caused by the dam, both upstream (flooding of land for the reservoir) and downstream (alteration to river flow, especially affecting fishers and floodplain farmers). One study estimated that there are at least 470 million "river-dependent people" living downstream of large dams and probably affected by hydrologic and ecological changes.[15] Another study found that, in India, communities downstream of large dams tended to experience net benefits from the water security provided by the dam, but communities upstream showed increased volatility in their agricultural productivity and experienced net increases in rural poverty from large dam projects.[16]

5.2. Dam Safety

Dams hold back huge volumes of water, and when they fail, they can cause extensive damage. By far the most devastating dam failure on record occurred in China in 1975, when high rainfall led to the overtopping and destruction of the Banqiao and Shimantan Dams, along with up to sixty-two smaller dams. At the time, this event was largely hidden from the outside world, but a later Human Rights Watch report estimates that 85,000 people were killed by the initial flood wave, and another 145,000 people died from starvation and disease in the weeks following as floodwaters covered thousands of square kilometers.[17]

The WCD report points out that the global record on dam safety has improved over time, with a failure rate of 2.2 percent for dams built before 1950 and less than 0.5 percent for dams built since then, even as dams continue to get larger. Certainly, engineers have learned more about the stresses on dam structures and how to best withstand them. Still, dam integrity is dependent on two factors that are unique to each site and difficult to assess: the geology that supports the dam and the hydrology that determines how much water it must be able to handle. In addition, the Banqiao case shows that single events can have catastrophic impacts, so average dam failure rates may be less important than the magnitude of the rare events that do happen.

An additional safety factor to consider is the issue of **reservoir-induced seismicity**. When dams are built in seismically active areas, the pressure of massive amounts of water, combined with water's lubricating effects, can lead to intensified earthquake activity. The 2008 earthquake in Sichuan Province, China, which killed about 80,000 people, is estimated to have been hastened by tens to hundreds of years by the 320 million tons of water in the reservoir of the nearby Zipingpu hydroelectric dam.[18]

Despite the well-deserved attention paid to catastrophic large dam failures, most dam-safety incidents involve small dams. For the 799 incidents reported to the US Association of State Dam Safety Officials between 2010 and 2020, the median dam height was 7 m; these incidents caused only two fatalities, but at least 190,000 people were evacuated from their homes.[19] Given the prevalence of small dams on the landscape, many people unknowingly live in *dam breach inundation zones*: areas that would be flooded if a dam were to fail. Small dams are often less well maintained than large ones; in extreme cases, dams that were constructed decades ago may not have an identified owner or may have an owner who is unaware that they are responsible for maintaining it. In some states, sellers are not required to disclose the existence of a dam on the property, and buyers have been known to unwittingly purchase poorly maintained dams and the associated liabilities.[20]

Dams in the United States are classified based on the consequences (not the likelihood) of dam failure. *High-hazard* dams are those where dam failure (or misoperation that leads to unscheduled water releases) is likely to cause loss of life; *significant hazard* implies no loss of life but significant economic damage, and *low hazard* means low economic damage. Property development in the dam breach inundation zone can increase a dam's hazard class without any changes to the dam itself. There are over 14,000 high-hazard dams in the NID database, of which almost 2,000 are rated as being in poor or unsatisfactory condition. Most dams in the United States are regulated by state dam safety agencies, which are highly variable in their regulations, staffing levels, and inspection capacity. Many states require high-hazard dams to undergo regular inspections and to have Emergency Action Plans outlining emergency response procedures.

Dams have a limited lifespan before they need significant repairs or decommissioning. The median age of dams in the NID is fifty-seven years (as of 2022), and many dams will need significant investments to maintain their safety and function. The American Society of Civil Engineers gives dams in the United States an overall grade of D and estimates a rehabilitation funding need of $28 billion for federal dams and $66 billion for nonfederal dams.[21] Are the benefits provided by dams fleeting, with large costs on either end (construction and decommissioning), or can we find ways to safely keep this generation of dams functional for many years to come?

6. Building the Right Dams and Managing Them the Right Way

Several of the previous sections have made the same point in different ways: Dams vary a great deal in their impacts as well as their benefits. In this section, we think about which dams are best to build (and how to decide) and discuss ways to manage existing dams to mitigate their negative effects.

6.1. Building Better Dams

Broadly speaking, "good" dam sites—ones that are likely to result in high benefits and low impacts—share certain characteristics, such as having low population density, high power density, low water retention time, low sediment loads, and few critical natural habitats.[22] Beyond these general principles, it is essential to understand the social and environmental costs of each proposed dam project and weigh them against the benefits before deciding to move forward. But how good are we at predicting these benefits and costs?

An analysis of dam costs for a representative sample of 245 large dams found that, on average, the dams cost 96 percent more to build than anticipated (i.e., almost double), with three quarters of the dams experiencing some degree of cost overrun.[23] (The authors did not evaluate nonmonetary costs, such as social and environmental impacts.) The average cost overrun was much higher than for most other types of large projects except nuclear power plants. In addition, about half of the dams studied would have had benefit/cost ratios less than 1 if their true costs had been known at the time the benefit–cost analysis (BCA) was conducted. In other words, half of these large dams make no economic sense, even if we ignore social and environmental costs (as most BCAs do).

But if we don't trust the BCA process to help us decide which dams to build, are there other ways to evaluate tradeoffs? One tool for this purpose is *multicriteria decision making*, in which different criteria are explicitly evaluated without conversion to the common currency of dollars.

An example will help illustrate. Hydropower dams are rapidly being built in the Mekong Basin in Southeast Asia, including in three significant tributaries: the Sre Pok, Se San, and Se Kong, collectively known as the 3S tributaries. The 3S basin generates 10 percent of the Mekong's flow but 25 percent of its sediment, indicating its importance for sand availability in the lower river and the Mekong Delta. Full buildout of hydropower potential in the 3S would reduce sediment loading from the basin by 97 percent,[24] which would threaten the delta and its 17 million inhabitants with inundation and salinization.[25]

Can we maximize the benefits of hydropower development in the 3S while minimizing the impacts of sand retention? Since each potential dam site has a unique combination of hydropower potential and sand retention, it should be possible to

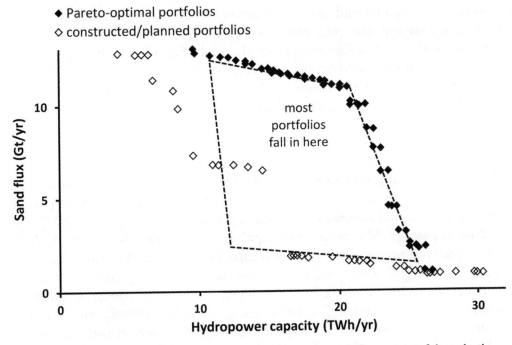

Figure 9-7. Dam portfolios for the 3S basin. Each point is a different mix of dams built or not built at forty-two potential dam sites. The Pareto-optimal portfolios are those that provide the optimal value on one axis given a specific value on the other axis. Most of the 16,638 portfolios analyzed are not shown but fall into the area outlined by the dashed lines. Redrawn from Schmitt et al. (2018).

evaluate individual dams—or, better, dam portfolios—against both criteria. A recent study did just that, examining 16,638 different dam portfolios for the 3S, with each portfolio consisting of a different combination of dams built or not built at forty-two potential dam sites. Estimating the hydropower capacity and sand flux for each portfolio allowed the authors to identify the **Pareto-optimal** portfolios (Figure 9-7): the portfolios that maximize the sand flux for a given hydropower capacity. Note that there is no attempt here to put the two axes into the same units, as one would do in a BCA.

A graph like that shown in Figure 9-7 does not necessarily imply any particular choice among the Pareto-optimal solutions; depending on how one values hydropower versus sand connectivity, one might choose different solutions.[26] But it clearly makes no sense to choose portfolios that are not on the Pareto frontier; these Pareto-inferior solutions provide less hydropower than they could for a given reduction in sand loading (or, equivalently, they reduce sand more than they need to for a given amount of hydropower production).

In practice, though, decision making in the 3S has been far from optimal. The portfolios that have actually been built (open symbols, Figure 9-7) are clearly doing more damage than they need to. The reasons for this involve both financial and political considerations but fundamentally indicate that dam builders and funders don't value sand connectivity. Studies in the Amazon have shown similar results; the portfolio of dams that is being built has greater environmental impact (greenhouse gas emissions in this case) than is necessary for a given amount of hydropower. Analysts are also starting to go beyond one environmental metric and analyze tradeoffs between hydropower and multiple environmental and social impacts.[27]

Some argue that all this just proves that the goal of building only good dams is unrealistic and that we should oppose all new dams. Without going that far, I would say at a minimum that environmental and social review processes need to be beefed up significantly instead of being the rushed, pro forma checklists that they too often have become.

6.2. Managing Dams Better

Can we manage existing dams to minimize their impacts? In addition to environmental flow releases, which we will discuss in Chapter 11, probably the most important mitigation measures are ***sediment passage*** and ***fish passage***.

Sediment passage: Minimizing the trapping of sediment behind dams is desirable for several reasons: to preserve reservoir capacity, to prevent downstream channel scouring by "hungry" water, and to replicate natural turbidity conditions downstream. Various techniques have been developed to maximize passage of sediment through the area of the dam, including bypassing turbid flood flows around the dam structure, moving turbid flood flows quickly through the reservoir without settling, and drawing down the water level in the reservoir to mobilize previously deposited sediments. The latter two approaches require a low-level outlet in the dam structure that can accommodate large amounts of water and sediment. Where sediment can't be effectively moved downstream, dredging the reservoir (or excavating the sediment after water drawdown) can be used to restore sediment volume, but it can be very expensive. Dealing with excessive erosion in the watershed is a prerequisite for effective reservoir sediment management.

Fish passage: Various structures can be installed to allow fish to pass around a dam, ranging from a fish ladder, in which fish move up from one small pool to another, to nature-like fishways, which are designed to mimic natural stream channels. Fish passage structures are generally designed to pass a small set of species, and their effectiveness is often low. There are also more surprising ways to move fish around obstacles, including truck and barge transport,[28] fish elevators,[29] and the Salmon Cannon.[30] In the Pacific Northwest, agencies such as the Bonneville Power Administration invest significant resources in salmon transport, but this artificial offstream migration has proven to be no substitute for the real thing.

7. Dam Removal

There has been great interest in the "selective removal of dams that don't make sense."[31] Environmentally, the goal is to restore river ecosystems by reversing the various ecological impacts associated with dams. Economically, many of the dams that are being removed are no longer serving their original purpose and need significant repair; dam removal is sometimes the cheapest option.

Dam removal has taken place around the world but has perhaps been best documented in the United States, where at least 2,025 dams have been removed as of 2022.[32] A significant milestone was reached in 2011, with removal of the 64-m-high Glines Canyon Dam on the Elwha River in Washington and four large dams on the Klamath River in California and Oregon being removed in 2023. Dam removal advocates are pushing for still more ambitious targets, including four dams on the lower Snake River in the Columbia Basin and, perhaps quixotically, the Glen Canyon Dam in Arizona and the Hetch Hetchy Dam in California.

How well does dam removal accomplish its goal of ecosystem restoration? Two important lessons have been learned from dam removal studies.[33] First, dam removal represents a short-term disturbance to the river system, especially when large volumes of reservoir sediments are released quickly, resulting in high turbidity and downstream sedimentation. Second, the long-term ecological outcomes of dam removal are usually very positive and can include the recovery of culturally and ecologically important fish populations. Of course, dam removal is unlikely to restore the riverine ecosystem to its exact predisturbance condition; the legacy effects of the dam, the presence of other stressors (e.g., water quality, climate change), and the nature of ecosystem dynamics all mean that the goal of dam removal is improved ecological health, not a specific state that is identical to some imagined predam condition.

Of the different components of ecological health, the quickest to respond are hydrology and connectivity; dam removal can quickly restore flow regimes and fish passage. It may take much longer for suspended sediment, river geomorphology, and floodplain ecology to stabilize.

Some of the issues that need to be addressed in planning for dam removal include the following:

- **Sediment dynamics**: Large volumes of sediment tend to accumulate behind dams, and thought must be given to the fate of this sediment after dam removal. Some of this sediment will be flushed downstream by dam removal, which may cause turbidity pulses and downstream sedimentation. In addition, scientists are increasingly realizing that stream systems in certain regions (e.g., eastern North America) are still experiencing the sedimentary legacy of thousands of small dams that breached naturally over the past few centuries, in the form of ***incised streams*** that flow through (and gradually erode) deep sediment deposits.[34] This makes it clear that dam removal in itself does not restore the floodplain habitats that existed before damming. Ironically, beaver dams (see Box 7-2) may help restore incised streams and reconnect them to their floodplains.[35]
- **Contaminants**: Sediments must be tested for contaminants to ensure that dam removal doesn't redistribute pollutants. The infamous 1973 removal of the Fort Edward Dam on the upper Hudson River resulted in the unwitting mobilization of sediments containing tens of thousands of kilograms of polychlorinated biphenyls (PCBs) dumped into the river over decades by General Electric, significantly complicating the eventual environmental cleanup.
- **Opposition**: There may be opposition to dam removal from people who use the reservoir or see the dam structure itself as historically significant. Not all dams are appropriate for removal, but an informed conversation about pros and cons is a good place to start.
- **Watershed perspective**: Dam-removal projects must be prioritized based on how many river miles they open up for fish passage, as well as whether other environmental impacts are likely to compromise the success of the project.
- **Permits**: Obtaining federal and state permits for dam removal can be a time-

consuming and frustrating process. The same safeguards that we have adopted to prevent the construction of poorly vetted, environmentally damaging dams also make it hard to remove environmentally damaging dams.
- **Ownership and liability**: Especially for larger projects, dam removal may entail significant financial risk that dam owners may not want to take on (although, as noted above, dam ownership also involves significant risk). To ease these concerns, coalitions of environmental groups and tribes can take on ownership and liability for dam removal.
- **Replacing dam functions**: Although many of the older, smaller dams removed to date have not been actively providing benefits, the same is not necessarily true of the next generation of dam removal. Managers and activists need to seriously consider whether and how to replace the storage, flood control, and power production being lost.

8. Aqueducts

Aqueducts are often used in conjunction with dams and reservoirs to move water from a dam site to where it is needed. While long-distance water transfer goes back to the Romans and earlier, the twentieth century saw a massive increase in the number and length of aqueducts, as well as three fundamental societal changes: We are now willing to use energy to pump water uphill, we are willing to invest in bringing water to agricultural fields (not just cities) that are far from water sources, and we have made long-distance water transfer cheap enough that building massive cities and agricultural complexes in the desert has become financially attractive.

The largest aqueducts operate at scales that are hard to comprehend. Shumilova et al. (2018) define water transfer megaprojects as those that meet one of the following criteria: construction cost $1 billion or more, length 190 km or more, or annual water volume 230 MCM or more.[36] Based on these criteria, there are currently thirty-four mega-aqueducts globally, of which six are in California. These mega-aqueducts cumulatively snake over 13,000 km and transfer over 200 km^3 of water per year. Projects currently under construction, including China's South–North Water Transfer and India's River Linking Project, will dramatically increase these numbers, to 39,000 km and 520 km^3/yr. Most of these aqueducts move water to cities and farms, although some, such as the James Bay Project, are primarily hydropower projects. Many of these aqueducts are ***interbasin transfers*** that move water from one watershed to another, often with severe effects on both the donating and receiving basins.

In addition to these mega-aqueducts, the cumulative impact of many smaller aqueducts is also impressive. A mapping effort by The Nature Conservancy estimated that the aqueducts operated by the world's large cities carry 184 km^3/yr of water (of which 30 km^3/yr represents interbasin transfers) and have a total length of about 27,000 km.[37]

Large aqueducts are vital for the survival of many urban areas and irrigation projects. In that sense, they also represent a significant vulnerability; what happens to Los Angeles if its aqueducts are disrupted by nature (earthquake, drought) or bad actors (terrorism)? In addition, of course, aqueducts contribute to the vicious cycle of supply; a new water supply begets more growth, which begets more water demand, which requires a new water supply. Some cities—including Los Angeles—have started to break this pattern by refocusing on local sources (Chapter 15). But other regions, such as the Lake Chad basin, are still floating ideas for new large-scale water transfers.

Chapter Highlights

1. Dams and reservoirs are used for many purposes, including instream uses and water storage for offstream uses. Satisfying these different uses requires careful reservoir operation.
2. We have built an astonishing number of dams; even the US National Inventory of Dams, with more than 90,000 structures, is an undercount. In the United States, the era of dam building ended in the late twentieth century, in part because of changing societal attitudes toward dams. In contrast, many low- and middle-income countries are rapidly building large dams, some of which are encountering significant local and international opposition.
3. Despite continued dam construction, reservoir storage volume—a metric of our resilience to hydrologic variability—has declined on a per capita basis due to population growth and *reservoir sedimentation*.
4. Probably the most significant global impact of dams has been the displacement of tens of millions of people, often river-dependent Indigenous communities, who have been forcibly resettled on marginal lands without adequate compensation.
5. Other negative impacts of dams include flooding of cultural, archeological, economic, and ecological resources; damage to rivers, fish, and river-dependent lifestyles; risk of catastrophic dam failure and *reservoir-induced seismicity*; and greenhouse gas emissions.
6. Data on existing large dams suggest that dam planners tend to underestimate costs and overestimate benefits, to the point that perhaps half of all the large dams we've built don't make economic sense, even if we ignore social and environmental costs.
7. Dams have widely varying distributions of benefits and costs. If we are going to build dams, we should build only the best dams, but in practice, we continue to build *Pareto-inferior* dams.
8. Some of the impacts of dams can be mitigated by management for environmental flows, *sediment passage*, and *fish passage*, although fish passage techniques have proven less effective than hoped.
9. Dam removal is growing in scope and scale, for environmental, social, and financial reasons. Dam removal initiates a complex process of geomorphic change but often succeeds in reestablishing a free-flowing river and associated fish runs.
10. Large aqueducts are used in tandem with dams to move water large distances, often across basin boundaries, enabling large cities and irrigation projects to sprout up in arid regions that don't have the local water to support them.

Notes

1. Major dams are those with heights greater than 150 m, or dam volumes larger than 15 MCM, or reservoirs larger than 25 BCM, or installed hydroelectric capacities greater than 1,000 MW.
2. The minimum power pool elevation for Lake Mead was long considered to be 1,050 feet, but modifications in 2016 lowered it to 950 feet.
3. Doyle and Patterson (2019).

4. Some of the traditional methods include sequent peak analysis, ripple mass curve analysis, and the Gould–Dincer approach. Modern dam planning uses automated versions of these methods.

5. Nover et al. (2019).

6. https://energyhistory.yale.edu/sierra-club-grand-canyon-dam-advertisements-1966/.

7. One of the commissioners, Medha Patkar of Narmada Bachao Andolan, signed the main document but also added a comment to note some of the deficiencies she saw in the report.

8. Briscoe (2015).

9. Root and Jones (2022).

10. This is the median value for large reservoirs in Wisser et al. (2013), Table 9-1. The same table cites a capacity-weighted mean of 0.66 percent/yr, and the WCD uses a value of 0.8 percent/yr, so my chosen value is on the low end.

11. This is consistent with Perera et al. (2023), who estimated that 16 percent of global reservoir storage has been lost to sedimentation, with loss expected to reach 26 percent by 2050.

12. Annandale (2013).

13. https://www.nytimes.com/2019/10/12/world/asia/mekong-river-dams-china.html.

14. Scudder (2005).

15. Richter et al. (2010).

16. Duflo and Pande (2007).

17. Human Rights Watch (1995).

18. Ge et al. (2009).

19. Calculated from the database at https://damsafety.org/incidents.

20. https://ctexaminer.com/2019/09/30/old-dams-and-new-problems-for-connecticut-home owners/.

21. ASCE (2021a).

22. Ledec and Quintero (2003).

23. Ansar et al. (2014).

24. Schmitt et al. (2018).

25. Kondolf et al. (2018).

26. Schmitt et al. (2018) argue that, within the Pareto-optimal solutions, the breakpoint at the top right of Figure 9-7, where additional increases in power come at increasing sand-retention costs, would be a good place to stop building additional dams.

27. Null et al. (2021).

28. Lusardi and Moyle (2017).

29. https://www.hged.com/community-environment/barrett-fishway/default.aspx.

30. Arguably the best take on the Salmon Cannon is John Oliver's, available at https://www.youtube.com/watch?v=l9qA8c-E_oA&feature=youtu.be.

31. American Rivers (1999).

32. https://www.americanrivers.org/2023/02/dam-removals-continue-across-the-u-s-in-2022/.

33. Foley et al. (2017).

34. Walter and Merritts (2008).

35. Pollock et al. (2014).

36. Shumilova et al. (2018).

37. McDonald et al. (2014).

Part III
Water Governance: Allocation and Reallocation, Cooperation and Conflict

While Parts II and IV of the book deal with specific uses of water (instream and offstream, respectively), Part III links them together by asking the basic question: Given a limited supply (as discussed in Part I), how can we best allocate water between competing uses? We also go beyond the question of allocation to other water governance issues, such as cooperation, coordination, and planning.

We start in Chapter 10 by describing the goals and challenges of water allocation, introducing economic efficiency, the tragedy of the commons, game theory, and common-pool resource management, followed by a description of current water allocation law in the United States. Chapter 11 picks up the central challenge posed by Chapter 10—how can we modify existing water allocation schemes to better achieve justice and sustainability?—and explores a variety of market and nonmarket tools for reallocating water and for improved coordination and planning. Chapter 12 addresses questions of conflict and cooperation over water, especially in international basins.

10 Water Allocation: Sharing the Common Pool

- How should we allocate water between different users, and how do we actually allocate it?
- How can we avoid the tragedy of the commons in water use?

The basic question of "who gets how much water?" arises in many contexts and at multiple scales, from a small group of farmers sharing an irrigation ditch to large basins shared by multiple countries. Both the difficulty and the importance of allocation are magnified in situations of scarcity, when there is not enough water to provide everyone all that they want.

This chapter addresses the ways in which water is allocated between users. We start the chapter by touching on economic principles for water allocation, which leads us to explore the common-pool nature of water and the ways that traditional water-sharing arrangements have overcome the tragedy of the commons. We then explore and evaluate the range of approaches used for water allocation in the United States.

There is no one perfect formula for allocation. Indeed, a great variety of approaches have been used in different times and places, some of which will be surveyed in this chapter. Before we start, however, it may be useful to set out goals for an allocation system. Ideally, our system for dividing up this scarce and precious resource should be:

- Economically efficient: maximize utility;
- Stable but adaptable: respect historical use but also allow for adjustment over time in response to changes in water availability, the economy, and societal values;
- Sustainable: maintain the resource, including its ecological value, for the long term;
- Just: divide up the resource in an equitable manner;
- Clear: avoid conflict by specifying how the resource should be shared, especially in times of scarcity.

These goals are in tension with each other, and not all of them can be fully met in most situations. Still, as we review allocation methods throughout the chapter, we will evaluate them against this wish list.

1. Water as a Common-Pool Resource

In this section, we explore the economics of water allocation and introduce one reason why water allocation is so complicated: Water is often a ***common-pool resource (CPR)*** and is thus susceptible to the ***tragedy of the commons***.

1.1. Economic Efficiency in Water Allocation

One of the goals we identified above for water allocation is economic efficiency, so we should probably define that a little more precisely. An efficient allocation is one in which the total net benefits (benefits minus costs) from the resource are maximized. The efficient allocation of a river's flow optimizes the value created by all the uses of the water, including both instream uses such as fishing and recreation and offstream uses such as irrigation and drinking water.

To understand allocation efficiency a bit better, let's look at Figure 10-1. You may remember from Chapter 1 that a user's marginal benefits from water use decrease as more water is available. (If you don't, you may want to review Box 1-2). In Figure 10-1, we plot *net* benefits (benefits minus costs), but the pattern is similar. In the simple case where we are allocating water between two users with identical demand curves (Figure 10-1a), it clearly makes sense to give half the water to each user, since this maximizes the total net benefit by satisfying the most valuable uses of each user and forgoing the least valuable uses. Put another way, an efficient allocation occurs when the marginal net benefits of water use are equal for the two users (Figure 10-1b). This result is generalizable beyond two identical users; water is allocated efficiently when the marginal net benefits are equalized across all users (each with their own demand curve). Pay attention to what this means: When efficiency is our goal, users who have a higher demand curve (i.e., those who value water more, maybe because they have water-intensive landscaping) should get more of the water.

This same logic can be applied to the intertemporal question of how quickly to draw down a nonrenewable water stock such as an ***aquifer***. For now, let's sidestep the question of shared aquifers and imagine a nonrenewable aquifer that is totally owned by you alone. You face the decision of how much water to use now and how much to save for later. The economist gives you advice: Maximize your net benefits by setting your marginal net benefits equal in the two time frames. Assuming that "future you" will use water much like you do now, your future demand curve looks much like your present one but must be ***discounted*** to account for the fact that those benefits occur later. Thus, you will use somewhat more than half the water now and save the rest for later, with the exact amount depending on your ***discount rate***. The same framework can be applied to other nonrenewable resources as well, including oil and other fossil fuels.

This is a good time to return to the topic of global and regional groundwater depletion, which we discussed in Chapter 5. As a reminder, we noted that some 10 to 20 percent of current global groundwater use is nonrenewable, and we looked at the Central Valley in California as one place where water tables are dropping due to

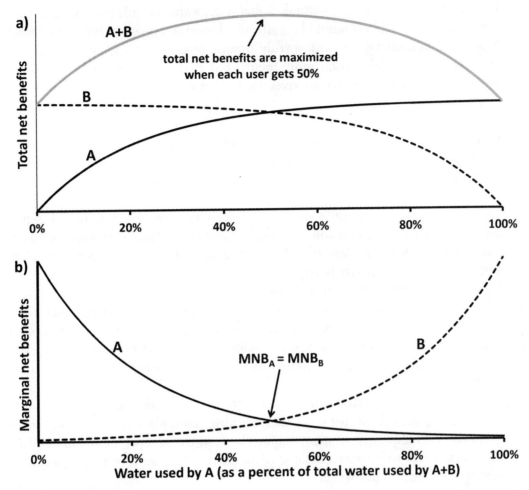

Figure 10-1. Total (a) and marginal (b) net benefits for two (identical) users as a function of how much water is used by each. Water used by B is 100 percent minus the water used by A. MNB, marginal net benefits.

overuse. It should be clear now that, from an economist's perspective, this is not necessarily a bad thing; just because a resource is nonrenewable doesn't mean it shouldn't be used. Thus, economists tend to reject the criteria of physical sustainability or safe yield (i.e., limiting groundwater use to capture rather than storage) that we raised in Chapter 5. A quote from a water resource economics textbook captures the argument:

> Instead of thinking about the repeatability of groundwater, why not emphasize what we really care about—the repeatability of human welfare? In the resource and environmental economics literature, inquiries about sustainability understand that depletable [i.e., nonrenewable] natural resources are always in decline due to their exploitation. At least until recently, each generation has been better off than the one preceding it. Historically, the depletion of resources has been overcome in some way(s). A primary way is that some economic activities create physical capital (machines, equipment, buildings, and infrastructure) and human capital (knowledge, technology, and experience) that have, up to recent times

anyway, offset the sacrificed natural capital (e.g., water, petroleum, soil, and timber resources).... There would be a stronger basis for a preservationist argument if we were addressing something truly unique (e.g., a species, an irreplaceable resource, or a special natural monument), but here we are merely considering groundwater, a substance that survives use and for which surface water is an excellent substitute.[1]

From my perspective, this argument has some merit—in a time of scarcity, it doesn't make sense to totally forgo nonrenewable groundwater use and just leave that water in the ground—but needs to come with three important caveats:

- Groundwater is not merely "a substance that survives use." Groundwater depletion can often have irreversible and undesirable side effects that must be taken into account, including loss of habitat, land **subsidence, saltwater intrusion**, and loss of storage capacity (Chapter 5).
- Groundwater depletion is a time-limited activity. We need to plan for what happens when the resource is gone, to prevent the economic and social shocks that can come from sudden changes in water availability (shocks that we have seen play out in the case of other nonrenewable resources). If human and physical capital can substitute for groundwater, they need to be there when the groundwater is gone.
- Groundwater is not as substitutable as other nonrenewable resources. Wind turbines can replace oil, but for some uses, nothing can replace water. Surface water is already fully exploited in many regions and, in the absence of adequate planning, is unlikely to be *more* available at the point when we turn to it as a substitute for depleted groundwater.

1.2. CPRs and the Tragedy of the Commons

If we understand what we mean by *efficient allocation*, we can turn to the question of how to achieve that allocation. For many resources, markets do a good job of achieving an efficient allocation. Markets ensure that the resource is sold to the users with the highest net benefits, since they are willing to pay the most for the resource. In addition, the market price of the resource will vary with changes in supply or demand, and thus it is an important signal of the scarcity of the resource and controls how much of it will be consumed.

Importantly, though, markets tend to work best for ***private goods***, defined as resources that are ***rival*** (my use precludes your use) and ***excludable*** (I can easily prevent you from accessing the resource) (Table 10-1). For other types of goods, markets may not lead to the socially efficient result, a situation called market failure. Market failures often result from some kind of ***externality***, where the costs (or benefits) of an activity are not borne solely by the person carrying out the activity.

The misalignment of private incentives and social goals is common for both public goods and CPRs. ***Public goods*** are those that are nonexcludable and nonrival, such as river habitat for an endangered snail: Everyone can benefit from knowing that the species is being preserved, and my benefit doesn't interfere with your benefit. If a factory destroys the snail's habitat, the public good will be lost, but the cost of that loss

Table 10-1. Classification of goods by rivalry and excludability, with examples given (in parentheses) for each category. For goods classified as nonexcludable, property rights systems can be designed that will limit the number of users, but from a purely physical perspective this is difficult to do.

		Rival?	
		Yes	No
Excludable?	Yes	Private good (car)	Club good (cable TV)
	No	Common-pool resource (open-ocean fishery)	Public good (streetlights)

to the factory owner is small, so the habitat will be undervalued relative to the socially optimal level. This is clearly a situation that calls for government regulation to ensure that the public good is not destroyed, perhaps by internalizing the externality through pollution fees or other mechanisms.[2]

CPRs—resources that are rival but not excludable, like most water sources—face a slightly different problem: Individuals acting in their own self-interest have an incentive to overuse the CPR, to the detriment of the resource and ultimately all its users. Imagine, for example, a shared aquifer with no rules on extraction; each user has an incentive to pump as much as possible from the shared resource before other users do, but this behavior quickly leads to groundwater depletion. This phenomenon has been referred to as the ***tragedy of the commons***, although this term is somewhat problematic, both because it misrepresents the way that traditional commons functioned (see next section) and because it was popularized by ecologist Garrett Hardin as part of a racist anti-immigrant agenda.

A word on terminology: Some economists prefer the term *open-access resource*, rather than *common-pool resource*, since it is the unfettered access to the commons that creates the tragedy. In our context, we will use *common-pool resource (CPR)* to refer to a physical characteristic of the system—it is physically difficult (but not impossible) to exclude people from using it—and reserve the term *open access* to refer to a particular type of property rights system that can be attached to a CPR, namely, that it is treated as public property. As we will discuss below, not all CPRs are open access; depending on the legal system, CPRs can also be treated as government property, private property, or common property.

1.3. Ostrom and Community Management

Two solutions are often proposed to the tragedy of the commons: privatization and government intervention. Here we discuss them briefly before turning to the third alternative: communal management.

Privatization of the commons—breaking it into individually owned parcels—changes the incentive structure and may make each landowner more likely to use it sustainably. More broadly, assigning enforceable private property rights to the commons (i.e., making it excludable; see Table 10-1) converts it to a private good and means that it can be allocated in markets. However, this approach ignores the fact that the commons as a whole may have a value that is greater than the sum of its parts. Privatization also tends to disenfranchise less privileged strata of society; indeed, the enclosure of the commons in England in the seventeenth through nineteenth centuries was notorious for dispossessing people of their traditional rights. Similarly, in the United States, the 1887 Dawes Act allowed the federal government to convert communally owned Native American reservations into allotments to individual tribal members; this privatization ignored the long tribal traditions of communal land management and led to loss of land from tribal control.[3] In the case of water, privatization is made considerably more complex by the fugitive nature of the resource. How do you privatize a shared aquifer when water flows invisibly across property boundaries toward the biggest pump?

As an alternative to privatization, government can impose rules to prevent the tragedy of the commons. In the aquifer example, the government can set and enforce a pumping limit on users. The problems with this approach may include poor understanding by central governments of the social and physical realities on the ground and the difficulty of enforcement (especially with a widely distributed *fugitive resource*).

Economist Elinor Ostrom pointed out that there is a third way that has been practiced for a long time in many places around the world: A group of users can own and manage the resource as *common property*, devising and enforcing their own rules for individuals' resource use (referred to as appropriation) and their contributions to operations and maintenance (referred to as provision). Often, she argued, these user-derived rules will be superior to government-imposed rules because they will be more deeply informed by local realities and more accepted by the users themselves.

Ostrom studied both successful and failed examples of community-based CPR management around the world and identified eight design principles that often characterized the successful cases (Table 10-2), although she emphasized that these characteristics were neither necessary nor sufficient to ensure success.[4]

While Ostrom's work applies to various kinds of CPRs (e.g., forests, fisheries), she and others have identified many examples in the water sector, most commonly involving communities using surface water for irrigation, where there is a substantial advantage to joint management of infrastructure. Community-based small-scale irrigation systems include huertas in Spain, subaks in Bali (Box 10-1), zanjeras in the Philippines, and acequias in Spain and former Spanish colonies, including the US Southwest. Small-scale local irrigation systems still manage a significant fraction of irrigated land in many countries.

The appropriation rules used by local irrigation systems vary widely in their specifics, but they often involve some form of proportional sharing (each farmer receives water in proportion to the amount of land irrigated, thus achieving some measure of equity) and rotation (each farmer receives the full flow of the local irrigation ditch for a period of time rather than continuous partial flow). Rotation of flow has advantages for both social reasons (it is easier for farmers to ensure they get their turn) and physical ones (full flow in a given canal reduces sedimentation and water loss to seepage and ensures that flow is sufficient to reach the tail end of the system). Note that proportional sharing presupposes design principle 1A (a limited, well-defined set of users; see Table 10-2), since proportional sharing in a system that allows unlimited new users will end up providing each user with a smaller and smaller—and ultimately insufficient—quantity of water.

Many of these local irrigation systems have proven resilient to social and ecological

Table 10-2. Design principles for successful common property management of CPRs, as proposed by Ostrom (1990) and reformulated by Cox et al. (2010). Asterisks indicate the principles that were identified as necessary for success in Ma'Mun et al.'s (2020) meta-analysis of sixty-two irrigation systems.

*1A. User boundaries: Boundaries between legitimate users and nonusers must be clearly defined.

*1B. Resource boundaries: Clear boundaries are present that define a resource system and separate it from the larger biophysical environment.

2A. Congruence with local conditions: Appropriation and provision rules are congruent with local social and environmental conditions.

2B. Appropriation and provision: The benefits obtained by users from a CPR, as determined by appropriation rules, are proportional to the amount of inputs required in the form of labor, material, or money, as determined by provision rules.

3. Collective-choice arrangements: Most individuals affected by the operational rules can participate in modifying the operational rules.

*4A. Monitoring users: Monitors who are accountable to the users monitor the appropriation and provision levels of the users.

*4B. Monitoring the resource: Monitors who are accountable to the users monitor the condition of the resource.

5. Graduated sanctions: Appropriators who violate operational rules are likely to be assessed graduated sanctions (depending on the seriousness and the context of the offense) by other appropriators, by officials accountable to the appropriators, or by both.

6. Conflict-resolution mechanisms: Appropriators and their officials have rapid access to low-cost local arenas to resolve conflicts among appropriators or between appropriators and officials.

*7. Minimal recognition of rights to organize: The rights of appropriators to devise their own institutions are not challenged by external governmental authorities.

8. Nested enterprises: Appropriation, provision, monitoring, enforcement, conflict resolution, and governance activities are organized in multiple layers of nested enterprises.

change over hundreds of years, but they face new challenges in the twenty-first century: climate change, population growth, increased demand for food, and new technologies. Perhaps most significantly, they face decollectivization pressure from governments that are eager to bring them under centralized control while not really understanding how they work (thus challenging design principle 7), along with private companies that see an opportunity to market their own expertise and technologies. While community irrigation systems may benefit from modern technologies such as drip irrigation, wholesale restructuring of governance and ownership may lead to system collapse (Box 10-1). We need to find ways to support local governance systems that are working well, while also ensuring equitable and affordable access to appropriate technology.

The cooperative allocation systems we have discussed in this section can work well in facilitating water sharing among a similar set of users, especially groups of farmers who all need water for irrigation. How well these systems function at larger scales, with a greater number of participants and a more diverse array of uses, has been less well studied. We turn now to the contemporary United States—where the legal system must balance water demand from cities, farmers, industry, hydropower, and others—and discuss the ways that water is allocated in practice.

> **Box 10-1. Bali's Subaks**
>
> On the Indonesian island of Bali, a complex social–religious–agro-ecological irrigation system has persisted for over 1,000 years. A subak is a jointly managed surface irrigation system for rice paddies (flooded rice fields), with each subak serving about 50–500 farmers, often from different villages. The island's more than 1,000 subaks are interconnected in a nested system that manages water flow through a complex network of rivers and diversions. Water management in this system is closely tied to religious ritual, with field flooding and planting schedules determined by a multilevel hierarchy of water temples, all ultimately deriving their authority (and their water) from the highest temple, on the shores of Lake Batur, the crater lake from which water flows downhill to the subaks.
>
> Anthropologist Stephen Lansing believes that one underlying dynamic driving the subak system is the tension between water scarcity—exacerbated when multiple subaks are on the same cropping schedule—and pest control, which is best done by coordinating cropping schedules between multiple subaks so they can engage in simultaneous postharvest flooding or burning of fields to eliminate pest refugia. In Lansing's view, this tension creates a self-organizing system—expressed in the water schedules organized by the water temples—that maintains cooperation and productivity over time.[a]
>
> As part of the **Green Revolution**, the Indonesian government encouraged Balinese farmers to abandon traditional temple-determined cropping schedules and shift to continuous cropping of high-yielding varieties of rice. What followed was chaos in water scheduling and an explosion of pest populations, which helped reveal the previously hidden functioning of the subak system and led to its restoration and recognition by the Indonesian government and, ultimately, by UNESCO, which has designated it a World Heritage Site.
>
> **Note**
> [a]Lansing and de Vet (2012), Lansing et al. (2017).

2. Water Allocation in the United States

Water allocation in the United States is generally controlled by states, not the federal government, with some exceptions that we will address below. Each state has its own idiosyncrasies, but states in the eastern part of the country generally use some version of the riparian doctrine, while states in the West use some version of the prior appropriation doctrine. Before delving into these doctrines, it is worth noting that both are consistent with the American reluctance to involve government directly in resource allocation—a reluctance that is not shared around the world. Even leaving aside centralized economies such as China's, several capitalist democracies—perhaps most notably Australia and Israel—give the government a much more direct role in water allocation.

2.1. The Riparian Doctrine

The **riparian doctrine**, common in states east of the Mississippi River, shares some of the features of the community allocation mechanisms described in Section 1.3. In particular, the riparian doctrine treats the river as common property, a resource to be shared by all those who own land bordering the river. No riparian owns the water itself, but each riparian has an equal right to use water from the river (a usufructuary right). The original riparian doctrine prohibited users from significantly altering the "natural flow" of the stream, which worked well when demand was low and much of the use was nonconsumptive (e.g., mills). However, as demands on the resource have increased, especially for municipal and industrial use, the riparian doctrine has generally abandoned the natural flow requirement and moved to requiring only that each use be *reasonable* relative to other uses. The original riparian doctrine also prohibited users from moving water from riparian parcels to unconnected land or land outside the watershed, but many states have abandoned this requirement.

Under the riparian doctrine, scarcity (due to drought or increasing demand) is shared equally by all users, following the same criteria of reasonableness. While all users are considered equal, the riparian doctrine does give some preference in times of scarcity to "natural" uses, particularly water for domestic uses and livestock.

How does the riparian doctrine fare on the evaluation criteria we identified at the beginning of the chapter?

- **Efficiency**: Riparianism has only a weak mechanism ("reasonableness") for ensuring that more water goes to the users who will derive the greatest benefit from it.
- **Stability and adaptability**: Riparianism arguably leans too far to the side of adaptability; as new riparian users come on board, they can affect the water available to previous users, which makes it hard to maintain a stable water supply for industry and cities.
- **Sustainability**: Riparianism makes no explicit provision for environmental flows.
- **Justice**: The riparian doctrine is equitable in the sense of treating all riparians equally, but it is not in the sense of ensuring that nonriparians (e.g., residents of cities) have enough water.
- **Clarity**: Riparianism can lead to conflict because of the lack of clarity on how to share shortfalls; persistent conflicts are resolved in the courts.

2.2. Regulated Riparianism

Because of the limitations of the pure riparian doctrine, most riparian doctrine states have moved toward a modified system called **regulated riparianism**, in which riparian users must apply for permits for water use. These permits are issued by a government agency, ideally based on a comprehensive review of water supply and demand in the basin, along with criteria such as efficiency and environmental needs. Permits are issued for a defined period, after which they must be renewed; this gives the opportunity for periodic reevaluation while still allowing secure water rights for a reasonably long time.

The promise of regulated riparianism lies in the possibility of comprehensive watershed planning, with government agencies weighing different needs—including environmental flows—and achieving an allocation that balances efficiency, equity, and environmental protection. This contrasts with the piecemeal allocation that is typical of riparianism (and, as we will discuss below, prior appropriation), in which there is no opportunity for holistic review of individual decisions accumulating over time.

Regulated riparianism is not a panacea. For it to succeed, the relevant government agency must have clear authority, sufficient resources, the ability to collect the data it needs, and clear communication channels with users.

2.3. Prior Appropriation

When large numbers of settlers moved into the western United States during the gold rushes of the late 1840s and 1850s, they found the riparian doctrine ill suited to the more arid landscape, where there was little riparian land (since there were fewer streams) and where the main uses of water (mining and irrigation) required moving it away from rivers to mining claims or suitable agricultural land. They also found that survival in these difficult conditions required more of an assured water supply than the riparian doctrine could provide, and they found that allowing all users equal rights to a limited water supply could leave each user with insufficient water to make a living. They ultimately settled on the ***prior appropriation*** doctrine, which derived from similar rules regarding mining claims, and now dominates most western states in some form or other, although some states incorporate both riparianism and prior appropriation into a hybrid model.

There are two pillars to the prior appropriation doctrine:

- **"First in time, first in right"**: Water use rights are acquired by physically diverting water from a river (including to nonriparian lands and even lands outside the watershed) and putting it to some type of ***beneficial use***, such as mining or irrigation. These requirements were meant to encourage water development while discouraging speculation and hoarding of water rights. Each water right has an associated quantity (the amount of water put to beneficial use) and date (the date the user began the legal or physical process of diverting the water). In times of scarcity, when there is not enough water available to satisfy all claims, water rights get filled in order of seniority, with more senior users (earlier dates) getting their full allocation before more junior users get any water.
- **"Use it or lose it"**: Unlike riparian rights, prior-appropriation rights are subject to forfeiture if they are not used for a certain period (as short as four years in New Mexico).

To understand prior appropriation better, let's look at the state of Colorado, which presents one of the purest examples of prior appropriation in action; appropriation is often called "the Colorado doctrine," and the Colorado Supreme Court's 1882 decision in *Coffin v. Left Hand Ditch Co.* is perhaps the most widely cited rejection of riparianism. Shortly after statehood in 1876, Colorado established courts to adjudicate water claims and passed legislation requiring claimants—many of whom had common-law appropriative rights from the decades before statehood—to submit evidence to those courts. In those early adjudications and others since, the courts sorted out competing claims and established a priority list (dates, amounts, and locations) for each basin in the form of water-rights decrees, which are considered transferable property rights for the use of water. (In other prior-appropriation states, administrative agencies, rather than courts, issue permits, rather than decrees.)

When there is insufficient water in a Colorado river to satisfy all the demands on it, a downstream senior user can place a priority call on the river, requiring upstream junior users to stop withdrawing water. Administration of this system—which is in the hands of local water commissioners under the authority of the state engineer—can be quite complex, since most rivers in Colorado have many appropriators, each with a different priority date.

A Colorado water right can be transferred from one use (or user, or location) to another while maintaining its priority date—but only if other users (including junior ones) are not harmed by the transfer. This means that the amount of water that can be transferred is only the amount that was used consumptively by the original user; any return flows from the original user must remain in place to avoid harming other downstream users. Thus, a transfer of water rights often involves a detailed analysis of current and historical water-use patterns to allow calculation of consumptive use. One exception to the no-reduction-in-return-flow requirement is when a user maintains the original use—say, agriculture—but reduces the return flow by increasing efficiency; such a change would generally be allowed. A second exception applies to imported water (sometimes called developed water), which is water brought in from another basin; the user who brought in the water may change the location or type of use (including reducing the return flow), since that water is considered entirely consumed from the perspective of the originating basin.

How does the prior appropriation system fare on our evaluation criteria?

- **Efficiency**: Prior appropriation fails the efficiency test in two ways: (1) Just because certain users (often mining and irrigation interests) have senior water rights does not mean that those are necessarily the best uses of water in the modern economy; and (2) although there is a requirement for "reasonably efficient" use of water, there is little incentive in the law for farmers to use more efficient irrigation methods or switch to less water-intensive crops. The incentive is the opposite: to continue to use as much water as possible to maintain one's claim into the future. At the same time, junior users in the same basin may have only ***paper water rights*** (rights that will rarely be filled) even if they could use the water very efficiently.
- **Stability and adaptability**: Prior appropriation is strong on stability (senior users know their rights will be protected) but weak on adaptability to changing economies (e.g., urbanization) and societal values (environmental flows, tribal rights). There is a great need to shift water from senior agricultural users to cities, rivers, and tribes; the next chapter will discuss some ways that this is being done.
- **Sustainability**: Prior appropriation, in its strictest incarnation, has no provision for environmental flows since water rights require a diversion.
- **Justice**: In many cultures, there is an inherent sense of fairness to the idea of "first come, first served." However, prior appropriation can also perpetuate preexisting inequalities of opportunity. In addition, prior appropriation failed to see that the water being appropriated was already in use by tribes, whose mostly instream uses (fishing, transportation, culture) were invisible to the capitalist culture that displaced them.
- **Clarity**: In theory, prior appropriation provides clear guidance for how to share water, but in practice, neither the law nor the water rights data are as clear as they should be, thanks to the accretion of layers of regulation, along with poorly documented conflicting claims.[5] Powerful special interests can take advantage of the complexity and opacity of water law—sometimes with the help of ***regulatory capture***—to manipulate the system for their own benefit.

Just as the riparian doctrine in the East has gradually evolved into regulated riparianism, prior appropriation has evolved over the last century, although perhaps in subtler ways. First, the increasing dominance of urban economies and populations has given urban water suppliers greater power and produced a reluctance to enforce strict priority rules against large cities. Second, the construction of large, federally funded, multipurpose infrastructure (dams and **aqueducts**) has reduced the need for short-term priority calls and has meant that water allocations are determined in part by contracts with state and federal agencies rather than individual water rights. Third, increasing awareness of the value of instream flows has led many states to change their definitions of beneficial use to include the protection of aquatic habitat through keeping water in rivers.

Before we leave prior appropriation behind, it is worth noting one important feature that it has in common with riparianism: Under both doctrines, users can claim a water right simply by using it, without paying anything for the privilege. Yes, developing the infrastructure to use that water can be expensive, but users rarely pay for the right to use the water. This essentially means that we are assigning the water itself a value of zero, a phenomenon that is at the heart of the undervaluing of water and the creation of scarcity through wasteful use. Actually, it's worse than that. Many irrigators in the West rely on highly subsidized large federal dam and aqueduct projects, so they are not even paying the costs of infrastructure, much less the value of the water.

2.4. Groundwater Allocation

Groundwater presents a tougher allocation challenge than surface water, for several reasons:

- Groundwater is widely accessible on the landscape, including on private property, making its use difficult to regulate.
- Basic data on groundwater (recharge rate, pumping rates, accessible volume, water quality) are much harder to obtain than for surface water.
- Groundwater can be extracted at rates greater than the rate of recharge, which presents questions of when and whether to allow groundwater mining.
- Groundwater use often affects surface water flows, but in complex, time-lagged ways that can be very hard to sort out.

For all these reasons, groundwater regulation has lagged behind surface water allocation, with states often implementing groundwater regulations only in response to a crisis. This neglect has resulted in the widespread depletion of aquifers (Chapter 5), although economists have found that groundwater use rates are often closer to the economically optimal rates than one might expect.

Common-law doctrines for allocating groundwater include the **rule of capture**, in which there are no limits on how much groundwater a landowner can pump; prior appropriation, in which an annual pumping volume is specified for each user based on seniority; versions of riparianism (correlative rights), in which pumping rights are shared by the aquifer "riparians" (those with land overlying the aquifer); and the "reasonable use" doctrine. In most states, these doctrines have been modified through legislation, which often includes permitting requirements and limits on groundwater pumping. For example, Arizona and California passed groundwater laws thirty-four years apart, but both aim to achieve "safe and sustainable yield" in the most overdrafted aquifers.

States have naturally focused on regulating the most overdrawn aquifers, but this creates incentives to deplete water elsewhere. In Arizona, the stark difference in the level of regulation between Active Management Areas (AMAs) and other regions has prompted water depletion in more rural parts of the state, as large farming operations move into unregulated areas for the same reason that Willie Sutton robbed banks: "because that's where the money [water] is." For example, large dairy operations are mining groundwater in Cochise County,[6] and in La Paz County, alfalfa is being grown with fossil groundwater and shipped to Saudi Arabia for feeding cattle.[7] Thus, while water levels in AMAs are recovering, water levels outside AMAs are dropping quickly, often drying up smaller household wells in rural areas.[8] Data illustrating the problem are rather limited, since neither monitoring of water levels nor metering of groundwater use is routinely required outside of AMAs.

The key to successful groundwater management is recognition of three basic facts:
- Groundwater pumping on private property can affect the water table on adjacent properties and the flow in adjacent surface water, so it should be subject to government regulation.
- Where groundwater is unregulated, it is subject to rapid depletion by intensive agricultural use.
- As discussed in Chapter 5 (Section 4.1), effective groundwater regulation needs to distinguish between hydrogeologic contexts. Groundwater pumping that affects surface water must be co-managed with surface rights, while depletable groundwater must be managed as a nonrenewable resource whose user should take into account future users as well as current ones.

2.5. Reserved Rights

An additional set of water rights, especially relevant in western states, is **reserved water rights**: implicit tribal and federal rights whose authority comes from outside state law.

The **Winters doctrine** (dating to the 1908 Supreme Court ruling in *Winters v. United States*) states that when the federal government establishes an Indian reservation through treaty or executive action, it also establishes an implicit water right for the reservation, since the reservation would not be able to fulfill its purpose without a water supply.[9] A reserved water right is different from standard prior-appropriation water rights in several ways. First, it is a federal right and thus supersedes any state water claims (although the McCarran Amendment of 1952 allows federal water rights to be adjudicated in state courts). Second, the priority date for the water right is either "time immemorial" or the date the reservation was established[10]; in either case, it is usually senior to all other water rights. Third, the water right does not require putting the water to use and is not lost by nonuse.

The size of the Winters water right is the amount necessary to fulfill the purpose of the reservation, which is often based on practicable irrigated acreage (i.e., the amount of water needed to irrigate the arable land on the reservation), on the assumption that the original purpose of the reservation was settled agriculture. However, most tribes have not yet adjudicated or developed their water rights, and the amounts involved are subject to various interpretations. The volume of potential outstanding Indian water claims in the West has been estimated at about 57 km^3/yr (three times the flow of the Colorado River), although Sanchez et al. suggest that this is probably an overestimate.[11] This is a source of great anxiety to non-Indian water managers in some western states; if tribes were to start using the large amounts of water that they have rights to, an already-stressed supply system would get even tighter.

An additional category of reserved rights is known as **Winans rights** (deriving from the 1905 case *United States v. Winans*), which are the rights of tribes to exercise their traditional practices, including hunting and fishing, on lands that the tribes gave up under coercion as part of the treaties establishing reservations. In the Pacific Northwest, for example, tribes have rights under the Stevens Treaties to fish at traditional river fishing sites—which also implies a right to have water and fish available at those sites. In *Washington v. United States* (2018), the Supreme Court affirmed a lower court decision that used the logic of Winans to compel the state of Washington to replace road culverts that were impeding salmon passage.

The complex landscape of water rights—tribal and federal reserved rights as well as claimants under state law—has led some states to carry out **general stream adjudications**, in which state courts sort out all the conflicting claims on a river basin. Besides quantifying reserved rights, these adjudications are opportunities to clean up the mess of poorly recorded appropriation claims, many of which date to the period before proper permitting and data-management systems were implemented. These adjudications can be extremely lengthy and expensive; the Snake River Basin Adjudication in Idaho, for example, took twenty-seven years and over $93 million in state funds to adjudicate 151,604 claims.[12]

2.6. Interstate Water Allocation

When John Wesley Powell appeared before Congress in 1890, he presented a now-famous map showing the western United States organized by watershed. The colorful prop reinforced one of Powell's key messages: In the West, where water scarcity is the primary limit on development, governance structures should reflect water's natural boundaries. Powell envisioned an irrigation society organized hydrologically into a few hundred "natural districts," each with its own small reservoir drawing on local water resources.

Of course, Powell lost this battle, defeated by the American tendencies to impose uniform structures across a diverse landscape and to prefer large over small. Instead of natural watershed boundaries, we have state boundaries, often drawn either as straight lines or along river courses (thus splitting the watershed).[13] Instead of small local reservoirs, we have large mainstem dams, tied to aqueducts that transport water large distances across state lines and over watershed divides.

The implications of these decisions are still with us: Since state boundaries do not follow watershed lines, water is often an interstate resource that states must work out how to share. The potential complications are numerous: Can an upstream state deplete all the water in a shared river? If downstream appropriators are senior, can they make a priority call on users in an upstream state? What if one state follows the riparian doctrine and another relies on prior appropriation? What if groundwater use

> **Box 10-2. Equitable Apportionment**
>
> In water disputes between states, the Supreme Court has original jurisdiction (i.e., it is the first court to hear the case), often appointing a special master to collect evidence and prepare recommendations. The Court decides these cases based on the common-law doctrine of "equitable apportionment," which, at its most basic, means that the states are on equal footing before the Court. Where both states use the same water allocation doctrine, the Court may decide to allocate based on that doctrine, as in the Laramie River case of *Wyoming v. Colorado* (1922)—both prior appropriation states—in which the Court limited the withdrawals of Colorado (the upstream state) to protect prior withdrawals in Wyoming. However, the Court will also apply a set of seven broad criteria:
>
> 1. Physical and climatic conditions;
> 2. Consumptive use of water in the several sections of the river;
> 3. Character and rate of return flows;
> 4. Extent of established uses and economies built on them;
> 5. Availability of storage water;
> 6. Practical effects of wasteful uses on downstream areas;
> 7. Damage to upstream areas compared to the benefits to downstream areas if upstream uses are curtailed.
>
> Notice the attention to protecting existing uses (4), to economic efficiency (6 and 7), and to working out ways that all uses might be satisfied (3 and 5).
>
> While equitable apportionment has typically been applied to shared surface water, the 2021 Supreme Court ruling in *Mississippi v. Tennessee* shows that it should be applied to shared aquifers as well, in this case the Middle Claiborne Aquifer, which extends under multiple states.

in an upstream state depletes river flows into a downstream state? What if the federal government is also involved—as it usually is—as water supplier (e.g., Bureau of Reclamation–operated irrigation infrastructure) or user (e.g., on federal lands)? These questions touch on core issues of state sovereignty and federalism.

There are three ways that water can be allocated between different states: litigation in the Supreme Court, which will apply the doctrine of equitable apportionment (Box 10-2); federal legislation specifying how much water each state gets; and a negotiated interstate compact, which must be approved by Congress. Litigation is expensive and time-consuming and can only resolve existing disputes (not plan future allocations), while legislation is divisive and rarely used, leaving compacts as the most desirable solution. Of course, compacts don't eliminate the threat of litigation, since states can sue over compact interpretation and implementation. For case studies of interstate water allocation, see the companion website.

Chapter Highlights

1. Economic theory suggests that efficiency in water use—defined as attaining the maximum aggregate net benefits—can be achieved by allocating water so that marginal net benefits are equal across users (and across time periods, after discounting). However, water is usually not allocated efficiently, in part because it is often a **common-pool resource** that is subject to the **tragedy of the commons**.
2. Economic efficiency is not our only goal for water use. We also want our water

allocation systems to be fair, sustainable, and clear, and we want it to find the right balance between stability and adaptability.
3. Three approaches are commonly used to avoid the tragedy of the commons: government regulation of water use, allocation of private property rights in water, and community management of water as ***common property***. Successful common-property management of irrigation systems often follows certain principles, such as well-defined user boundaries, effective monitoring, and proportional sharing.
4. Water allocation law in the United States is complex, opaque, and multilayered, and consequently water rights are often unclear and subject to litigation. This complexity also creates room for manipulation by powerful insiders.
5. Surface water allocation under state law tends to rely on the ***riparian doctrine*** in eastern states and ***prior appropriation*** in the drier western states.
6. Riparianism ties water rights to land ownership and relies on the sharing of water as common property based on reasonable use. Urbanization, economic development, and increased consumptive use have driven a shift to the greater clarity provided by the permit system known as ***regulated riparianism***.
7. Prior appropriation allocates limited private rights in water use based on the concepts of ***beneficial use***, seniority, "use it or lose it," and transferability (under certain conditions). Prior appropriation has been criticized for its rigidity and its failure to achieve efficiency, justice, sustainability, and clarity.
8. Groundwater allocation doctrines vary by state and can include elements of ***rule of capture***, correlative rights, and prior appropriation. A sensible groundwater law should distinguish between aquifers that are connected to surface water and those that aren't. Groundwater "water grabs" are happening with increasing frequency, driven by water scarcity and lack of regulation. At the same time, sustainable, cooperative groundwater management is emerging in some places.
9. In response to changing economies and values, we need to reallocate some water from agricultural users (who tend to have senior rights under prior appropriation) to urban users and environmental flows.
10. Indian reservations and tribal fisheries have ***reserved water rights*** under federal law that predate and supersede claims under state law, but these rights are often poorly quantified and enforced.
11. In the United States and other federal countries, interstate basins and aquifers present complex, multiscale coordination problems. Interstate water allocation is best achieved by negotiation of a compact but can also be worked out in the Supreme Court.

Notes

1. Griffin (2016).
2. Another example of the issues posed by public goods is a beach cleanup day. Volunteers spend their time providing a public good—a clean beach—but that good can be enjoyed by everyone, including *free riders* who choose not to participate in the cleanup.
3. Similarly, laws around tenancy in common have led to large losses of land ownership among African Americans, who have frequently been forced to sell jointly inherited property (referred to as heirs' property) if even one of the heirs (or a speculator who bought an heir's share) demands a sale.
4. Ostrom (1990).

5. Hanemann and Young (2020) point out that prior appropriation was much easier to implement for mining claims, where the resource consisted of small, adjacent plots of land, than for water, which is harder to measure and moves over large distances than can't be visually monitored.

6. https://www.nbcnews.com/video/water-mining-by-industrial-farms-leaves-wells-dry-in-rural-arizona-residents-say-69145669779.

7. https://revealnews.org/blog/debate-spreads-about-saudi-dairy-drilling-wells-in-arid-arizona/.

8. https://www.azcentral.com/in-depth/news/local/arizona-environment/2019/12/05/unregulated-pumping-arizona-groundwater-dry-wells/2425078001/.

9. Rulings in a case involving the Agua Caliente Band of Cahuilla Indians in Palm Springs, California, suggest that reserved rights apply to groundwater as well as surface water.

10. When the reservation documents (e.g., treaty or executive order) aim to protect aboriginal tribal activities (e.g., hunting), the water right will be considered to date from time immemorial; in contrast, when those documents envision the tribe engaging in a new type of water use, such as irrigated agriculture, the water right will date to the time of reservation establishment. It is worth noting here the racist attitude embedded in the vision of "civilizing" Indians by turning them into proper farmers (even on landscapes where farming was unsustainable).

11. Sanchez et al. (2020).

12. MacDonnell (2015).

13. Popelka and Smith (2020) calculate that globally 17 percent of subnational boundaries follow large rivers (excluding those that follow coastlines), compared with 23 percent of national boundaries and 12 percent of second-level (e.g., county) boundaries.

11 Reallocation and Coordination for Improved Water Governance

- How can we move toward more efficient and equitable water allocation?
- What role should markets play in reallocating water?
- How can we protect and restore environmental flows and tribal water rights?
- How can we better coordinate different water uses and plan for a changing future?

In Chapter 10, we discussed the existing rules for allocating water between users and concluded that changes in the economy (urbanization) and societal values (recognition of the needs of Indigenous peoples and aquatic ecosystems) mean that water allocations need to change as well. The first half of this chapter discusses approaches for reallocating water toward cities, the environment, and tribes, and the second half broadens our lens toward other water governance issues in the United States, particularly coordination and planning within and across state lines.

1. Reallocation for Cities, Tribes, and the Environment

Water allocation, particularly in the American West, is often described as using nineteenth-century institutions (prior appropriation) to manage twentieth-century infrastructure (dams and aqueducts) in the twenty-first century. The rules we have inherited are indeed far from ideal. Some might want to throw out these rules and start fresh, but that is probably neither realistic nor fair. The question is, How can we tweak our water allocation rules—or supplement them with other measures—to better meet current and future needs, in particular to reallocate water toward urban, environmental, and tribal needs?

There are at least three mechanisms by which water can be reallocated:

- Water markets, in which water rights are transferred through voluntary sales or leases, are popular in many corners and will be the topic of our first section below.
- Negotiations between parties—the topic of our second section—are probably the best way to reallocate water, as long as the resulting outcomes are fair and don't

hurt third parties (including the environment). Negotiations in which the two sides are vastly different in power can produce problematic outcomes.
- Government actions can reallocate water rights in various ways. In more centrally managed water economies (e.g., those of Australia or Israel), reallocation can be fairly simple legally, although it may be quite difficult politically. In the United States, where individuals have acquired a form of private property rights in water, changes are often incremental, in the form of state laws that directly or indirectly modify existing allocation systems. Federal laws and court rulings, such as the Endangered Species Act and the **Winters doctrine**, also can be used to reallocate water rights, as we will see below.[1]

1.1. Water Markets

Some have seen water markets as a promising solution to the inefficiency and inflexibility of the prior appropriation system. The idea is simple: The sale (or short-term lease) of water rights allows water to flow to the highest-valued uses, just like markets in other goods. Farmers who know their water has market value will increase their water-use efficiency and sell their extra water to urban users who are willing to pay high prices for it. Markets will increase total societal benefits by transferring water to users with higher marginal benefits until marginal net benefits are equal across all users and total net benefits are maximized (see Figure 10-1). The prior appropriation system would appear to be well positioned to take advantage of the power of markets, since it already grants transferrable private property rights for water use.

However, there are at least five types of problems with water markets, stemming from the unique physical and legal characteristics of water.

First, the transfer of water from agricultural to urban areas can devastate rural economies and ecologies. Once a few farmers sell their water, the burden of maintaining irrigation infrastructure falls more heavily on the remaining farmers. Slowly at first, and then quickly, the local ag-centered economy—the seed and fertilizer stores, the farm equipment sellers, the local restaurants—gets hollowed out and collapses. The abandoned fields dry up and generate toxic dust storms. Rural communities are understandably fearful of the **buy and dry** phenomenon and wary of the seemingly insatiable thirst of urban centers. In response, several states have passed "area of origin" laws to protect current and future water users from excessive interbasin transfers.

Second, property rights in water—even under prior appropriation—are often uncertain and poorly quantified. Where water rights have not been fully adjudicated, it may be hard to sell a water right of uncertain seniority. Similarly, the ambiguity of water law raises the risk that a farmer who leases her water to another user for a few years may unexpectedly find that she has permanently lost her right because of "use it or lose it" provisions.

Third, because of the interconnected nature of water systems, water rights are not truly private property rights. As discussed in Chapter 10, water transfers must satisfy certain legal criteria, such as causing no harm to other users and maintaining the timing and location of return flows. Besides the difficulty of actually meeting these criteria, the time and expense involved in proving it (to a state agency or court) can be daunting. An analysis of transaction costs (specifically, legal and hydrologist fees) for water sales in Colorado found that they varied greatly but on average exceeded the value of the water itself.[2]

Fourth, the physical costs of water transfers can be large as well. That is, infrastructure to move water from seller to buyer must be built and operated, which can be expensive given the high weight-to-value ratio of water. For this reason, the most successful water markets to date are in places where this infrastructure already exists.

Last, markets transfer water to those who are willing to pay the most for it. Many people object on moral grounds to the idea that money would determine who has access to this most basic and vital of resources. All humans and aquatic ecosystems deserve some water, even if they can't compete in a marketplace with deep-pocketed corporate interests. Despite these concerns, many environmental groups have welcomed water markets as an opportunity to buy water rights and convert them to environmental flows.

These issues mean that water markets will function effectively only in certain settings, and even there they will need significant regulatory oversight to ensure that they are operating efficiently and equitably. Active water markets have developed in the US West, as well as in Australia and Spain, where institutional and physical conditions favor their development.

California accounts for about half the water traded in the United States,[3] with annual volumes growing rapidly in the 1980s and 1990s and remaining roughly stable over the last twenty years at about 1.5 MAF/yr (out of a total statewide use of about 39 MAF/yr), though with significant year-to-year variability. In recent years, long-term leases (ranging from two to 100 years) have dominated the market, but permanent sales and short-term (annual and subannual) leases also have significant volume. As expected, most of the sellers are farmers, and most of the buyers are cities and environmental uses. However, there is also a large volume of ag-to-ag trade, which on the whole moves water from other agricultural regions to the water-scarce San Joaquin Valley.

In addition to water trading, California also allows water banking: Someone who artificially recharges groundwater into a designated groundwater bank can reserve that water for their own future use. Groundwater banks are a useful tool for storing water from wet to dry periods, but the legal and accounting mechanisms for managing them have lagged behind the engineering, resulting in manipulation of the system and extended legal battles.

Studies in Australia and the United States suggest that markets do improve the economic efficiency of water use. In the southern Murray–Darling Basin in Australia, water trading is estimated to have reduced the economic impact of the Millennium Drought by $4.3 billion over a five-year period.[4] In western US water markets, the water lease price paid by cities is about three to four times higher than the price paid by farmers, suggesting substantial gains in value through transfers.[5] In the lower Rio Grande, water markets have shifted water to higher-value crops, especially during drought years.[6] Increased ag-to-ag trades in California could reduce the costs of agricultural water scarcity by $360 million per year.[7]

1.2. Negotiating Ag-to-Urban Reallocation

Around the world, as cities face increasing water demand (from population growth) and shrinking supply (from climate change and water pollution), they are turning to agricultural water as a solution. A review of water transfers from rural users to cities found 103 such transfers over the last century, with the pace of transfers accelerating in the last twenty years.[8] These transfers served sixty-nine cities with a total population of 383 million and transferred a total of 16 km^3/yr over a cumulative distance of 13,000 km. Only 11 percent of the transfers took place in markets, but other types of transfers (e.g., through negotiation or government decree) often involved monetary compensation.

As the pressures to transfer agricultural water to cities continue to grow, it will be even more important to develop alternatives to "buy and dry," in the form of arrangements that build long-term relationships between farming communities and cities, allowing both to thrive.

Landmark agreements in southern California are arguably models that could be replicated elsewhere. These agreements have some features of markets—they involve cities buying water from farmers—but they are broader and deeper than simple one-time exchanges in a marketplace. The primary players are a pair of irrigation districts with senior water rights (Palo Verde and Imperial) and a pair of urban water suppliers looking to diversify their portfolio (Metropolitan and San Diego). Despite a complex maze of agreements, the underlying exchanges are simple: Urban centers pay irrigation districts and individual farmers to implement water-saving measures, with the saved water going to the cities and money flowing to rural areas. Limits on the amount of land that is fallowed help ensure that the agricultural economy is maintained. The mostly successful implementation of these agreements, despite institutional barriers and a long history of distrust between rural and urban areas in California, is a testament to the power of scarcity (and external pressure from other states and the federal government) to produce agreement.

1.3. Environmental Flow Requirements

Throughout the book, we have noted repeatedly the damage that anthropogenic changes to flow regimes have done to aquatic ecosystems, and in Chapter 5 we briefly mentioned **environmental flow requirements (EFRs)** as one way to ensure that aquatic ecosystems are protected or restored. This section describes ways to quantify EFRs, and the next section covers implementation.

There is not one perfect EFR that will ensure ecosystem health while allowing continued human water use. Any amount of hydrologic alteration is likely to have some impact on the health of aquatic ecosystems, and those impacts will increase with the degree of alteration. There may be thresholds beyond which impacts are increasingly severe, but those thresholds are often site-specific, hard to quantify, and dependent on which aspects of ecological integrity are examined. Ultimately, society must judge what level of degradation is acceptable for a particular river, given the other demands on those water flows and the uniqueness of the ecosystem. Thus, the most useful EFR methods are those that recognize that different rivers should be subject to different standards and that many rivers will experience some level of impact.

Having said that, we should also recognize that there have been missing voices in societal discussions of the balance between river health and human use. The missing voices often include those of people living and working by the river, people whose lives are organized around the river's daily and seasonal rhythms but whose concerns are drowned out by water planners who want to satisfy a seemingly never-ending human thirst. And, of course, the voices of the rivers themselves have been missing. In trying to find the right balance between human and ecosystem uses, it may be helpful to subtly change the question, from "How much water do we have to leave in the river?" to "How much water is it okay to take?" That is, it may be helpful to recognize the river itself as the core user, with human uses limited to what is available once that core has been satisfied (recognizing, as noted above, that the core may be defined more or less protectively for different rivers).

To illustrate the range of approaches toward defining EFRs, Table 11-1 summarizes four EFR methods of different levels of complexity. In designing EFRs, there is an inherent tension between complexity and ease of implementation. In a hypothetical world where we had perfect understanding of every river's ecology—along with a consensus on how to balance human and ecological needs—we could design an EFR for each river that would maximize its ecologic integrity and human utility. In the real world, information is expensive, and we need methods that can be applied quickly without decades of intensive data gathering. It may be better to have an implementable standard that is slightly less protective rather than an ideal standard

Table 11-1. A sample of EFR methods, in rough order from simpler to more complex. MAF, mean annual flow (under natural conditions); MMF, mean monthly flow (under natural conditions).

Name	*Description*
Sustainability boundary approach (Richter et al. 2012)	For each day, flow must be within 10 percent (more protective class) or 20 percent (less protective class) of naturalized flow. This is meant as a "presumptive standard" for streams for which more detailed studies have not been done. This standard is used as the safe Earth system boundary for surface water.[a]
Tennant (Tennant 1976)	This is one of a group of related methods that calculate monthly flow requirements separately for low-flow and high-flow months. Tennant defines low-flow months as those whose MMF is less than or equal to the MAF. The EFR for low-flow months is 20 percent of the MAF, and the EFR for high-flow months is 40 percent of the MAF.
Ecological limits of hydrologic alteration (ELOHA) (Poff et al. 2010)	Rivers within a region are placed into groups based on hydrologic similarity. For each group, studies are done to determine the relationship between hydrologic alteration and ecological responses. These relationships are used as the basis for stakeholder discussions aimed at producing standards that reflect ecological reality and human values. ELOHA analyses can be quite complex but can cover multiple rivers at the same time.
Instream flow incremental methodology (IFIM) (Bovee et al. 1998)	Knowledge of habitat needs for specific fish species (water depth, velocity, substrate) is used to elucidate how habitat conditions change as a function of flow. These relationships are used as the basis for a stakeholder process to determine the acceptable balance between habitat and water use.

[a]Rockström et al. (2023).

that can't be implemented because of lack of data. At the same time, recognizing our incomplete understanding of the complexity of aquatic ecosystems should make us err on the side of caution in protecting these treasures from irreparable harm.

One way to get more bang for your buck from environmental flows is a ***pulse flow***, a one-time release of a large volume of water. This pulse can mobilize sediment, restore ecosystem connectivity, raise public awareness, and reconnect river communities to their waterways. Perhaps the most famous pulse flow was the Colorado River flow of 2014, in which 130 MCM (about 1 percent of the Colorado's historical annual flow) was released from the Morelos Dam as part of a US–Mexico agreement. This allowed the Colorado to reach the ocean for the first time in many years and provided ecological and social benefits along the way.

How do water managers go about meeting EFRs? Most often, EFRs are applied in settings where water managers are already controlling flow rates, so meeting them typically involves reoperating a dam or modifying withdrawal rates. In situations where groundwater withdrawals are affecting streamflows, meeting surface water EFRs may require limits on pumping, although these must take into account the time lag between pumping and effects on surface water. Introducing EFRs as a constraint can increase the complexity of managing water systems for multiple benefits and can result in lost water supply or electricity generation.

1.4. Legal Tools for EFRs

In this section, we review several approaches to implementing environmental flows within the US legal framework; other relevant tools not covered here (but discussed in the companion website) include the Clean Water Act, the Endangered Species Act, Federal Energy Regulatory Commission relicensing, and state EFR laws.

APPROPRIATING INSTREAM FLOWS

Under the prior appropriation system, the obvious way to implement environmental flows is to obtain an ***instream flow right*** (a right to keep water in the river), similar to offstream rights claimed by other users. However, under traditional appropriation requirements, a water right is claimed through diversion and beneficial use, neither of which is applicable to instream flows, especially when beneficial use is defined in narrow economic terms. Fortunately, most states have changed their requirements to include habitat (especially for harvestable fish) as a beneficial use and to eliminate the requirement for a diversion. Still, these issues point to a fundamental mismatch between the logic of prior appropriation (you have to put in work to claim water) and the philosophy of instream flows (we should do less work on rivers, not more).

One ongoing manifestation of this mismatch is reflected in the question of who can claim or own an instream flow right. If states were to allow any entity to claim instream flows, what would prevent an environmental group from immediately claiming all the unappropriated flows in the state, resulting in the water hoarding that the beneficial use and diversion requirements were meant to prevent?[9] The answer in most states is that there are specific rules on who may claim a new instream flow and where and how they can do so.

In any case, claiming unappropriated flows as instream flow rights will seldom be enough to achieve reasonable environmental flows, given the high level of depletion that already exists in many rivers in the western United States. Thus, states have made provisions for buying up existing water rights and converting them to instream flow rights (without losing their priority dates). The funds for these purchases come from state governments and from environmental nongovernment organizations (NGOs). Ideally, state agencies and NGOs work closely together to prioritize and acquire instream flows, since NGOs are sometimes able to build trust with farming communities more easily than state agencies.

The Public Trust Doctrine

The **public trust doctrine** reflects Emperor Justinian's declaration, "By the law of nature these things are common to mankind: the air, running water, the sea." In other words, although government may recognize individual rights to use water, government is also responsible for upholding the general public's interest in—and ownership of—the waterways that are their natural inheritance, an interest that clearly should include ensuring adequate flow to protect instream values.

In most states, the public trust doctrine has not played a significant role in water reallocation, but an important case in California suggests that it could be interpreted broadly. The context was the ecological devastation of Mono Lake due to water withdrawals by the Los Angeles Department of Water and Power. In 1983, the California Supreme Court, in *National Audubon Society v. Superior Court*, ruled that the public trust doctrine applies even to previously granted appropriative rights and that water supply benefits must be weighed against the public-trust ecological and recreational values of Mono Lake. This led to a 1994 decision by the State Water Resource Control Board that limited LA's water diversions from Mono Lake tributaries based on water level in the lake, with the hope of restoring the unique Mono Lake ecosystems and mitigating the air quality impacts of the exposed lakebed.[10]

Federal Reserved Flows

In 1963, the Supreme Court, in *Arizona v. California*, extended the Winters doctrine (Chapter 10) to other (non-Indian) federally reserved lands, such as national parks. This doctrine was the basis for a 1976 ruling (*Cappaert v. United States*) that restricted groundwater pumping affecting water levels in Devils Hole, Nevada (part of Death Valley National Park), to protect an endangered pupfish (*Cyprinodon diabolis*).

As in the case of Indian reservations, the reserved water rights associated with federal land are limited to the amount necessary to fulfill the primary—but not secondary—purpose of the land designation. Thus, for example, despite the US Forest Service's proud proclamation that "Water is one of the most important natural resources flowing from forests," and "National forests are the single most important source of water in the US,"[11] water is generally considered a secondary purpose of Forest Service lands, and thus those lands are not eligible for reserved rights.

Wild and Scenic Rivers

Predating the Clean Water Act and the other environmental laws of the 1970s, the Wild and Scenic Rivers Act (1968) established a mechanism for protecting the environmental values of the nation's most pristine rivers. Once a river reach is designated as wild or scenic under the Wild and Scenic Rivers Act (by Congress or the interior secretary), new impoundments and diversions are severely restricted, making this a strong tool for protecting environmental flows as well as water quality. To date, 21,500 km of rivers in forty-one states has been protected as wild or scenic.

RIGHTS OF NATURE
Another approach to protecting environmental flows (and other aspects of aquatic-ecosystem integrity) is emerging as part of the **rights of nature** movement. This movement, which traces its origins to Christopher Stone's 1972 article "Should Trees Have Standing?," tries to expand our understanding of nature from property that can be owned (and therefore can be destroyed) to an entity with rights, thus evoking a human responsibility to protect those rights.[12] This movement has taken different forms in different places, ranging from recognizing the human right to a healthy ecosystem, to asserting the rights of ecosystems to exist and evolve, to granting legal personhood to specific natural features. These approaches generally allow any local resident to sue on behalf of nature or appoint legal guardians to represent the ecosystems.

In Ecuador, the 2008 constitution guaranteed both the human right to *buen vivir* ("good living")—the Spanish translation of the Indigenous term *sumak kawsay*, which implies living in harmonious relationship with nature—as well as the right of Mother Earth to "integral respect for its existence and for the maintenance and regeneration of its life cycles, structure, functions and evolutionary processes."[13] In 2011, in response to damage to the Vilcabamba River from a road-widening project, two American citizens living along the river used these constitutional provisions to sue successfully on behalf of the river.

In India, the High Court in the state of Uttarakhand declared in 2017 that the highly polluted Ganges and Yamuna Rivers—both holy in Hinduism—were legal persons with the status of minors, and the court appointed two officials to represent their interests, including presumably being free of pollution. However, the state government successfully appealed this decision to the nation's Supreme Court, arguing, among other things, that legal personhood for the rivers opened up the possibility that the rivers could be sued for flood damages.

In the United States, several local governments have passed rights of nature ordinances, but their effectiveness in protecting river flows has not yet been fully tested.

1.5. WaterBack: Decolonizing Water Management
This section discusses ways to increase Indigenous access to, and control over, rivers and other water bodies. There are at least two reasons that this is important. First, it is a clear moral imperative to partially right the historical injustices inflicted on Indigenous people by settlers and colonial cultures, including the theft of land and water resources. Second, many Indigenous peoples have long traditions of sustainable water management, refined over thousands of years of cultural coevolution with particular places; we sorely need to learn from these traditions.

As noted in Chapter 10, US law in theory recognizes the seniority of tribal water rights (Winters reserved rights) over other water users in prior-appropriation states. However, as of 2019, only fifty-six out of 226 reservations in the western United States had had their rights quantified through an adjudication process (twelve through court decrees and forty-four though negotiated settlements), and these adjudications took an average of twenty-one years to complete, in part because each adjudication involved, on average, thirty-three parties (mostly nontribal water users in the same basins).[14] One troubling dynamic impeding the quantification of tribal rights is that nontribal water users are often invested in the status quo and can afford to delay and obstruct, while tribes have limited resources to invest in a legal battle that does not help meet their pressing short-term needs. State water managers, and the state courts where these cases are heard, also have an interest in delaying tribal adjudications, because tribal allocations count against state allocations in interstate basins such as the Colorado.

Tribal water rights can be adjudicated through either legal proceedings or negotiated settlements. The latter approach has several advantages. First, settlements can avoid the costs and animosity associated with litigation. Second, settlements can, at least in theory, proceed more quickly, providing all parties with more certainty on water rights (although, as noted above, some parties may prefer delay). Most important, settlements provide a tribe not just with ***paper water rights*** but also with federal funding to develop the infrastructure to use those rights. Still, progress on settlements has been slow, in part because some states have used the process to try to force non–water-related conditions on the tribes.[15] The recent Supreme Court decision in *Arizona v. Navajo Nation*—which rejected the Navajos' request that the federal government quantify their Winters rights—leaves the tribes at the mercy of the states' delaying tactics.

Once water rights are adjudicated, what the tribes do with that water should be decided by each tribe, but the federal government often sets conditions as part of its funding. Although the Winters doctrine foresees primarily agricultural use, tribes may choose to use their water for other forms of economic development (although a troubling court ruling in Wyoming found that the Wind River Tribes could not use their Winters rights for instream flows). Some tribes with adjudicated water rights are leasing that water to other users (as in the case of the Gila River Indian Community) or accepting compensation to not use that water in time of drought (as in the case of the Colorado River Indian Tribes, who have senior rights to over 650,000 AF/yr)[16]; legal restrictions sometimes stand in the way of these types of transactions.[17] Most fundamentally, tribes need water to provide their members with the basic human right to water and sanitation, a need that is not currently being met in many reservations; we will return to this issue in Chapter 15.

Because of legal barriers and lack of infrastructure, even tribes with settled water rights often use or lease less than half of their allocated water. The forgone revenue from this unused water (estimated at up to $1.6 billion per year) feeds the vicious cycle of underfunding, in which tribes lack the resources to develop the infrastructure that would allow them to claim their water, a necessary precondition for economic development.[18]

Beyond the quantification of offstream water rights and the development of infrastructure to use them, there is also a need for greater tribal involvement in management of rivers and watersheds to support traditional cultural values and healthy ecosystems. In short, we need a water counterpart to LandBack, the movement that is trying to reclaim Indigenous sovereignty over land and other resources.[19]

One tool for WaterBack is tribes' Winans rights to fishing and other traditional activities (Chapter 10). Three court cases in the Klamath River Basin in California and Oregon show the potential power of this tool. *United States v. Adair* (1983) established that the tribes' fishing rights imply an instream water right and "the right to prevent other appropriators from depleting the streams waters below a protected level in any area where the [instream] right applies." *Parravano v. Babbit* (1993) ruled that nontribal commercial ocean fishing of salmon can be curtailed to ensure adequate return of salmon for both tribal fishing and spawning. Finally, *Baley v. United States* (2017) found that when irrigation deliveries were reduced in 2001 to protect salmon for tribal fishing, irrigators did not suffer **regulatory takings** because their water rights were inferior to the tribes' "time immemorial" instream water rights.

A second tool for WaterBack is the rights of nature approach discussed in the previous section, particularly in situations where Indigenous leaders with long cultural connections to the river are appointed to speak for the river as its guardians. The most prominent example is probably New Zealand's recognition of Te Awa Tupua (the Whanganui River) as a legal entity, driven in large part by a process of reconciliation with the various Maori iwi (tribes) that traditionally lived along the river. The 2017 law is notable in trying to incorporate Indigenous conceptions of the river as a living being: "Te Awa Tupua is an indivisible and living whole, comprising the Whanganui River from the mountains to the sea, incorporating all its physical and metaphysical elements. . . . The iwi and hapū [subtribes] of the Whanganui River have an inalienable connection with, and responsibility to, Te Awa Tupua and its health and well-being." While a tribal guardian (alongside a nontribal guardian) will be responsible for representing the interests of Te Awa Tupua, the personhood law specifically protects existing nontribal water use rights. Around the world, not all tribal leaders view the rights of nature approach as suitable, noting that it operates within the Western tradition of emphasizing individual rights rather than communal responsibilities.

In the United States, the Sauk-Suiattle Indian Tribe recently sued the city of Seattle in tribal court on behalf of salmon (Tsuladx^W in the Lushootseed language), whose migration is blocked by three water-supply dams with no fish passage. The lawsuit asks the court to:

> a. Declare Tsuladx^W within the territory of the Sauk-Suiattle Indian Tribe is protected and possesses inherent rights to exist, flourish, regenerate, and evolve. . . .
> b. Declare the Plaintiff's tribal members possess a right and public trust, or trust responsibility, to protect and save Tsuladx^W. . . .
> d. Declare that Defendants individually and collectively knew or should have known that obstructions to Tsuladx^W's way of life was undertaken without the free, prior, informed consent of Tsuladx^W as sentient beings and without the free, prior, informed consent of or consultation with the [Sauk-Suiattle Indian Tribe] or other tribes who owe a duty to Tsuladx^W.

In the US Southwest, the Pueblo Action Alliance is fighting for WaterBack in a broad sense, including not just increases in tribal water allocations but also paradigm shifts toward water relationships built on respect and reciprocity rather than the market. Their WaterBack Manifesto starts with the following statement:

Within the Pueblo perspective, rivers are viewed as mothers and women are the keepers of water. Waterways are spiritually governed by our mothers and have provided sustenance for what grows on the land. The resurgence of Indigenous feminist water management strategies is the only way to protect our water resources and ensure clean water for the future.[20]

From these words, it should be clear that WaterBack is not just about reallocation but instead seeks a larger shift in values, attitudes, power dynamics, and governance systems.

2. Coordination and Planning

In the last chapter and the first half of this one, we have focused mostly on rules for water allocation. Here we turn from rules to institutions, and we move from a narrow focus on allocation to broader questions of water management. Specifically, we ask how agencies and organizations of various kinds (both public and private) interact to formulate and implement rules for water management. What scales and institutional forms are best suited for coordinating the diversity of instream and offstream water uses and values? We examine this first at the state level and then at larger scales.

2.1. State Coordination and Planning

To achieve water security for their populations, states must balance a variety of water uses and manage other impacts (pollution, land use, climate change) that affect water quantity and quality. Since water touches on so many aspects of our lives, there are typically a number of agencies charged with administering state and federal law related to water. The existing configuration of agencies and interactions is the outcome of a complex history and is rarely ideally suited for current challenges. Yet these agencies have typically developed strong cultures and power centers that are often resistant to change.

Just as each state is different in the rules it uses for allocating water, states also differ in the structure of their water management agencies, with some states exerting more centralized control and others devolving most decision making to counties, municipalities, and special districts.

Many states—recognizing the changes that will be facing their water systems over the coming decades—have undergone planning processes to produce state water plans to guide water policy and management. Close examination of three of these—California (updated 2018), Colorado (2015), and Connecticut (2018)—will help us understand the issues facing state-level water governance.

All three water plans have elements that research[21] has identified as critical to adaptive and resilient state water planning:

- **Science and data**: The plans emphasize the importance of being guided by the best science. All three states recognize that their information systems for water supply and demand data have significant shortcomings. Especially for Colorado and Connecticut, the planning process was used as an opportunity to compile data to help understand water availability and use in different areas of the state.
- **Nonstationarity**: All three plans take into account the (uncertain) effects of

climate change over their planning horizons. This is not true of all state water plans; five out of eleven state water plans reviewed by the Interstate Council on Water Policy did not address climate change.[22]
- **Adaptive management**: The plans recognize the importance of continuous adaptation in the face of changing conditions. California focuses on evaluating the outcomes of state investments, performance tracking, and coordinating climate science and monitoring efforts. Colorado and Connecticut use scenario planning to identify short-term no-regrets actions, along with time horizons for future decision making in response to the effectiveness of those actions and the emerging impacts of climate change.
- **Multilevel collaboration**: The plans were produced through a collaborative process between state agencies, water users, environmental groups, tribes, consultants, academics, and other stakeholders. The states have committed to continued collaboration in implementing and updating the plans.
- **Clear water allocation rules**: Here the plans diverge somewhat:
    - California does not directly address water-rights issues in its plan. However, there are certainly active discussions in the state about how to adapt its allocation system. The Planning and Conservation League recently convened a group of experts who proposed several changes to California water law, mostly focused on improving the state's ability to verify, monitor, adjudicate, and enforce water rights and water use.
    - Colorado is very proud of its prior appropriation system. The main sore spot in water allocation is the phenomenon of transmountain diversions, in which rapidly growing cities on the eastern slopes of the Rockies divert water from the less populated and wetter western slopes. Since 2005, the regional tensions created by these diversions have been managed through nine regional roundtables and an Inter-Basin Coordinating Committee. As part of the state water-planning process, the committee produced a conceptual framework to guide future transmountain diversions.
    - The primary water rights problem in Connecticut is the issue of *zombie registrations*: water rights that were grandfathered in during the transition to regulated riparianism and are no longer active (but could be reactivated at any time). The water plan compiles data on the potential volume of these registrations by basin and calls for further work to investigate and remove unused registrations.

Beyond these basic features, the three plans share many commonalities:
- **Interagency coordination**: The plans look at ways that different state agencies can work more smoothly together in managing water.
- **Supply and demand**: One key focus for all three plans is ensuring adequate water supply to meet demand into the future.
- **Conservation**: The plans all call for reductions in demand to fill the supply–demand gap, although Connecticut, as the wettest of the three, acknowledges that the state traditionally has lacked a water conservation ethic.
- **Drought**: All three plans address the challenge of managing water during drought (and all three states also have separate drought planning processes).

- **Tribal water**: All three plans pay some attention to tribes' need for increased water allocations, although this is least prominent in Connecticut.
- **Funding**: The plans all call for increased funding to support implementation.
- **Land use**: The plans all recognize the link between land use and water quality and quantity, and they call for integration of water goals into land-use planning.
- **Agriculture**: All three plans address the role of agriculture in water use and the need to balance efficient use with maintaining a viable agricultural economy.
- **Ecosystems**: The plans all call for greater attention to protection and restoration of water-dependent ecosystems while recognizing the need to balance ecological protection with human uses.
- **Education**: All three plans emphasize the importance of educating children and adults about water issues in their state.
- **Groundwater**: The plans all recognize the complexities of groundwater management.
- **Flooding**: All three plans examine ways to better manage flooding.
- **Values**: The plans all contain value-based goals for improved water management. California uses sustainability as its framework, Connecticut invokes the triple bottom line (environmental, social, and economic), and Colorado outlines three goals: "(a) a productive economy that supports vibrant and sustainable cities, viable and productive agriculture, and a robust skiing, recreation, and tourism industry; (b) efficient and effective water infrastructure promoting smart land use; and (c) a strong environment that includes healthy watersheds, rivers and streams, and wildlife."[23]

At the same time, each plan has unique areas of emphasis that are absent or less prominent in the other plans, reflecting the varying physical and legal geographies of these states; a nonexhaustive overview of some of these topics is provided below.

California
- **Disadvantaged communities**: Given the large number of Californians who lack access to safe drinking water, the plan focuses on "engag[ing] proactively with disadvantaged community liaisons," where disadvantaged communities are defined as those with median incomes less than 80 percent of the state average.
- **Wildfire**: As the site of some of the country's most destructive recent wildfires, California understandably emphasizes wildfire mitigation and resilience more than the other states.
- **Operational flexibility**: Given the state's heavy reliance on a complex, interconnected network of dams and aqueducts, it makes sense that one of the plan's five goals is "strengthen resiliency and operational flexibility of existing and future infrastructure."

Colorado
- **Growth**: As the fastest-growing state of the three, Colorado has special concern over the impacts of population growth on water demand.
- **Storage**: As a headwater state, Colorado can increase its water supply by increasing storage to capture some of the water that now flows downstream. In its plan, Colorado sets a specific goal of 400,000 acre-feet of additional storage by 2050.[24]

CONNECTICUT
- **Regionalization**: Connecticut has a large number of small water suppliers and is interested in potentially increasing efficiency through consolidation.
- **Class B water**: Connecticut is unique among the fifty states in not allowing discharges of treated wastewater into any streams designated for actual or potential drinking-water supply. This means that most of the large rivers in the state are not available for future supply. Although there is broad consensus that this restriction should remain, there is interest in allowing use of Class B waters for nonpotable uses.

2.2. Interstate Coordination and Planning

About twenty-five of the world's nations—including the United States, India, Australia, Brazil, Germany, and Mexico—have federal forms of government, where responsibility for water management is shared in complex ways between the national government and state or provincial governments. *Federal river basins* are defined as domestic river basins that span multiple states or provinces within a federal country and international basins where at least one of the riparians is a federal country.[25] This includes some of the world's largest and most important river systems.

Federal rivers pose unique coordination challenges, both within a given level of government (e.g., between two states) and across levels (between the states and the national government). A bewildering array of agencies and rules may apply to different river segments or to different aspects of the same river segment (e.g., fish vs. water quality), leading to confusing and disjointed management unless the agencies can harmonize their goals and operating rules. In this section, we will focus on interstate water management in the United States, but many of the lessons are relevant to other federal countries.

Traditionally, water management has mostly been a state responsibility, but over time the federal government has come to play a larger and larger role, starting with navigation and extending into flood control, water supply, and environmental protection. The federal role has included both regulation and funding, with the latter making the former more palatable to states. The large federal water agencies—the Bureau of Reclamation and the US Army Corps of Engineers—came to the forefront of water management in the first half of the twentieth century, when construction of federally funded infrastructure brought them power and influence. Both agencies still manage a large fraction of the country's river flow and water delivery, but the decline in new infrastructure construction has shrunk their budgets and left them searching for new missions and new tools to induce state cooperation.

Since at least the mid-nineteenth century, the watershed, or river basin, has often been seen as the ideal scale for water management.[26] The philosophies underpinning this perspective have included regionalism, rational planning, high modernism, and ecosystem-based management, but have all shared the sense that the river basin is the "natural" unit of management. In reality, as we have seen in previous chapters, not all water management problems are amenable to solutions at the basin scale. Still, there are certainly many situations where basin-scale approaches are called for.

Institutionally, the focus on river basins—and the fact that political jurisdictions don't align with those basins—has sparked interest in the creation of **river-basin** *organizations (RBOs)* that work across political boundaries to manage water problems holistically. Depending on the setting and the type of problem, RBOs may take different institutional forms.

In Chapter 4, we described the federal government's various efforts at river-basin planning, starting with Teddy Roosevelt's Inland Waterways Commission in 1907 and continuing with the "308 reports" in the 1920s–1940s and the creation of the TVA in 1933. Additional attempts at basin coordination continued through most of the twentieth century. Starting in 1943, Federal Inter-Agency River Basin Committees were formed in six basins, with the goal of coordinating the actions of existing agencies without creating a new layer of bureaucracy. The 1965 Water Resources Planning Act created a Water Resources Council and six "Title II" federal–state river basin commissions, although these were eliminated by President Reagan in 1981. Both the interagency committees and the Title II commissions were generally considered ineffective, in part because funding and power tended to flow through existing agencies, undercutting the authority of these nascent RBOs.

In contrast to the top-down federal planning attempts, bottom-up RBOs and other coordinating arrangements have flourished where they are needed to solve real problems. The Interstate Council on Water Policy identifies fifteen institutional forms that have resulted from these efforts.[27] In addition to dealing with water allocation, these RBOs also take on topics such as water quality and flood control. For analyses of some of these RBOs, see the companion website.

Given Ostrom's design principle 2A (rules should be congruent with local conditions; see Table 10-2), perhaps it is not surprising that a diverse set of regionally specific cooperative arrangements has had greater success than the relatively uniform interagency commissions and Title II committees of the mid-twentieth century. Of course, these diverse institutions are operating within a set of uniform national laws and regulations that provide important guiderails. Finding the right balance between stability and innovation, between consistency and specificity, is an elusive and ongoing task.

Chapter Highlights

1. To achieve a water allocation that is more just, sustainable, and efficient, we need to reallocate some water toward tribal, environmental, and urban users. Ultimately, this means that the water footprint of agriculture will have to shrink, especially in water-scarce basins, but this transition must be gradual and must consider the needs of farmers and farm workers.
2. Within prior appropriation systems, water markets can increase efficiency by shifting water to more valuable uses. But markets in water come with significant problems, including externalities and transaction costs; appropriate regulation can help alleviate these problems.
3. As urban water scarcity increases, water reallocation from agricultural to urban uses is increasingly common. These transfers can be carried out through markets, decrees, or negotiations. Negotiated agreements between cities and farmers are a good option for shifting some water to urban centers while avoiding ***buy and dry*** and ensuring the continuing viability of rural communities.
4. Methods for quantifying environmental flow requirements—meant to protect

natural flow regimes—vary in their ease of implementation and the level of protection that they provide.
5. Many types of policy instruments are being used to return water to the environment, including reserved rights, appropriation of **instream flow rights**, the **public trust doctrine**, and **rights of nature** laws.
6. Tribal reserved rights are a powerful tool for remedying historical inequities in water use, but many tribes still have not had their water rights recognized and quantified through legal proceedings or negotiated settlements.
7. WaterBack demands more than water reallocation; it requires changes in governance to allow tribes more water management authority and to incorporate Indigenous voices and values into basin-level organizations. In some cases, tribes are being named as legal guardians of rivers or are suing on behalf of culturally important aquatic species.
8. State-level water governance typically involves a complex network of agencies and relationships between local, county, state, federal, and nongovernment actors.
9. In response to changes in water supply and demand, states are actively engaging in water planning processes. These planning processes take different forms in different states but share some common threads.
10. Interstate water management seems to work best when states form **river-basin organizations** around issues of common concern (ranging from flooding to environmental protection) rather than when the federal government imposes top-down river basin planning.

Notes

1. Marston and Cai (2016).
2. Womble and Hanemann (2020a).
3. Schwabe et al. (2020b).
4. Grafton and Horne (2014).
5. Grafton et al. (2012).
6. Debaere and Li (2020).
7. Arellano-Gonzalez et al. (2021).
8. Garrick et al. (2019).
9. To be clear, I am not arguing that claiming unappropriated flows on behalf of the environment would necessarily be a bad thing in many basins, just that states understandably don't want to open up all unappropriated flows to be immediately claimed as instream flows.
10. https://www.latimes.com/california/story/2022-04-15/l-a-gets-less-water-from-mono-lake-due-to-declining-levels.
11. https://www.fs.usda.gov/managing-land/national-forests-grasslands/water-facts.
12. Stone (1972).
13. Constitution of the Republic of Ecuador, 2008. https://pdba.georgetown.edu/Constitutions/Ecuador/english08.htm.
14. Sanchez et al. (2020).
15. https://www.hcn.org/issues/55.7/indigenous-affairs-colorado-river-how-arizona-stands-between-tribes-and-their-water-squeezed.
16. https://www.tribalwateruse.org/?page_id=569; https://apnews.com/article/arizona-science-government-and-politics-business-environment-and-nature-3d27c894e028629e87a14d7a41ddd5db.

17. https://www.circleofblue.org/2020/world/colorado-river-indian-tribes-take-another-step-toward-marketing-valuable-water-in-arizona/.

18. Sanchez et al. (2022).

19. For more on Land Back, see Treuer (2021), https://landback.org/, and #landback.

20. https://www.puebloactionalliance.org/water-back.

21. Kirchhoff and Dilling (2016).

22. Interstate Council on Water Policy (2021).

23. Arguably, point (b)—water infrastructure and land use—should be identified as a means toward a goal, not a goal in itself.

24. Colorado's goal of increased storage has drawn attention from downstream states. In one of the more bizarre developments, Nebraska has allocated funding to begin constructing the $500 million Perkins County Canal, which would transfer water to Nebraska from the South Platte River in Colorado. What makes this bizarre is that (a) Colorado's storage plans on the South Platte—which at this point are just a list of potential projects—would not interfere with Colorado's ability to deliver to Nebraska the water it is required to deliver during the irrigation season under the South Platte Compact; and (b) the canal would be allowed to transfer water only during the off season, and even then it would have low-priority rights under the compact, so it is unlikely to provide any additional water to Nebraska. Nebraska's plan has been widely derided as a political stunt. See https://somachlaw.com/policy-alert/a-reality-check-on-nebraskas-proposed-south-platte-river-canal/ and https://grist.org/politics/nebraska-south-platte-river-canal-colorado/.

25. Garrick and De Stefano (2016) also include a third category: intrastate rivers in a federal country.

26. Molle (2009).

27. Interstate Council on Water Policy (2020).

12 Transboundary Water Management: Conflict and Cooperation

- Are "water wars" real, and is scarcity making them more common?
- What rules govern the allocation of shared water resources between countries?
- What strategies can be used to shift transboundary water interactions from conflict to cooperation?

As a vital and often-scarce resource, water can be the focal point of conflict between countries, states, or user groups, but water can also bring countries and communities together to reap the benefits of cooperation and shared management. This chapter focuses on conflict and cooperation over water, mostly in the context of transboundary waters: rivers and aquifers that cross state or national borders. It continues our discussion in Chapters 10 and 11 of how to share water between groups, but it broadens the lens to include international water bodies.

For several reasons, this larger scale complicates the task of reaching the type of **common-property** management discussed in Chapter 10.[1] First, the larger number (and greater cultural diversity) of participants—at nested scales ranging from individual users to national governments—makes it harder to agree on collective rules. Second, the sovereignty of national governments poses challenges to binding participation in collective action and generally requires unanimous agreement between parties before action can proceed. Third, at these larger scales, water is closely linked to other common-pool resources such as energy and climate, further raising the stakes and complicating the necessary management regime.

Note that the two topics of water conflict or cooperation and transboundary water management are overlapping but not identical; water conflict can take place in non-transboundary situations, and transboundary water management involves not just conflict management but also other governance and infrastructural issues. Still, it is convenient to focus this chapter primarily on the area of overlap—conflict and cooperation over transboundary waters—while also touching on related topics.

We open the chapter with an overview of water conflict and then turn to an analysis of conflict and cooperation in international basins before briefly discussing nonstate water conflicts. Throughout the chapter, we try to understand what drives water conflict and how to achieve more peaceful and equitable transboundary water management. Several case studies are provided on the companion website.

1. Introduction to Water Conflict

Conflict over water is as old as civilization, but the increasing stresses on water systems in modern times (e.g., increased demand, climate change, pollution, flooding) have led to growing concern over this issue.[2] This concern has sometimes found voice in neo-***Malthusian*** "water wars" rhetoric—the idea that "the wars of the twenty-first century will be fought over water"[3]—which, as we will see later in the chapter, is not supported by the data. Still, lower-level water conflicts at a variety of scales are certainly happening.

The Pacific Institute provides a periodically updated Water Conflict Chronology, which covers events at international as well as subnational scales and includes 1,297 events from 3000 BCE to 2022.[4] Most of these events occurred after 1900, mostly because of a paucity of information for earlier time periods. These water-conflict events are classified into three types: water as a trigger of conflict, as a casualty, and as a weapon. The remainder of this chapter will focus mostly on water as a trigger, probably the most common meaning of the phrase "water conflict," but Boxes 12-1 and 12-2 address water as a casualty and weapon, respectively.

Recent years have seen increased attention to water as an issue of national security, part of a larger trend toward identifying environmental issues (e.g., climate change) as posing security threats (e.g., climate refugees). In the United States, a 2012 Intelligence Community Assessment, titled "Global Water Security," identified four water issues as having potential national security implications: internal instability in other countries, international conflict, global food issues, and global economic impacts from scarcity and pollution.[5] A July 2020 memo from the National Intelligence Council voiced similar concerns and predicted increased water conflict in the next two to three decades.[6]

A 2021 National Intelligence Estimate on climate change highlighted water as an area of concern[7]:

- "We judge that transboundary tensions probably will increase over shared surface and groundwater basins as increased weather variability exacerbates preexisting or triggers new water insecurity in many parts of the world." Particular basins identified as worrisome include the Nile, Mekong, and Indus.
- "We judge that cross-border migration probably will increase as climate effects put added stress on internally displaced populations already struggling under poor governance, violent conflict, and environmental degradation. Triggers for increased migration are likely to include droughts, more intense cyclones—with accompanying storm surges—and floods."

What are we to make of the intelligence community's interest in water issues? On one hand, increased attention at the highest levels of government may lead to greater resources being devoted to managing these critical problems. For example, the awareness that global water issues have national security implications played a role in the

> **Box 12-1. Water as a Casualty of War**
>
> Many of the military conflicts of the twenty-first century—from Yemen to Syria to Ukraine—have involved attacks on densely populated urban areas. The stark images of destroyed buildings and cratered roads have a less visible counterpart in the destruction of water and sanitation systems, whether targeted or collateral damage. This destruction leaves civilian populations without access to safe drinking water and sanitation services, potentially leading to cholera epidemics (as in Yemen in 2016–2017) and other waterborne diseases. Effects on children are particularly severe and tragic.[a]
>
> In just the first three months after the February 2022 Russian invasion of Ukraine, at least forty-nine conflict-related incidents affected Ukrainian water systems: fifteen disruptions to treatment plant operations, eight disruptions to water transfers, seven disruptions to centralized water supplies, six incidents of water pollution, six cases of mine flooding, six incidents of damage to dams and hydropower stations, and one case of bacterial pollution.[b] As I wrote in June 2023, the world is still absorbing the devastation being wrought by the destruction of the Kakhovka Dam, including short-term effects (flooding of downstream cities) and longer-term impacts (reductions in power production and water supply).
>
> A startling example of water infrastructure as a casualty of war comes from the US campaign against ISIS in Syria. According to a *New York Times* report from January 2022, the US military attacked the ISIS-controlled Tabqa Dam on the Euphrates River on March 26, 2017. The strike—which apparently used three 2,000-pound bombs, including a "bunker buster" that fortunately failed to detonate—violated military guidelines, which had included the dam on a no-strike list, noting that dam failure could potentially kill thousands of civilians downstream. The attack seems to have targeted ISIS forces who were using the dam's towers as a base, but instead they damaged the dam's water release mechanisms, causing reservoir water levels to rise rapidly, which could have led to catastrophic overtopping and potential dam failure. Fortunately, Turkey was able to reduce water releases from upstream dams to mitigate pressure on Tabqa, and a group of engineers was able to open a floodgate the following day and avert catastrophe. *The New York Times* describes the dramatic scene:
>
>> Less than 24 hours after the strikes, American-backed forces, Russian and Syrian officials and the Islamic State coordinated a pause in hostilities. A team of 16 workers—some from the Islamic State, some from the Syrian government, some from American allies—drove to the site, according to the engineer, who was with the group. They worked furiously as the water rose. The distrust and tension were so thick that at points fighters shot into the air. They succeeded in repairing the crane, which eventually allowed the floodgates to open, saving the dam.[c]
>
> **Notes**
> [a] Morris-Iveson et al. (2021).
> [b] Shumilova et al. (2023).
> [c] https://www.nytimes.com/2022/01/20/us/airstrike-us-isis-dam.html.

2014 passage of the Senator Paul Simon Water for the World Act, which expanded the US Agency for International Development's role in improving global water and sanitation access. On the other hand, defining an issue as a national security concern can be used to justify unilateral, unprecedented, and illegal measures, a process called *securitization*. Securitization of water issues does not necessarily bode well for the prospect of reaching equitable and peaceful solutions to international water conflicts.

2. International Basins and Aquifers

We start this section by exploring the concept of international water resources and discussing the international legal framework governing use of these waters. We then

> **Box 12-2. Water as a Weapon of War**
>
> During its campaign to control territory in Iraq, the terrorist group ISIS used water as a weapon in numerous ways[a]:
>
> - ISIS controlled the Fallujah Barrage on the Euphrates River for over two years, from April 2014 to June 2016. On several occasions, ISIS manipulated water flow through the Fallujah Barrage and the nearby Nuaimiyah Dam to flood cropland and Iraqi troop positions or to deprive farmers of irrigation water.
> - In August 2014, ISIS briefly held the 113-m-high Mosul Dam, cutting off water supplies to nearby towns and raising fears that the group would release a flood of water that could threaten Baghdad downstream. Adding to the concerns was the poor state of the dam (sometimes called the most dangerous dam in the world), which needs constant maintenance (grouting of cracks in the **karst** rock underneath the foundation) to avoid catastrophic failure.
> - In a September 2014 battle, ISIS forces reportedly deployed chlorine gas, obtained from the chlorine supply of water treatment plants it had seized.
> - In October 2014, ISIS cut off the electricity supply to water wells in the village of Talkhaneim and extorted payment from local villagers to restore the supply.
>
> These examples illustrate the adage that water is power, especially in arid regions, and highlight the vulnerability of water infrastructure to terrorism.
>
> The vulnerability of water infrastructure was also demonstrated by a February 2021 hacking attack on the drinking water treatment plant in Oldsmar, Florida (population 15,000). Using the system's remote access software, an unknown cyberintruder changed the concentration of sodium hydroxide from 100 ppm to 11,100 ppm, a change that would have made the water highly caustic. Fortunately, the hacking was detected and no harm was done, but this event (and others like it) should serve as a warning call to all water utilities, especially small ones that may not have the resources to pay much attention to cyber security. Computerized, automated control systems have dramatically changed the operations of water utilities, allowing fewer staff to operate treatment plants more efficiently but also increasing vulnerability to malicious or accidental software malfunctions. "Cyber hygiene" services, provided by the federal Cybersecurity and Infrastructure Security Agency at no cost,[b] can help water and wastewater utilities reduce their vulnerability. As of March 2023, the Environmental Protection Agency requires water utilities to include cybersecurity assessments in their periodic sanitary surveys (or an alternative mechanism).
>
> **Notes**
> [a]Strategic Foresight Group (2014).
> [b]https://www.cisa.gov/cyber-hygiene-services.

turn to the central questions of this chapter: Does water tend to drive conflict or cooperation (or both), and how can we shift water interactions toward more peaceful and equitable paths?

2.1. The Extent of Transboundary Waters

International basins—***river basins*** that span more than one country—are estimated to occupy 47 percent of the world's land area (excluding Antarctica) and house 52 percent of the world's population.[8] Some international basins are shared by only two ***riparian*** countries (e.g., the Colorado is shared between the United States and Mexico), while others are shared by multiple nations (up to nineteen in the case of the Danube). In some cases, the basin is shared evenly between riparians, and in others most of the basin is in one country. For example, the Mississippi River Basin is arguably an international basin in name only, given that only 1.6 percent of its area is in Canada.

We can distinguish several patterns in how rivers flow between countries, which may produce different types of geopolitical dynamics. The two simplest arrangements are as follows:

- **Successive (border-crossing)**: The river flows from an upstream country to a downstream country.
- **Contiguous (border-defining)**: The river serves as the border between two countries. Rivers are convenient borders—approximately 23 percent of international borders (excluding oceans) are large rivers[9]—but from a watershed perspective, those borders split the basin in half.

A convenient unit of analysis for international basins is the basin–country unit (BCU). For example, the Colorado–US BCU is the area of the Colorado River Basin that lies within the United States, and the Colorado–Mexico BCU is the portion of the basin that is in Mexico. Although a focus on international basins and BCUs can be useful, we should note two shortcomings to this approach:

- It implies that the only relevant players in water conflict are nation-states. In fact, there may be nonstate actors that are of equal importance, whether those are insurgency groups, nongovernment organizations, or multinational corporations. In addition, countries are, of course, not unitary actors; a country's behavior in international relations is inevitably driven by internal politics.
- It implies that the relevant geographic unit is the basin. In fact, there are multiple situations in which the relevant "problem-shed" is not identical to the hydrologic basin. This is especially true at the international scale, where countries may share multiple basins (and other transboundary resources) and may find it easier to resolve problems by expanding the frame beyond the individual basin.

Just like surface water, groundwater can also be a transboundary resource, although these systems are much more complex and poorly understood than their surface water counterparts. In addition, there is arguably a spectrum of transboundary character, depending on how much (and how quickly) groundwater pumping in one country affects water levels in another country—which depends on complex hydrogeologic characteristics as well as pumping locations and rates.

Despite these definitional and methodological issues, the International Groundwater Resources Assessment Centre has mapped over 300 transboundary aquifers, involving most countries on the planet.[10] The UN's Transboundary Waters Assessment Programme Groundwater (TWAP Groundwater), which focused on the 199 largest aquifers, found that transboundary aquifers were particularly important in Africa (40 percent of the land area, 30 percent of the population) and South America (39 percent and 21 percent, respectively).[11] TWAP Groundwater also found that most groundwater depletion is currently taking place in domestic, not transboundary aquifers, although they also identified some hotspots of potential international conflict over groundwater.[12]

2.2. International Law of Transboundary Waters

There are many situations in which countries must figure out how to share transboundary water resources. What guidance does international law give on how to do so? Compared with the well-defined water allocation doctrines we discussed in Chapter 10, we are dealing here with a vaguer set of legal principles governing relations between sovereign states. And, of course, enforcement of international law is weak at best. Still, there is a body of international water law that has developed over the past century or so,[13] mostly in the form of customary law, periodically summarized in resolutions by international legal groups such as the International Law Association, whose Helsinki Rules of 1996 and Berlin Rules of 2004 have been particularly influential.

In 1997, drawing on the Helsinki Rules, the UN adopted the Convention on the Law of the Non-Navigational Uses of International Watercourses (the **UN Watercourses Convention**). The convention entered into force in 2014, when Vietnam became the thirty-fifth country to ratify it; as of 2022, only two additional countries (i.e., thirty-seven in total) have ratified the convention. Thus, twenty-five years after its initial adoption, the convention is still not binding on most countries. Still, case law in the International Court of Justice largely reflects the same principles as the convention.

Several key provisions are articulated by the convention:

- **Equitable and reasonable utilization** (Article 5). This is the cornerstone of international allocation: Each riparian has a right to use the waters of the basin, but it's not an absolute right. Riparians must use water in ways that are equitable and reasonable relative to other users, reflecting the notion of "limited territorial sovereignty" (Box 12-3). Of course, defining what is equitable and reasonable is tricky; Article 6 of the convention provides a list of factors to be considered.[14] Broadly speaking, reasonableness can be thought of as a test of the use itself (is it appropriate to the setting? is it wasteful?), while equity can be thought of as a test of the use relative to the other riparians (does the balance of uses between riparians reflect a fair outcome?).

- **Equitable and reasonable participation; the duty to cooperate** (Articles 5 and 8). In order to achieve an equitable and reasonable outcome, the convention calls for each riparian to participate with the others in managing the basin and even articulates (in Article 8) an "obligation to cooperate." Similarly, Article 3 calls for the creation of treaties for individual basins that "adjust the provisions of the present Convention to the characteristics" of the basin.

- **No harm** (Article 7). The principle that one country's use should not cause harm to another country is in tension with the principle of equitable and reasonable use, since some degree of cross-border harm will often occur when a country develops its water resources (e.g., when an upstream country starts diverting some amount of water for its own use). The convention does not forbid the causing

> **Box 12-3. Limited Territorial Sovereignty**
>
> As a fugitive resource, water challenges notions of state sovereignty: What does it mean to own a river when its water flows into another country? This conundrum is reflected in the variety of historical doctrines for water allocation on international rivers. Upstream countries often invoke the doctrine of ***absolute sovereignty***,[a] in which a country has complete right to the flow of any rivers within its borders. Downstream countries often invoke the doctrine of ***absolute river integrity***, which states that a downstream riparian has the right to receive the natural flow into the country from upstream, undiminished in quantity or quality. Clearly, these doctrines are incompatible, and international law has settled on an intermediate position, called ***limited territorial sovereignty***, in which all riparians have limited rights of use.
>
> **Note**
> [a]This is often called the Harmon doctrine, named for US attorney general Judson Harmon, who defended the rights of farmers in the New Mexico Territory to deplete the flow of the Rio Grande, to the harm of downstream farmers in Mexico.

of harm but requires countries to "take all appropriate measures to prevent the causing of significant harm" and, where significant harm is caused, to "discuss the question of compensation."

- **Data sharing** (Article 9). Some degree of joint data availability is crucial to successfully manage an international river. Data sharing promotes technical collaboration and allows countries to operate from a common understanding of reality.
- **Notification and consultation** (Articles 11–19). The convention emphasizes the importance of avoiding unilateral actions by establishing procedures for countries to notify and consult on planned changes (e.g., dam construction).
- **Environmental protection** (Articles 20–23). The convention urges riparians to individually and jointly act to protect the environmental values of international rivers.
- **Dispute resolution** (Article 33). Successful cooperation requires some mechanism to peacefully resolve the disputes that do arise. The convention provides detailed guidance on third-party mediation of disputes.

2.3. Water Wars or Water Peace?

Predictions of future water conflict—even wars—have been a media staple over the last few decades. Thoughtful scholars, too, have identified a "new geography of conflict"[15] in which the central driver of international conflict is not ideological differences but scarcity of water and other essential resources. But are these predictions justified? What can analysis of international interactions over water teach us about the prevalence of water conflict and cooperation?

One set of answers to these questions has emerged from Aaron Wolf and colleagues at Oregon State University (OSU), who have used the "event data" approach to compile a database of international interactions over water,[16] with each event coded along a fifteen-point scale (known as the Basins at Risk scale) from highly conflictual to highly cooperative. For the period 1948–2008, 66 percent of events showed some degree of cooperation, and only 29 percent of events were conflictual. Almost all the conflictual events were minor, mostly ranked as –1 or –2. Of course, just because there have been no water wars in the last sixty years doesn't mean there won't be any in the future. But the track record of cooperation over water is heartening.

The OSU approach operates on the assumption that conflict and cooperation are opposites and that international interactions can be neatly coded into one or the other category. Other scholars have suggested that conflict and cooperation are not mutually exclusive and indeed often come together. Or, to put it another way, if there are no tensions over shared water, there is no need to cooperate. The Transboundary Water Interaction Nexus operationalizes this insight by using a matrix in which bilateral relations over water are categorized by their level of conflict *and* their level of cooperation.

Another critique of the OSU approach is that the absence of conflict over water does not mean that water is being shared fairly; it may instead indicate that a powerful country is taking more than its share of water and using its power to suppress dissent. Mark Zeitoun and colleagues have articulated a *hydro-hegemony* framework, which analyzes power asymmetries and how they affect transboundary water management.[17] In this framework, regional hegemons—countries with some combination of economic, political, and military power, favorable position in the basin (i.e., upstream), and advanced technological capacity to exploit water resources—exert their power in a variety of ways, potentially leading them to consolidate control over water and reinforce their dominant position. Alternatively, hegemons may also show positive leadership and bring the basin to a position of shared control and equitable cooperation. However, at least in the Middle East (Nile, Jordan, and Tigris/Euphrates basins), the hydro-hegemony analysis suggests that hegemons (respectively, Egypt, Israel, and Turkey) have tended toward domination, with "power asymmetries influencing an inequitable outcome—at the expense of lingering, low-intensity conflicts."[18]

2.4. Drivers of Water Conflict

Shared water resources can clearly be the focus of both conflict and cooperation. What factors tend to drive particular situations toward positive or negative outcomes? Many scholars have tried to isolate the physical, political, social, and economic factors that lead to conflict or cooperation, often with contradictory results. A review article summarizes twelve factors that have been identified as important (Table 12-1).[19]

One of the best-studied drivers of water conflict has been water scarcity (reflected in Table 12-1 as "precipitation"), since scarcity forms the basis of the neo-Malthusian "water wars" argument discussed above. Here, too, results have not been totally consistent. For example, one study found, "Everything else being equal, a river-sharing dyad in which at least one member suffers from water scarcity has a 41% higher risk" of a militarized dispute,[20] while another study found an inverted-*U*-shaped relationship between scarcity and cooperation, with cooperation most common at intermediate levels of scarcity.[21]

The upstream–downstream dynamic (Table 12-1) can be extremely important. The asymmetry created by the directional flow of water and pollutants—in which the downstream country is affected by the upstream country's actions but not vice versa—tends to create "malign" problems, where the interests of the countries are fundamentally at odds, as opposed to "benign" ones, where coordination can benefit

Table 12-1. Summary of variables identified by Bernauer and Böhmelt (2020) as affecting the likelihood of conflict. In their analysis, trends for cooperation are considered to be the opposite of trends for conflict.

Variable	*Description*
Territory in basin	Basins that are dominated by one country have less conflict.
External water dependence	Countries with higher dependence on upstream countries for their water supply have more conflict.
Precipitation	Higher precipitation leads to less conflict.
Income levels	Richer countries have less conflict.
Population	Basins with larger populations have more conflict.
Form of government	Country pairs have less conflict when both are democracies.
Security alliance ties	Country pairs with existing security ties have less conflict.
Intergovernmental organization memberships	Country pairs that are in the same intergovernmental organization have less conflict.
Legal system similarity	Country pairs with similar legal systems have less conflict.
Distance and contiguity	Country pairs that are closer and share a longer boundary have more conflict.
Number of riparian countries	Basins with more riparians have more conflict.
Upstream–downstream configuration	The "successive" arrangement leads to more conflict than "contiguous" or other arrangements.

both countries.[22] In some situations, the advantages of the downstream position—access to deep-water ports and seawater for desalination—can help balance the scales.

Construction of new water infrastructure, especially large dams, can disrupt the status quo and create hydropolitical tensions that can lead to water conflict. Scholars have identified certain factors that are likely to exacerbate these tensions and other factors—often referred to collectively as ***institutional resilience***—that are likely to ameliorate them (Table 12-2). Climate change and population growth can contribute to rising risk by increasing flow variability and decreasing per capita water availability

Table 12-2. Exacerbating and ameliorating factors for hydropolitical tensions, from de Bruin et al. (2023).[a] Each factor is calculated at the scale of the basin–country unit (BCU).

Exacerbating Factors
Flow variability is high.
Per capita water availability is low.
The country is highly dependent on water from the basin.
Educational levels are low.
Governance is weak.
The country has low per capita income.

Ameliorating Factors (Institutional Resilience)
The BCU is involved in at least one water treaty.
The treaty has a conflict resolution mechanism.
The treaty has a mechanism for managing flow variability.
The treaty incorporates the principles of "no significant harm" and "equitable and reasonable utilization."
The BCU is involved in a river basin organization.

[a]This study builds on the framework from De Stefano et al. (2017).

(Table 12-2). A recent analysis using this framework projected that by 2050, under a business-as-usual scenario, over 500 million people will live in BCUs with a very high risk of conflict, but this number could be reduced to 7 million with a high-ambition risk-reduction program involving improved governance, higher water-use efficiency, and transboundary dam management.[23]

2.5. Building Institutional Resilience

What factors make some international basins more resilient to the stresses posed by dam construction and climate change? And what can we do to shift basins toward more resilient, cooperative, and equitable solutions? Table 12-2 has prepared us for this conversation by identifying two key components of institutional resilience: effective treaties and **_river-basin organizations (RBOs)_**.

Although institutional resilience is often identified with the presence of regional water treaties, not all treaties create resilient institutions. For example, the Mekong River Commission—created in 1995 by a treaty between Cambodia, Laos, Thailand, and Vietnam—has been described as "an amputated RBO with its tributaries pruned and its headwaters lopped" because upstream China is not party to the treaty, and the commission has little power to affect the building of dams on tributaries.[24]

A 2014 analysis found 250 treaties in 113 transboundary basins, covering some 70 percent of the area of transboundary basins.[25] However, these treaties vary widely in their specificity and effectiveness. Only a minority of treaties include all the riparians within the basin. Groundwater is mentioned in only 14 percent of treaties, although recent treaties are more likely to include it. One of the few treaties focused specifically on groundwater is the 2010 Guaraní Aquifer Agreement between Argentina, Brazil, Paraguay, and Uruguay, which entered into force in 2018.

Treaties are more likely to be effective if they include certain features:

- Equitable and clear allocation mechanisms that explicitly address variability in water availability;
- The creation of a joint institution (RBO) with sufficient funding and decision-making power;
- Mechanisms for enforcement and conflict resolution;
- Provisions for data collection and sharing;
- Inclusion of all riparians;
- Provisions for adjusting the treaty over time.

You may notice that many of these design features have been incorporated into the Watercourses Convention (Section 2.2).

How do we move the needle toward effective treaties and resilient cooperative institutions? Several strategies have been identified.

Side payments. Side payments are a recognized strategy in game theory, in which player A offers player B compensation for playing in a way that benefits player A. Assuming that the value to player A of the favorable game outcome is higher than the value of the side payment, everyone is better off. For a water-specific example of a side payment, we can look to the Rhine River, where the Netherlands, as the downstream country, has been suffering for many years from high chloride inputs from upstream countries, especially France and Germany. Solving this problem ultimately required (among other measures) a side payment in the form of Dutch financial contributions to chloride abatement in upstream countries, despite the polluter-pays principle.

Issue linkage. Linking a transboundary water conflict to unrelated issues between the same countries can change the leverage points and help reveal ways that both sides can benefit from cooperation. For example, in the Aral Sea Basin, the upstream countries are water-rich but poor in energy resources (other than hydropower), while the opposite is true for the downstream countries. For the upstream countries, it makes sense to release water from reservoirs in the winter in order to generate hydropower for home heating, but the downstream countries would prefer that water be stored in the winter and released in the summer for irrigation. The obvious solution is to link the water and energy issues, with the downstream countries exporting cheap energy to the upstream countries in the winter in exchange for summer water releases. When all five countries were part of the Soviet Union, these transactions were centrally managed, but since then the countries have been negotiating these relationships as independent states, with periods of greater and lesser success.

Benefit sharing. In many water situations, the total benefits of cooperation[26] (summed across all countries involved) are larger than the total benefits of uncoordinated action; thus, even if some countries will lose out from cooperation (e.g., having less water available for their own use), redistribution of benefits can create a win for all countries. There are several reasons that cooperation is often economically beneficial. First, integrated management of water infrastructure allows more efficient water storage, delivery, and use, effectively expanding the pie of available water; in the Aral Sea Basin, for example, transboundary cooperation in the timing and volume of water delivery could increase agricultural output by about $1.75 billion per year.[27] Second, cooperation allows countries to undertake joint infrastructure projects to make better use of their shared waters and often attract outside investment in these projects, although we should be careful to avoid the trap of building environmentally damaging megaprojects purely to increase cooperation. Third, conflict has its own costs to the countries involved, whether military costs or forgone opportunities for cooperation in other areas.

From rights to needs. Aaron Wolf has argued that successful water negotiations move from conflicting claims based on *rights* to water to recognition of mutual *needs* for water.[28] As noted above, rights in transboundary waters are inherently disputed and relative: Fixed rules, such as prior appropriation or proportional sharing based on land area or flow contribution, simply don't make sense in the international context. Needs, on the other hand, can provide some guidance: Given the inherent equality of all people, there is a logic to allocating water based on the number of people dependent on the resource or the amount of irrigable acreage.

Technical cooperation. Cooperation at the technical level (e.g., data collection) between water managers in different countries can be a first step toward building trust and establishing working relationships that can be maintained even when higher political echelons are saber-rattling for their own reasons. A classic example of technical cooperation is provided by the "picnic table talks" between Israeli and Jordanian water managers, who met periodically, even when their two countries were officially at war, to coordinate management of the Jordan River.

3. Nonstate Water Conflicts

Conflicts over water occur not just between nations but also between groups within a country and between local groups and transnational corporations. In contrast to the rich research on international water interactions—which draws on strong foundations in international relations and geopolitics—nonstate water conflicts are relatively understudied. This section will briefly discuss some types of nonstate water conflicts.

Farmer–herder conflicts. Conflicts between agriculturalists and pastoralists go back to ancient times, as reflected in the biblical story of Cain and Abel. In modern times, these tensions are exacerbated by population growth and climate change, both of which can destabilize traditional patterns of coexistence between different groups. Although many factors play into these conflicts—religious and tribal identities, land access, historical injustices, the mismatch between traditional territories and modern borders—disputes over water often play a role. In particular, droughts and upstream water diversions tend to draw different groups into closer proximity around the remaining viable water sources and grazing lands. In Nigeria, for example, nomadic Fulani herders have been migrating farther south in search of water, bringing them into conflict with local farmers; eighty-three people were killed in farmer–herder violence in April 2021.[29] Water management disputes between ranchers and fishers can also descend into violence, as in northern Cameroon in December 2021.[30]

Antigovernment protests. When a government (or utility) fails to provide its citizens with the water solutions they want and need, protests can erupt that may escalate into violence, especially when the government responds aggressively. In 1999–2000, the famous "water war" in Cochabamba, Bolivia, brought masses of people out in protest against water privatization, leading to the death of a protestor shot by police. In 2018, citizens of Basra, Iraq, took to the streets to protest poor water service after more than 100,000 people were hospitalized due to poor drinking-water quality; government buildings were burned, and twenty protestors were killed.[31] In 2020, protests by Mexican farmers against water deliveries to the United States from the La Boquilla Dam resulted in two deaths, illustrating the role that international arrangements can play in internal tensions. In 2021, cities and villages around Iran saw fatal protests centered on the water shortages that have affected households, farmers, and power availability. Many observers believe that water issues played a role in initiating the Syrian Civil War, as drought-stressed farmers, angry over the government's water management, migrated into cities (Chapter 3).[32] All these examples demonstrate that providing water is central to any government's responsibilities, and failure to provide water can make citizens question the government's legitimacy.

Water protectors. Proposals for new hard-path projects—especially dams, mines, and oil pipelines—have often been the impetus for local communities to organize to protect their waters, lands, and livelihoods. Despite the almost uniformly nonviolent nature of these "water protectors," they are often met by physical violence and other forms of intimidation. These fights are often one-sided—the national and corporate interests that will benefit from these projects have much greater political, legal, and financial power than local communities—but the passion of local leaders and their international allies has sometimes prevailed. One notable example is the success of antimining organizers in El Salvador, whose efforts led to the world's first countrywide ban on metal mining, passed in 2017.[33] However, Marcelo Ramirez, an early leader of the antimining coalition in northern El Salvador, was tortured and murdered in 2009, and although the government officially blamed gangs, many suspect that promining interests were involved. In Honduras, indigenous leader Berta Cáceres was murdered in 2016 after she led a successful campaign against dams on the Gualcarque River. The organization Global Witness, which has been tracking killings of environmental defenders since 2012, documented 227 such killings in 2020, with more than half taking place in Colombia, Mexico, and the Philippines.[34] In North America, antipipeline movements led by Indigenous peoples have often been met by governmental and corporate intimidation and violence.

Western US water conflicts. In the American West, the (re-)allocation of water between consumptive and environmental uses often provokes strong feelings, leading to what are sometimes referred to simplistically as fish–farmer conflicts, even though they involve complex values and identities. These conflicts often take place in the context of the larger urban–rural divide, with liberal urban environmentalists pitted against farmers and ranchers. Representatives Cathy McMorris Rodgers and Dan Newhouse, both Republicans from Washington State, expressed this divide in their response to a 2019 study on removing the Snake River dams, "This privately-funded study is a slap in the face of our state's agricultural economy. It is another example of Seattle-based interests seeking to disrupt our way of life in Central and Eastern Washington."[35] In reality, the diverse stakeholders in most water conflicts in the West include not just environmentalists and farmers but also tribes, power users, urban water utilities, barge companies, and recreationalists. Resolving these complex conflicts requires patience, creativity, time, and a commitment to listening and building relationships.

Chapter Highlights
1. Water can be a trigger, casualty, or weapon of conflict.
2. Where river basins cross international boundaries, they pose a challenge to national sovereignty and an opportunity for cooperative management.
3. There are no clear allocation criteria for international basins, but the ***UN Watercourses Convention*** includes the principle of ***limited territorial sovereignty***, reflecting the tension between the right to use and the duty not to cause harm.
4. Countries have both positive and negative interactions over shared waters, but cooperation is generally more common than conflict.
5. Factors thought to increase the likelihood of international conflict over water include rapid changes (e.g., dam construction, climate change), water scarcity, groundwater depletion, upstream–downstream arrangements, and other factors.
6. Factors thought to calm conflict include joint institutions (river-basin organizations) and treaties with clear but flexible allocations and dispute resolution mechanisms. Not all water treaties and joint institutions are effective or fair.
7. Strategies to move toward resilient cooperation include side payments, issue linkage, benefit sharing, technical cooperation, and a focus on needs.
8. Nonstate actors—from terrorist groups to international nongovernment organizations to multinational corporations to water protectors—are often engaged in water-related conflict.

Notes

1. Ostrom et al. (1999).
2. Gleick and Iceland (2018).
3. This phrase originates from a 1995 speech by Ismail Serageldin, in which he said, "The wars of this century have been on oil, and the wars of the next century will be on water . . . unless we change the way we manage water."
4. https://www.worldwater.org/water-conflict/.
5. Office of the Director of National Intelligence (2012).
6. National Intelligence Council (2020).
7. National Intelligence Council (2021).
8. McCracken and Wolf (2019). Previous analyses also calculated that international rivers had 60 percent of global river flow (e.g., Wolf 2007). The areas identified as international basins have changed somewhat over time, because of political change (e.g., the dissolution of the Soviet Union) and methodological issues (e.g., the resolution with which basin boundaries are identified).
9. Popelka and Smith (2020).
10. https://ggis.un-igrac.org/view/tba.
11. https://www.un-igrac.org/news/population-and-areal-statistics-199-transboundary-aquifers.
12. UNESCO-IHP and UNEP (2016).
13. Interest in shared use of rivers for navigational purposes goes back farther than the nonnavigational uses discussed here. In Europe, the Act of the Congress of Vienna in 1815 ensured freedom of navigation on shared rivers for all riparian states. The first resolution to deal with nonnavigational uses was the 1911 Madrid Declaration of the Institute of International Law.
14. The seven factors listed in article 6 are "(a) Geographic, hydrographic, hydrological, climatic, ecological and other factors of a natural character; (b) The social and economic needs of the watercourse States concerned; (c) The population dependent on the watercourse in each watercourse State; (d) The effects of the use or uses of the watercourses in one watercourse State on other watercourse States; (e) Existing and potential uses of the watercourse; (f) Conservation, protection, development and economy of use of the water resources of the watercourse and the costs of measures taken to that effect; (g) The availability of alternatives, of comparable value, to a particular planned or existing use." Note again the competing principles we discussed in Chapter 10: efficiency, equity, stability, adaptability, and sustainability.
15. Klare (2001).
16. https://transboundarywaters.science.oregonstate.edu/content/international-water-event-database.
17. Zeitoun and Warner (2006), Warner et al. (2017).
18. Zeitoun and Warner (2006).
19. Bernauer and Böhmelt (2020).
20. Furlong et al. (2006).
21. Dinar and Dinar (2017).
22. Dieperink (2011) articulates the malign–benign distinction in the water context. Young (2011) argues that the malign–benign distinction is not a good predictor of the effectiveness of international environmental governance.
23. de Bruin et al. (2023).
24. Lebel and Garden (2008).
25. Giordano et al. (2014).
26. Pohl et al. (2017) prefer the phrase "costs of non-cooperation" to "benefits of cooperation," arguing that human psychology sees costs as more tangible than benefits.
27. Pohl et al. (2017). Cooperation in Aral Basin reservoir management would also reduce damage from floods and mudslides, but those benefits have not been quantified.

28. Wolf (2007).
29. https://www.economist.com/middle-east-and-africa/2021/05/22/a-nigerian-plan-to-reconcile-farmers-and-herders-is-not-working.
30. https://www.voanews.com/a/cameroon-says-new-clashes-kill-at-least-10-displace-hundreds-/6342586.html.
31. Human Rights Watch (2019).
32. See, for example, Gleick (2014).
33. This story is told in powerful detail in Broad and Cavanagh (2021).
34. https://www.globalwitness.org/en/campaigns/environmental-activists/last-line-defence/.
35. https://mcmorris.house.gov/posts/mcmorris-rodgers-newhouse-reject-flawed-privately-funded-dam-study.

Part IV
Offstream Water Use: Cities and Farms, Mines and Factories

The final six chapters of the book examine how water is used in various contexts, from households to farms. Our goal is to understand how offstream water use is currently managed while also examining emerging trends that might move us toward a more just and sustainable future. We explore questions related to water quantity (how can we decrease use and increase supply to alleviate scarcity?), water quality (how can we provide users with safe water and ensure that their use doesn't contaminate water for others?), infrastructure (how well is our current water infrastructure working?), water governance (how are drinking water quality and wastewater discharges regulated?), and funding (how do we balance conservation incentives with affordability?).

Chapter 13 returns to our earlier theme of scarcity as a mismatch of supply and demand, examining potential supply-side solutions that might be less harmful than dams and aqueducts: desalination, wastewater reuse, water harvesting, and aquifer storage. Demand management does not appear in Chapter 13 but features prominently in later chapters.

Chapter 14 is the first of three chapters on household and urban water use, and it focuses primarily on drinking-water quality, including a critical evaluation of water supply and sanitation systems in the United States. Chapter 15 deals with urban water-management issues, including water-supply reliability, water conservation, pricing, and stormwater management. Chapter 15 also explores the One Water concept, in which different urban water flows (drinking water, wastewater, stormwater, rainwater) are managed holistically to simultaneously alleviate scarcity, flooding, and pollution. Chapter 16 turns to low- and middle-income countries, exploring the successes and failures of the global development agenda around WASH (water, sanitation, and hygiene); we use the Sustainable Development Goals as our basic framework for this discussion.

Chapters 17 and 18 deal, respectively, with industrial and agricultural water management. In Chapter 17, we examine water quantity and quality issues associated with energy production, mining, and manufacturing, and turn a critical eye toward corporate water stewardship. Chapter 18 explores how the largest user of water—food production—affects water quantity and quality and how better agricultural water management (and better diet choices) could help improve the sustainability and productivity of agriculture more broadly.

13 Beyond Dams: Old and New Solutions for Water Supply

- Are there alternatives to dams and aqueducts for water supply?
- Will desalination solve all our water problems?
- Can we safely reuse our wastewater to get more mileage out of a limited supply?
- Can we "harvest" water in sustainable ways by using traditional and modern technologies?

Previous chapters have presented two stark realities that together pose one of the central dilemmas of modern water management:

- Increased demand for water, combined with more frequent droughts driven by global change, is driving water scarcity in many regions (Chapter 5).
- The standard hard-path approach for dealing with scarcity—building dams and aqueducts to modify the spatiotemporal availability of water—results in often unacceptable social and environmental impacts (Chapter 9).

Does rising scarcity mean we have no choice but to build more large dams? Do we have to choose between healthy rivers and a reliable water supply? Or are there other ways to manage scarcity without building more hard infrastructure?

The solutions to scarcity include both increases in supply and decreases in demand. Ways to decrease demand in each water-using sector will be discussed in upcoming chapters; this chapter focuses on innovative and traditional nondam alternatives for creating resilient water supplies.

The first four sections of this chapter cover the pros and cons of four types of water supply: desalination, wastewater reuse, water harvesting, and aquifer storage. Each of these technologies will also come up in more specific contexts in later chapters.[1] The chapter closes with an analysis of energy requirements for water supply and a comparison of the four supply options.

1. Desalination

Desalination—the removal of salts from saltwater to produce freshwater—has generated a great deal of interest, with its promise of a limitless, reliable supply of water. Will desalination be the silver bullet that solves our water scarcity problems?

The idea of desalination goes back to ancient times, but given its high energy requirements, it was only with the ready availability of fossil fuels in the modern era that desalination was practiced at significant scale. The first generation of modern desalination plants used distillation, also called thermal desalination, in which water is evaporated by heating under vacuum and then condensed; several stages of evaporation–condensation lead to freshwater. Most newer desalination plants use the more efficient ***reverse osmosis (RO)*** process, in which water is forced under pressure through semipermeable salt-rejecting membranes. RO is part of a family of rapidly improving ***membrane filtration*** techniques that remove various substances from water depending on the pore sizes used (Table 13-1).

Thanks to these technological improvements, desalination has made remarkable advances over the last several decades. As of 2020, there are over 16,000 operational desalination plants around the world, with a production capacity of about 35 km³/yr (increasing by about 7 percent per year).[2] This represents less than 1 percent of global water withdrawals and about 3% of municipal and industrial withdrawals—statistics that understate the importance of desalination for certain countries, especially water-scarce Middle Eastern countries, which were the first to move into this space and which continue to rely heavily on desalination. As of 2018, the United States has about 400 desalination plants in thirty-five states, producing about 1.9 km³/yr of drinking water; Florida, Texas, and California account for most of these facilities.[3] San Diego's Carlsbad plant, which opened in 2015 after a three-decade legal battle, produces 70 MCM/yr, making it the largest in the country.[4]

One advantage of desalination is that it is "drought-proof": It frees its users from the vagaries of the hydrologic cycle and the uncertain impacts of climate change. In addition, it sidesteps the ***tragedy of the commons*** (since it is controlled by one actor) and can potentially ease water conflict by alleviating regional scarcity (Box 13-1).

The two big drawbacks of desalination are cost and energy use, both of which have decreased dramatically over the last several decades but are often higher than those of other water supply options (where those exist). The cost of desalinated water can vary significantly depending on the size of the project, the salinity of the source water, the technology used, the local costs of energy and land, and other factors. The high cost of desalinated water means that it generally goes to "high-value" users, particularly cities and industry, rather than farmers, who can't afford to pay high prices for water.[5]

Table 13-1. Membrane filtration processes. Modified from Capodaglio (2021).

Process	Pore Size (mm)	Operating Pressure (bar)	Substances Removed
Microfiltration	0.1–10	0.1–2	Most bacteria
Ultrafiltration	~0.003–0.1	0.2–5	Bacteria, some viruses, proteins, polysaccharides
Nanofiltration	~0.001	5–20	Bacteria, viruses, organic matter, multivalent ions
Reverse osmosis	<0.001	10–100	Bacteria, viruses, organic matter, ions, molecules with molecular weight >200

> **Box 13-1. Desalination and Hydropolitics**
>
> Given that scarcity and variability can drive water conflict (Chapter 12), it is often posited that desalination can reduce conflict by increasing and stabilizing a region's water supply. Desalination is expensive, but so is conflict; once the economic losses from conflict (e.g., decreased trade, increased military spending) reach the cost of desalination, the conflicting actors would do well to stop fighting and just build a desalination plant—if they were rational actors and water had no emotional valence.
>
> Desalination may also open up opportunities for collaboration in the form of jointly managed desalination plants, financing agreements, or energy–water exchanges. One notable development is a 2021 agreement between the United Arab Emirates, Jordan, and Israel, under which the United Arab Emirates (rich in cash and solar technology) will help Jordan (rich in desert land ideal for solar power) build a 600-MW solar power plant, with the renewable energy being sold to Israel, which has ambitious renewable energy goals but little land for solar and wind power. In return, Israel (rich in desalination technology and coastline) will construct a desalination plant that will provide Jordan with 200 MCM/yr of water that it sorely needs. This "Green Blue Deal" was first advanced by the trilateral (Israeli, Jordanian, and Palestinian) environmental nongovernment organization EcoPeace Middle East and represents a success of "track II diplomacy" in advancing both a peace-making and an environmental agenda.
>
> On the other hand, desalination may not always reduce conflict and could even lead to increased conflict over issues such as unilateral construction, brine disposal, or access to newly valuable brackish waters; desalination plants could also present new targets for military or cyberattacks.[a] Desalination can also shake up hydropolitics in potentially unpredictable ways, such as reversing the usual upstream–downstream power dynamics.
>
> **Note**
> [a] Katz (2021).

Desalination's unique combination of drought resilience and high cost means that some desalination plants essentially serve as expensive insurance policies. In response to the Millennium Drought in Australia, several cities—including Adelaide, Gold Coast, Melbourne, and Sydney—built expensive desalination plants that have been in standby mode for most of their lifetime because of the return of wetter conditions.[6] Still, for wealthy cities in drought-prone areas, this insurance premium may be worth paying.

Energy represents about one third to one half of the costs of desalination. Distillation plants use both electrical energy (mostly to generate vacuum) and thermal energy (to evaporate water). RO plants are more energy-efficient but still use about 2–4 kWh of electricity per square meter of water produced, mostly to pressurize the seawater. This is a significant reduction from earlier numbers and is approaching the theoretical thermodynamic limit of about 1 kWh/m^3. Desalination of brackish water (water with a salinity between that of seawater and freshwater) uses less energy and is often cheaper.

Along with the high energy consumption of desalination comes a correspondingly high production of greenhouse gases. The carbon footprint of a cubic meter of desalinated water is estimated to be 0.4–6.7 kg CO_2 for seawater RO[7]; this range is larger than the energy footprint range, in part because of the different energy sources used in different plants (e.g., coal vs. natural gas). Global CO_2 emissions from desalination plants were estimated at 76 Mt in 2016 (0.2 percent of global CO_2 emissions), projected to grow to 218 Mt by 2040.[8]

In response to this conundrum, there is increasing interest in powering desalination plants with renewable energy from the grid or using wind or solar energy to power desalination plants directly. One problem that must be overcome is the mismatch between the intermittency of renewable energy and the need for continuous flow through the desalination plant. In addition, the land footprint needed for solar energy is much larger than the footprint of the desalination plant itself.[9] Another relatively sustainable option is cogeneration of power and water, in which waste heat from electricity generation is used as a heat source for thermal desalination. Emerging technologies may ultimately allow affordable, small-scale passive solar desalination units that could be deployed at the household scale.[10]

Beyond energy use, desalination has other environmental and social drawbacks, although these are generally less significant than those of dams or aqueducts:

- **Intakes**: Like other water intakes in coastal areas, desalination plants can lead to mortality of fish and other organisms that are sucked into the intake pipes. Also, in settings where sewage or oil spills often pollute near-shore waters, intakes may need to be placed farther offshore.
- **Brine disposal**: The other output from a desalination plant, besides freshwater, is a concentrated salt solution, referred to as a ***brine*** (sometimes called a concentrate). The volume of the brine exceeds the volume of freshwater produced by about 50 percent[11]; this discharge also contains various chemical additives and cleaning solutions. For coastal desalination plants, discharge back into the ocean is usually an acceptable solution, especially if the discharge is far enough offshore that it is rapidly diluted; still, monitoring to confirm minimal effects on aquatic life should be required. For inland facilities (e.g., those desalinating brackish groundwater), disposal options are more problematic.
- **Boron toxicity**: Boron, which is found in seawater at concentrations of 4–7 mg/L, is not fully removed by desalination, and some desalinated waters may exceed the World Health Organization drinking-water guideline of 0.5 mg/L. Many newer facilities use specific processes to ensure adequate boron removal, which can add significantly to the expense.
- **Low ion concentrations**: Ironically, desalinated water may be too clean to drink and too pure for our pipes. The low salt content of desalinated water means that it is low in essential nutrients and, in addition, tends to cause pipe corrosion. To solve the latter problem, many desalination plants carry out "remineralization," which typically adds calcium but not magnesium. A study of Israelis drinking desalinated water found an elevated risk of heart disease, probably due to magnesium deficiency.[12] Iodine deficiency is also an emerging health issue associated with drinking desalinated water.

Desalination is far from a silver bullet. Although it can serve as a crucial backstop source for coastal cities, it is still too expensive and energy-intensive to play a large role in global water supply.

2. Wastewater Reuse

A second potential "new" source of water is the reuse of wastewater, especially municipal wastewater.[13] Interest in wastewater reuse stems partly from water scarcity but is also part of a larger movement toward a circular economy, in which once-through resource flows are eschewed and "waste" is recognized as a valuable resource for further use.

There are many variants on water reuse, sometimes referred to by different names. We will use the term *water reclamation* to refer to the treatment process that allows wastewater to be reused by another user; we will call the entire process *water reuse* or *water recycling*. Water reuse can be categorized as centralized or decentralized, direct or indirect, potable or nonpotable, and planned or unplanned, but the most important distinctions have to do with whether the wastewater is adequately treated. Figure 13-1 provides a visual guide to several common centralized reuse scenarios that we will explore in this section and future chapters; we will discuss decentralized reuse in Chapter 15.

Scenarios A and B in Figure 13-1 involve unplanned, indirect reuse of inadequately treated sewage and carry high levels of risk. Scenario A involves the reuse of wastewater-rich river water by downstream farmers. The benefits to farmers are clear: a constant supply of water containing high levels of plant nutrients. The risks of disease—to farm workers, surrounding communities, and consumers of crops—are also very real. A 2017 study estimates that some 36 Mha of agricultural land (about 17 percent of total irrigated land) is irrigated with river water that contains more

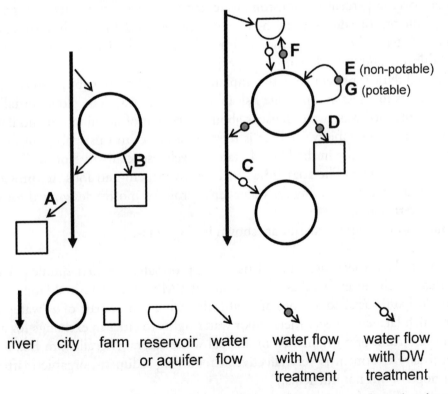

Figure 13-1. Various water reuse scenarios discussed in the text, indicated with capital letters. DW, drinking water; WW, wastewater.

than 20 percent wastewater; 29 Mha was in countries where less than 75 percent of wastewater is treated, suggesting high risk of exposure to untreated (but somewhat diluted) wastewater.[14]

Scenario B also involves farmers using untreated urban wastewater—but this time directly, without dilution, so the risks are even higher. This practice is common in low- and middle-income countries, with about 4–6 Mha globally irrigated directly with untreated wastewater and an additional, unknown amount of agricultural land being used for application of untreated fecal sludge (i.e., undiluted human waste, often emptied from pit latrines).[15]

The drivers behind scenarios A and B include a lack of alternatives for sewage disposal, along with farmer preferences. Farmers who have no other water source and can't afford commercial fertilizer are highly dependent on wastewater and fecal sludge for water and nutrients, and they have been known to oppose construction of wastewater treatment plants for that reason. The World Health Organization has published water quality and treatment guidelines for wastewater reuse in irrigation, but these are rarely met in most low- and middle-income countries. For example, a meta-analysis found that farm workers using untreated and partially treated wastewater, and their families, were more likely than their neighbors to suffer from diarrheal diseases and worm infections.[16] Risk of disease is also high for consumers of produce (especially raw produce) irrigated with poorly treated wastewater.[17]

Scenario C is another form of unplanned indirect reuse, in which a downstream city uses river water containing a significant fraction of treated sewage from upstream communities. This type of reuse is common in the United States and elsewhere; for example, over 14 percent of Houston's water supply consists of treated sewage.[18] The safety of the practice depends on the level of treatment used for both wastewater and drinking water and on the distance and travel time in the river, which affords additional dilution and breakdown.

Planned reuse is preferable to the unplanned reuse discussed above, since it allows policymakers to assess and manage risk in a systematic way. The global installed capacity for planned wastewater reuse is about twice that of desalination and has been growing even more rapidly.[19] China is the largest player in this space, but reuse has been increasing rapidly in the United States as well, especially in Florida, Texas, Arizona, and California. The 2021 Infrastructure Investment and Jobs Act provided $1 billion in new federal funding for water reuse projects, mostly designated for water-scarce western states.

Four planned-reuse scenarios are shown in Figure 13-1:

- Scenario D involves agricultural use of appropriately treated municipal wastewater, which currently takes place on about 1 Mha of agricultural land.[20] Where physical and social conditions are right, this reuse can be part of a "water swap," in which farmers transfer their senior water rights to cities in exchange for a guaranteed stream of nutrient-rich municipal wastewater.[21] (Recall from Box 8-1 that secondary treatment is very effective at removing sediment, organic matter, and pathogens, but not nutrients.)
- Scenario E involves wastewater reuse within a city for nonpotable uses, such as landscape irrigation, street cleaning, sewer flushing, and industrial uses. This

scenario is arguably the most sustainable in terms of involving minimal additional treatment while posing low risk.
- Scenario F involves indirect potable reuse (IPR), in which treated wastewater is discharged to an *aquifer* or reservoir, where it is diluted and undergoes further microbial, chemical, or photochemical breakdown before being withdrawn and sent to a drinking-water treatment plant and incorporated into the city's water supply.
- Scenario G involves direct potable reuse (DPR), in which treated wastewater is used directly as part of a city's drinking-water supply, often after blending with other water sources.

In California (where good data are available because recycled water users are required to report usage to the state), the end uses for recycled water in 2020 consisted of agricultural irrigation (28 percent), landscape irrigation (27 percent), IPR (24 percent), industrial uses (15 percent), and other (6 percent).[22] California does not currently allow DPR, although it is moving in that direction.

These various uses have different water quality requirements and will involve different levels of treatment (and expense). Some uses can directly tap standard secondary-treated wastewater, while others will involve additional (tertiary) treatment. In its 2012 Water Reuse Guidelines, the Environmental Protection Agency refers to this as the "fit for purpose" principle. Although the agency does provide general guidance on water reuse, states are responsible for regulating the practice and are rapidly trying to determine the risk of exposure for different uses and set appropriate standards. The resulting state-to-state variability in definitions and standards illustrates both the differential risk tolerance across states and the large degree of scientific uncertainty associated with these risk assessments.

The most controversial aspect of water reuse has been the idea of potable reuse, which has been disparagingly called "toilet to tap." One notable facility producing potable recycled water is the Groundwater Replenishment System (GWRS) in Orange County, California, which is in the middle of an expansion that will increase its capacity to 130 MGD (180 MCM/yr), driven in part by California's water scarcity problems and in part by the desire to avoid building a second ocean outfall for treated sewage. In the GWRS, effluent from a secondary wastewater treatment plant undergoes three further treatment steps: membrane filtration, reverse osmosis, and a chemical oxidation process using ultraviolet light and hydrogen peroxide. The treated water is then pH-adjusted through removal of excess CO_2 and addition of lime and is used for groundwater recharge and ultimately extracted for water supply (IPR). This multibarrier process is effective at removing salts, nutrients, pathogens, and microcontaminants (Figure 13-2).

What about DPR? Windhoek, Namibia, has been practicing DPR since 1968, with no evidence of harmful effects. In 2013, Big Spring, Texas, opened a $14-million, 2-MGD reclamation plant that treats wastewater via microfiltration, reverse osmosis, and ultraviolet disinfection. This water is drinking-water quality but is delivered (together with raw water) to the town's water treatment plant for further treatment. El Paso is designing a reclamation facility that will bypass the final step, producing drinking water that will go directly into the distribution system.

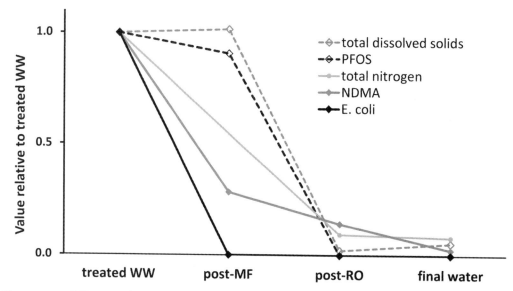

Figure 13-2. Water quality parameters at various stages in water treatment at the GWRS: input water (i.e., secondary-treated wastewater [WW]), after membrane filtration (MF), after reverse osmosis (RO), and final water. Each parameter is scaled relative to its concentration in input water, so all parameters start at 1. *Escherichia coli* is effectively removed by membrane filtration, as are suspended solids (not shown). Total dissolved solids and PFOS (a perfluorinated "forever chemical") are effectively removed by RO. NDMA (an uncharged, small-molecule disinfection byproduct) requires chemical oxidation, as well as the first two steps, for effective removal. Data from Orange County Water District (https://www.ocwd.com/gwrs).

The drawbacks of wastewater reuse fall into five broad categories:

- **Human health**: The health risks of reuse are highly dependent on the level of treatment and the type of reuse. The burden of disease from the informal reuse of untreated wastewater in agriculture is undoubtedly high. Well-regulated potable reuse has very low risk of transmitting infectious disease, but there are legitimate concerns over emerging contaminants in wastewater, especially uncharged small molecules that are not well removed by RO. Still, a well-designed multibarrier treatment system should be able to reduce these risks to very low levels (Figure 13-2).
- **Soil health**: Also worrying in the long run is the effect of wastewater reuse on soils. Studies of Israeli farmland irrigated with wastewater suggest a gradual increase in salts and organic contaminants in the soil.[23] Given that these problems have already emerged in the short time that these areas have been practicing wastewater reuse, are these practices sustainable for any length of time?
- **River health**: When wastewater is reused rather than being discharged into a river, the river's flow may be substantially reduced. Of course, rivers receiving large volumes of wastewater are clearly not pristine in their water quality or their flow regime, but they may still be better off ecologically with the (treated) wastewater than without. More broadly, whenever a new water source is being

evaluated—including wastewater recycling—it is important to think about where that water is coming from and what its value is in situ.
- **Energy use**: Energy requirements for wastewater reuse depend on the level of treatment needed and the distance and elevation that water must be moved. The most energy-intensive option, IPR, can use up to about 2 kWh/m^3 (still lower than desalination) when water is treated by RO, transported to a reservoir or aquifer, and then treated again. DPR uses less energy, which could soon be under 1 kWh/m^3 with some minor technical improvements.[24]
- **Consumer and farmer acceptance**: Water customers can sometimes be strongly opposed to potable reuse, and both consumers and farmers may be hesitant about agricultural reuse.[25] However, increased awareness of scarcity, greater familiarity with water reuse,[26] and a transparent public process can help alleviate public concerns.

3. Water Harvesting

The same basic factor that drives dam building—the need to overcome the temporal variability in water availability—has also, for millennia, driven a variety of smaller-scale approaches to capturing and storing rainwater and runoff, collectively called ***water harvesting (WH)***.

Examples of WH techniques include the collection of rainwater in cisterns or rain barrels, the diversion of hillslope runoff into ponds or infiltration basins, the use of small check dams in wadis to slow runoff, and many others. These approaches differ from large dams in their scale and their underlying logic: The goal of WH is to retain water locally for later use, not to capture and distribute water over a large region. Therefore, WH generally focuses on rainwater or local runoff rather than large rivers.

Two additional characteristics stem from this difference in scale between dams or aqueducts and WH. First, WH differs from dam construction in the tremendous diversity of techniques that are used, depending on local conditions and traditions. Second, WH is less prone to corruption, capture by elites, and imposition of solutions onto local communities; decision making tends to be local and benefits tend to flow to the local community, although the distribution of benefits may mirror and exacerbate preexisting inequities within that community.

WH has two additional benefits besides alleviating scarcity. First, it can reduce flood damages by slowing water down and storing it instead of letting it all run off at once. Second (and related), WH can reduce the environmental impact of stormwater, especially in urban streams, where ***impervious surfaces*** lead to flashy and polluted flows. When the primary goal is ecological protection or flood management rather than scarcity alleviation, these practices are often called stormwater management (Chapter 15) rather than WH, but the logic is the same: slow water down and keep it in the watershed rather than the river.

Given the diversity of WH practices around the world, we turn now to examples of specific practices. WH is most fully developed in countries with highly seasonal water availability, such as India and Peru, where storing water for the dry season is vital.

India: India, perhaps the global center of WH, has a great variety of regionally specific WH practices.[27] Many Indian states have rainwater harvesting requirements, especially for new construction,[28] which typically involve using rooftop drainage to recharge groundwater. In rural areas, WH practices focus on slowing water down as it moves across the landscape, allowing it to accumulate in ponds and recharge groundwater for use in the dry season. The cumulative effect of many small-scale practices is often a wholesale transformation of the landscape into a catchment for WH.

In the southern Indian state of Tamil Nadu, thousands of small reservoirs, called tanks or eris, were constructed in the precolonial era and still dot the landscape. Simple structures formed by crescent-shaped embankments at their downstream end, the reservoirs are designed to capture surface runoff during the monsoon season. On average, each tank irrigates a rice paddy of about 20 ha, although there is large variation around this average and there are other uses not included in this number, such as productive use of tank margins and of the tank itself after it is dry.[29] This system requires extensive coordination between tanks, as well as frequent maintenance (repairs to embankments and sluices, removal of silt from tanks and channels), which is traditionally done collectively through a system known as *kudimaramath*. When responsibilities for maintenance are not clearly delineated and enforced (as is often the case with mandated urban WH systems[30]), effectiveness can drop rapidly.

In the desert state of Rajasthan, one traditional practice that is being revived is the khadin. During the rainy season, runoff from a rocky hillslope is captured in the valley bottom by a low earthen embankment. This water infiltrates into the soil, recharging both soil moisture and downgradient wells. The water stored in the soil within the khadin area can support a dry-season crop of wheat or millet, as well as fruit trees along its upper edge. The khadin also accumulates fine sediment, nutrients, and organic matter, which increases the soil's productivity and water-holding capacity. The khadin captures runoff from a catchment area that is some fifteen to fifty-six times as large as the planting area, so even low rainfall amounts generate enough runoff to grow crops.

Peru: Lima, Peru, on the arid Pacific coast of South America, has a population of almost 10 million people and is the second driest capital city in the world (after Cairo, Egypt). It obtains much of its water from the Andes, including an interbasin diversion from the eastern slopes (Amazon Basin) through a tunnel. There is great interest in supplementing this imported, high-impact water supply with more sustainable local sources. In fact, Peruvian water utilities are required to spend a portion of their revenues on nature-based solutions.[31]

An indigenous WH system was recently restored in Huamantanga, an Andean community (elevation 3,300 m) about 100 km from Lima.[32] The system, which dates to about 1,400 years ago, diverts water from small, seasonal streams and moves it to small ponds and infiltration areas, where it sinks into the soil and ultimately emerges in downhill springs. A dye study found that water takes about seven weeks to travel through the subsurface and reemerge, which allows the community to extend water availability into the dry season. The study calculated that widespread implementation of these techniques would supplement the current use of gray infrastructure (dams and reservoirs) and help mitigate Lima's water scarcity by increasing dry season flows by 7.5 percent.

Urban water managers in Peru are also turning their attention to the water retention functions played by wetlands within the Andean **water towers** that provide most of the water for the arid coast. In particular, peatlands known as *bofedales* appear to be important for regulating downstream water flows and are under threat from climate change, peat harvesting, and other stressors. Bofedales are now being more widely understood as complex socioecological systems that have been managed for centuries by herders, and there is interest in using **payments for ecosystem services** to retain their hydrologic functions for both local and downstream users.

Peru is also famous for its fog catchers, which we turn to now in the context of WH from air.

Vapor capture: The idea of collecting water from air probably has a long history, although the details of that history are poorly understood. Given the large amounts of water in the atmosphere—an order of magnitude more than in rivers—there is renewed interest in capturing that water through a variety of approaches, from traditional and low-tech to innovative and high-tech.

The optimal setting for water collection involves air with relative humidity at or close to 100 percent, where water vapor has condensed into **fog**. Fog fences or **fog catchers** are simple structures with a variety of designs that use a high surface area to mimic the function of cloud forests (Chapter 3). Fog catchers are used in the periurban hillslopes around Lima, Peru, areas that are typically occupied by low-income informal communities without secure land title and with limited water access. During the winter, these hillslopes are shrouded in fog, and a mesh fog catcher ($24\ m^2$) can produce 200–400 L of water per day despite the lack of rainfall.[33] The nongovernment organization Peruanos sin Agua (Peruvians without Water) has been installing hundreds of fog catchers in Lima and elsewhere, in cooperation with local community leaders, who are trained to maintain them and collect funds for ongoing expenses; the fog catchers are generally taken down in the summer to extend their lifetime. One fog catcher is typically shared between two families and provides supplemental water to allow small-scale vegetable farming. While the amount of water is negligible compared with the needs of the city as a whole, these local solutions do provide families with a more reliable and cheaper source than the water trucks that they must otherwise depend on.

Where air is not water-saturated (i.e., relative humidity is less than 100 percent), water capture is more challenging, since water must be condensed from vapor to liquid form. Traditional approaches include simple rock structures meant to provide a cool surface for morning dew condensation, as well as more complex but still passively functioning "air wells."

Recently, there has been growing excitement about active water-from-air devices, with more than thirty companies offering products that can provide high-quality drinking water in remote locations without the need for infrastructure.[34] Despite the hype, a close examination of these technologies suggests that they are too expensive and energy-intensive to be appropriate for most settings. Just because something is labeled as "water harvesting" doesn't mean that it is truly a low-impact, sustainable water solution.

Cloud seeding: For precipitation to occur, it is not enough for air to be saturated with water vapor; that water vapor must condense into precipitable (i.e., large) droplets or ice crystals, a complex process that is determined by temperature, saturation state, cloud type, and the number and type of nuclei on which water molecules can condense. Since at least the 1940s, there has been interest in facilitating this process by injecting various substances (e.g., silver iodide, dry ice, hygroscopic salts) into clouds to increase the precipitation rate, an approach called ***cloud seeding***.

Several countries (and several US states) have ongoing cloud seeding programs, mostly involving silver iodide introduced to clouds by planes, ground-based projectiles, or combustion at the surface; some of these programs are aimed primarily at reducing hail damage (by converting it to smaller crystals), but many aim to increase snow or rainfall. Despite decades of research, it has been hard to definitively prove that cloud seeding is successful, and its efficacy is probably highly variable. However, many agencies that use the practice claim that they are seeing results, and their continued financial commitment suggests that the investment is worth it. A recent study in Idaho used a combination of radar and snow gauges to definitively show an increase in snowfall associated with seeding, although the amounts involved were quite small.[35] Still, cloud seeding is cheap, and increased scarcity is likely to bring more efforts in this direction.

Potential negative effects of cloud seeding include dispersal of toxic silver into the environment, but the amounts of silver involved are probably too small to worry about. Downwind effects—the depletion of moisture from a cloud that would otherwise have dropped precipitation in a downwind state or country—may be a slightly larger concern, if only because they have the potential (justified or not) to lead to conflict.

Although cloud seeding may be too high-tech and invasive to qualify as WH in many people's minds, its low cost and high potential yield make it an approach that managers facing scarcity probably can't afford to ignore.

4. Aquifer Storage

Artificial groundwater recharge has come up repeatedly in the last two sections. Storage of water in aquifers, rather than aboveground in reservoirs, can play a key role in many modern water supply strategies, whether the water being recharged is treated wastewater, harvested rainwater, or a more conventional water source such as a river. Advantages of belowground storage include the absence of evaporation losses, the smaller impacts on aquatic and terrestrial ecosystems, and the small aboveground footprint needed. Importantly, groundwater recharge can help mitigate some of the problems associated with groundwater overpumping, such as ***land subsidence*** and ***saltwater intrusion***.

Still, belowground storage is not without its risks, especially in terms of water quality. It is important to ensure that the recharged water is clean enough to avoid contaminating existing groundwater sources and to confirm that the interaction of water chemistries doesn't mobilize subsurface contaminants such as arsenic. In addition, the characteristics of the subsurface formation—which are often poorly understood—will affect important parameters such as recharge rate, total available storage, degree of mixing, and ease of recovery. Aquifer recharge by percolation is most suitable when the aquifer is overlain by highly porous soil; when that is not the case, injection wells, rather than infiltration basins, can be used to add water directly to the aquifer without traveling through the soil.

Different terms are used for different types of aquifer storage. The general process of artificially recharging an aquifer is called **managed aquifer recharge (MAR)**. When aquifer recharge is accomplished with floodwaters, it is sometimes called *Flood-MAR*. When wastewater is recharged for water quality improvement, it is often called *soil aquifer treatment*. When the recharged water will later be recovered from the same location, it is often called *aquifer storage and recovery (ASR)*. ASR can even be used to store freshwater as a "bubble" within a brackish or saline aquifer.

Some interesting examples of MAR in the western United States are worth highlighting:

- In Orange County, California, the Talbert Seawater Barrier has been fighting saltwater intrusion since 1976. After an expansion between 1998 and 2006, the barrier now consists of thirty-six injection wells, which collectively use 42 MCM/yr of drinking water–quality recycled water from the GWRS (see Section 2). This creates a groundwater mound in the vicinity of the barrier and controls the hydraulic gradient to prevent saltwater from contaminating the highly productive Orange County groundwater basin.
- In California's Central Valley, farmers are experimenting with using their fields as recharge basins (*Ag-MAR*) during the winter when crops are dormant and excess surface water is often available.[36] Alfalfa fields, for example, can withstand winter flooding without damage, as long as soils are sufficiently permeable.[37] The wet winter and spring of 2023 provided a unique opportunity to scale up this practice, but they also revealed substantial regulatory barriers, leading to an emergency order suspending several regulations.[38]
- In San Antonio, Texas, water managers pump water from one aquifer (the Edwards), only to inject it into another (the Carrizo) for storage. Although this seems counterintuitive, it is justified by the unique characteristics of the Edwards, a heavily used **karst** aquifer that is vital for several endangered species and that responds quickly to drought (a common feature of karst aquifers). The Carrizo ASR facility allows San Antonio to stop pumping from the Edwards when conditions require it and instead retrieve the stored Edwards water from the Carrizo.

5. Energy for Water Supply

One theme running through this chapter has been that different water supply alternatives vary substantially in their energy needs. This brief section—part of our discussion of the **water–energy nexus** throughout the book—highlights this topic and puts it in the context of the energy needed for the whole anthropogenic water cycle.

Every step of the water supply cycle—from water extraction through treatment, distribution, use, and disposal—may use energy. For that reason, conserving water also saves energy. Water infrastructure varies tremendously in its energy intensity,

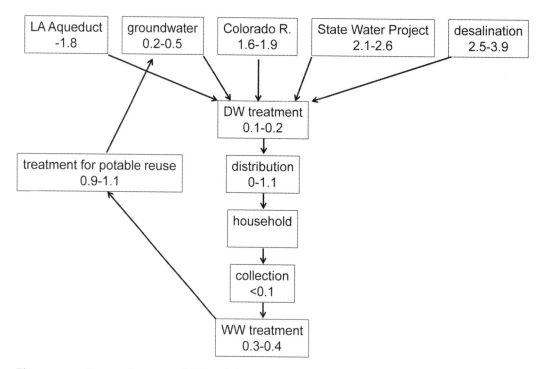

Figure 13-3. Energy intensity (kWh of electricity equivalent needed per m³ of water used) for various components of the urban water supply cycle in southern California. DW, drinking water; WW, wastewater. Household energy use (e.g., for water heating) is not included. Distribution energy intensity varies depending on the need for pumping and pressurization (e.g., the elevation of the highest floor that water must reach). The negative value for the Los Angeles Aqueduct reflects hydropower production. Data from Porse et al. (2020).

with the biggest energy demands coming when water must be moved or lifted large distances or must be intensively treated to remove pollutants or salts.

Figure 13-3 illustrates the energy intensity of various steps in the urban water cycle in southern California. The region's various water sources vary quite a bit in their energy intensity, with local surface water and groundwater using less energy than desalination or imported water (except for the Los Angeles Aqueduct, which is a net producer of energy since hydropower is generated as water drops from the high-elevation Owens and Mono Lakes to LA). Of course, the low-energy sources (both local sources and the LA Aqueduct) are limited in how much water they can supply. IPR occupies a middle ground, as it uses more energy than other local sources but is less energy-intensive than imported water or desalination (although, as noted above, DPR would be even better from an energy perspective). Note that the energy saved by water conservation is the energy saved by the marginal water supply source (the one that provides the last increment of water), which—if energy were the only factor—would be desalination or the State Water Project. In practice, legal, financial, and operational considerations mean that this is not always the case.

Figure 13-3 also shows that the energy intensity of the remaining steps in the southern California water supply process—drinking water treatment, distribution, collection, and secondary treatment—can range from about 0.4 to 1.7 kWh/m³, which is significant but generally less than for supply. This does not include water-related energy used inside homes or industrial facilities, since much of this energy is used for heating and cooling purposes, not for the water cycle per se. Estimates that do include heating and cooling can be very large (e.g., 19 percent of California's electricity use is water related,[39] and 47 percent of US energy is used for "water-related purposes"[40]) but should be interpreted with care.

6. Summary

Table 13-2 provides a simplified summary of the four water-supply strategies we have discussed in this chapter, along with more traditional strategies (dams and aqueducts). Each strategy has drawbacks, and different strategies will be best in different situations. Maintaining a diverse portfolio of strategies can increase water supply resilience by reducing dependence on any one source (Box 13-2). However, the emergence of desalination has highlighted an inherent tension between diversification and energy use: A concern for the water–energy nexus would suggest that desalination plants should be built only when other sources are completely tapped out, whereas a concern for supply diversification would suggest that desalination can provide much-needed resilience.[41]

In some settings, resilience can also be increased—and costs reduced—by regional collaboration around water supply solutions. For example, Las Vegas and Arizona are planning to help fund a new wastewater reclamation plant in southern California; if

Table 13-2. Summary of traditional and alternative water supply strategies. For more on the difference between supply and storage, see Chapter 9, Section 1.4.

Strategy	New Supply or Storage?	Co-Benefits	Drawbacks	Most Appropriate Setting
Dam	Storage	Hydropower, flood control, recreation	Social and ecological impacts (Chapter 9)	See Chapter 9
Aqueduct	Supply	Hydropower (sometimes)	Energy use (often), impacts on source area	A water-abundant area is located near (and uphill of) a high-demand area
Desalination	Supply	Alleviates conflict? (Box 13-1)	Energy use, cost	Coastal setting with brackish water, high-value uses, and no other alternatives
Water reuse	Supply	Alleviate wastewater pollution, close the nutrient loop	Health risks, public acceptance, long-term soil health	Water swap between cities and nearby farms, strong governance and monitoring capacity to minimize risk
Water harvesting	Storage and supply[a]	Alleviate flooding and stormwater pollution	Uncertainty over cumulative effects	Strong local tradition of water harvesting
Aquifer storage	Storage	Reduce evaporation, minimize impacts of groundwater depletion	High data needs	Suitable hydrogeology

[a] Although water harvesting is primarily a storage technique, it can also increase blue water availability by capturing water that would have otherwise infiltrated into the soil and undergone evapotranspiration.

> **Box 13-2. Singapore's Four National Taps**
>
> Over the last few decades, Singapore has worked to diversify its water supply and reduce its dependence on water imports from Malaysia. Singapore's water utility, the Public Utilities Board (PUB), has developed four water sources (Table 13-B1):
>
> - Imported water: Singapore imports water from the southern Malaysian state of Johor through a pipeline that crosses the strait separating the two countries.
> - Local runoff: Stormwater from almost 90 percent of Singapore's land area is directed into a series of reservoirs.
> - Desalination: Singapore opened its fifth RO desalination plant in April 2022.
> - NEWater: Singapore has built five water-reclamation plants, which use membrane filtration, RO, and ultraviolet disinfection to produce water that is used mostly for industrial purposes but also for IPR. To increase public acceptability of NEWater, PUB has invested heavily in education and awareness raising.
>
> **Table 13-B1.** Singapore's four national taps. Data from Irvine et al. (2014), Vincent et al. (2014), and https://www.pub.gov.sg/watersupply/fournationaltaps.
>
> | | Current Fraction | Future Plans | Energy Intensity (kWh/m^3) |
> |---|---|---|---|
> | Imported water | ~30% | Contract ends in 2061, may or may not be renewed | 0.25 |
> | Local runoff | ~30% | Not much room for expansion | 0.25 |
> | Desalination | ~10% | 30% by 2060 | 3.5 |
> | NEWater | ~30% | 50% by 2060 | 0.95 |
>
> To its credit, Singapore has also worked hard to reduce per capita household use and system leaks. Still, as the city-state continues to grow and tries to wean itself from imported water, it will increasingly rely on desalination and NEWater, both of which are quite energy-intensive. Water supply (including wastewater treatment) used 840 GWh in 2012 (2 percent of the country's energy use), and this use may double by 2030.[a] Over the long term, PUB is hoping to reduce the energy intensity of desalination to 1 kWh/m^3 by using new techniques, but it is not clear that such efficiency gains are possible.
>
> **Note**
> [a]Vincent et al. (2014).

all goes according to plan, the plant would reduce the LA region's use of imported Colorado River water, allowing Las Vegas and Arizona to increase their withdrawals from the river.[42] Such a swap makes sense financially and ecologically and is facilitated by existing hydrologic and institutional ties.

One last thought on water-supply solutions: We must not let the promise of new supplies distract us from working to control demand. If we respond to scarcity as we have in the past—by building new supplies to maintain the illusion of abundance—we will remain stuck in an endless supply loop, at great financial, ecological, and social expense. We should use the tools in this chapter to take the edge off of scarcity and allow essential uses of water to continue, but at the same time we need to work hard at efficiency, conservation, and prioritization. Luckily, great progress is being made in these areas, as we will see in the next few chapters. Importantly, though, this progress can be sustained only if the public knows that there are no magic supply-side solutions, that scarcity is not going to disappear, that every drop of water is precious.

Chapter Highlights

1. As water scarcity grows and the drawbacks of dams and aqueducts become more obvious, water managers are devising new water supply solutions and rediscovering old ones. Each of these solutions has its pros and cons, and different solutions will be appropriate in different places. Some of these approaches are aimed at increasing water availability, while others provide water storage to increase resilience to temporal variability.
2. Desalination has experienced great growth and technological improvement over the last few decades and is already an important water source in some parts of the world. The main drawbacks of desalination are cost and energy use.
3. Wastewater recycling—the reuse of treated municipal wastewater for potable and nonpotable uses—is emerging as a major "new" water source in some regions. Wastewater reuse has the potential to alleviate water scarcity while reducing pollution and closing nutrient loops, but it can also pose significant threats to public health and soil quality. The key to safe and sustainable wastewater reuse is finding the appropriate level of treatment to reduce the risk to people and soils while not wasting energy by overtreating.
4. ***Water harvesting (WH)*** is a catch-all term for a variety of small-scale, low-impact approaches to increasing local water supply, including both traditional practices such as khadins and emerging technologies such as ***cloud seeding***. Some energy-intensive and expensive technologies, such as active water-vapor capture devices, are masquerading as WH approaches.
5. Water storage in aquifers has advantages over reservoir storage, including eliminating the need for dams and avoiding evaporation losses. Where local hydrogeology is suitable, ***managed aquifer recharge*** can be an important component of water supply strategies that involve wastewater recycling and WH.
6. In evaluating water-supply options, it is important to aim for diverse supplies and to consider the energy impact of the entire water supply cycle from source to disposal.

Notes

1. Virtual water transfer—the importation of water-intensive food crops to water-scarce areas—is also sometimes considered a water-supply solution, but we will discuss it in the context of global agriculture in Chapter 18.
2. Eke et al. (2020).
3. Mickley (2018); this includes facilities that desalinate brackish water, seawater, and wastewater.
4. The facility's webpage, at https://www.carlsbaddesal.com/, offers a "pop-up tour of the amazing" plant and notes that "Desalinated water costs half a penny per gallon," which works out to $1.30/m^3, a cost that is higher than that of most desalination plants and much higher than many other water sources.
5. Jones et al. (2019) found that 1.8 percent of desalinated water globally goes to irrigation.
6. The Adelaide plant has been at 10 percent production since 2012, the Gold Coast plant is in "hot standby," the Melbourne plant was not used until 2017 despite being completed in 2012,

and the Sydney plant, which was taken out of production in 2012, was restarted in 2019 because of a drop in reservoir levels. Perth, on the other hand, now operates two desalination plants, which provide half of its water supply.

7. Cornejo et al. (2014).
8. https://www.waterworld.com/water-utility-management/energy-management/article/16202012/look-to-windward-the-case-for-wind-powered-desalination.
9. Roth and Tal (2022).
10. Xu et al. (2020).
11. Jones et al. (2019).
12. Shlezinger et al. (2018).
13. Industrial effluents often contain toxics, and agricultural return flow is often already being used by downstream farmers.
14. Thebo et al. (2017).
15. Jimenez et al. (2010).
16. Adegoke et al. (2018).
17. Quansah et al. (2020).
18. Turner et al. (2021).
19. Global Water Intelligence and International Desalination Association (2022).
20. Drechsel et al. (2022b).
21. Drechsel et al. (2022a).
22. https://www.waterboards.ca.gov/water_issues/programs/recycled_water/volumetric_annual_reporting.html.
23. Tal (2016).
24. Tow et al. (2021).
25. Suri et al. (2019), Dery et al. (2019), Savchenko et al. (2019).
26. Savchenko et al. (2019).
27. For an overview of many of these practices, see https://www.cseindia.org/rainwater-harvesting-1272 and https://www.indiawaterportal.org/topics/rainwater-harvesting.
28. https://www.cseindia.org/laws-and-policy--1161.
29. Mosse (2003).
30. https://www.hindustantimes.com/cities/gurugram-news/majority-of-84-rainwater-harvesting-systems-checked-by-admin-team-choked-101627842203376.html.
31. https://www.bbc.com/future/article/20210510-perus-urgent-search-for-slow-water.
32. Ochoa-Tocachi et al. (2019).
33. Abel Cruz, Peruanos Sin Agua, personal communication.
34. https://www.aquatechtrade.com/news/water-treatment/zero-mass-water-why-all-the-hate/.
35. Friedrich et al. (2020).
36. Levintal et al. (2023).
37. Dahlke et al. (2018).
38. https://www.gov.ca.gov/wp-content/uploads/2023/03/3.10.23-Ground-Water-Recharge.pdf?emrc=642515541dbb2; https://water.ca.gov/News/News-Releases/2023/May-2023/Putting-Flood-Waters-to-Work-State-Expedites-Efforts-to-Maximize-Groundwater-Recharge.
39. California Energy Commission (2005).
40. Sanders and Webber (2012).
41. Williams (2018).
42. https://www.watereducation.org/western-water/drought-shrinks-colorado-river-socal-giant-seeks-help-river-partners-fortify-its-local.

14 Drinking Water, Sanitation, and Health: Water 3.0

- What are the links between water and infectious disease?
- How have urban water and sanitation systems developed historically?
- What steps are involved in providing safe drinking water and sanitation?
- How well does the US water and wastewater system work?

Drinking water and other household uses account for only a small fraction of global water consumption, but these uses are among the most basic and essential of all. In many *low- and middle-income countries (LMICs)*, lack of access to safe water and sanitation services has devastating health consequences; providing clean water to these households is a moral imperative of the highest order. At the same time, households in rich countries have access to water-supply and sanitation systems that are much safer, but they still have their problems. In addition, the ready availability of inexpensive, high-quality water in high-income countries often leads to wasteful, excessive use; this waste is no longer acceptable in a time of growing scarcity. Compounding these drinking-water issues, flows of sewage and stormwater can contaminate drinking-water sources and create significant environmental problems, especially in cities.

These interlinked issues—drinking water, health, sanitation, stormwater, and urban water supply—are the subjects of this chapter and the next two. This chapter covers drinking water and sanitation in high-income countries, especially the United States, with a focus on water quality issues. Chapter 15 is about urban water management, covering water supply and scarcity, as well as urban stormwater. Chapter 16 broadens our lens to examine global issues of water, sanitation, and health, especially in lower-income countries.

We start this chapter with an introduction to the links between water and infectious disease, followed by a discussion of the history of urban drinking water and sanitation. We then turn to a description of the current drinking-water and sanitation system in the United States, followed by an evaluation of its strengths and weaknesses.

1. Water and Infectious Disease

Despite the dominance of chemical toxicity in the average American's fears about water contamination, water-related infectious disease—caused by a wide variety of bacteria, viruses, protozoa, and **helminths** (parasitic worms)—has been, and continues to be, a greater threat globally. Five types of **water-related diseases** can be identified (Figure 14-1):

- *Waterborne diseases*—which include cholera, dysentery, typhoid, enteric viral infections, and other diarrheal diseases—are transmitted when feces from an infected person contaminates the household or environment, leading a susceptible person to ingest an infectious dose of the pathogen. Drinking water contamination is one of the most common fecal–oral pathways, but pathogens can also be transmitted by flies, food, and other pathways. Soil transmission (e.g., walking barefoot on feces-contaminated soil or eating food grown in contaminated soil) is particularly important for helminths, which affect about 1.5 billion people in LMICs. Breaking the cycle of transmission for waterborne diseases requires addressing all the pathways shown at the top of Figure 14-1, by providing water, sanitation, and hygiene, often referred to collectively as **WASH**.
- *Water-washed diseases*—eye and skin infections such as trachoma, body lice, and scabies—are those whose transmission can be prevented by adequate washing of hands, faces, and bodies (hygiene).
- *Water-based diseases* are caused by pathogens whose life cycles require an intermediate water-based host; adequate sanitation is key in breaking the cycle of transmission. **Schistosomiasis**, for example, is spread when feces or urine from an infected person ends up in a water body, where the helminth eggs hatch into a life stage that infects snails and matures inside the snail into a stage that is released into the water and can penetrate human skin.
- Some insect-borne diseases, including malaria (caused by a protozoan carried by mosquitoes), are considered water-related, since environmental water management activities (shown as E in Figure 14-1)—such as eliminating stagnant, shallow waters—can reduce the populations of the insect disease vectors.
- *Environmental pathogens*, such as *Legionella* bacteria and nontuberculous mycobacteria, spend a large part of their life cycle outside of human hosts, so they can be commonly found in soil and water even where there are no direct human sources of infection. These pathogens are often found in recreational water (e.g., hot tubs) and in building plumbing systems, and people can be exposed from drinking, inhalation of water vapor, or dermal (skin) exposure. Prevention is primarily through management of building water, including not just drinking water but also heating, ventilation, and air conditioning (HVAC) systems and recreational water.

Preventing these diseases requires providing safe and adequate water for household use (including hygiene) and ensuring that human waste is adequately separated from water sources. Historically, this was hardest to do in urban agglomerations, so these locations were where innovations in water and sanitation originated. We turn to this history next.

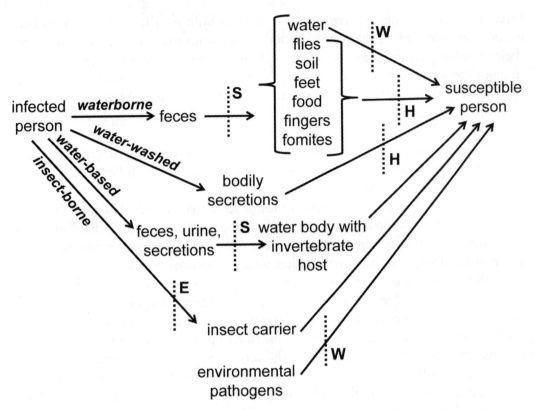

Figure 14-1. Schematic of disease transmission pathways for waterborne, water-washed, water-based, and insect-borne diseases. Dashed lines show the barriers to transmission that are provided by clean water (W), sanitation (S), hygiene (H), and environmental water management (E).

2. History of Urban Water and Sanitation

In Chapter 4, we discussed the early history of urban water provision, ending with Rome, which built sophisticated water systems to bring in clean water, often from some distance, and distribute it to city residents. In this section, we add in the two other important components of urban water systems: wastewater and stormwater.

2.1. Ancient and Medieval Sanitation and Drainage

Archeological evidence suggests that, as early as the third and fourth millennia BCE, some city dwellers in Mesopotamia were using simple indoor toilets that were connected to a sloped channel that drained to the street or to a pit supported by a stack of perforated ceramic rings; the pit (referred to as a cesspit or cesspool) leaked liquid waste into the soil but retained solid waste, presumably requiring periodic cleaning or relocation.[1] In the Indus Valley cities of Harappa and Mohenjo-Daro (fourth to second millennia BCE), some dwellings had indoor baths and water-flushed toilets. Wastewater from these facilities flowed into covered sewers that also carried stormwater.

Rome—a city built on a wetland—had a drainage system that grew increasingly complex over time. The Cloaca Maxima, which ultimately became the city's main sewer, was originally an open artificial stream meant to help drain the swampy land.

Over time, the channel was covered, the city was built over it, and numerous connections were made into the system, including openings where residents could empty their chamber pots (buckets for collecting human waste). Direct household connections to the sewer were uncommon, so the main alternatives to chamber pots were pay-for-use public toilets that flowed into the sewer. The Cloaca Maxima also carried groundwater and rainwater, but the bulk of the water flowing through it consisted of the discharge from the fountains and other water sources in the city, most of which flowed around the clock. This constant flow of water through the streets and sewers helped flush debris and human waste into the Tiber River, which kept the city clean but led to poor water quality in the river.

After the decline of the Roman Empire, its water systems fell into disrepair. Throughout the Middle Ages and beyond, European cities generally used local water sources, which were often contaminated, since the main mechanisms of human waste disposal were cesspools or the dumping of chamber pots into the streets. Because those cities lacked the Rome-style continuous flow of clean water through the streets and abounded in other waste sources (e.g., animal manure, household trash, and waste from butchers, tanners, and dyers), medieval European cities were highly polluted and odoriferous. When cesspools filled up, the ***fecal sludge*** was collected and removed (usually at night, hence the term *night soil*), and sometimes used as agricultural fertilizer. In China and Japan, where less animal manure was available, agricultural markets for night soil were more fully developed (and cities probably didn't smell quite as bad).

2.2. Epidemics and the Sanitary Revolution

Water-related diseases, in both endemic and pandemic forms, were probably common from the earliest days of urban settlements, although we lack a good understanding of their impact on public health. In recent decades, paleo-parasitologists have examined soils and coprolites (preserved fecal material) from historic toilets and burial grounds and have found helminth eggs and the molecular signatures of protozoa, including those that cause dysentery, malaria, and giardiasis.[2] Pathogenic bacteria and viruses are not well preserved, so we know less about their historical abundance.

The early nineteenth century saw the first pandemic of cholera, a diarrheal disease transmitted through contaminated water. Cholera outbreaks were recurrent features of life in many European and American cities until the early twentieth century, and they still pop up today around the developing world. In hindsight, the cause of these nineteenth-century cholera outbreaks seems clear: the inadequate separation of human waste and water sources in rapidly growing cities. However, the dominant explanation for epidemics in the early nineteenth century was the miasma theory, which held that "bad air" from decomposing organic matter was the source of these diseases. While this led some cities to focus on cleaning up organic wastes, these measures were ineffective given the lack of attention to water contamination.

A major turning point toward modern germ theory—and thus modern water systems—came during the 1854 cholera epidemic in London, when physician John Snow demonstrated a high prevalence of cholera among those getting their drinking water from the Broad Street well—which, as it turned out, was directly adjacent to a cholera-contaminated cesspool. Snow convinced local authorities to remove the pump handle, preventing further use of the contaminated water and leading to a decline in neighborhood cholera cases.

The idea that disease can be spread by contamination of drinking water with sewage—and thus can be prevented by the provision of clean water and effective sanitation—has been termed the **Sanitary Revolution** (Figure 14-2). In several European cities, the Sanitary Revolution led quickly to a trio of measures that drastically reduced waterborne disease: the development of water supplies brought from outside the city and delivered to individual homes, the treatment of drinking water through filtration before delivery to consumers, and the use of flush toilets (water closets) connected to new or newly expanded sewers.

The nineteenth century saw increasing recognition that provision of water and wastewater services was vital to the health and success of cities and their residents. In addition to protecting public health, an urban water supply could help fight fires and produce a sense of civic pride and identity. The emergence of the private bathroom often led to the decline of publicly available water and baths, changing the nature of water services in the public sphere. A household water supply and sewer connection—provided to private individuals as a public service—transformed what it meant to live in a city and even, on some level, what it meant to be human, as Carl Smith explains: "The miracle of the bathtub and the water closet was not just that by taking city water one could remove oneself from city life, but that they also made one's

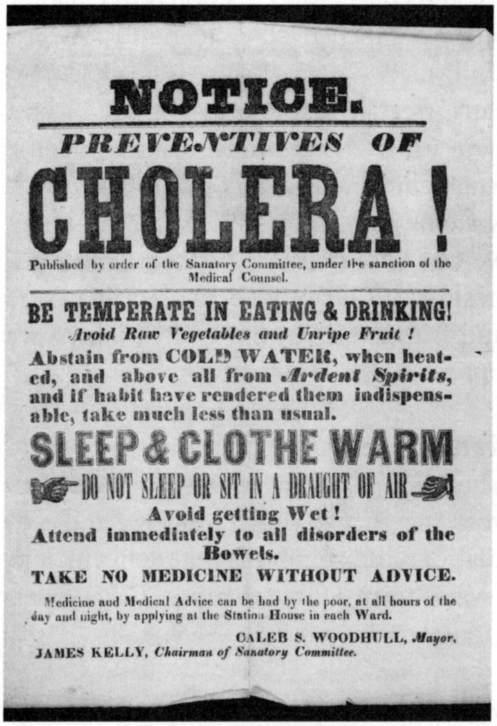

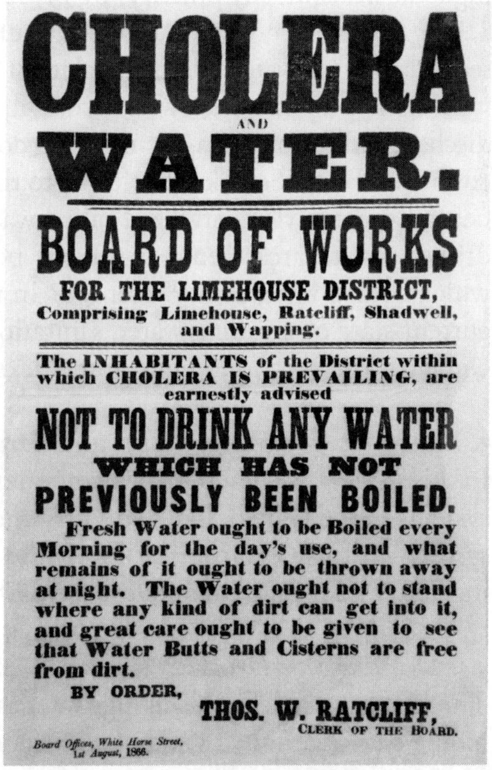

Figure 14-2. Cholera advice before and after the Sanitary Revolution: (a) New York, 1849; (b) London, 1866. The latter is more scientifically accurate. Images in the public domain.

natural body odor and waste—reminders that nature itself is 'dirty' and that existence is inescapably corporeal—seem to disappear."[3]

2.3. The American Experience
Cities in the United States underwent similar transformations to the European cities discussed above. Chicago, for example, originally used Lake Michigan for both water supply and waste disposal and struggled with repeated disease outbreaks until it took a series of steps to improve its water and sanitation: building a 2-mile tunnel to bring water from deeper parts of the lake, raising the entire low-lying city so a proper sewer could be built, reversing the flow of the Chicago River to send its wastewater to the Mississippi River rather than Lake Michigan, and building water and wastewater treatment plants.

Other American cities followed trajectories similar to Chicago's, with Philadelphia bringing in water from the Schuylkill River in 1801, New York from the Croton River in 1842, and Boston from Long Pond in 1848 (after a two-decade-long effort following the fire of 1825). In each case, the arrival of water from outside the city limits was celebrated with great excitement as a symbol of progress. In each case, the city soon outgrew its initial waterworks and developed additional supplies of increasing complexity and geographic reach. And in each case, the focus on importing water was followed by implementation of drinking-water treatment and ultimately sewage treatment as well. By 1940, practically all US cities had sewers and drinking-water treatment. Sewage treatment lagged a bit, with 57 percent of urban sewage undergoing treatment (mostly primary) by 1950 and 89 percent by 1970. As noted in Chapter 8, the 1972 Clean Water Act (CWA) ultimately compelled cities to build modern secondary wastewater treatment plants (WWTPs).

This approach—take water from upstream, treat it, use it, treat it again, and discharge it downstream—is now the dominant urban water model in high-income countries. David Sedlak has referred to this arrangement as **Water 3.0**, with Water 1.0 being Roman-style centralized piped water systems and sewers and Water 2.0 being drinking-water treatment.[4] Is this the best we can do? Are there further evolutions, which Sedlak refers to as Water 4.0? In the rest of this chapter, we will evaluate the current state of drinking water, sanitation, and stormwater in the United States; we come back to these questions in the next chapter.

3. Water 3.0: Drinking Water and Sanitation in the United States
In this section, we describe the components of modern drinking-water and sanitation systems in the United States and how they work together to protect human and ecosystem health. We start by exploring the regulations and standards that define drinking-water safety and then follow the path of water from source to tap to wastewater.

3.1. Defining Safe Drinking Water
The best way to define drinking-water safety is by absence; water should *not* contain biological, chemical, or radiological contaminants at levels that may cause harm. However, given the diverse array of potential contaminants (including naturally occurring ones), the goal of ensuring zero risk is probably unrealistic, and drinking-water regulation in practice involves balancing risks and costs.

In the United States, the Environmental Protection Agency (EPA) is responsible, under the Safe Drinking Water Act (SDWA) and its various amendments, for setting acceptable levels of contaminants in drinking water and establishing regulations to ensure that water suppliers provide clean water. None of these federal regulations apply to private drinking-water wells, which serve 13 percent of Americans.

The EPA sets two types of standards for each contaminant. A ***Maximum Contaminant Level Goal (MCLG)*** is an unenforceable public-health goal, set at a level that should protect against any adverse health effects. MCLGs for ***carcinogens*** are set at 0, and MCLGs for noncarcinogens are set below the threshold of harm (with a margin of safety). The second type of standard is a ***Maximum Contaminant Level (MCL)***, an enforceable standard that is often higher than the MCLG and is determined by balancing health risks against the cost and feasibility of controlling the contaminant.

As of 2023, EPA has established MCLs or the equivalent for four radionuclides, seven microbial groups (including turbidity), seven disinfectants and disinfection byproducts, sixteen inorganic contaminants, and fifty-three organic contaminants.[5] The universe of potentially harmful chemicals that might be found in drinking water is much larger than this, so it is critical for EPA to continue to evaluate data on the toxicity and prevalence of drinking-water contaminants in order to prioritize additional contaminants for regulation. Under the process established by the 1996 SDWA Amendments, EPA periodically compiles Contaminant Candidate Lists (CCLs) and then makes a regulatory decision (regulate or not regulate) for a subset of those contaminants, based on three factors: toxicity, the opportunity to meaningfully reduce risk through regulation, and presence in drinking-water supplies.

In four CCL rounds over the last twenty-three years, EPA has chosen to regulate only four compounds and has implemented no new MCLs.[6] In part, this reflects the fact that understanding how a chemical affects the human body, and what levels are safe, is a difficult task that requires a ream of chemical and toxicological studies. But it also reflects a lack of political will, along with a cumbersome process that is prone to manipulation by powerful interest groups who produce or use the chemicals in question and seize on scientific uncertainty to derail regulatory efforts. Even for chemicals that do have MCLs (such as nitrate and arsenic), the levels may not necessarily be adequately protective.

Some would argue that our drinking water should be completely free from any anthropogenic contamination. Yet there are real costs to implementing MCLs. Distasteful though it may be, we must make choices as to where we should spend money to best protect public health. Overly conservative MCLs may increase the cost of water to households, which may have its own negative health effects. A study using an approach called health-health analysis to evaluate health tradeoffs estimated that, for small water suppliers, the annual cost of meeting the new arsenic standard will be more than $400 per household; if these higher costs are passed on to households, the resulting economic distress is estimated to result in "diseases of poverty" that would undo about half the health benefit of the lower standard.[7] However, the solution in these cases may be to provide federal or state subsidies for meeting standards, rather than abandoning the standards.

How do drinking-water suppliers go about meeting the standards set by EPA? The

multiple-barrier approach involves protecting water quality from source to tap to disposal; the next four sections walk us through those steps.

3.2. Source Protection

Protecting drinking-water sources from contamination is far more effective than treating water after it is already contaminated. The surest way to protect drinking water is to ensure that water-supply watersheds remain in their natural state (e.g., forested) and are protected from future development. In many cases, though, some development in the watershed is inevitable, and the focus ought to be on excluding highly polluting industries and protecting the most sensitive areas (e.g., buffers along watercourses and reservoirs, steep slopes that are subject to erosion).

Since many cities obtain their water from sometimes-distant rural watersheds, urban water quality is vulnerable to land-use changes over which cities typically have little direct control. Urban water managers need to build strong relationships with rural communities and collaborate on solutions that protect water quality while minimizing the constraints on rural lifestyles; ***payments for ecosystem services*** can be an important part of the exchange. New York City's program to protect its water sources in the Catskills is an example of effective large-scale watershed protection in a rural agricultural setting.

Compared with surface water, groundwater is often assumed to be better protected from anthropogenic activities, since water percolating through the soil undergoes natural purification. ***Confined aquifers*** are even better protected than unconfined ones, since a low-conductivity layer protects them from activities at the surface. Still, even confined aquifers have a recharge zone (except for ***fossil aquifers***), and not all contaminants are effectively removed during percolation, so regulation of land use in recharge zones can be an important measure. Oil and gas extraction can also pose a threat to aquifers. Once aquifers are contaminated, they can be very difficult to remediate.

3.3. Water Treatment

The SDWA's Surface Water Treatment Rule (1989) requires that, unless a community receives a filtration waiver from EPA, all surface water supplies need to be treated by filtration to remove suspended sediment and its associated contaminants, followed by disinfection (e.g., with chlorine) to kill pathogens. Groundwater generally needs only disinfection. Water utilities usually add other chemicals as well, including anticorrosives and fluoride.

Several cities have obtained waivers from EPA to avoid building filtration plants for their surface water sources (but must still disinfect their water). Most prominently, Boston, New York City, Portland, Seattle, and San Francisco have been successful, at least so far, in their battles to avoid spending billions of dollars on new filtration plants and focus instead on source protection.

One of the trickiest issues in drinking-water treatment is the problem of ***disinfection byproducts (DBPs)***. On one hand, the use of chlorine for disinfection has dramatically reduced the incidence of waterborne disease. On the other hand, chlorine interacts with natural organic matter found in source water (especially surface water) to form small amounts of toxic byproducts. The goal, then, is to balance microbial and chemical risks by applying enough disinfectant to eliminate even the most resistant organisms while minimizing formation of DBPs. This goal is easier to achieve if the source water is cleaner to start with, in terms of pathogens, natural organic matter, and turbidity. The future success or failure of source protection programs in New York and other cities will probably hinge on their ability to keep their source water clean enough that they can achieve the dual goals of adequate disinfection and minimal DBP formation.

3.4. The Distribution System

Once source protection and water treatment have produced safe drinking water, that water still needs to travel through a network of pipes and pumping stations to homes. This underground water infrastructure suffers from a serious visibility problem: Since it is out of sight, people don't think about it as long as they can turn their taps on, making it hard to mobilize funding for pipe maintenance.

There are 2.2 million miles of drinking-water pipes in the United States, including both water mains (larger-diameter pipes running under streets) and service lines (smaller pipes running from mains to houses). Although they vary in age, many were installed in the early to mid-twentieth century and are nearing the end of their expected lifetime of 75–100 years, resulting in a distribution network that is aging and vulnerable to leaks and breaks. The American Society of Civil Engineers, which gives US drinking-water infrastructure a grade of C– (up from D in its previous assessment), notes that, nationwide, about 15 percent of treated drinking water is lost to leaks and that over 250,000 water main breaks occur each year.[8]

In addition to water loss, the aging distribution system can also lead to water contamination, particularly through corrosion of lead-containing pipes. The lead contamination crisis in Flint, Michigan, brought this issue to public attention, but this problem exists in many other cities as well. Lead is found primarily in service lines made of lead or galvanized steel, as well as in household plumbing fixtures, joints between pipes, and solder. Despite the 1986 ban on new lead pipes, many older pipes are still in place. In addition, pipes installed as recently as 2014 may contain up to 8 percent lead. Many utilities don't have an inventory of their lead pipes, and estimates of the total number of **lead service lines (LSLs)** in the United States range from about 6 to 10 million.[9] LSLs are more common in older homes and poorer neighborhoods.[10]

Although a well-planned corrosion control program can minimize lead leaching from LSLs, the ultimate solution may be to remove LSLs and replace them with copper lines. EPA estimates an average cost of about $4,700 per LSL (range $1,200–12,300),[11] suggesting a total cost of up to $50 billion. Despite the high cost, there are also great benefits of reduced lead exposure (increased IQ and earnings, decreased cardiovascular disease), and several analyses suggest benefit/cost ratios greater than 1 for complete LSL replacement.[12] The 2021 Infrastructure Investment and Jobs Act allotted $15 billion for this purpose, a significant investment but probably short of what is needed.

Some cities have already been moving ahead with LSL replacement. Newark has won kudos for its quick work, replacing its 22,000 LSLs in less than three years.[13] Several factors contributed to this success, including funding from the state and the county that allowed the city to do the work at no cost to homeowners, along with a provision that allowed the city to proceed without receiving homeowner permission.

In 2020, Chicago announced an initiative to replace all 400,000 lead service lines in the city, at an estimated total cost of $6–10 billion.[14] For low-income families with children or with documented high lead levels, the city will pay the entire cost of service line replacement. However, communities of color have expressed some hesitancy about participating, for reasons that include distrust of city authorities (for good historical reasons) and concern that accepting public assistance will affect their immigration status.[15]

Manny Teodoro points out that cities that have already completed their LSL replacement will not benefit from the federal funding in the Infrastructure Investment and Jobs Act and notes the moral hazard that this creates: Why should cities and states act proactively to invest their own resources in water infrastructure when the feds will eventually step in if the problem gets bad enough?[16] This issue goes beyond LSLs, of course, and speaks to the complex dance of regulation and funding between levels of government in the United States.

3.5. Urban Sanitation

Once clean water reaches the user, most of that water undergoes **nonconsumptive uses** such as bathing and toilet flushing and enters a second set of pipes, namely the sewer system. At that point, responsibility shifts from the SDWA—with its goal of protecting human health—to the CWA, with its goal of protecting aquatic ecosystems for ecological and human benefits. Still, it is helpful to think of these as two parts of one larger system; safe sanitation works together with safe drinking water to safeguard our drinking-water sources and protect us from waterborne disease.

The sanitation system in American cities has two components: the water-based flushing of waste through a network of pipes (about 800,000 miles of sewer mains and 500,000 miles of lateral sewers connecting homes and businesses to mains) and the treatment of that wastewater at WWTPs (more than 16,000 facilities). The flushing protects the user, but only treatment eliminates the threat to ecosystems and to downstream users. Wastewater thus presents a classic **externality** problem: Cities don't feel the full costs of their environmental contamination because they discharge it downstream, and regulation (i.e., the CWA) is necessary to ensure that cities treat their wastes.

There is little doubt that the CWA's emphasis on **secondary treatment** for municipal wastewater has resulted in improved water quality in aquatic ecosystems, which of course also serve as the source of most of our drinking water. But our wastewater system as a whole still has significant problems, as indicated by a grade of D+ from the American Society of Civil Engineers[17] and EPA's estimate of a twenty-year funding need of $271 billion.[18] One expensive problem facing many WWTPs in coastal communities is flooding due to **sea-level rise**, since WWTPs are typically built at the lowest elevation of the community to minimize the need to pump wastewater.[19]

3.6. Combined Sewers

Perhaps the toughest problem facing the wastewater system is the issue of **combined sewers**. Some 860 communities in the United States, with a total population of about

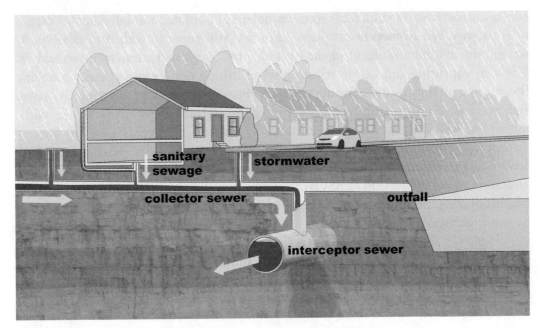

Figure 14-3. Schematic of a combined sewer. During normal conditions, both stormwater and sanitary sewage flow to the WWTP, but when the capacity of the sewer system is exceeded, the combined sewage discharges into the river from the outfall in a CSO event. Image courtesy of Justin Brown, Metropolitan Water Reclamation District of Greater Chicago, with labels added by the author.

40 million people, have sewers that are designed to carry both the sewage from people's homes and the stormwater that enters from streets, roofs, and so on; the remaining communities generally have two separate sewer systems, one for wastewater and one for stormwater. The advantage of combined sewers is that stormwater is treated at the WWTP before being released to rivers. The disadvantage is that during large rain events, the combined volume of stormwater and wastewater may exceed the capacity of the sewer pipe or the treatment plant, and the excess—which includes untreated human waste—is either discharged to a nearby stream (a combined sewer overflow [CSO]) or backs up into people's basements (Figure 14-3).

The CSO issue has proven to be an expensive and vexing problem. The obvious solution—separating the combined sewers into storm and sanitary sewers—has been implemented in some places but is very slow and expensive, since it involves large-scale digging up of underground sewer lines; sewer separation also creates a new storm sewer system that can generate its own pollution problems. Other options have been tried, including the construction of large underground tunnels for storing combined sewage until the storm has passed and then pumping the combined sewage to the WWTP. Chicago, for example, is near the end of its multidecade, $4 billion Tunnel and Reservoir Plan, which will use 109 miles of deep tunnels and three large reservoirs to provide 17.5 billion gallons of storage for combined sewage and floodwaters. Together with WWTP upgrades, the plan has already improved water quality in local waterways.[20]

Because of the long timeline for planning and implementing CSO solutions, these efforts are particularly vulnerable to the problem of outdated rainfall data.[21] Alternatives to **_gray infrastructure_** can help address CSOs more quickly and flexibly. As we will discuss in more detail in the next chapter, **_green infrastructure_** can address CSOs by capturing and slowly releasing water during rain events, thereby reducing the volume of stormwater running off **_impervious surfaces_** and entering the sewer system. However, regulators may be loath to approve these cheaper alternatives without strong evidence that they will substantially reduce CSO events.

The same phenomenon of insufficient capacity can also happen with sanitary sewers (sewers that, in theory, carry only wastewater), leading to a discharge called a sanitary sewer overflow (SSO). SSOs can occur for a variety of reasons, including pipe blockages and pumping station failures, but they often happen during rain events due to two processes:

- **Inflow**: the direct flow of rainwater or runoff into the sanitary sewer through connections such as roof leaders;
- **Infiltration**: the leakage of groundwater into sewer lines when water tables are high.

Solving the inflow and infiltration problem takes a great deal of painstaking work to disconnect nonwastewater connections and increase the integrity of pipes.

One emerging tool that has helped some communities with their CSO and SSO problems is the "smart sewer," in which a network of real-time sensors helps to optimize flow through different parts of the system in response to changing conditions. South Bend, Indiana, which has been using such a system since 2008, has saved hundreds of millions of dollars by improving wet-weather flow through its existing pipes and designing future improvements based on a spatially explicit analysis of past performance.

3.7. Rural Sanitation

Where population density is low, sewers are prohibitively expensive, so 20 percent of the US population, mostly in rural areas, instead use some type of decentralized wastewater treatment system, most commonly an underground septic tank, which is essentially a simple, miniature primary treatment plant. Household sewage travels by gravity to the tank, where pollutants are removed by settling, flotation, and anaerobic digestion. Effluent from the septic tank is usually distributed into the soil through a gravel drainfield and undergoes further treatment in the soil before reaching groundwater. Sludge slowly accumulates in the septic tank, so tanks must be periodically cleaned and the fecal sludge (material removed from the tank) disposed of properly at a centralized treatment plant.

A properly functioning septic tank with an adequate residence time and a sufficiently large drainfield can protect water quality, but septic systems can also cause significant contamination. Septic tanks and drainfields have an unfortunate combination of two properties: They are underground (so easy to ignore) and they are the responsibility of individual homeowners, who may not know or care much about what happens when they flush the toilet. In the absence of effective local regulations, maintenance and cleaning are often deferred until a problem occurs.

4. Evaluating Water 3.0

In discussing the components of our current water and sanitation system in the previous section, we have already encountered some of its shortcomings: the slow pace of standard setting, the underfunding of aging infrastructure, the ubiquity of lead in our plumbing, and the problem of CSOs. In this section, we step back and evaluate the performance of the system as a whole in several different arenas. We start by examining how well the system performs in achieving its primary goal: protecting public health.

4.1. Health Effects

The water and sanitation system in the United States has dramatically improved public health. Waterborne diseases such as typhoid and dysentery, which killed thousands of Americans every year at the beginning of the twentieth century, have been virtually eliminated. Cholera epidemics—which regularly brought COVID-like fear to American cities—are no longer a concern.

A small number of water-related disease outbreaks (clusters) do still occur in the United States, but the larger threat comes from low-level widespread exposure. A recent Centers for Disease Control and Prevention (CDC) study estimated that the total annual burden of water-related infectious disease (from drinking water, recreational water, and building HVAC systems) is about 7 million illnesses, 6,600 deaths, and $3.3 billion in healthcare costs.[22] The bulk of illnesses were minor (4.7 million cases of swimmer's ear), and the bulk of deaths and healthcare costs were caused by environmental pathogens (nontuberculous mycobacteria and *Legionella*), which primarily affect older adults and immunocompromised people.

The CDC has also documented an increase in Legionnaire's disease, with the incidence rate rising from about 0.5 cases per 100,000 in the 1990s to almost three cases per 100,000 in 2018. Over the period 2003–2018, the incidence rate for Black Americans was more than twice the rate for Whites. Cases were highest in the summer and fall, suggesting a role for seasonal cooling systems and exposure through ambient surface water.[23]

What about the burden of chemical toxicity from drinking water? A recent study estimates that the national average lifetime cancer risk associated with drinking water (based on average contaminant levels) is four in 10,000, roughly the same as the cancer risk associated with air pollution and much higher than the goal of roughly one in 1 million.[24] Most of the cancer risk comes from arsenic (especially in groundwater systems) and DBPs (especially in surface water systems), and the risk is generally higher in more arid western states.

4.2. Disparities in Access

Until now, we have focused largely on the *average* performance of Water 3.0. This section focuses on distributional issues, particularly the racial and class disparities in access to Water 3.0 in the United States.

Contrary to what many people assume, not everyone in the United States has access to the benefits of indoor water and toilets. There is no systematic count of those without plumbing access, but one estimate suggests that at least 2 million people, including about half a million unhoused people and 250,000 residents of Puerto

Rico, lack at least one element of indoor plumbing (hot and cold running water, a sink, a shower or bath, or a flush toilet).[25] Compared with White households, African American and Hispanic households are almost twice as likely to experience plumbing poverty, and Native Americans are about twenty times as likely. A peer-reviewed study found similar numbers (1.1 million people, not including the unhoused or residents of Puerto Rico) and similar racial disparities.[26] The water problems faced by these households are part of a larger pattern of substandard housing and inadequate infrastructure in low-income communities and communities of color.

The low level of water access among Native Americans—48 percent of homes lack access to reliable water and sanitation[27]—is especially shameful given the federal government's treaty and trust responsibilities to tribes. Existing government programs to help tribes build and maintain drinking-water infrastructure are underfunded, hard to access, and scattered across agencies.[28] The disproportionate toll of the COVID-19 pandemic on Native Americans has brought into stark relief the importance of a home water supply and has hopefully provided some momentum for government action.

In the 71,000 km² Navajo Nation reservation, about a third of residents lack access to drinking water and must haul water to their homes from collection points many miles away, at a cost that amounts to seventy times the cost of water in nearby suburban communities.[29] This water is sometimes contaminated with uranium from abandoned mines or naturally occurring arsenic. Since 2014, the Navajo Water Project, a community-managed "utility alternative" run by the nonprofit DigDeep, has been providing households with a water system that includes a 1,200-gallon buried water tank, a solar-powered water heater, and indoor plumbing. Tanks are refilled by a water truck that obtains water from safe local sources—usually wells with treatment that includes arsenic removal.

The *colonias* along America's southern border—informal periurban communities of low-income Latinx immigrants—are home to some 840,000 people, of whom 30 percent are estimated to lack water access.[30] Barriers to access include the cost of extending service—which often must be borne by the residents—and the informal nature of the settlements and the frequent lack of land title. Residents resort to drinking water from contaminated wells or buying bottled water.

Even for people who nominally have access to water services, the quality of that water is often affected by structural racism. For example, in the San Joaquin Valley in California, where intense agricultural use has contaminated groundwater with nitrate, salts, and pesticides,[31] water systems are more likely to exceed the NO_3^- MCL if a greater fraction of their customers are Latinx; these differences are largest for smaller systems.[32] Similarly, a nationwide study found that water systems with nitrate concentrations above 5 mg/L were more likely to have a higher percentage of Latinx customers.[33] Residents who receive unsafe water often end up paying twice: once for their nonpotable tap water and again for the bottled water they must bring home for drinking, cooking, and other uses.

One of the factors that perpetuates racial bias in water access is jurisdictional history. In the San Joaquin Valley, many Latinx people live in unincorporated communities that are either served by a water system that is out of compliance with the SDWA or not served by any water system. Many of these residents are in close proximity to in-compliance water systems, but those systems are not required to extend coverage into the unincorporated areas.[34]

A similar phenomenon has been documented in Wake County, North Carolina, where *extraterritorial jurisdictions*—areas that are outside municipal boundaries but are partially controlled by municipal councils—often lack water service, especially when their population is mostly African American. Water service often stops at the boundaries between White and African American neighborhoods, even when both neighborhoods are extraterritorial.[35]

In addition to water access problems, communities of color often face sanitation problems. A flush toilet may not do much good if it connects to a malfunctioning septic system or a "straight pipe" that discharges directly into your yard. Catherine Coleman Flowers has brought nationwide attention to this issue, starting with her work in mostly Black Lowndes County, Alabama, where clay soils render septic systems ineffective, low population density makes sewers expensive, and sanitation problems are part of a vicious cycle of criminalized poverty. A 2017 study found that a third of stool samples from Lowndes County had DNA from the hookworm *Necator americanus*, a parasite that is more typically associated with LMICs or earlier centuries.[36] In 2021, the Department of Justice opened a civil rights probe into how the state and county health departments have handled wastewater disposal in Lowndes County, noting, "Sanitation is a basic human need, and no one in the United States should be exposed to risk of illness and other serious harm because of inadequate access to safe and effective sewage management."[37]

4.3. Utility Structure

In this section, we examine two of the underlying determinants of the success or failure of Water 3.0: the size and ownership structure (public vs. private) of our utilities. Figure 14-4 presents a summary of drinking-water utilities in the United States. Study the data for a moment; what conclusions do you draw?

Perhaps the most important fact demonstrated by Figure 14-4 is simply the huge number of drinking-water utilities in the country—about 50,000, compared with about 3,200 electric utilities, 1,400 gas utilities, and 7,000 internet service providers.[38] This extreme fragmentation produces a complex and inefficient institutional landscape,[39] further complicated by a network of relationships between multiple water wholesalers and retailers in the same region and by the fact that wastewater utilities are often (but not always) separate from water-supply utilities.

The second thing to notice in Figure 14-4 is that the United States has both private and public water suppliers, but most Americans, especially in large cities, are served by publicly owned utilities.[40] The early history of urban water supply in the United States often saw a shift from private to public utilities, as fires and disease outbreaks made it clear that a safe, reliable water supply was a public good. The 1990s and early 2000s saw many cities in the United States (and elsewhere) turn their water systems (along with other utilities) over to private companies, a trend that was driven by a variety of factors: increasingly complex SDWA requirements, a desire to save money, and a 1997 law that provided tax-exempt status to municipal water system debt, even

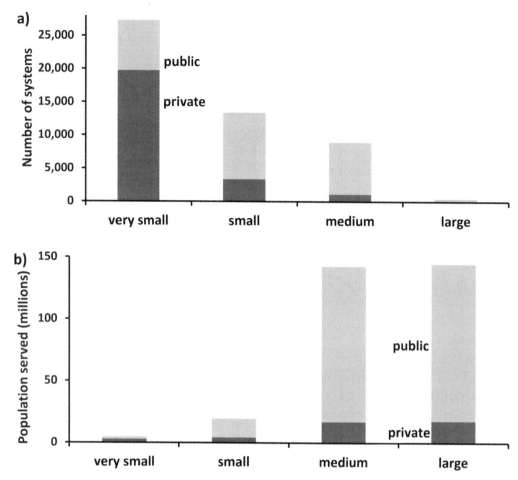

Figure 14-4. (a) Number of utilities and (b) population served, by utility ownership structure and size (very small, 25–500 customers; small, 501–3,300; medium, 3,301–100,000; large, more than 100,000). Data from GPRA Inventory Report (https://obipublic.epa.gov/analytics/saw.dll?PortalPages&PortalPath=/shared/SFDW/_portal/Public). Only community water systems (systems that provide water to the same population year-round) are included.

if that debt was issued by private operators. Underlying all this was the narrative that the private sector had the expertise to run water and sanitation systems more efficiently than overworked and undertrained public employees.

In fact, it is hard to make generalizations about whether private or public management of water supply is "better." Private utilities have slightly lower rates of SDWA violations than public ones.[41] And there are certainly public utilities that are poorly managed and unresponsive to customers.[42] However, in the last two decades, several high-profile failures of private management—high prices and low quality in Atlanta and Gary, Indiana, among others—have prompted a wave of remunicipalizations, which now includes at least fifty-eight cities in the United States and 126 cities in other high-income countries (perhaps most famously Paris).[43] In some cases, remunicipalization is driven by community mobilization around "our water" (i.e., by an ethical rather than a practical objection to the involvement of for-profit entities in providing this vital public service).

Our third and final observation from Figure 14-4 is that, although most Americans (92 percent) who get water from a utility are served by medium or large utilities, most water utilities in the United States (81 percent) are small or very small. These 40,000+ small or very small utilities, mostly in rural areas, face many of the same issues of water delivery that larger utilities deal with but don't benefit from economies of scale, resulting in much greater per capita funding needs to solve infrastructure and water quality problems. These smaller systems often lack the financial and managerial resources they need to adequately serve their communities, and they are more likely to violate SDWA requirements.[44] Similar problems face wastewater utilities in rural areas, where sewer and wastewater systems can be the largest item in a small town's budget.

Private wells (a dominant water source in many rural areas) may also be less safe. A US Geological Survey study of water quality in private wells found that 23 percent of the wells examined had at least one contaminant at concentrations higher than health guidelines.[45] A recent study in North Carolina found that children drinking private well water were more likely to have elevated blood lead levels.[46] A statewide study from California found that violations of arsenic, nitrate, and chromium MCLs were practically nonexistent in large systems but were common in domestic wells and smaller systems. This reflects an underlying power dynamic in which rural areas are seen as places for agriculture, not for people.

One solution to this problem may be the expansion of utilities into unserved areas and the consolidation of smaller utilities into larger, regional entities[47]; different types of institutional arrangements short of full consolidation can allow some economies of scale without loss of local control.[48] While voluntary consolidations should be the goal, California passed a law in 2015 allowing the State Water Board, under certain circumstances, to mandate consolidation or extension of a utility's service into unserved or poorly served areas,[49] a power that it has used several times, including requiring the Orosi Public Utility District to consolidate with the poorly served East Orosi system. Successful consolidation may require state funding.[50]

Besides small, rural utilities, another area of concern is the financial stability of water and wastewater utilities in **legacy cities**: cities such as Flint or Detroit or Newark that have suffered from industrial decline, population loss, redlining, and White flight. The Lincoln Institute of Land Policy identifies almost 100 legacy cities in the United States and notes, "Once drivers of industry and wealth, legacy cities experienced decline in the 20th and 21st centuries due to changing industries and government policies that steered investments away from communities of color. . . . Home to nearly 17 million people and a collective economy of $430 billion, these cities still possess the infrastructure and institutions that can launch an equitable revival."[51]

For legacy cities to launch that revival, many of them will need to invest substantial resources in water and wastewater systems that are increasingly stressed by past underinvestment, climate change, and CWA requirements such as addressing CSOs. Hartford, Connecticut, for example, has a CSO problem similar to Chicago's but has a much smaller and poorer customer base to pay for solutions. Jackson, Mississippi—whose ailing water infrastructure was brought to the nation's attention by crises in 2021 and 2022 that left residents without water for weeks—needs some $1–2 billion to fix that infrastructure, but the city has an annual municipal budget of only

$300 million and houses many state government properties that pay no taxes. It is no coincidence that these cities are also disproportionately home to people of color. Legacy cities will need to be creative in finding cheaper, greener solutions, but suburbs, states, and the federal government must also share the funding burden.[52]

4.4. The Rise of Bottled Water

Over the last thirty years, per capita consumption of bottled water in the United States has more than quadrupled, to the point where each of us (on average) buys the equivalent of one 500-mL bottle every day of the year. We collectively spend over $18 *billion* per year on a product that would cost about $20 *million* if drawn from the tap (even though some bottled water brands are in fact repackaged municipal tap water). Despite pressure from environmental activists and even some municipal bans on sales of single-use bottled water, the trend shows no signs of slowing down, although much of the recent increase appears to represent a shift from sweetened drinks to water.

While some bottled water use is driven by convenience (especially given the disappearance of public water fountains in recent years), much of it appears to be associated with an increasing distrust of tap water. People of color and lower-income people are less likely than their White and high-income counterparts to trust the safety of their water and are thus more likely to take on the financial and logistical burden of buying bottled water.[53]

Are these fears justified? On one hand, bottled water is less tightly regulated than tap water, and there is no reason to believe that it is generally safer than most American tap water. On the other hand, suspicion of tap water is certainly understandable given all the issues discussed in previous sections of this chapter, including high-profile incidents of contamination and chronic neglect of water infrastructure, both of which disproportionately affect communities of color. Indeed, bottled water sales tend to rise in response to well-publicized health-based drinking-water violations.[54]

Sadly, the bottled water industry has preyed on these fears, spending a great deal of money on advertising to make sure that bottled water is associated in our minds with health, safety, and beauty. On the tap water side, the counterpart to these flashy ads is the Consumer Confidence Report, an annual report that utilities must send to their customers detailing the results of water quality sampling. Utilities must find ways to make these reports more user-friendly and accessible while still retaining their primary function of educating the public about the sources and quality of their tap water.

The rise of bottled water is a deeply disturbing phenomenon. We need to invest in the long-term health of our public water systems for the benefit of all Americans, rather than providing profits to private companies selling an unsustainable, polluting product.

4.5. The Problem of Linear Flows

Perhaps the most far-reaching critique of Water 3.0 is that it relies on linear flows of water and materials, as opposed to the sustainable circular flows that are common in nature. Arguably the most absurd manifestation of this linearity is the fact that we take water from the environment, clean it to drinking-water quality, use it to flush our toilets, and then treat it as waste. Compounding the absurdity is the fact that wastewater contains high levels of nutrients and other valuable materials.

There is a long historical tradition of capturing the concentrated chemical value that urine and feces represent, but several factors ultimately led to the decline of this tradition of excreta reuse. First, the adoption of the water closet led to dilution of human waste with clean water, making it harder to recover its value. Second, the growing availability of other fertilizer sources reduced the value of fertilizer from human waste. Third, the Sanitary Revolution led to increased awareness of the dangers of growing food with poorly treated sewage and a desire to simply "get rid of it." Fourth, the emerging aesthetic of the "modern city" favored large, centralized sewers over informal, decentralized collection and reuse.

Given the scale of modern cities, there may be no way to return those nutrients to the soil they came from. But we should at least ask whether there are safe ways to start closing the loop. We have already discussed the potential of wastewater reclamation as a "new" water supply (Chapter 13); we will return to this theme in the next chapter.

Chapter Highlights

1. Water can transmit infectious diseases through multiple pathways; there are **waterborne**, **water-washed**, **water-based**, insect-borne, and **environmental pathogens**. Breaking the cycle of disease transmission requires access to safe water, sanitation, and hygiene (WASH), as well as better management of ambient surface water.
2. The historical evolution of urban water systems consists of three stages. In David Sedlak's formulation, Water 1.0 is the Roman model of centralized water delivery and drainage, Water 2.0 adds drinking-water treatment on the front end, and **Water 3.0**, which serves most Americans, also adds wastewater treatment, which is important for protecting aquatic ecosystems and drinking-water sources.
3. Even with water treatment plants in place, it is still important to protect water sources from polluting activities. Water providers are implementing effective source-protection programs in cooperation with local governments and residents.
4. Under the Safe Drinking Water Act, EPA sets standards for contaminants in water (**MCLGs** and **MCLs**) but has been slow to address new threats and to update existing standards.
5. Water 3.0 has led to a large decline in water-related infectious diseases in the United States over the past century. The remaining health threat is associated primarily with environmental pathogens such as *Legionella*, along with a poorly quantified risk from chemical contaminants, including arsenic, lead, and **disinfection byproducts**.
6. Some 2 million people in the United States—disproportionately Indigenous and people of color—do not have access to drinking water and sanitation. Even among those with access, there are significant racial and economic disparities in safety and quality of service. Prominent examples include the lead water crisis in Flint, Michigan, the lack of household plumbing in the Navajo Nation, the contamination of private wells in California's Central Valley, and the use of "straight pipes" in Lowndes County, Alabama.

7. Some of the weaknesses of US water systems include underinvestment in infrastructure maintenance; environmental contamination from combined sewer overflows (CSOs); a slow and reactive approach to new chemical threats; a reliance on potable water to flush away human waste; linear flows of water and nutrients; the large number of small, underresourced rural utilities, especially in low-income and minority communities; the high costs of maintaining and improving water infrastructure in *legacy cities*; and the increasing use of bottled water, which reflects, in part, public mistrust of the water-supply system.

Notes

1. McMahon (2015), Flammer et al. (2020).
2. Anastasiou and Mitchell (2015).
3. Smith (2013).
4. Sedlak (2014).
5. https://www.epa.gov/sdwa/drinking-water-regulations-and-contaminants.
6. Since the 1996 amendments, EPA has established five new standards for disinfection byproducts and microbes using rule-making processes separate from the CCL process.
7. Raucher et al. (2011).
8. ASCE (2021b).
9. GAO (2020b).
10. Ibid.
11. EPA (2019).
12. https://www.brookings.edu/blog/up-front/2021/05/13/what-would-it-cost-to-replace-all-the-nations-lead-water-pipes/.
13. https://gothamist.com/news/newark-nears-finish-lead-pipe-removal-record-time.
14. This is much higher than the cost per line cited above, in part because city and state regulations require that sewer lines be replaced at the same time.
15. https://www.circleofblue.org/2021/world/some-chicagoans-wary-of-lead-pipe-replacement/.
16. https://mannyteodoro.com/?p=2888.
17. ASCE (2021d).
18. https://www.epa.gov/cwns.
19. Hummel et al. (2018).
20. Pluth et al. (2021).
21. https://grist.org/cities/cities-are-investing-billions-in-new-sewage-systems-theyre-already-obsolete/.
22. Collier et al. (2021).
23. Barskey et al. (2022).
24. Evans et al. (2019).
25. DigDeep and US Water Alliance (2019).
26. Meehan et al. (2020).
27. https://tribalcleanwater.org/.
28. Tanana et al. (2021).
29. Bureau of Reclamation (2018), Chapter 5.
30. https://www.texastribune.org/2017/08/22/colonias-border-struggle-decades-old-water-issues/.
31. One of the main chemicals of concern in San Joaquin Valley groundwater is 1,2,3-trichloropropane (1,2,3-TCP), which is not itself a pesticide but was a manufacturing contaminant in two widely used fumigants. 1,2,3-TCP was banned in the 1990s but is still commonly found in drinking water wells in the valley and beyond.

32. Balazs et al. (2011).

33. Schaider et al. (2019).

34. London et al. (2018).

35. Gibson et al. (2014).

36. McKenna et al. (2017).

37. https://www.justice.gov/opa/pr/justice-department-announces-environmental-justice-investigation-alabama-department-public.

38. Further complicating the picture are another 18,000 or so non-transient non-community water systems (defined by EPA as "public water systems that regularly supply water to at least 25 of the same people at least six months per year," e.g., a hospital with its own water supply) and more than 80,000 transient non-community water systems ("public water systems that provide water in a place such as a gas station or campground where people do not remain for long periods of time").

39. Pincetl et al. (2016).

40. In reality, there is a spectrum of private involvement in water supply, from private ownership of public debt, through management contracts, to build–operate–transfer arrangements. In our discussion, any situation in which a private company is responsible for day-to-day operation is referred to as a private utility.

41. Allaire et al. (2018).

42. Baltimore is a notable example of a "recalcitrant" agency with "an astonishingly inept and unresponsive bureaucracy"; Lee (2021).

43. Kishimoto et al. (2015).

44. Allaire et al. (2018).

45. DeSimone et al. (2009).

46. Gibson et al. (2020).

47. https://www.theguardian.com/us-news/2021/feb/28/california-east-orosi-toxic-america-water.

48. Nylen et al. (2018).

49. https://www.waterboards.ca.gov/drinking_water/programs/compliance/.

50. https://www.southerncitymagazine.org/regionalization/.

51. https://www.lincolninst.edu/research-data/data-toolkits/legacy-cities.

52. Atlanta supplements user fees with a small sales tax dedicated to water infrastructure. Although a sales tax is a regressive instrument, it does have the advantage of shifting some of the cost to visitors and commuters in recognition of the burden they place on the city's water and sewer systems.

53. https://waterpolls.org/ap-gfk-flint-water-poll-2016/; Javidi and Pierce (2018).

54. Allaire et al. (2019).

15 Urban Water Management: Water Conservation, Stormwater Management, and Beyond

- How can cities prepare for drought and ensure water-supply reliability?
- How much water do we use in our homes, and how can we reduce this amount?
- How can better water pricing help support healthy utilities, promote water conservation, and ensure that water is affordable to all?
- What can cities do to make better use of local water flows, such as stormwater and wastewater?

Urbanization is a defining feature of our time, with both positive and negative environmental effects. On one hand, urbanization concentrates the human footprint, thus preserving biodiversity elsewhere, and city dwellers tend to use less energy for transportation and heating than people living in rural communities. On the other hand, cities are the endpoints of massive supply chains that bring food, energy, and goods from farms, factories, mines, and refineries around the world; the remote impacts of these supply chains are invisible to most city dwellers. The balance between the positive and negative effects of urbanization depends on how cities are built and managed.[1]

In this chapter, we take a water perspective on the impacts of urbanization, trying to remain cognizant of both local and remote impacts. In the former category, we spend some time thinking about urban runoff and wastewater and how they can be better managed to protect urban streams. In the latter category, remote impacts should include the industrial and agricultural water footprints of cities, but those are covered in Chapters 17 and 18, so we focus here on direct water use as the thread that connects cities with the rural areas that often provide their water. Minimizing water use through conservation and making better use of local water resources (rainfall, stormwater, and wastewater) can help reduce the amount of water that the city must import and thus provide water security to cities while reducing impacts on aquatic ecosystems and rural communities.

This chapter brings together, and expands on, multiple threads from previous chapters:

- Many cities face a looming scarcity crisis as demand continues to grow and supply becomes more variable (Chapter 5). Wastewater reuse and water harvesting can help alleviate this scarcity (Chapter 13), as can the reallocation of agricultural water to urban use (Chapter 11).
- Flooding problems are increasing in cities because of climate change, land-use change, and poor flood management (Chapter 7).
- Water infrastructure in the United States faces problems of deferred maintenance, funding gaps, inequitable coverage, loss of public trust, and degraded water quality (Chapter 14).
- The linear design of urban water systems—clean water in, wastewater out—may not be sustainable in a rapidly urbanizing world (Chapter 14).

These intersecting crises are birthing a new paradigm in urban water management, which aims to help alleviate both scarcity and flooding while also engaging urban residents and enhancing their quality of life. This paradigm, sometimes known as Integrated Urban Water Management or One Water, is built on an integration of previously separate realms—water supply, wastewater, drainage, flooding, land use, ecosystem health—into a more holistic approach to urban water management. The cornerstones of this approach are the following:

- **Reduce**: Instead of taking customer water demand as a fixed (or ever-rising) quantity that must be met, cities and utilities are working to reduce demand by using a variety of tools.
- **Retain and reuse**: Recognition of the value of other urban water flows—rainwater, stormwater, wastewater—is generating strategies that capture those flows for productive use while mitigating flooding and environmental degradation.
- **Restore**: Restoration of urban streams is creating better stormwater solutions, healthier ecosystems, and valuable recreational spaces.
- **Rebuild relationships**: Through all these measures, relationships are being nurtured: relationships between utilities and their customers, between city governments and their residents, between people and rivers, and between urban and rural areas.

While the main focus of this chapter will be the newer tools mentioned above, we shouldn't forget that the most basic responsibility of an urban water utility is to provide a reliable water supply to its customers; indeed these tools are meant to address that goal (and others). The chapter thus starts with a focus on reliability.

1. Defining and Achieving Reliability

The most basic responsibility of an urban water manager is to maintain an adequate water supply in the face of hydrologic variability. At the same time, it doesn't make sense to build expensive excess supply that is never used. This is the unenviable balancing act that water managers must perform: *Do* prepare for drought, but *don't* overbuild in a vain attempt to reduce risk to zero.

Central to this balancing act are the concepts of risk and reliability. Let's return to the idea that risk is the product of likelihood and consequence, a concept that we first introduced in the context of drought (Chapter 2). In the case of urban water supply planning, we are dealing with the likelihood that a water supply system will not be able to meet demand, with the consequence being mandatory restrictions on water use. *Reliability*, then, can be defined as the opposite of likelihood (i.e., the probability that a water supply system will be able to meet demand; a 1 percent likelihood of failure corresponds to a reliability of 99 percent).

Ideally, we'd like an urban water supply to have a reliability close to 100 percent. But in practice, achieving that reliability in the face of climatic variability can be prohibitively costly. Figure 15-1, a modified version of Figure 9-3, illustrates the problem: As the desired reliability increases from 90 percent to 95 percent to 99 percent, the amount of storage needed increases dramatically. At some point, it probably makes more sense to think about ways to reduce the *consequences* of low-probability shortages rather than trying to further decrease their likelihood.

To put it slightly differently, every water system has some level of drought at which it will experience shortage (i.e., will not be able to meet all the demand). Once we accept this, two questions follow:

- **What is an acceptable likelihood of shortage?** In the United States, urban water systems generally are expected to be able to meet demand during a *critical dry period*, usually defined as either a hundred-year drought or the drought of record. Of course, droughts more severe than the critical dry period do happen (especially with a changing climate), implying a nonzero probability of shortage. Regulators may require a margin of safety (an excess of supply over demand during the critical dry period[2]), which reduces but does not eliminate the risk of shortage.
- **How will shortage be managed?** Shortage is not necessarily catastrophic. Most urban water systems can achieve a significant short-term reduction in water use

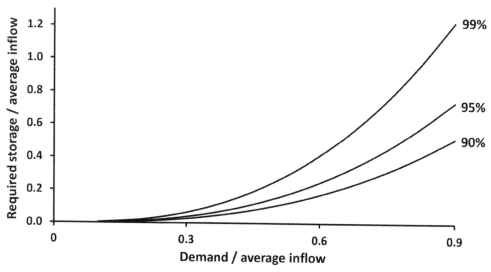

Figure 15-1. Effect of desired reliability on needed reservoir storage. Vertical axis: the amount of storage needed, expressed as a fraction of average inflow. Horizontal axis: water demand, expressed as a fraction of average inflow. Curves correspond to reliabilities of 90 percent, 95 percent, and 99 percent. Calculated using the same approach as in Chapter 9, Figure 9-3 (CV = 0.4 and skewness = 0.4).

without a correspondingly large reduction in quality of life by targeting nonessential uses, especially outdoor water use. In addition, drought planners should be prepared to activate backup or emergency supplies, typically water sources that are more expensive, have poorer water quality, or pose greater environmental impacts. Agricultural water can provide a useful safety valve for urban water supply during drought, but equitable agreements must be reached with farmers before droughts materialize. Planning for drought before it happens can help dampen what has been called the ***hydro-illogical cycle***, in which concern builds to panic during drought periods, but apathy follows as soon as it rains. Drought *will* recur, and we will handle it better if we are prepared. Drought planning is rendered more complex by the fragmentation of responsibility across utilities and levels of government. In addition, smaller utilities may have a harder time preparing for drought.[3]

2. Conservation

Reducing water use is a key tool for cities and regions facing scarcity, whether that scarcity is driven by drought or climate change or population growth. Even for regions that have abundant water supplies, reducing water use is helpful in reducing energy use and minimizing the environmental impacts of water withdrawal, treatment, and disposal.

2.1. How Much Do We Use?

To understand the potential for urban water conservation, we first need a better sense of how much water we use and for what purposes. This section explores the ways that households and cities use water and how much water those uses typically need.

To live a healthy life, people need water for drinking, as well as preparing and cooking food and cleaning ourselves, our homes, and our clothes; the minimum acceptable amount is often cited as somewhere between 20 and 50 L per person per day (Lpcd). In Chapter 4, we estimated current domestic water use in the United States (based on US Geological Survey data) at 500 Lpcd, much higher than these values. Is this a fair comparison, or have we ignored some important definitional issues?

You probably won't be too surprised to learn that definitional issues do in fact confound this comparison. While Americans certainly use more than the minimum amount of water, we don't actually use 500 Lpcd inside our homes, because the US Geological Survey data (and AQUASTAT domestic water-use data) include several categories in addition to indoor water use:

- Outdoor water use (e.g., for watering lawns and gardens and filling swimming pools) can account for more than half of total household water use in regions with large lawns and high potential evapotranspiration.
- Commercial, industrial, and institutional (CII) water use is water from the local utility that is used by businesses and institutions such as schools and hospitals.
- ***Nonrevenue water (NRW)***—water for which a utility does not receive payment—can be grouped into three subcategories: real losses (water lost to leaks in water mains and other infrastructure); apparent losses (water that is used by customers but is not billed because of metering inaccuracies or water theft); and authorized unbilled uses (e.g., water used for firefighting or pipe flushing).

All of these are included in the 500 Lpcd number that we are citing for US domestic use. In general, when comparing "municipal" or "domestic" or "residential" data from different sources, it is important to clarify whether CII and NRW are included. In this book, we use "municipal" or "domestic" interchangeably to include CII and NRW and specify "household" use when CII and NRW are excluded.

We can look in more detail at household and CII water use by using a database of monthly use (excluding NRW) in seventy-eight US cities from 2005 to 2017 (Figure 15-2). We note, first, that household use in these cities is, on average, somewhat larger than CII use. Second, CII and especially household use are highest in the summer, reflecting the importance of outdoor use. Outdoor water use is highly variable between cities and regions, with the highest values in the arid cities of the West (where grass needs constant watering during summer) and the lowest values in the cool, wet cities of the Northeast (Figure 15-2b). As a result, urban water demand tends to be highest where water supply is lowest. Per capita water use also tends to be higher in smaller, lower-density cities and suburbs, where houses have large lawns and gardens.[4] Likewise, water use tends to go up during droughts, when less green water is available to support outdoor uses.[5]

A final observation from this database is that both household and CII water use have gone down over time, dropping by 21 percent and 15 percent, respectively, over

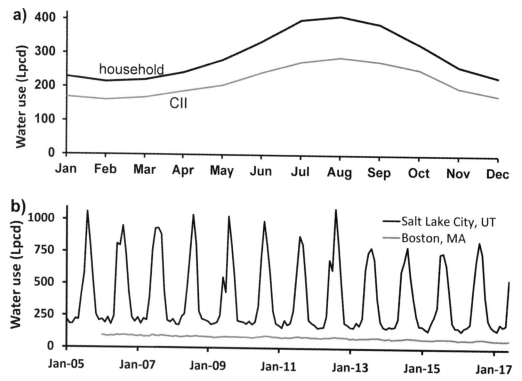

Figure 15-2. (a) Average household and CII water use in seventy-eight US cities by month and (b) monthly household water use for Boston and Salt Lake City. The difference in water use between Boston and Salt Lake City is driven in part by greater evaporative demand in Salt Lake City but also by much lower population density (i.e., more sprawl) in Salt Lake City. Data from Chinnasamy et al. (2021), including only the seventy-eight cities for which data were available for at least nine out of the thirteen years.

this short time period; the decrease has been largest in arid cities, and much of the decrease is associated with outdoor water use.

2.2. Tools for Conservation

How can we reduce household water use? In this section, we outline tools that individuals, utilities, and city planners can use to achieve that goal.

Outdoor water use: Given the high volume and fundamentally discretionary nature of outdoor water use, it is often a primary target of conservation programs. Landscape watering (including golf courses) is a particularly large water user in arid regions and in suburban households with large, manicured lawns. In arid regions, replacing lawns with drought-tolerant native species (xeriscaping) can eliminate most outdoor water use. During droughts, utilities and governments often impose mandatory restrictions on outdoor water use, although they can be hard to enforce. In a sign of the severity of the recent California drought, some utilities have resorted to installing flow-restricting devices when repeated fines don't deter wealthy homeowners from overwatering.[6] On the flip side, it is often appropriate to designate some water for irrigation of native, drought-tolerant vegetation in public parks, given the multiple benefits they can provide, including recreational opportunities and mitigation of the ***urban heat island effect***.

Leaks and fixtures: The average American "uses" 30–40 Lpcd in household leaks, so repairing these leaks represents a significant water saving opportunity. In addition, technological advances mean that newer plumbing fixtures and appliances do the same job with less water. In calculating payback periods for replacing older fixtures with more efficient ones, it is important to include not just the cost of the water saved but also the cost of the energy used to heat that water.

Behavioral change: Motivated people can save water through behavioral changes, such as shorter showers, fewer loads of laundry, or less lawn watering. However, behavioral change can work in the opposite direction as well. In particular, efficient fixtures don't always reduce water use by the expected amount because of the rebound effect; for example, when efficiency reduces the cost of operating a washing machine, people may do more laundry.[7] To create sustainable behavioral change in the right direction, people need to be aware of the impacts of their water use, but many people, even in water-scarce regions, are not aware of water scarcity or of how much water they are using.[8]

Incentives and awareness: Water utilities can support household conservation by providing discounted fixtures, water audits, turf buyback programs, and conservation-oriented pricing (see below). In addition, various behavioral tools can be used to raise awareness of water scarcity and conservation; these tools, such as bills that compare your use to your neighbors', are widely used by electric and gas utilities but are just starting to make their way into the water world.

Nonrevenue water: For older systems, NRW can amount to 20 percent or more of a utility's water use, which is a problem both for utility finances and because of the real water losses involved. These systems can benefit greatly from a concentrated effort to reduce NRW, which might include leak detection and repair, pipe replacement, and upgrades to water meters.

Population growth and land development: In areas that are experiencing rapid population growth and suburban sprawl, all the conservation measures discussed above may not be enough to counteract the upward trend in water demand. As more and more people move into an area and build large houses surrounded by huge expanses of grass, they may stress the water supply to its limit. ***Assured water supply*** laws are mechanisms to align land development with water availability, although loopholes sometimes limit their effectiveness.

3. Pricing

Appropriate water pricing is an important tool for conservation, but it is more than that. In this section, we identify three goals for the pricing of household water—cost recovery, conservation, and affordability—and discuss how to achieve these goals. This discussion focuses on pricing for water supply, but wastewater pricing involves similar considerations.

3.1. Cost Recovery

Some people believe that water should be free. After all, it falls from the skies as a gift from the hydrologic cycle and flows downhill to us without cost. Upon reflection, though, most people realize that what is being delivered from their taps is not the same as what falls from the sky or flows in a river. The water company must divert the water, store it, treat it, test it, and deliver it at a suitable pressure. Doing this requires the construction and maintenance of a complex system of infrastructure—reservoirs, pumping stations, treatment facilities, and pipes—that must be paid for. In fact, water utilities have a higher capital-to-revenue ratio than any other type of utility.

When water prices are too low, utilities may not be able to cover their costs and may have to cut back on their level of service or defer maintenance on essential infrastructure. Underfunded utilities may experience a vicious cycle of poor performance, where inadequate financing prevents the utility from taking steps—such as metering, bill collection, rate increases, and NRW reduction—that would put it on a more solid financial footing. Taken to the extreme, chronic underfunding leads to customers becoming reconciled to an unsafe and unreliable water supply and turning to bottled water or other private solutions (if they can afford them); this is the case in many cities in LMICs.

In other words, if we want a safe, well-managed water (and wastewater) system, we must be willing to pay for it. The next question, though, is *who* should pay for it. Economists tend to argue that costs should be borne by those who receive the service, to avoid **moral hazard** and increase accountability. On the other hand, water supply—and certainly wastewater treatment—has a public-good nature to it, so partial government funding may be necessary to achieve economically efficient levels of provision. One of the reasons for the success of the Clean Water Act was the massive federal spending that allowed communities to build wastewater treatment plants that they couldn't build on their own, providing two positive **externalities**: healthier rivers and cleaner source water for downstream communities.

Federal funding for water and wastewater infrastructure peaked in the late 1970s and by 2017 had fallen to 4 percent of the total, with the other 96 percent covered by states and consumers.[9] The remaining federal funding flows through a number of loan and grant programs:

- Two revolving loan programs—the Drinking Water and Clean Water State Revolving Funds—have replaced the federal grant programs of the 1970s and 1980s.
- The Water Infrastructure Finance and Innovation Act of 2014 provides loans and loan guarantees for large water and wastewater projects.
- The 2016 Water Infrastructure Improvements for the Nation Act provides funding for water systems in "small, underserved, and disadvantaged communities," as well as lead testing and abatement projects.
- The US Department of Agriculture's Water and Waste Disposal Loan and Grant Program provides funding to rural areas, small towns, colonias, and tribal land in rural areas.

The 2021 Infrastructure Investment and Jobs Act (IIJA) changed this picture dramatically by allocating substantial funding to water infrastructure, second in US history only to that provided by the Clean Water Act. In addition to $15 billion for lead pipe replacement, the IIJA provides $11.7 billion each to the two revolving funds, $10 billion for emerging contaminants (half through the revolving funds and half through the Water Infrastructure Improvements for the Nation Act), $3.5 billion for tribal water, sanitation, and solid waste infrastructure, and $1 billion for the Bureau of Reclamation's rural water infrastructure program.

Is this federal funding a long-overdue investment in the water sector, or will it create more problems than it solves by giving some utilities more money than they can responsibly spend, while well-managed utilities lose out because they have been maintaining their systems? The answer depends on how the money is targeted. The IIJA is a unique opportunity to level the playing field by preferentially directing funding to the sectors of society that have suffered from structural racism and systemic underinvestment and to the places that have the greatest gap between need and ability to pay. This includes the small, rural utilities and *legacy cities* that we identified in Chapter 14 as having the greatest problems.

States also need to step forward to prioritize funding for their neediest systems. California led the way in 2019 by creating a $1.4 billion fund to provide safe and affordable drinking water to vulnerable communities and live up to the promise of a 2012 state law declaring that "every human being has the right to safe, clean, affordable, and accessible water adequate for human consumption, cooking, and sanitary purposes."[10]

Still, despite federal and state support (which often comes in the form of loans that must be repaid), most water utilities in the United States rely on water fees from customers to pay for all or most of their expenses. Thus, ensuring a healthy utility generally means pricing water for *cost recovery*: charging customers at rates that reflect the total costs to the utility (including maintenance) divided by the total volume of water delivered (i.e., price = average cost).

3.2. Conservation

Besides maintaining fiscally sound utilities, water prices should also send a signal about the value of water. If utilities were to raise water prices to levels more closely approximating the true value of water, household water use would decline and water would be freed up from wasteful household uses and transferred to uses that provide greater societal value, such as keeping rivers from going dry.

To an economist, the right water price includes not just the costs of treatment and delivery but also the value of the natural water itself. The fact that water falls from the sky doesn't make it worthless, just as the fact that oil occurs naturally doesn't mean that it is without value. The value of oil rises and falls in response to changes in demand and supply, with the market price increasing as oil becomes scarcer; shouldn't the same be true for water? The ideal price of water, then, should equal the marginal cost of water (the cost of the last increment of water), including both supply costs (the costs incurred by the utility) and the value of the raw water (often expressed as the ***opportunity cost***, the value of the water for other possible uses). Note that the goal of economic efficiency (price = marginal cost) can conflict with the goal of cost recovery (price = average cost).[11]

Yet water is different from oil in several ways that make pricing it at its true cost very difficult:

- Water is essential for life. Applying cold-hearted economic calculations to a basic necessity seems wrong to many people. Yet economists would point out that food and shelter are also essential for life, and we nonetheless set prices for those goods in the marketplace.
- Water delivery is a natural monopoly. It doesn't make sense for multiple companies to lay parallel water pipes to serve the same area. Thus, water prices can't be set by market competition between suppliers. As is the case for other natural monopolies, government regulation of water rates is critical, and indeed water utilities in the United States must generally have their rates approved by a state agency of some kind.
- Water has uses that are hard to value. The value of water to support instream habitat is undoubtedly greater than zero, but it is hard to put a number on.
- Water is publicly owned. For cultural, historical, ethical, and emotional reasons, there is a consensus in most societies that private ownership of water sources (as opposed to use rights) is wrong.

Despite these issues, it seems clear that water prices should be higher, especially in water-scarce locations, to encourage conservation and more efficient use. Economists argue that price increases can achieve the same conservation targets as regulatory tools (e.g., limits on landscape watering) but can do so with greater efficiency, since customers make their own choices about how much they value different uses of water. If your garden is really important to you—and you can afford it—you will pay the high marginal prices for your water use, while your neighbor, who would rather spend the money on a vacation, will let her garden wither, making both of you happier than if the city told you how often you are allowed to water.

How much water could be saved by price increases? The standard measure of the responsiveness of demand to price is the *demand elasticity*, defined as the percentage change in demand that results from a 1 percent change in price. Residential demand elasticity in the United States varies depending on the type of use, the location, the pricing structure, and other factors, but it is often about −0.4 (i.e., a 1 percent increase in price results in a 0.4 percent decrease in demand).[12]

3.3. Affordability

The case for higher water prices that we made in the previous sections runs into one obvious problem, already alluded to above: High prices may make this basic human right unaffordable for low-income customers. It seems self-evident that nobody should be priced out of using a basic amount of water. An often-used threshold for affordability is that households should spend no more than 2 percent of their income on water and no more than 4.5 percent of their income on water plus wastewater services. By these metrics, water is unaffordable to at least 12 percent of Americans (and rising rapidly)[13] and to the lowest-income households in every census tract in eleven out of twelve US cities.[14]

A survey of twelve American cities by Circle of Blue found that over 1.5 million American households have water bills that are past due, with a total of $1.1 billion owed, including fees for late payment.[15] In Philadelphia, 36 percent of residential accounts are past due, with a median debt of $663. Part of the problem is that water bills are often opaque, and utilities are often unresponsive to customers who complain about incorrect charges or inconsistent billing.[16] In many jurisdictions, unpaid water bills, even small ones, can result in tax foreclosures. Water shutoffs, in which utilities deny water service to people who are behind on their bills, have rightly attracted significant anger from activists and the general public. Shutoffs contribute to a vicious cycle of poverty, since a lack of water at home requires households to spend more money and time getting water in other ways.

The fundamental solution to this problem lies not just in the water realm but in alleviating poverty more broadly. In many cases, the real problem in poor communities is not that water prices are too high but that people are too poor. But given the realities of poverty both globally and in the United States, how should we deal with water pricing for poor households?

Two principles should guide our response to this question. First, affordability should not be an excuse for lowering water rates across the board and giving up on the goals of cost recovery and conservation. Second, ensuring that essential water is affordable for all is a high-priority goal that serves the public good by increasing public health and community integrity.

There are at least two ways to square these competing principles. First, we can make a basic amount of water free or very cheap for everyone and then charge much higher prices to higher-volume users in order to cover the cost; this would have the advantage of creating a large conservation incentive across the board. This solution suffers from two problems: providing free water to rich people and having to define what a "basic" amount of water is. (Given the large variation in the number of people in a household, a uniform per-household limit is clearly inequitable, but many utilities in the United States do not have the information needed to implement per-person limits.)

Instead of subsidizing basic use for everyone, a second approach (more common in the United States) is to subsidize all water use for low-income residents, similar to low-income assistance programs for other utilities, such as heating fuel or electricity. Several American cities have recently enacted affordability programs, such as Philadelphia's Tiered Assistance Program, which caps water bills based on residents' income and makes up the difference with a small surcharge on all customers. A similar program in Baltimore, passed in 2019, was not implemented until 2022, showing how hard it is for some water utilities to take on this nontraditional role of social utility. In Detroit, a proposal to tie water rates to income (among other measures) was rejected by voters in August 2021. In California and some other states, publicly owned water utilities are explicitly prohibited from using water rates to cross-subsidize different groups of customers.

In response to the COVID-19 pandemic, Congress created a federal Low Income Household Water Assistance Program, although it appears to be having trouble reaching those who need help.[17] Analyses of existing low-income customer water assistance programs find significant problems, including burdensome paperwork, insufficient outreach, flawed eligibility criteria, and exclusion of renters from eligibility.[18]

3.4. Rate Structures

To summarize, we want our water prices to do several things at once: ensure cost recovery, provide an appropriate signal of value, and be affordable to all. Each utility tries to satisfy these different goals in its own way, leading to a bewildering variety of water rates and rate structures.

There are several components to a typical residential water bill:

- **Fixed charge**: This is a fixed fee for service regardless of how much water is used. For households that don't have a meter, this may be the whole bill, but for households with a meter, there is also a . . .
- **Water-use fee**: This charge depends on how much water is used, based on one of several approaches:
    - **Uniform volumetric rate**: The same per-unit rate is applied regardless of water use.
    - **Decreasing block rates**: After a certain volume of water use, additional water is charged at a lower rate. This option—which was meant to reward high-volume customers—is becoming less popular as water becomes scarcer.
    - **Increasing block rates**: After a certain volume of water use, additional water is charged at a higher rate. In the extreme case, the first block, perhaps meant to cover essential uses, is free. Many advocates believe that increasing block rates are beneficial from both conservation and affordability perspectives, although economists generally prefer uniform volumetric rates.

The balance between the fixed charge and the variable water-use fee is a critical issue in rate setting. Putting more of the charge into the variable fee is beneficial in terms of both conservation (customers are more motivated to reduce water use) and equity (low-income customers can better achieve affordable water bills by minimizing their water use). However, because so much of the utility's costs is fixed, a reduction in water use can quickly lead to significant shortfalls when fixed charges are low.

One last thought on water pricing: For complex reasons (including the low cost of water and the resistance to treating it as an economic good), the water sector lags behind other utilities both technologically and institutionally. For example, most renters don't pay for water, but they do pay for electricity and gas. Electricity, gas, and internet are typically paid monthly, whereas water is often paid quarterly. Practically all electric and gas connections are metered, which is not the case for water. Some electric utilities are starting to use "smart meters" and dynamic pricing (in which the cost of electricity fluctuates in response to changes in supply and demand), a move that most water utilities can only dream of. Improving the capacity of water utilities to implement innovative pricing structures will ultimately improve both water supply and ecosystem health.

4. Reuse and Decentralization

The previous sections have focused on urban water supplies: ensuring their reliability (Section 1), reducing use to alleviate scarcity (Section 2), and pricing (Section 3). We turn now to other urban water flows, specifically wastewater and stormwater. In Chapter 13, we discussed the shifting attitude toward these water flows, from "get rid of them as quickly as possible" to "use them as water sources." Here we return to this theme and look at some specifically urban examples of wastewater reuse (in this section) and stormwater harvesting (in the next section). Since we have already discussed centralized water reuse (Chapter 13), this section focuses on decentralized wastewater reuse. (Besides water, there is also the potential to recover nutrients, energy, and information from sewage flows; those topics are covered on the companion website.)

The centralized water systems that dominate modern cities have both strengths and weaknesses (Table 15-1). Some have argued that the future of urban water lies in decentralized water systems that are more sustainable, flexible, and resilient; water harvesting and wastewater reuse would feature prominently in these systems.[19] Decentralization could solve many of the problems we identified in Chapter 14 by creating small, circular water systems that are more adaptable, less dependent on large-scale infrastructure, and more responsive to user needs.

A key component of safe decentralized reuse is the separation, at the source, of the different types of wastewater:

- **Graywater:** Water draining from sinks, dishwashers, clothes washers, and showers is not highly polluted, but it can contain soaps, sediment, grease, dander, personal

Table 15-1. Summary of the strengths and weaknesses of centralized urban water infrastructure (Water 3.0).

Strengths	Weaknesses
- Well-established engineering requirements for treatment, distribution, and collection	- Requires an extensive network of underground infrastructure, which is expensive and hard to monitor, maintain, and replace
- Known public health effectiveness	- Uses potable water for flushing human waste
- Amenability to centralized operation, regulation, and monitoring	- Treats resources (rainwater, stormwater, human excreta) as wastes to be disposed of
- Economies of scale	- Vulnerable to terrorist attacks, natural disasters, and energy supply disruptions
	- Lasts for multiple decades, limiting resilience to climate change
	- Not adaptable to changes in population size
	- Failures can lead to large-scale problems

care products, and even pathogens at low concentrations. This ***graywater*** can be reused within the household—most commonly for landscape irrigation or toilet flushing—with low risk. Graywater reuse can range from jerry-rigged approaches (a bucket in the shower that is then used to flush the toilet) to more sophisticated systems involving pumps, storage tanks, and filters.[20]

- **Blackwater**: Water from toilets—referred to as ***blackwater***—is highly pathogenic and much riskier to reuse at the household scale. Modular ***membrane bioreactors*** can be used to treat blackwater (or combined graywater and blackwater) from apartment buildings or clusters of houses, although the costs are usually high. Composting toilets have long been in use at backcountry sites and are being tried in office buildings as well, though not always successfully.[21]
- **Urine and feces**: Because of the difficulties of managing blackwater—and the desire to avoid using water to flush toilets—there is increasing interest in ***urine-diverting dry toilets (UDDTs)***, which provide further separation at the source:
  - Urine is nutrient rich and not highly pathogenic and can be disposed of in various ways, including use as fertilizer.
  - Feces is highly pathogenic, but when separated from urine and flushing water, it is low in volume and easier to dry and treat.

Water for flushing or anal cleansing should not be used in UDDTs, although a small amount of water is often necessary to keep the device clean. UDDTs are often part of an approach called *ecological sanitation*, whose primary goal is to close the loop and capture the value of human excreta.

There are many permutations of decentralized reuse; Figure 15-3 identifies three scenarios.

Scenario A: Use nontraditional sources within the existing grid. San Francisco is a good example of the emerging interest in incorporating decentralized reuse and water harvesting into a predominantly centralized water system (Figure 15-3a). Since 2015, the city has required new buildings over 250,000 square feet to capture graywater, rainwater, and foundation drainage (groundwater collected to prevent seepage into the foundation) and use that water to meet toilet-flushing and irrigation needs. San Francisco's regulations require treatment that depends on the type of reuse. For example, buildings that reuse graywater for toilet flushing must reduce the concentration of enteric viruses by six orders of magnitude to avoid aerosolization of viruses during flushing.

Scenario B: Get off the grid. Although the San Francisco buildings are an opportunity to showcase new technologies and approaches, the fact that they are still connected to the centralized water and sewer systems undercuts some of the advantages of decentralization. The ultimate goal would be a truly off-the-grid home or building complex, using water harvesting and reuse to meet all its water needs (and some of its energy needs) while generating little if any waste (Figure 15-3b). However, this is rarely practical in areas that are already served by water systems, because of the sunk costs and technological lock-in associated with current systems. Social lock-in is an important limitation too; most people in US cities are content with the current system, where they can turn on a tap and flush a toilet without knowing or thinking

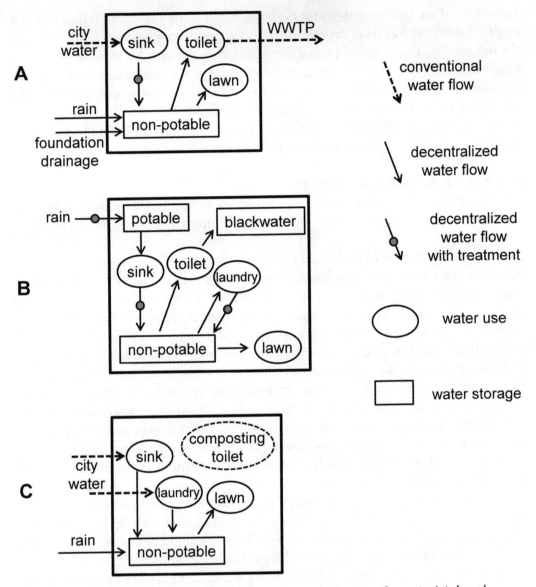

Figure 15-3. Scenarios for decentralized urban reuse or harvesting. Scenario A is based on San Francisco and scenario B is modified from Rabaey et al. (2020). In scenario B, blackwater can be treated for nonpotable reuse, discharged to a septic tank, or stored and picked up as needed.

much about the water systems involved. Decentralization would also require a paradigm shift for traditionally conservative utilities and regulators. All these factors mean that innovative, decentralized systems may be more appropriate in rapidly growing cities in low-income countries, where resource constraints and astronomical growth rates mean that simply extending current systems is not a viable option; we will return to this issue in Chapter 16.

Scenario C: Use some city water but eschew sewers. For urban households that already have a safe centralized water supply, it probably makes sense to continue using that source for drinking, but there are still opportunities to minimize impact on the larger system. In Figure 15-3c, for example, we imagine a household that uses city water for drinking but relies on rainwater and graywater for many nonpotable uses. We also imagine the household using a composting toilet or other waterless solution, which reduces water use, eliminates the need for a sewer connection, and captures the value of the excreta. The viability of this scenario is heavily dependent on the availability of a safe, affordable, reliable, and effective waterless toilet. There are several such models at different stages of development, using treatment approaches that range from membrane filtration to combustion to composting.

5. Urban Stormwater Management

The last section of the chapter discusses urban stormwater management, a key tool for improving stream health and increasing water supply.

5.1. Stormwater and the Urban Stream Syndrome

Although some cities have **combined sewers** (which present their own problems, as noted in Chapter 14), stormwater in most cities is carried to rivers by a system of drainage pipes called a **municipal separate storm sewer system (MS4)**. MS4s do not carry sewage, but the stormwater they carry can still be a problem, since it carries pollutants from urban surfaces, including petroleum products from roads, parking lots, and gas stations; salt from winter road treatment and concrete weathering; nutrients from lawn fertilizers, septic systems, pet waste, and atmospheric deposition; metals from roofs and car wear; thermal pollution (heat) from hot **impervious surfaces**; and bacteria from pet waste and septic systems. Urbanization increases pollutant loads in two ways: by increasing sources of pollution in the landscape and by shifting the pathways of water flow, which decreases the opportunities for pollutants to be removed as water moves through soils.

The problems with urban stormwater go beyond traditional pollutants. As noted in Chapter 4, urbanization—particularly the combination of impervious surfaces and MS4s—short-circuits watershed flow pathways, leading to flashier streams. This flashiness also affects stream geomorphology: Higher stormflows cause incision and bank erosion, while the rapid decline in flows after storms can lead to instream sediment deposition, especially if watershed erosion is high (e.g., from construction sites). Add to this the direct physical modifications to urban streams (e.g., levees, riprap, straightening, burying, culverts) and, sometimes, the impacts of CSOs or SSOs, and you have a phenomenon that has been called the **urban stream syndrome**: the complex but consistent pattern of ecological degradation of urban streams. Since these same urban streams are the most frequent experience of nature for many people, their degradation leads directly to a lower quality of life and a disconnection from the natural landscapes that are so critical to our well-being.

5.2. Urban Stream Restoration

What can we do to better manage urban watersheds and restore urban streams? Given the numerous insults to these ecosystems, improving their health requires a variety of different approaches targeted at different scales, ranging from the stream itself to the larger watershed.

Stream Channel

Stream stability and habitat quality can be improved by various restoration techniques, including dam removal, ***daylighting*** of buried rivers, replacement of streambank riprap with vegetation, improvement of instream habitat (e.g., by adding woody debris), and beaver restoration. These measures require a deep understanding of how rivers function geomorphically and biologically. Restoration of physically degraded rivers must intelligently increase channel complexity, adjust channel capacity to new hydrologic regimes, and work with, rather than against, the power of water. Instream restoration should also include improvement of water quality by removal or mitigation of pollutant sources such as CSOs, point source discharges, and contaminated sediments. However, instream restoration is unlikely to be fully successful if attention is not paid to floodplain and watershed processes as well.

Floodplain

The riparian zone should be a particular target for protection and restoration. When left as an intact vegetated system, this part of the landscape provides room for the river to flood, slows down stormwater en route to the stream, and removes pollutants through biological and physical processes. The functions of the riparian zone will be undermined if storm sewers carry water under the riparian zone directly to the stream, so restoration efforts must minimize the flow of water through MS4s directly to streams.

Watershed

Successful urban stream restoration must address the watershed and its role in generating stormwater. The basic goal of urban stormwater management is to slow water down and retain it on the landscape rather than having it run off quickly to the stream. This requires a fundamental rethinking of urban design, since traditional engineering practices are centered around getting water off the landscape as quickly as possible. There are three primary benefits to slowing water down:

- Streams regain a more natural, less flashy flow pattern, which improves ecological health and reduces flood damages along stream corridors.
- Stresses on the sewer system (MS4 or combined sewer) are reduced, decreasing the likelihood of basement flooding and CSO events; this can minimize the need to invest in expensive, hard infrastructure such as tunnels.
- Water is retained in soils for supporting plant growth and recharging groundwater, with the latter potentially increasing the water available for human use.

The first two factors (streams and sewers) tend to drive stormwater management in wetter areas, while the latter (water harvesting) is more important in arid regions. Regardless of the primary goal, the practices involved are similar and are referred to broadly as stormwater best management practices (BMPs) or green infrastructure (GI); some practices are summarized in Table 15-2. The common thread that ties these together is the effort to undo the hydrologic effects of urbanization by minimizing impervious area or disconnecting it from waterways.

Table 15-2. Some green infrastructure (GI) practices for urban stormwater and the scale at which they operate (individual lot, street, or neighborhood).

Scale	Name	Description
Lot	Rain barrel	A barrel that collects water from rooftops (usually through gutter downspouts); that water is later used for landscape irrigation or other uses.
Lot	Rain garden	A depression that receives flow from surrounding pervious and impervious areas, planted with vegetation that can withstand occasional standing water as runoff infiltrates into the soil.
Lot	Green roof	A vegetated rooftop that absorbs water rather than shedding it.
Lot	Minimizing structural imperviousness	Taller buildings with lower rooftop area.
Lot or street	Minimizing paved imperviousness	Clustered housing, shared driveways, narrower streets, use of pervious pavers that allow water infiltration.
Street	Bioswale	A ditch or depression that receives runoff from the street and is designed to infiltrate water while pollutants are removed by plants, soil, and sunlight.
Street	Infiltration gallery	A gravel infiltration trench installed under the street, with stormwater from catch basins directed to it.
Neighborhood	Detention basins and retention ponds	Depressions that collect stormwater flow and slowly release it, infiltrate it into the soil, or treat it through sediment deposition and nutrient uptake. Detention basins are designed to be dry between storm events, whereas retention ponds have water most of the year.
Neighborhood	Constructed wetlands	Vegetated systems designed to treat stormwater.
All scales	Pollution prevention	Minimizing pollutant inputs through reduced lawn fertilizer use, pet waste cleanup, and other practices.
All scales	Education and maintenance	Improving the uptake and functioning of other practices.

These techniques are being broadly implemented in urban areas to reduce stormwater problems and recharge groundwater. China has designated thirty "sponge cities" that are meant to have 80 percent of their area capturing runoff by 2030. Compared with hard stormwater solutions such as sewers or floodwalls, GI is generally cheaper, more flexible, more modular, and easier to implement. At the same time, engineers may hesitate to rely on GI because its performance is more site-specific and harder to model than traditional **gray infrastructure**. The small scale of GI contributes to this problem, since one rain garden is unlikely to measurably affect runoff and it is hard to measure (or predict) the cumulative effect of a large number of unevenly distributed GI practices.

When evaluating GI for stormwater, we should also be aware of the other benefits that it can provide, although many of these have not been well studied,[22] and these co-benefits are highly dependent on the specific type and location of the GI.

- **Carbon sequestration**: Vegetated BMPs (e.g., bioswales and constructed wetlands) take up carbon dioxide and convert it to organic matter, which is stored in

the vegetation and soils, although this is likely to be insignificant in a city's carbon budget.
- **Wildlife**: GI provides patches of habitat in a landscape that is otherwise dominated by concrete and asphalt.
- **Heat reduction**: Vegetation can reduce the urban heat island effect by providing shade and transpiring water.
- **Human well-being**: GI can be calming and provide people with a daily dose of nature. Healthier rivers can become valuable recreational resources instead of hazardous eyesores.

On the other hand, GI is not a panacea and can come with significant drawbacks:

- **Water use**: In water-scarce cities, some GI may need to be irrigated to survive and may not an appropriate choice.
- **Pests**: GI can breed mosquitoes, especially when poorly maintained.
- **Litter**: Plastic bags, bottles, and other litter may accumulate in GI depressions. Of course, this is litter that otherwise would probably have ended up in rivers, so this can be seen as GI doing its job by trapping pollutants and preventing them from entering storm sewers.
- **Safety concerns**: Stormwater BMPs can sometimes be viewed as "abandoned" spaces and may become a magnet for illegal activity.
- **Gentrification**: GI can accelerate neighborhood gentrification.

Many of these problems can be solved by investing resources in maintenance and cleaning and by ensuring local management and control.

5.3. Legal and Financial Tools for Stormwater Management

The stormwater practices described in the previous section provide clear benefits to the public and the environment, but they also require leadership, commitment, and funding. What legal and financial incentives are there for cities to implement these practices?

First, most urban stormwater is regulated under the Clean Water Act. Despite the fact that it originates from diffuse sources over a large area, urban runoff is considered a ***point source***, since it flows through storm sewer pipes and discharges to streams from distinct outfalls. As a result, over 7,000 regulated entities (cities, towns, counties, special districts, universities, prisons, and state agencies) are required to obtain stormwater ***National Pollutant Discharge Elimination System (NPDES)*** permits for their MS4s.[23] These permittees are divided into Phase I MS4s (about 850 cities of more than 100,000 people that have been regulated since 1990) and Phase II MS4s (a larger number of smaller entities in urbanized areas that have been regulated since 1999). Phase II MS4 permits are often *general permits*; that is, there is one MS4 permit for the entire state, and each regulated entity must comply with its conditions. There are also permit requirements for stormwater runoff from construction, industrial sites, and transportation corridors.

To meet MS4 permit requirements, the town (or other MS4 owner) must create a stormwater management plan, which is meant to provide a framework for tackling the complex problem of reducing the quantity and improving the quality of stormwater runoff. Stormwater management plans have six required components, including public education, implementation of GI, and elimination of illegal connections to storm systems; monitoring of stormwater outfalls is often required as well. Still, most of the requirements are quite general, and enforcement is usually weak.

In cases where stormwater is a significant contributor to a stream's impaired status, ***total maximum daily loads (TMDLs)*** have been used to manage these impacts. However, the complex, multifaceted nature of stormwater impacts means that the traditional TMDL approach, which focuses on one pollutant, is not a perfect fit. Several TMDLs have tried more innovative approaches (e.g., setting limits on effective impervious cover), with mixed success.

The requirements to manage stormwater and comply with MS4 permits often fall as an unfunded mandate on cash-strapped municipal budgets. State governments and nonprofits can offer substantial technical assistance. In Connecticut, the Department of Energy and Environmental Protection provides maps of impervious areas and impaired waters for each town,[24] and the University of Connecticut's NEMO (Nonpoint Education for Municipal Officials) program provides substantial training and guidance on how to meet the various requirements.[25]

In terms of paying for stormwater management, close to 2,000 municipalities around the United States (as of 2020) have created ***stormwater utilities***: local or regional agencies that manage the stormwater system and charge customers (i.e., homeowners) a fee for this service.[26] Fees are typically based in some way on the amount of impervious surface on the property, and there are often credits for BMPs that disconnect impervious surfaces from the stormwater network. The median stormwater fee is about $5 per month for a single-family residence, but it can be much larger for businesses or institutions with large parking lots. In some places, the idea of a stormwater utility has been met with strong public opposition, often with "stop the rain tax" as a rallying cry. Of course, the fee is not for the rain itself but for what happens to the rain when it hits your driveway. For both political and legal reasons, advocates for stormwater utilities need to make sure that the fee is clearly tied to the service provided, a difficult task given that most homeowners do not understand that stormwater flowing off their property must be managed by the city.

Another way to pay for stormwater management is to put the burden on developers. Many cities require new developments to retain a given amount of stormwater on site (often the runoff from the first inch of rainfall). These programs can help with combined sewers, too; in New Haven, a "green redevelopment" requirement to retain on-site the runoff from a two-year, six-hour storm has reduced the number of annual CSO events by seven and the volume by 3.3 million gallons.[27]

An innovative new program in the Chicago area addresses the spatial disconnect between redevelopment areas—often Whiter, richer neighborhoods—and the areas with the greatest stormwater flooding problems, which are often poor communities of color. This program, called Stormstore, allows developers to buy credits from stormwater management programs elsewhere in the same watershed and will hopefully allow funding to flow toward the communities that need it the most.[28]

Chapter Highlights
1. Urban water managers are starting to think in an integrated way across various water flows (drinking water, wastewater, rainwater, stormwater, urban streams) to

alleviate scarcity, flooding, and pollution and improve the health of urban waterways and communities.
2. Urban water suppliers should strive to achieve a high level of reliability, but they must recognize that they may not be able to satisfy 100 percent of the demand 100 percent of the time. Advance planning can help cities weather drought and avoid the **hydro-illogical cycle**.
3. Urban water use includes indoor and outdoor household use, as well as commercial, industrial, and institutional use and **nonrevenue water**. Water demand, especially for outdoor use, tends to be highest in dry regions and during dry periods.
4. Water conservation programs have had great success in reducing urban water demand, but most cities still have more conservation potential. The greatest focus should be on reducing outdoor use, fixing leaks, and encouraging adoption of efficient appliances.
5. Higher water prices are a critical tool for encouraging conservation and ensuring financially sustainable water utilities, but there must be mechanisms to ensure that lower-income households can afford the water they need for basic uses.
6. Urban use of rainwater, stormwater, and wastewater as water supplies is growing, as it simultaneously addresses scarcity, stream water quality, and flooding. There is growing interest in combining wastewater reuse and water harvesting to create small-grid and nongrid alternatives to Water 3.0.
7. Urban streams suffer from a variety of insults, not all of which fit well with the tools available under the Clean Water Act. Stormwater flowing through **municipal separate storm sewers (MS4s)** is considered a point source, but effective stormwater management requires watershed-wide changes rather than end-of-pipe treatment.
8. Stormwater managers are increasingly using green infrastructure to reduce effective imperviousness, recharge groundwater, and help restore natural hydrologic pathways.

Notes

1. Simkin et al. (2022).
2. For example, the Connecticut Department of Public Health recommends a supply/demand ratio of 1.15 during the critical dry period (i.e., a margin of safety of 15 percent).
3. Mullin (2020).
4. Chinnasamy et al. (2021).
5. https://www.nytimes.com/2022/03/25/us/water-conservation-california.html.
6. https://www.nytimes.com/2022/06/03/climate/california-water-restrictions.html.
7. At the extreme, the rebound effect may be strong enough that efficiency leads to more use; this is referred to as the *Jevons paradox* and is commonly discussed in the energy context.
8. Brown (2017) finds that making water sources and their depletion more visible can help change attitudes toward conservation. Water managers need to walk a fine line between raising awareness of drought or scarcity and reassuring the public that they are doing their job of providing a secure water supply.
9. Congressional Budget Office (2018).
10. https://www.waterboards.ca.gov/water_issues/programs/hr2w/.
11. As a city builds out its water system, the marginal costs of water typically go down through economies of scale. However, when it has exhausted the cheap, nearby sources of water and must

turn to importing or desalinating water, marginal costs will go up. At some point, marginal costs may become higher than average costs, meaning that the economically optimal price will lead to a large profit for the utility, something that regulators frown upon. The economist's preferred solution—bill for water at the marginal cost but rebate the excess profit to customers or apply it to conservation—is often politically unpalatable, and prices in this situation are usually much lower than the marginal cost of water.

12. Bruno and Jessoe (2021).

13. Mack and Wrase (2017).

14. Colton (2020).

15. https://www.circleofblue.org/water-debt/.

16. https://www.npr.org/2019/02/08/691409795/a-water-crisis-is-growing-in-a-place-youd-least-expect-it.

17. https://mannyteodoro.com/?p=1521.

18. Vedachalam and Dobkin (2021).

19. Larsen et al. (2016), Rabaey et al. (2020), Leigh and Lee (2019), Hoffmann et al. (2020).

20. Hydraloop (https://www.hydraloop.com/) is an example of a high-end water reuse solution that is probably overkill. The household-scale version—which costs several thousand dollars and is designed for use in homes with conventional water supply and sewers—incorporates six-stage treatment of graywater, including ultraviolet disinfection, after which the water is used for toilet flushing, laundry, and landscape irrigation.

21. https://bullittcenter.org/wp-content/uploads/2021/03/The-Bullitt-Center-Composting-Toilet-System-FINAL.pdf; https://www.sierraclub.org/sierra/2021-4-fall/stress-test/gritty-truth-about-multistory-composting-toilets.

22. Pataki et al. (2011).

23. https://www.epa.gov/npdes/stormwater-discharges-municipal-sources.

24. https://portal.ct.gov/DEEP/Water-Regulating-and-Discharges/Stormwater/Municipal-Stormwater.

25. https://nemo.uconn.edu/ms4/index.htm.

26. Campbell (2020).

27. Greater New Haven Water Pollution Control Authority (2018).

28. https://www.metroplanning.org/work/project/48.

16 Water, Sanitation, and Health in Low- and Middle-Income Countries: Leaving No One Behind

- What are the health and economic impacts of inadequate access to water, sanitation, and hygiene (WASH)?
- How do we define acceptable levels of WASH access?
- Has WASH access improved over the last several decades, and how far do we have to go?
- How much funding will it take to solve WASH problems, and where will this funding come from?

In this chapter, we take a global view of water, sanitation, and hygiene (WASH), with a particular focus on low- and middle-income countries (LMICs) that are struggling to achieve WASH access for all their residents. We start by examining the links between WASH and development, including the health and economic burdens imposed by inadequate WASH and the role of WASH in the **Sustainable Development Goals (SDGs)**. We then use the SDG framework to analyze the current state of each of the three subsectors (water, sanitation, hygiene). The latter part of the chapter turns to solutions to WASH problems, focusing on the need for funding and the potential of unconventional solutions.

1. Water and Sustainable Development

In this section, we examine the critical role that water plays in global development. We start with the problem—the ways that inadequate WASH contributes to poverty and poor health—and then outline the solution envisioned in the SDGs: adequate WASH for all.

1.1. The Burdens of Inadequate WASH

Although there are many reasons to work toward adequate WASH for all, one of the most salient is the disease burden that inadequate WASH imposes on people in LMICs. Accurately estimating this disease burden requires knowing not just how many people die of WASH-related diseases, such as diarrhea, but also what fraction of this could be prevented by better WASH, a parameter called the population-attributable fraction (PAF).[1] Table 16-1 provides estimates of the WASH-related disease burden for different diseases, expressed both in deaths and in ***disability-adjusted life years (DALYs)***, a measure of the impact of disease on years of healthy life.[2]

These numbers lend urgency to this topic, as they reveal over 1.2 million deaths and 75 million DALYs per year attributable to inadequate WASH, mostly from diarrheal disease and respiratory infections, and mostly in children under the age of five. An additional 355,000 deaths and 30 million DALYs per year from malaria are attributable to environmental water management.

Table 16-1 represents an underestimate of the WASH disease burden, in at least two ways:

- Because of insufficient data, the analysis excludes chemical toxins (e.g., arsenic, fluoride), as well as some potentially significant infectious diseases (e.g., dengue, Legionnaire's disease).
- The PAFs shown in Table 16-1 are based on studies comparing disease incidence in households receiving WASH interventions (e.g., chlorine packets for disinfecting water collected from an unsafe source) with "control" households that did not receive the intervention. These comparisons are notoriously difficult for a variety of statistical, logistical, and ethical reasons. One could argue that if we were to succeed in truly solving WASH problems and providing high-quality water and sanitation to all, then childhood diarrheal deaths (and schistosomiasis) in LMICs would drop to near zero, as they have in the United States and other high-income countries; this suggests that almost all 1.4 million diarrheal deaths could be attributed to WASH.[3]

Although the burden of water-related disease is still much too high, it is important to acknowledge that progress has been made. The share of global DALYs that are

Table 16-1. Population-attributable fraction (PAF) and water-attributable disease burden in 2016. Using the first row as an example, there were about 1,400,000 deaths from diarrheal disease in 2016, of which 60 percent (PAF) are attributed to water, sanitation, and hygiene (WASH), resulting in 829,000 WASH-attributable diarrheal deaths. DALY, disability-adjusted life year. Data from Prüss-Ustün et al. (2019).

		Water-Attributable Disease Burden	
Disease	*PAF*	*Deaths (thousands)*	*DALYs (millions)*
Diarrhea	0.60	829	49.8
Acute respiratory infections	0.13	370	17.3
Malnutrition	0.16	28	3.0
Schistosomiasis	0.43	10	1.1
Soil-transmitted helminths	1.00	6	3.4
Trachoma	1.00	0	0.2
Total WASH		1,243	75
Malaria	0.80	355	29.7

attributable to WASH dropped from 14 percent in 1990 to 5.5 percent in 2019.[4] (The numbers for children under 10 are 25.2 percent and 17.3 percent, respectively.) On the other hand, climate change may undermine some of this progress, since both drought and flooding can drive increases in diarrheal disease.[5]

While undoubtedly important, health is not the only reason to work toward WASH access for all:

- Having convenient water and sanitation frees people from the time burden of meeting those needs, allowing those hours to be spent in other ways, such as education and employment.
- Water access opens up possibilities for increasing one's economic security, such as growing food in a garden plot or opening a small business.
- Acceptable water solutions are often much cheaper than the alternatives, such as buying water from a street vendor.
- Current water and sanitation options often pose unacceptable safety risks, especially for women and girls.

In short, lack of WASH access has real economic costs, which have been estimated at $260 billion per year, or 1.5 percent of global gross domestic product (GDP); for some regions, such as sub-Saharan Africa, these costs amount to over 4 percent of regional GDP.[6] Inadequate WASH access is both a symptom and a cause of poverty, playing a central role in a vicious cycle that perpetuates global and local inequalities. To address global poverty, we must ensure WASH access for all, but poverty often robs citizens of the political and economic power needed to achieve WASH access. Reversing this cycle is a top priority for global development.

1.2. WASH Goals

Given the importance of WASH for health, dignity, and economic advancement, the international community has increasingly recognized that achieving universal WASH access should be a global goal.

Global goals for WASH have been part of both the Millennium Development Goals (MDGs) and the SDGs[7]:

MDG Target 7.C: Halve, by 2015, the proportion of the population without sustainable access to safe drinking water and basic sanitation.

SDG Target 6.1: By 2030, achieve universal and equitable access to safe and affordable drinking water for all.

SDG Target 6.2: By 2030, achieve access to adequate and equitable sanitation and hygiene for all and end open defecation, paying special attention to the needs of women and girls and those in vulnerable situations.

As we discussed in Chapter 1, the SDGs for water are substantially more ambitious than the MDGs, and this pattern is clearly reflected in the statements above. Relatedly, the MDG for drinking water (though not for sanitation) was achieved by 2015, while the universal access called for by the SDGs clearly will not be met, at least in the strict sense, by 2030. Still, these ambitious and specific goals have their place in refocusing our efforts on trying to reach everyone.

2. Drinking Water

In this section, we address the global status of SDG 6.1 (drinking-water access). In order to do so, we first need to define what we are measuring.

2.1. From Goals to Indicators

Given the lofty SDG goal of "universal and equitable access to safe and affordable drinking water," we need to define a concrete, measurable indicator that will track progress toward that goal. The task of defining and monitoring water and sanitation indicators under SDGs 6.1 and 6.2 falls to the Joint Monitoring Programme (JMP), a collaboration of the World Health Organization and UNICEF. The JMP indicator for SDG 6.1 is "6.1.1. Proportion of the population using safely managed drinking water services," which in turn requires us to define "safely managed." JMP uses a five-rung ladder to define levels of water access, including "safely managed" (Table 16-2).

A key feature of the drinking-water ladder is the distinction between unimproved and improved technologies for drinking water. Improved technologies are those that are presumed to provide safer water and include the following:

- Piped supplies
    - Onsite: a tap in the home or yard
    - Public standpipe: a public tap where users can fill their water jugs
- Nonpiped supplies
    - Tubewells: narrow-bore drilled wells
    - Protected wells and springs: dug wells or springs, on the condition that they are protected from contamination (e.g., the well is covered)
    - Rainwater: usually collected from roofs
    - Bottled water: purchased in bottles or jerrycans
    - Trucked water: delivered periodically to households, often stored in a home tank.

The advantage of using these types of technological distinctions is clear: A survey can quickly determine whether a household is using one of the improved technologies. The disadvantage is also clear: Just because you are using one of these technologies

Table 16-2. Drinking water ladder used by the Joint Monitoring Programme. Only rung 5 is considered to fully satisfy SDG 6.1. The ladder used during the MDG period was similar but lacked the fifth rung; basic access was considered sufficient to achieve the drinking water MDG.

Rung	Description
1. Surface water	Untreated surface water
2. Unimproved	Unprotected wells and springs
3. Limited	Improved technology, but too far from home (>30 minutes round trip)
4. Basic	Improved technology, within 30 minutes round trip
5. Safely managed	Improved technology, on premises, available 12 hours/day, contaminant-free

does not mean that your water is actually safe and sufficient. Protected wells can run dry or be contaminated with arsenic. Piped supplies may be contaminated or may only run for a few hours a week. Water trucks may come infrequently, charge exorbitant prices, or draw their water from an unprotected source. Even if water is clean at the source, it can easily be contaminated during household storage. (Remember from Chapter 14 the importance of protecting drinking water along the whole pathway from source to tap.) And the burden of water carrying can be substantial, even if the source is within the acceptable distance of thirty minutes (Table 16-2); this burden falls disproportionately on women and girls.

The shortcomings of the purely technological metrics led the JMP to add a fifth rung to the ladder. "Safely managed" adds three requirements beyond basic access:

- The source must be on the premises (house or yard).
- The source must be available for at least twelve hours per day (to avoid the need for storage).
- The water must meet water quality standards for some of the most common contaminants: *Escherichia coli* (or other pathogen indicator), arsenic, and fluoride.

Unfortunately, this fifth rung is much more data intensive to monitor, and the most recent JMP report has "safely managed" data for countries representing only 45 percent of the world's population.[8] Hopefully, one benefit of adding the fifth rung will be a gradual improvement in countries' monitoring abilities. However, most funders would rather pay for a new water source than for monitoring to determine where a new water source is most needed.

2.2. Where We Stand

Figure 16-1 summarizes the state of drinking-water access globally for 2000 (the year that monitoring began) and 2020 (the most recent year available), and for the least developed countries (LDCs) for 2020 only.[9] It is clear that there has been some progress over the last twenty years: Even as population has grown, 12 percent more of the world's population has gotten access to a safely managed water source (now at 73 percent), and the percentage of people using surface water or other unimproved sources has been cut by more than half, to 7 percent. Nonetheless, some 770 million people (10 percent of the world's population) still don't have access to at least basic water services. In LDCs, about a third of the population still lacks basic access.

There are reasons to be somewhat skeptical of the claim that 73 percent of the world's population has truly safe drinking water. First, as noted above, data on safely managed water are incomplete, so the statistics in Figure 16-1 are extrapolated from limited data. Second, although a treated piped water supply (similar to that in US cities) is often assumed to provide safe water, this is not necessarily the case in many LMICs; in one review, more than half of studies found fecal contamination in at least some piped water samples.[10] In many cases, piped water may indeed be safe when it leaves the treatment plant (and thus technically qualify as safely managed) but becomes contaminated in distribution or storage.

One common cause for contamination of piped water supplies is **intermittency**. In many cities in poorer countries, households—even in richer parts of the city—receive

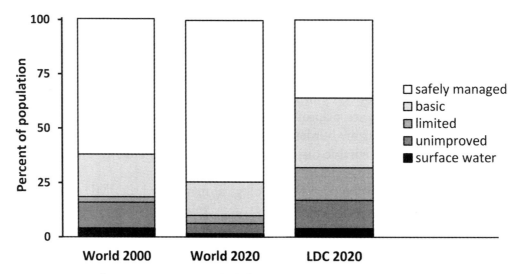

Figure 16-1. Population percentages with different levels of drinking water access. Least developed countries (LDCs) are the forty-six countries that the UN has identified as "low-income countries confronting severe structural impediments to sustainable development" (https://unctad.org/topic/least-developed-countries). Data from WHO/UNICEF (2021).

tap water for only a few hours a day, a consequence of inadequate water supply, insufficient pressure, or a desire by water managers to reduce water loss from leaking pipes. In India, for example, 93 percent of households with piped water reported having intermittent water availability.[11] Intermittency has several negative effects: inadequate pressurization allows contaminated water to enter the pipes; households must store water, leading to contamination during storage; repeated cycles of stagnant and flowing water in pipes can lead to release of biofilms, sediments, and pipe materials into the water; and households may discard stored water when taps come back on, leading to increased waste.

When piped water supplies are contaminated or unreliable, those who can afford it adapt by installing filters and water tanks or buying bottled water.[12] These understandable private responses can lead to acceptance of the status quo and disinvestment in the public supply. At the same time, many lower-income urban residents—especially those living in slums—lack basic water services and must resort to scrounging for water from local low-quality sources or paying high prices for trucked water (often of equally poor quality).

This discussion highlights the large difference in water access that often exists between rich and poor residents of the same city. Given the SDGs' emphasis on universal access, the JMP tries to disaggregate data to reveal the variability that averages can mask. For example, the JMP notes that in almost every LMIC, the richest quintile is more likely to have water access than the poorest quintile, and city dwellers are more likely to have water access than rural residents.[13]

Vulnerable and marginalized subgroups are likely to have particularly poor WASH access. The Equitable Access Score-Card, developed by the United Nations Economic Commission for Europe, grades existing water systems and policies on how well they serve the needs of groups for whom water access is both particularly important and particularly challenging, including people with special physical needs, imprisoned people, refugees, unhoused people, travelers and nomadic communities, and users of health facilities, educational facilities, and retirement homes.[14] However, most countries lack data on the extent to which vulnerable groups are left out of progress toward water access.

2.3. Affordability

> Anyone who has ever struggled with poverty knows how extremely expensive it is to be poor. —James Baldwin

SDG 6.1 calls for drinking water to be affordable as well as safe, but no indicators of affordability are routinely monitored. Despite the lack of monitoring, the available data suggest that water is often unaffordable to the poorest sectors of society.

Poor households spend a larger percentage of their income (and time) on water than rich households do. More surprising, perhaps, is that poor households often pay much higher rates per unit of water than richer households. This is especially true in underserved urban slums, where people must buy water from water trucks or other private suppliers, who charge much more than the utilities that provide piped supplies to richer neighborhoods. For example, in Kaula Bandar, a nonnotified (i.e., unapproved) slum in Mumbai, the median water price in 2012 was 135 Indian rupees (INR)/m^3 (almost \$3/m^3), whereas the price for piped water in nonslum areas was 5 INR/m^3 and the price for water from public taps in notified slums was 3 INR/m^3.[15]

One proximate factor driving up water prices in urban slums is the dominance of water cartels, sometimes called water mafias.[16] In cities from Mumbai to Karachi to Nairobi, these groups control local water sources, including illegally constructed boreholes and unauthorized diversions from the formal water supply, and make a significant profit selling this water to people who have few other options. The most brazenly illegal parts of these operations often take place at night, but the cartels also gain tacit approval from authorities through bribes and intimidation. When cartel water is taken from the public water supply, it will be categorized as nonrevenue water and will perpetuate the institutional and financial weakness of the public utility. Still, it is important to understand that the larger context is what allows water cartels to operate. If cities were to provide a reliable piped water supply to underserved areas, the cartels would lose their power.

More broadly, corruption is common in the water sector, due in part to the physical and institutional complexity of water and sanitation delivery and the information asymmetry posed by the inability to ascertain water quality without expensive testing. Various estimates suggest that at least 20 percent of global investment in the water sector is dissipated through corruption at various levels of government; in some regions, this fraction can be greater than half.[17]

3. Sanitation

The second part of the WASH triad—sanitation—gets less publicity than water but is equally important. A safe sanitation system should achieve three goals:

- Ensure the health, safety, and dignity of the user;
- Ensure the health, safety, and dignity of workers involved in the sanitation industry;

- Keep the environment free from contamination with human waste, in order to protect people, water sources, and ecosystems.

3.1. Sanitation Solutions

Before we look at JMP data on sanitation, we need a fuller understanding of the variety of sanitation solutions used around the world. Broadly speaking, a sanitation system consists of up to five components (Figure 16-2a): a user interface, waste storage, conveyance, treatment, and disposal. For a typical urban household in the United States (Figure 16-2b), the user interface is a flush toilet, conveyance is the sewer system, treatment is at a wastewater treatment plant (WWTP), and disposal is to water and land; there is essentially no storage. In a ***pit latrine*** (Figure 16-2c), one of the simplest and most common sanitation systems around the world, the user interface is typically a squat hole, and the collection and storage device is a simple pit dug in the ground. The size of the pit and the number of users determine how frequently it fills with fecal sludge. When the pit does fill, there are two options: Top the pit with a layer of soil and seal it, or remove the fecal sludge and transport it, ideally to a treatment facility.

There are many variations on the pit latrine, including:

- Ventilated improved pit latrine: A passive ventilation system is installed to reduce odors and flies.

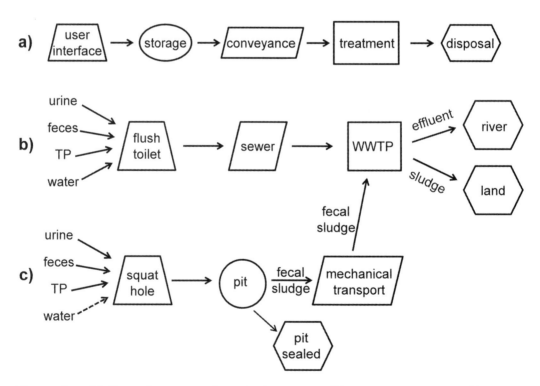

Figure 16-2. (a) General structure of a sanitation system; (b) sanitation system used by a typical urban household in the United States; (c) sanitation system using a pit latrine. TP, toilet paper; WWTP, wastewater treatment plant. The dashed line indicates that water may or may not be used in a pit latrine.

- Offset latrine: The pit is located adjacent to, rather than directly under, the latrine structure. A small pipe directs waste from the squat hole to the pit, reducing odors and allowing the latrine to be located inside the home, with the pit outside.
- Pour-flush latrine: Water is poured down after the waste to clean the pipe and (with a U-bend in the pipe) provide a water seal to prevent vapors from moving up the pipe.
- Twin-pit latrine: Waste can be directed to two different pits. When one fills up, it is sealed and waste is directed to the other pit. By the time the second pit fills, waste in the first pit has dried and degraded and is safer to remove and potentially reuse.
- Double dehydration vaults: A ***urine-diverting dry toilet*** is used to separate urine from feces. Urine is disposed of in the soil or reused, while feces is stored in a sealed, ventilated pit for drying. Addition of ash or lime can speed drying and increase pH, allowing safe reuse in about six months, so two pits with sufficient capacity can be alternated indefinitely.[18]

3.2. From Goals to Indicators

As we discussed above for water access, definitions are key to assessing progress toward universal sanitation. The metric used by the JMP to monitor SDG 6.2 is Indicator 6.2.1, "proportion of population using safely managed sanitation services, including a hand-washing facility with soap and water." We will come back to hand washing in Section 4, but for now we focus on defining "safely managed sanitation services."

JMP's sanitation ladder (Table 16-3) is structured similarly to the drinking-water ladder. Improved technologies include the following:

- Networked sanitation
    - Flush toilets connected to sewers: This is the system familiar to most urban users in high-income countries. A manual flush system (pour-flush) is also considered acceptable.
- Onsite sanitation
    - Flush (or pour-flush) toilets connected to septic systems (as used by most rural Americans).
    - Pit latrines, as long as there is a hard slab on the floor of the latrine (usually made of concrete or plastic) that can be cleaned. Pit latrines without a slab are unacceptable, since users are likely to contaminate their feet and hands during use.
    - Composting toilets

Besides slabless pit latrines, other unacceptable technologies are *hanging latrines*, in which the waste falls to the ground or water below, and *bucket latrines*, in which the user must empty the bucket manually (often onto the ground or into a water body).

The third rung on the ladder—"limited"—refers to toilets that are technologically acceptable but are shared by two or more households. The logic behind this distinction is twofold: Shared toilets may not be safe, especially for women and especially at night; and shared toilets may not be cleaned regularly.[19]

Table 16-3. Sanitation ladder used by the Joint Monitoring Programme. Only rung 5 is considered to fully satisfy SDG 6.2. The ladder used during the MDG period was similar but lacked the fifth rung; basic access was considered sufficient to achieve the sanitation MDG.

Rung	Description
1. Open defecation	Defecation in fields or streets
2. Unimproved	Unimproved technology
3. Limited	Improved technology, shared among multiple households
4. Basic	Improved technology, single household
5. Safely managed	Improved technology, single household, safe disposal

As was the case for drinking water, the reliance on technological distinctions has some shortcomings. Probably the most important is that the definitions are focused on the first two steps of the sanitation service chain (user interface and storage) and not on the final three (conveyance, treatment, and disposal). Put another way, these criteria are focused on protecting users and not on protecting workers or the environment; an improved technology such as a pit latrine or sewer system can lead to major environmental contamination if the waste is not treated properly.

To deal with the later stages of the sanitation service chain, JMP introduced the fifth rung, "safely managed," in which the waste is safely disposed of by one of the following methods:

- For networked sanitation, wastewater is treated properly before being discharged into the environment.
- For onsite sanitation, either the fecal sludge is emptied and treated properly, or the pit is sealed and poses no hazard to the environment.

Similar to the drinking water situation, data are less available for this fifth rung, although the most recent JMP report reported "safely managed" data from countries representing an impressive 81 percent of the world population. Unfortunately, even the fifth rung ignores one important aspect of safe disposal: the safety of workers in the sanitation industry. Latrine emptying is a necessary but hazardous job, as are other jobs in the sanitation industry, and workers do not routinely have adequate personal protective equipment. In addition, these tasks often fall to poorly trained, underpaid workers from disadvantaged backgrounds.[20]

3.3. Where We Stand

Figure 16-3 shows progress toward sanitation goals for the world as a whole and for the LDCs. The sanitation situation clearly lags behind that of drinking water, with 44 percent of the world having less than basic sanitation in 2000 (compared with 18 percent for water), increasing to 22 percent in 2020 (compared with 10 percent for water). Some 1.7 billion people still lack access to basic sanitation, including 63 percent of people who live in LDCs.

Slower progress on sanitation compared with water is unsurprising; a new standpipe or well is more attractive to donors and politicians than an improved pit latrine

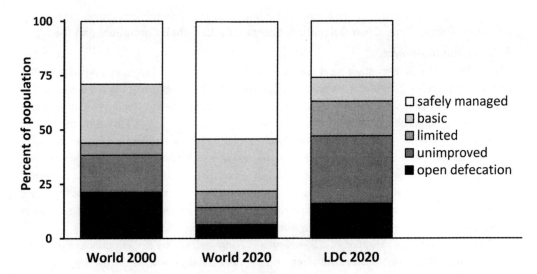

Figure 16-3. Population percentages with different levels of sanitation access. LDCs, least developed countries. Data from WHO/UNICEF (2021).

or even a new sewer line. And the greater complexity of sanitation makes it harder to figure out what solutions are best. But providing water without sanitation is clearly an incomplete solution.

Still, there has been some progress in moving people up the ladder. On the low end, there has been a particular focus on ending open defecation (OD), which is explicitly mentioned in SDG 6.2 and is often associated with widespread environmental contamination and poor health outcomes. Progress on OD varies from country to country, but most SDG regions are on track to eliminate OD by 2030, with the notable exception of sub-Saharan Africa and Oceania.[21] Five countries—Cambodia, Ethiopia, India, Mozambique, and Nepal—saw OD reductions of more than 10 percentage points in just five years (2015–2020), although there is some skepticism about reported progress in India (Box 16-1).

On the upper end of the ladder, almost half the world lacks access to safely managed sanitation (Figure 16-3). In other words, half the excreta generated by humanity is being discharged to the environment with essentially no treatment, through one of the following mechanisms: open defecation, latrines that do not safely contain excreta (e.g., hanging latrines, inappropriately constructed or sited pit latrines), pit latrines or septic tanks that are emptied but the fecal sludge is disposed of inappropriately, or untreated wastewater from sewers.

To help clarify the complex fate of human waste, the Sustainable Sanitation Alliance uses a tool called the ***shit flow diagram (SFD)***. SFDs have been prepared for over 100 cities around the world.[22] An example SFD for Lima, Peru, in 2016 is shown in Figure 16-4. Peru is an upper-middle-income country, and most of Lima has modern sanitation, with flush toilets and a sewer network. Despite this, as of 2016, only 46 percent of Lima's excreta were safely disposed of, because most of the wastewater collected either was discharged directly to a waterway or was delivered to a WWTP but was not effectively treated; WWTP upgrades are expected to improve

Box 16-1. Deeper Dive: Open Defecation, Community-Led Total Sanitation, and the Swachh Bharat Abhiyan

Open defecation (OD) is mostly a rural practice; the Joint Monitoring Programme estimates that 13 percent of people living in rural areas practice OD, compared with only 1 percent of city dwellers. It is useful to distinguish two reasons that people practice OD: lack of alternatives and preference for OD. The latter may sound odd to Western ears, but surveys of rural Indians suggest that they often prefer OD even when latrines are available, for a variety of reasons: a sense of freedom and connection to nature, appreciation of the opportunity to leave the house and walk with others, disgust over having a toilet in the home, or a feeling that diseases of poor sanitation are preferable to the diseases of modernity and the city, such as cancer.[a] Some of these concerns about latrines can be addressed—for example, by ensuring that toilets are clean, well lit, and well ventilated, and pits are deep enough to avoid rapid filling—but others are deeply held traditional values.

Community-led total sanitation (CLTS) emerged in Bangladesh in the early 2000s as a response to programs that provided villagers with free or subsidized toilets, which were then largely unused because the toilets were ill suited for what the villagers wanted. CLTS is based on two underlying ideas:

- Sanitation is a communal, not an individual, issue, both in terms of health impacts (using a toilet won't protect you from disease transmission if the rest of your village is practicing OD) and in terms of attitudes (you are unlikely to want to use a toilet if the rest of your village believes that OD is superior).
- Attitudes are as important as infrastructure. Instead of "build toilets and they will come," CLTS preaches "make them want toilets and they will build them." This philosophy includes an opposition to subsidies for toilets, although this position has softened over the years.

The actual practice of CLTS is centered around a "triggering" process, in which the village is taken through a process meant to elicit disgust over OD and a communal desire to find better solutions. CLTS has been criticized for sometimes being coercive and for not providing enough support for constructing infrastructure. CLTS has been adopted in dozens of countries, with great diversity in how it is practiced. A review paper found that "the evidence base on CLTS effectiveness . . . is weak."[b]

The global reduction in the number of people practicing OD has been driven in large part by rural India. Several government programs over the years have tried to reduce OD in India, but Narendra Modi's Swachh Bharat Abhiyan (SBA, Clean India Mission), initiated in 2014, has had the greatest results, at least on paper. In October 2019, Modi announced that the entire country was OD-free, having built 110 million toilets in five years to serve 600 million previously unserved people. Although sanitation experts appreciated the concentrated focus and funding that the SBA brought to bear on the issue, there is widespread skepticism over the conclusion that India is OD-free. Many of the toilets are apparently not in use, whether because they are locked or dirty or used for other purposes (e.g., storage) or because locals still prefer OD. In addition, the SBA led to significant shaming, coercion, and punitive measures, especially against Muslims and people from lower castes.

How do we balance the public-health goal of ending OD with recognition of cultural values and respect for individual choices?

Notes

[a] Yogananth and Bhatnagar (2018), Clair et al. (2019).
[b] Venkataramanan et al. (2018).

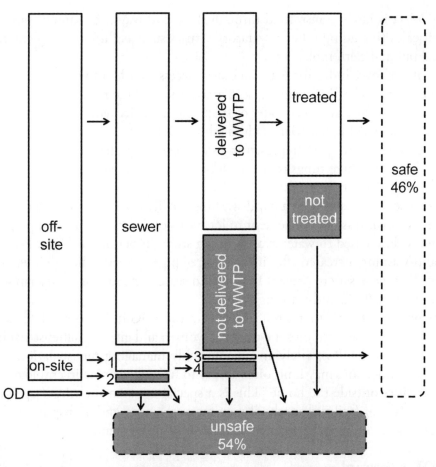

Figure 16-4. Shit flow diagram (SFD) for Lima, Peru, for 2016. Dark shapes represent unsafe pathways. Numbers refer to the following pathways: 1, contained; 2, not contained; 3, fecal sludge never emptied (considered safe); 4, fecal sludge emptied but not treated (disposed of in dump or waterway). Five unsafe disposal pathways—open defecation (OD), uncontained onsite sanitation, untreated fecal sludge, sewage that is not delivered to a wastewater treatment plant (WWTP), and sewage that is delivered but not treated—account for 54 percent of Lima's excreta. Redrawn from Furlong (2016).

the situation. In addition, 8 percent of Lima's population lived in periurban slums without access to sanitation. Only 1 percent practiced OD, but 6 percent used onsite sanitation that was uncontained or was contained but the fecal sludge was not treated when emptied. As shown in Figure 16-4, there are five distinct pathways by which excreta can contaminate Lima's soil and water.

4. Hygiene

The SDGs also call for universal access to hygiene, which is an important tool for reducing disease transmission. Hygiene behaviors include hand-washing, especially after defecating and before eating or preparing food; washing of face, body, and hair; cleaning of dishes and food preparation areas; oral hygiene; and menstrual hygiene. In the early days of the COVID-19 pandemic, hand and surface hygiene took center stage in public health recommendations. Although it later became clear that

SARS-CoV-2 is rarely transmitted through these pathways, the same is not true for other diseases (including other respiratory viruses such as flu), and hygiene remains a critical public-health tool.

As noted above, Indicator 6.2.1 includes "access to a hand-washing facility with soap and water." The JMP hygiene ladder consists of three rungs: no hand-washing facility, a facility lacking either water or soap, and a facility with both water and soap (or equivalent cleaning material, such as ashes). Data for this ladder are available for only seventy-nine countries (representing 50 percent of global population), but they suggest that those three rungs represent 9 percent, 21 percent, and 71 percent of the world's population, respectively.

Of course, the existence of a hand-washing facility does not mean that it is being used, especially if it is not convenient to locations of defecation or food preparation. Hygiene is closely tied to water availability; a study in periurban Lima, Peru, found that hand-washing increased after installation of piped water, although it was still less frequent than necessary to prevent fecal–oral disease transmission, and soap was used in only about half of hand-washing events.[23]

One important topic that is not included in the SDG indicators—despite the call for "paying special attention to the needs of women and girls"—is menstrual hygiene management. In LMICs, safely and comfortably managing menstruation can be a significant burden and may limit one's ability to fully participate in employment and other activities outside the home. This is a special concern for school-age girls, for whom the lack of facilities at school may mean shaming and a drop in attendance, contributing to the gender gap in education and employment.

5. The Way Forward

How do we move forward to achieve safe WASH for all? The solution may seem obvious: Replicate *Water 3.0* throughout the world. We know how to build and operate treatment plants, pumping stations, pipes, and sewers for urban areas; we know how to drill wells and install septic tanks for rural homes. Shouldn't those solutions—which have turned WASH into something that most Americans don't have to think about—be available to every person on the planet?

Actually, there are several reasons why bringing Water 3.0 to all may not be the right answer:

- The scale of the challenge is unprecedented, especially in the rapidly growing cities of the developing world. One analysis of sixty cities found that, in order to extend sewers to their entire population by 2030, most of these cities would need to construct sewers at rates of 10,000–100,000 people connected per month, but past projects have typically reached only 100–1,500 people per month.[24]
- The most rapidly growing parts of these cities are urban slums, where Water 3.0 faces several serious problems. First, city governments don't want to sanction informal settlements, and they worry that providing water and sanitation will encourage more in-migration. Also, these areas are often unsuitable for traditional pipes because they are steep, crowded, or flood-prone. Finally, low-income residents can't afford to pay for new pipes and connections, so water utilities don't see the opportunity to make a profit or even recover costs.

- Incremental approaches that are well short of Water 3.0 can provide benefits right away. Given the central role of WASH in health and economic advancement, a water-first approach to poverty alleviation may be called for, even if the water solutions implemented are cheaper and less technologically advanced than those used in high-income countries. The urgency of the moment supports this view; people in urban slums need WASH solutions every day, and they can't afford to let the perfect be the enemy of the good.
- Water 3.0 is actually far from perfect, as we noted in Chapter 14. Do we really want to replicate the centralized, inefficient systems of high-income countries at the very same time that the sustainability of those systems is being seriously questioned? Are there other solutions out there that can close the loop while still avoiding environmental contamination and disease transmission? Can LMICs leapfrog over the developed world's mistakes and lead the way to more sustainable water and sanitation systems?
- The wholesale application of Water 3.0 to the entire world ignores important differences in geography, climate, culture, resources, and even the way that water problems are defined. There is a long history of aid groups coming into areas and providing technologies that are not appropriate to local conditions, perhaps because spare parts and maintenance training aren't available or because the technologies don't actually solve a problem experienced by local populations.

These issues have driven a move toward innovative water and sanitation solutions that don't simply replicate Water 3.0. (For more detail on these approaches, see the companion website.) These solutions are often decentralized, both because that makes them easier to implement and because centralization is viewed by many as one of the weaknesses of Water 3.0. Some of these solutions are best thought of as short-term approaches that are better than nothing but are not sustainable or adequately protective in the long term, while others may have the potential to serve as true substitutes for Water 3.0. It will take time to sort out which is which, and we will no doubt need a mix of approaches in various places.

A few thoughts to keep in mind when evaluating WASH solutions:

- It is critical to work with communities to figure out what people want and what solutions would be acceptable rather than imposing externally designed solutions on them. Cultural, geographic, and capacity issues may be opaque to outsiders, but they are very real factors in the success or failure of water and sanitation projects.
- In many LMICs, Water 3.0—the flush toilet in particular—is seen as the best solution, if only because it represents modernity and wealth. In the long run, changing this perception will require solutions that look just as good and function just as smoothly as flush toilets—solutions that people in rich countries would also be happy to use.
- Solutions to water problems involve not just technologies but also management systems and institutions. Even the best technology will not be sustainable if there is no plan to maintain it, fund it, and distribute its benefits in a clear, equitable way. Strengthening communal institutions and capacity must be part of addressing WASH.

6. WASH Funding

Achieving WASH for all will require innovative technologies and improved governance, but it will also require funding. Figure 16-5 shows estimates of the annual capital investment needed to reach universal basic services (rung 4 in the water and sanitation ladders) and safely managed services (rung 5). There is a great range in the possible costs, depending on which options are chosen (e.g., simplified sewers are about half the cost of traditional sewers), but the number is in the ballpark of $30 billion per year for basic WASH and an additional $85 billion per year for safely managed services (on top of the cost of maintaining existing infrastructure). Globally, an annual investment of $30 billion is a small amount (about 0.02 percent of global GDP), and **benefit–cost analyses** consistently show that this investment will more than pay for itself, but the need is greatest in the countries that can least afford to pay for it. Indeed, a survey of twenty LMICs suggests a funding gap of about 61 percent, suggesting that current investment needs to increase threefold.[25]

There are three basic ways that water and sanitation can be paid for: tariffs (user fees), taxes (funding from local or national governments), and transfers (grants or loans from richer countries, charities, or international funders such as the World Bank). In addition, when water services are poor or nonexistent, households often invest significant monetary, physical, and time resources in self-supply, investments that may not be captured in surveys.[26] For thirty-five countries for which data were available, 66 percent of WASH funding came from households, 22 percent from governments, and 12 percent from international aid (3 percent from grants and 9 percent from loans).[27]

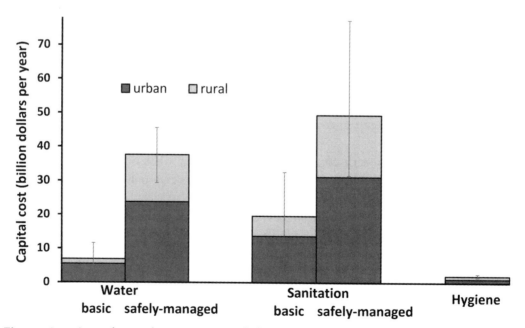

Figure 16-5. Annual capital investment needed to meet Sustainable Development Goal targets by 2030. Costs for safely managed water and sanitation (rung 5 on the ladders) are additional to the costs for basic services (rung 4). Data from Hutton and Varughese (2016). A 2020 follow-up study (for sanitation only) showed very similar results (Hutton and Varughese, 2020).

6.1. Tariffs
Some WASH practitioners argue that user fees should cover most or all of the costs of WASH:

- Unless user fees can support water services, the utility is doomed to the vicious cycle of underfunding and poor performance. Reliance on external funding makes the utility unsustainable in the long term.
- People value what they pay for. If you give people latrines for free, they will undervalue and underuse those latrines.
- Willingness to pay is a good measure of community values. If people aren't willing to pay for something, it means that what you are offering is not the right solution for their problems.

These arguments suggest that the solution to WASH problems lies not in providing subsidies but in generating greater demand through activities such as WASH education and CLTS (see Box 16-1).

Despite these arguments, I believe that subsidies do make sense in many situations, for two basic reasons:

- Appropriate WASH solutions may simply be too expensive for poor people to afford, no matter how much they want them. If we want to break the cycle of poverty, we need to collectively fund basic infrastructure to serve all citizens.
- Water and sewer infrastructure should be considered a public good, because the benefits it affords extend beyond the individuals using it to encompass their neighbors and the environmental health and livability of the entire region. Indeed, much of the WASH infrastructure in the United States and other high-income countries was built with public money for this very reason.

In an ideal case, public investment in initial infrastructure construction for unserved neighborhoods will allow residents to advance economically and ultimately pay the ongoing costs of service provision.

Research on the role of subsidies has had mixed results, but at least some studies have shown a positive effect. For example, one study in rural Bangladesh found that subsidies for construction of hygienic latrines led to a 22-percentage-point increase in latrine ownership, with the effects even spilling over to unsubsidized neighbors, who saw an 8.5-percentage-point increase in latrine ownership.[28] In contrast, two other interventions—a community motivation program and technical assistance in buying and installing latrines—did not significantly increase latrine ownership.

6.2. Taxes
If local residents can't necessarily pay for WASH solutions, where should the funding come from? The public-good argument above suggests that city and national governments (i.e., taxpayers) should be willing to pay a significant portion of these costs, as they have in the United States and other high-income countries. This is the position taken by the organization Public Finance for WASH[29] and supported by some WASH practitioners, especially with respect to high- and middle-income countries.

In LDCs, however, governments have scarce resources and many competing needs and are unlikely to be able to bear the costs involved, suggesting that international aid must be part of the answer.

6.3. Transfers

In recognition of the need—and obligation—for richer countries to help poorer ones achieve WASH access, one of the SDG indicators specifically addresses foreign aid (known as official development assistance [ODA]):

> **SDG Target 6.a.** By 2030, expand international cooperation and capacity-building support to developing countries in water- and sanitation-related activities and programmes. . . .
>
> **SDG Indicator 6.a.1.** Amount of water- and sanitation-related ODA that is part of a government coordinated spending plan.

The UN has been tracking ODA for WASH but has not been able to assess the extent to which that ODA is part of a spending plan. Note also that there is no specific target for WASH ODA, although rich countries are expected to spend 0.7 percent of their GDP on total ODA. Only six countries met or exceeded this goal in 2021: Luxembourg, Turkey, Norway, Sweden, Germany, and Denmark. The United States, at 0.18 percent of GDP, is well below the target and well below the average for countries the Organisation for Economic Co-operation and Development's Development Assistance Committee member countries (0.33 percent).

International aid for WASH has increased more than threefold over the last two decades, but it is still far short of what is needed to close the financing gap. In addition, international aid does not necessarily go to the countries and sectors that need it the most. For example, 59 percent of WASH aid in 2019 went to water supply and only 35 percent to sanitation (despite the greater need in the sanitation sector).[30] Aid flows to certain issues and countries for many political, logistical, and cultural reasons, often understandably bypassing the neediest countries because of poor governance, corruption, and conflict. The 2014 Paul Simon Water for the World Act, which aimed to increase the effectiveness of US water aid, acknowledged this tension in "direct[ing] the Administrator to ensure that USAID projects and programs are designed to achieve maximum impact and long-term sustainability by prioritizing countries based upon: (a) the population using unimproved drinking water or sanitation sources; (b) the number of children younger than five years of age who died from diarrheal disease; and (c) the government's capacity and commitment to work with the United States to improve access to safe water, sanitation, and hygiene."[31]

Given the dire need for funding, is there a role for the private (for-profit) sector in building out WASH infrastructure in LMICs? The 1990s–2000s wave of privatization seen in the United States also played out in LMICs, driven by at least three factors: the need for large upfront infrastructure investments, multinational water companies' quest for new markets, and the neoliberal philosophy of international lenders. Indeed, privatization—of the water sector as well as other sectors—was often central to the ***structural adjustment programs*** demanded by the International Monetary

Fund and the World Bank. Even when water utilities were not privatized outright, they were expected to emulate the private sector in their functioning, including pricing for full cost recovery regardless of customers' ability to pay. This emphasis on cost recovery from customers led utilities to cherry-pick high-income neighborhoods for service and continue to neglect urban slums.

The collision of profit-seeking water conglomerates with the complex realities and limited resources of LMIC cities has led to a series of privatization failures and high-profile battles. In the public discourse, the privatization argument has hardened into two camps—pro-public and pro-private—each pointing to the failures of the other side to provide reliable, high-quality water supplies to all residents. This argument, of course, taps into larger discussions about capitalism, the role of government, corruption, and human rights.

To my mind, there may be a role for private capital and expertise to help communities and governments achieve the goals of SDG 6. Critically, though, regulators must set out and enforce strict guardrails for private involvement—including requirements to provide service to the poorest parts of the community and not to undermine existing arrangements if they are working well.

Chapter Highlights

1. The global burden of disease associated with inadequate WASH has declined over the past several decades, but it is still unacceptably high.
2. The SDG goal of ensuring universal access to safe water, sanitation, and hygiene is critical for human health, economic development, environmental protection, human dignity, and gender equity.
3. Translating SDG WASH goals into specific, measurable indicators is a complex task and in the past has relied heavily on technological distinctions that may not always be relevant.
4. The JMP ladders are a way of monitoring progress toward improved WASH; they show progress, but there is a long way to go.
5. When urban slums do not have access to water infrastructure, poor households must pay more for water (in both money and time), exacerbating inequality.
6. Ending open defecation is high on the development agenda, but methods to achieve this goal are sometimes coercive and ineffective.
7. People around the world use many different types of sanitation solutions, often involving a *pit latrine* of some type. Depending on how they are managed, both *networked* and *onsite* sanitation solutions can result in either safe or unsafe disposal of human waste; *shit flow diagrams (SFDs)* are useful tools for understanding the complex fate of fecal matter.
8. Hygiene is now included in the SDGs, but menstrual hygiene has been largely neglected.
9. A central question for the future of water is whether LMICs should (or will be able to) replicate Water 3.0 for all their citizens, or whether there are decentralized or alternative models that may be more appropriate in the short and long terms.
10. Successful WASH interventions must take into account local physical and cultural conditions and avoid creating other problems.
11. There are three basic sources of funding for WASH: tariffs, taxes, and transfers.

Where WASH solutions are unaffordable to local residents, taxes and transfers (as well as tariff cross-subsidies) are absolutely necessary. Meeting WASH goals is unaffordable locally but affordable globally and has a high benefit/cost ratio.

Notes

1. The PAF depends on what counterfactual WASH scenario is used for comparison; see Prüss-Ustün et al. (2019).

2. A DALY is the sum of two components: the number of years lost to premature death (relative to life expectancy) and the number of years living with a disability multiplied by the weighting given to that disability (e.g., blindness = 0.19, where 0 is perfect health and 1 is death). Some readers may be troubled by the implication that a blind person's life is worth 81 percent of a sighted person's, but DALYs are widely used in the public-health field.

3. Of course, children in high-income countries often benefit from improved health care and diet in addition to better water services; it is the desire to isolate the effects of WASH that leads to the use of PAFs in the first place. Still, a small-scale intervention study is likely to underestimate the effects of wholesale availability of adequate water and sanitation to an entire society.

4. Murray et al. (2020).

5. Dimitrova et al. (2023).

6. Hutton (2013).

7. https://www.un.org/sustainabledevelopment/water-and-sanitation/; https://www.sdg6data.org/en/node/1; https://www.un.org/millenniumgoals/environ.shtml.

8. WHO/UNICEF (2021).

9. Sanitation data for LDCs for 2000 are incomplete, so I have not included 2000 LDC data in either Figure 16-1 or Figure 16-3.

10. Bain et al. (2014).

11. Kumpel and Nelson (2016).

12. Consumers in India spent over $1 billion on water purifiers in 2015 (Rabaey et al. 2020).

13. WHO/UNICEF (2021). However, the urban–rural gap in water access may partly reflect the limitations of technology-based definitions, as "unprotected" water sources in rural areas (surface water or open wells) are arguably more likely to be safe than similar sources in cities.

14. https://unece.org/environment-policy/water/areas-work-protocol/equitable-access-water-and-sanitation.

15. Subbaraman et al. (2015).

16. https://www.reuters.com/article/women-cities-kenya-water-idINKCN0JA0P620141126? https://thewaterstory.com.au/2019/11/03/the-water-mafia-of-mumbai/.

17. Jenkins (2017).

18. https://sswm.info/sswm-solutions-bop-markets/affordable-wash-services-and-products/affordable-technologies-sanitation/dehydration-vaults.

19. However, Exley et al. (2015) showed that surfaces in shared toilets were no more contaminated with *E. coli* than surfaces in private toilets.

20. In India, the task of "manual scavenging" of human waste from public areas and toilets is often assigned to lower-caste people (https://www.hrw.org/report/2014/08/25/cleaning-human-waste/manual-scavenging-caste-and-discrimination-india).

21. WHO/UNICEF (2021).

22. https://sfd.susana.org/.

23. Oswald et al. (2014).

24. Öberg et al. (2020).

25. WHO (2019).

26. Danert and Hutton (2020).

27. WHO (2019).
28. Guiteras et al. (2015).
29. https://www.publicfinanceforwash.org/.
30. WHO (2019).
31. https://www.congress.gov/bill/113th-congress/house-bill/2901.

17 Industrial Water Use: Our Invisible Water Footprints

- How much water is used in the energy, manufacturing, and mining sectors?
- How do various industrial activities affect water quality?
- What laws are used in the United States to control chemical production, disposal, and cleanup?
- How can corporations be better water stewards?

Industry—a term covering a wide variety of activities—is second only to agriculture in the volume of water used globally (Figure 4-4). This chapter will explore the water quantity and quality impacts of the three primary water-using industrial sectors: energy, manufacturing, and mining. We close the chapter with a discussion of the recent interest in water stewardship on the part of large corporations.

For many of us, energy production, manufacturing, and mining can seem far removed from our daily lives. But we each use energy, manufactured goods, and mined materials every day; understanding (and reducing) the water footprints and water-quality impacts of those activities is one way to take responsibility for our actions.

1. Water for Energy

Previous chapters have laid out various aspects of the ***water–energy nexus*** (see Chapter 1, Section 2.6). We now turn to the final piece of the puzzle: the water impacts of energy use. We address this topic in three parts: water use for electricity, water use for transportation, and water quality issues associated with energy production and use.

1.1. Water for Electricity

Electricity can be generated from a variety of primary energy sources, with the most important being natural gas, coal, and nuclear, followed by the renewables: hydropower, wind, and solar. Depending on the energy source, some amount of water is used at each stage of the power cycle: production (e.g., mining coal), processing (removing impurities), transportation (to a power plant), conversion (use of the fuel to produce energy), and postconversion (ash disposal). For oil and gas, the need for water in the production step has proven problematic in some places, especially with the rise of hydraulic fracturing, commonly known as *fracking*, in which large volumes of water (mixed with sand and chemicals) are used to create fractures in low-permeability rocks (e.g., shale) to allow oil or gas extraction. But for the energy sector as a whole, the conversion step—in particular the use of water to cool **steam turbines** in thermoelectric power plants—is by far the largest water user. Electricity sources that don't involve steam turbines, such as solar photovoltaics, wind, and natural gas turbines, do not need cooling and thus have much lower water use.

Different types of cooling systems have different water needs:

- Older power plants often use **once-through cooling** systems, which withdraw large volumes of water from a river, lake, or coastal water body, use it to condense steam, and then return the warmed water to the environment (with minor losses to evaporation), usually near where it was withdrawn.
- The combination of increasing water scarcity and more stringent environmental regulation has led to an increase in **recirculating cooling** systems, which recycle cooling water within the power plant.[1] In these systems, the cooling water—after being warmed during steam condensation—dissipates its heat to the environment through evaporation and heat exchange in cooling towers or similar structures and then returns to condense more steam. Water withdrawals are thus limited to the *makeup water* needed to compensate for evaporation and *blowdown* (water that is removed from the cooling system to maintain water quality and prevent scale buildup). Compared with once-through cooling, recirculating cooling has a much lower water withdrawal but a much higher evaporation or consumption rate. Thus, the shift to recirculating cooling protects power plants from having to shut down during low-flow conditions, but it also exacerbates downstream scarcity.[2]
- Dry cooling systems use air rather than water for cooling, either by directly condensing steam with air or by using a closed-water recirculating system in which the cooling water is cooled without loss to evaporation.

As of 2021, 43 percent of US electric generation capacity involved steam turbines and thus required cooling; these cooling systems were dominated by recirculating cooling (62 percent of capacity) and once-through cooling (34 percent).[3] Given its heavy reliance on cooling water, our electricity supply may be vulnerable to drought and climate change. A US Department of Energy study found forty-three incidents between 2000 and 2015 in which US power plants were at risk because insufficient water was available, the intake water was too warm for efficient cooling, or the discharge water was so warm that it posed a risk to aquatic life.[4,5]

To sum up, then, electricity-related water use is a function of both energy source and cooling technology (Figure 17-1). Several features of Figure 17-1 are worth noting:

- Water use is expressed in units of L/MWh (i.e., liters of water used per megawatt-hour of electricity generated). For reference, the average per capita residential electricity use in the United States is 0.012 MWh/day, so if your electricity comes from a once-through-cooled coal plant (for example), your daily electricity use at home is responsible for about 1,600 L of water withdrawal (130,000 L/MWh × 0.012 MWh = 1,560 L) and 8 L of water consumption (680 L/MWh × 0.012 MWh = 8.2 L).

- Combined-cycle natural gas has lower water use (both withdrawal and consumption) than either coal or nuclear power, because of its more efficient production of electricity. Simple-cycle natural gas (not shown in Figure 17-1) has even lower water use, because gas turbines do not need cooling water.
- Among the renewables, **concentrating solar power (CSP)** has high water consumption associated with cooling the turbine, washing the solar reflectors, and producing the specialized materials used in power plant construction. As one report points out, "One of the most significant challenges associated with CSP development in the Southwest is that those areas with the most consistent and direct sunlight are also some of the most water scarce regions in the country."[6] Many new CSP plants use dry cooling, but even so, water consumption is still higher than that of other renewables. Photovoltaics also need water for cleaning the panels (as frequently as once per week in desert environments[7]), although electrostatic dust removal systems may one day reduce the need for water.[8]
- Hydropower has remarkably high water consumption due to evaporation from reservoirs, although this number is hard to estimate and varies by orders of magnitude between reservoirs depending on surface area, temperature, and power production.[9] Two recent studies using different methods and different sets of reservoirs produced average values that differ by a factor of six (Figure 17-1).

1.2. Water for Transportation

Most of the energy used in transportation comes from gasoline and other petroleum products such as diesel and jet fuel. However, the transportation sector is increasingly turning to two nontraditional energy sources: **biofuels**, especially ethanol produced from corn (Box 17-1), and electricity, either as a supplement to gasoline engines (hybrid vehicles) or as the sole energy source (electric vehicles). The production and consumption of each of these traditional and alternative fuels requires water. Given what you know about water use in agriculture and electricity production, how do you think the three main options for passenger vehicles—gasoline, ethanol, and electricity—compare in terms of water withdrawal and consumption?

The answer is shown in Figure 17-2. Ethanol has the highest **blue water** use (both withdrawal and consumption) of the three options, since it takes a lot of water to grow corn.[10] On the other hand, powering an electric vehicle also results (on average) in very high water withdrawal for cooling water (as discussed above). By these metrics, traditional gasoline comes off looking pretty good! Of course, this does not take into account greenhouse gas emissions, air pollution, or water pollution. From a broader environmental perspective, it's probably safe to say that electric vehicles are the best option, especially if your electricity comes from a low-carbon source, ideally one with low water use as well. (Actually, driving less is the best option!)

Not all biofuels are made from irrigated crops. The (blue) footprint of ethanol made from rainfed corn is about 1 percent of that shown in Figure 17-2,[11] although of course the **green footprint** is still high. In Brazil, bioethanol is made mostly from

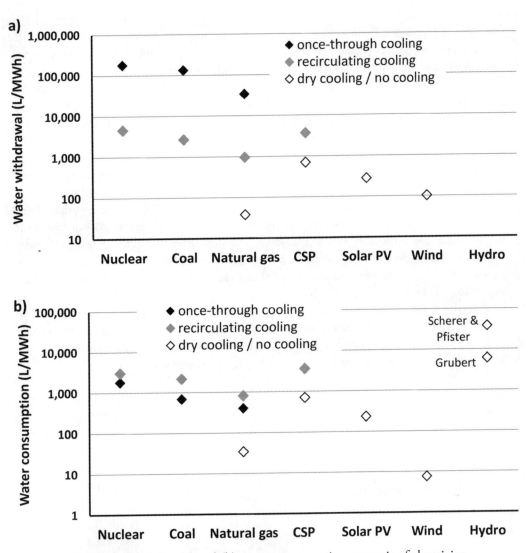

Figure 17-1. (a) Water withdrawal and (b) water consumption per unit of electricity generated, illustrating the role of various energy sources and cooling technologies. Concentrating solar power (CSP) generally does not use once-through cooling. Solar photovoltaic (PV) and wind don't need cooling. Nonhydropower data are medians from Meldrum et al. (2013) and include the entire power cycle from fuel production to postconversion as well as water use in power-plant construction and decommissioning. Natural gas is combined cycle; CSP is for a power tower. Hydropower withdrawal is zero, and two estimates are shown for hydropower-allocated net reservoir evaporation: the median in a global study by Scherer and Pfister (2016) and the average in a US study by Grubert (2016). Note that both y-axes use log scales but cover different ranges.

rainfed sugarcane, with a low blue footprint, and scores favorably on other sustainability metrics as well.[12]

1.3. Pollution from Energy Production

In addition to the water quantity issues discussed above, the use of fossil fuels can cause serious water quality problems.

> **Box 17-1. Biofuels and Food**
>
> US biofuel policy provides an example of unintended consequences within the food–energy–water nexus. Renewable Fuel Standards mandate increased use of bioethanol and biodiesel in transportation fuels, with the goal of reducing dependence on petroleum products. Most US bioethanol comes from diverting corn from the food supply, and most biodiesel comes from vegetable oils that could otherwise be used for human consumption. Besides the questionable ethics of burning food products to run our cars, ethanol mandates have increased the water footprint of transportation fuels and led to expansion of corn production and the associated fertilizer and pesticide use. Taking into account these secondary effects, ethanol's carbon footprint is probably as high as gasoline's,[a] but the political power of the corn lobby means that US lawmakers of both parties have lined up to support ethanol mandates.
>
> The biodiesel mandate has also had unforeseen effects on global food markets. Perhaps most notoriously, Indonesian palm oil has taken the place of US-produced vegetable oils that have been diverted to biodiesel production; palm oil production is highly destructive environmentally and socially and produces large emissions of greenhouse gases from forest burning. Palm oil is now the world's top vegetable oil by production volume and is used as a cooking oil and as an ingredient in both food and nonfood products.
>
> Given the water and land constraints facing global agriculture, prime agricultural land should be used to produce food, not fuel. Next-generation biofuels—made from algae, nonedible parts of food crops (e.g., corn stalks), or nonfood crops grown on marginal land—will hopefully be better at solving the complex problems of producing both food and energy with limited land and water.
>
> **Note**
> [a]Lark et al. (2022).

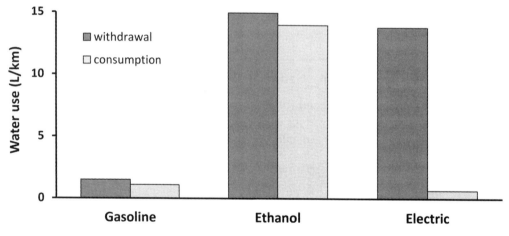

Figure 17-2. Blue water withdrawal and consumption for different passenger vehicle fuels, in liters of water per kilometer driven. Data (from Scown et al. 2011) are totals for the four stages of the transportation power cycle: feedstock production (e.g., oil extraction), feedstock transportation (from the wellhead to a refinery), fuel production (conversion of oil to gasoline), and fuel transportation (from the refinery to your local gas station).

Coal

Generally speaking, coal is the "dirtiest" of the fossil fuels, both in greenhouse gas emissions per unit energy and in impurities such as sulfur and toxic metals. After combustion in a power plant, many of these impurities end up concentrated in coal combustion residuals, more commonly known as **coal ash**, which US power plants produce at an astonishing rate of 130 million tons per year.[13] Storage, treatment, and disposal of coal ash have led to several environmental disasters. A 2008 dike collapse at a Tennessee Valley Authority power plant in Kingston, Tennessee, released over a billion gallons of coal ash slurry, caused extensive damage to the surrounding land and water, and cost around $1 billion to clean up; people who worked on the cleanup are still suffering health effects.[14]

Another vulnerability in the coal fuel cycle is the use of toxic chemicals for processing of coal before use in power plants; the 2014 accidental release of methylcyclohexanemethanol, a coal-cleaning chemical, into the Elk River in West Virginia contaminated the water supply serving 300,000 people in the Charleston area.

Oil

Every step in the petroleum lifecycle—from extraction to processing to transportation to use—can release toxic oil products into the environment. Catastrophic releases, such as the Exxon Valdez spill in Prince William Sound (1989) or the Deepwater Horizon blowout in the Gulf of Mexico (2010), can cause short-term devastation and long-term impacts for both ecosystems and human communities. Less publicized but more common are the everyday, low-level releases of petroleum products during processing and use. Toxic air pollution is still common in "Cancer Alley," along the Mississippi River between Baton Rouge and New Orleans, where over 100 refineries and petrochemical plants convert crude oil into gasoline, plastics, and other products. And nearly every gas station has the potential to contaminate surface water and groundwater due to leaking underground storage tanks, leaking pipes and connections, or the cumulative effect of losses during vehicle filling.

Fracking

The practice of oil and gas fracking has attracted significant opposition, including outright bans in places such as the Delaware River Basin. Several distinct fracking impacts are of concern:

- Methane emissions: Natural gas wells, whether fracked or conventional, can be significant sources of atmospheric methane, a strong **greenhouse gas**.
- Groundwater contamination: The rock formations that are subjected to fracking sometimes have drinking water **aquifers** above them, and there have been cases where gas, petroleum, or fracking chemicals have contaminated these aquifers, although definitively tying methane contamination to fracking activities has proven difficult. The likelihood of contamination is probably low in cases where the well is properly constructed and maintained, where there is significant vertical distance between the fracked formation and the aquifer, and where surface spills are prevented or promptly contained, but none of these factors are a given.
- Produced water: Oil and gas production, whether conventional or fracked, also bring up from the subsurface large volumes of water, referred to as **produced water**, which can be naturally high in salts, toxic metals, and radionuclides.[15] The low permeability of shale formations means that produced water from fracking can't be reinjected into the formation (as is commonly done in conventional oil

production) and must be used to frack new wells or disposed of in some other way. There is interest in using produced water for other purposes, but significant treatment is usually needed to remove salts and other chemicals.[16]

Tar Sands

Bitumen is a viscous petroleum product that is produced from tar sands through mining or steam extraction. To flow through pipelines, bitumen is diluted with lighter petroleum products to form ***dilbit*** (diluted bitumen). The diluted product retains much of the acidity and corrosivity of the initial products, and pipeline leaks are common. Spills of dilbit into water bodies can be even more problematic than spills of conventional petroleum, since the bitumen tends to sink to the bottom because of its high density, especially after the lighter diluents evaporate. A 2010 spill in Michigan released more than a million gallons of dilbit from a ruptured pipeline into a tributary of the Kalamazoo River, leading to widespread environmental degradation that has cost more than $1 billion to remediate. Bitumen is also highly problematic from a climate-change perspective, both because of the vast quantities of it available and because its extraction is energy intensive.

Pipelines

The United States is criss-crossed by oil and gas pipelines: more than 80,000 miles of pipeline transporting crude oil from wellfields to refineries, more than 62,000 miles transporting refined products (mostly gasoline) to markets, more than 70,000 miles transporting natural gas liquids (small hydrocarbons used primarily as feedstocks for manufacturing plastics and other chemicals), more than 300,000 miles transporting natural gas to utilities, and more than 2 million miles of gas distribution lines to homes and businesses.[17] Each of these pipelines poses a potential threat of soil and water contamination. New pipelines are continuing to be built despite the need to transition to a cleaner energy economy, and many of these pipelines have been the targets of protests and lawsuits, often based on water issues. These protests have had both successes and failures:

- In 2016, the Standing Rock Tribe of the Oceti Sakowin (Sioux) led a protest against construction of the Dakota Access Pipeline, which runs under Lake Oahe and the Mississippi and Missouri Rivers, carrying shale oil from the Bakken formation in North Dakota to an oil terminal in Illinois. The protest camps attracted large numbers of sympathizers and drew international attention with the slogans #NoDAPL and Wni Miconi ("Water Is Life" in Lakota), but in early 2017 the Trump Administration allowed the project to move forward.
- After more than a decade of protests against the Keystone XL pipeline—which would have traversed ecologically and culturally sensitive areas to bring dilbit from Alberta to the Gulf Coast—Indigenous and environmental activists scored a victory in 2021, when the Biden Administration denied a permit, leading TC Energy to abandon the project.
- Another recently completed dilbit pipeline project—Enbridge's expanded Line 3, which runs through Anishinaabe reservations and treaty territory in Minne-

sota—poses serious risks to wetlands and culturally vital wild rice (Manoomin) and is still being opposed by a coalition of water protectors.[18] In 2021, Manoomin was named as a plaintiff in a "rights of nature" lawsuit opposing the pipeline in White Earth Nation tribal court.
- In 2021, community groups in Memphis successfully derailed the Byhalia Pipeline project, which would have transported crude oil through an aquifer recharge area and across mostly Black neighborhoods that were described by an agent associated with the project as the "point of least resistance."
- The Mountain Valley Pipeline, which will transport natural gas through Virginia and West Virginia, faced significant opposition, including a lawsuit challenging West Virginia's Clean Water Act §401 certification of the project (Chapter 8), but was forced through by Congress as part of the 2023 debt crisis negotiations.

The battle over pipelines is also taking place at the national scale, particularly with reference to the permit required for pipeline construction under §404 of the Clean Water Act. Rather than individually permitting thousands of projects a year, the US Army Corps of Engineers typically relies on one nationwide permit for "linear" utility projects, including oil and gas pipelines, fiber-optic cables, and electric lines. However, Nationwide Permit 12, issued in 2017, was vacated by a Montana court in 2020, on the grounds that it sidestepped the biological consultations required under §7 of the Endangered Species Act. The US Army Corps of Engineers announced in April 2022 that it is reviewing the permit.

2. Water for Manufacturing

Every manufactured product—from automobiles to toilet paper—has a chain of water use and pollution behind it. In this section, we discuss the water quantity and quality impacts of manufacturing.

2.1. Direct Water Use

For a first look at industrial water use in the United States, we can turn to the US Geological Survey data collected every five years since 1950. Figure 17-3 shows that per capita industrial water withdrawal has dropped by almost 80 percent over the last sixty-five years. This decline has more than kept pace with population growth, so total industrial withdrawal has declined by over 50 percent. The drop in industrial water use is even greater (more than 90 percent) when water use is expressed per unit of industrial production, since (contrary to many people's assumptions) the United States now manufactures more than it ever has. These declines suggest that manufacturing has become more water efficient over time, a trend driven by improved technology, stricter effluent treatment requirements, public pressure, and increased water scarcity.

However, there are several methodological issues that limit the conclusions we can draw from Figure 17-3. Most importantly, the mix of products manufactured in the United States has changed over time, making it hard to know whether the decline in water use represents efficiency gains or structural changes in the economy.

We can unpack this question with the help of water consumption data for various manufacturing sectors, collected by Landon Marston and coauthors.[19] Average

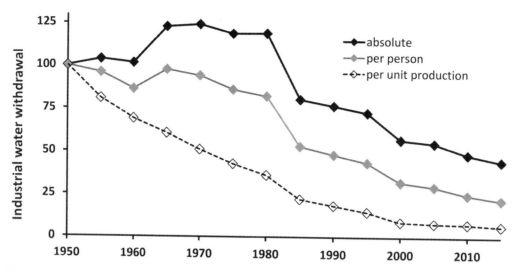

Figure 17-3. Industrial water withdrawal (absolute, per capita, and per unit of industrial production) in the United States. All curves are scaled to a value of 100 in 1950. Water withdrawal data are from US Geological Survey reports (see Chapter 4 for details), and industrial production manufacturing index is from the Federal Reserve (https://fred.stlouisfed.org/series/IPMANSICS).

water footprints by sector are provided in Table 17-1, in units of liters per dollar (since monetary value allows us to compare across different types of products). Table 17-1 makes it clear that "manufacturing" includes a large range of products with different water footprints, from high-value, low-water products (e.g., electrical equipment and computers) to low-value, water-intensive products (e.g., steel and concrete). An examination of Federal Reserve data shows that US production of the most water-intensive products in this table has lagged considerably behind other types of industrial production (and in the case of metals has actually declined over the last fifty years). Other, less water-intensive sectors (e.g., computers) have taken up the slack and contributed more and more over time to US industrial production, helping to drive down US industrial water use. This is not to say that the various sectors have not grown more water efficient—they clearly have—but the decline of more than 90 percent shown in Figure 17-3 overstates those efficiency gains, since it is driven in part by a shift to less water-intensive manufacturing sectors.

Are there still water efficiency gains to be achieved in manufacturing? A recent study suggests that US manufacturing could reduce its direct water use by over 1 km^3/yr, second only to agriculture in the potential for water savings.[20] This study used a benchmarking analysis, in which the distribution of water footprints was analyzed and water savings were calculated by assuming that all facilities within a given climate zone and sector (e.g., motor vehicle manufacture) could be upgraded to match the water footprint of the top 25 percent of facilities in that zone and sector.

2.2. Upstream and Downstream Water Use

The full water footprint of an object includes not just the water used directly in manufacturing but also upstream water use in the supply chain, which is often larger than direct water use. This upstream water use can include other manufacturing

Table 17-1. Direct blue water footprint (liters per dollar of value) for various manufacturing sectors, shown by North American Industrial Classification System (NAICS) code. Data from Marston et al. (2018).

NAICS Code	Manufacturing Sector	Water Footprint (L/$)
323	Printing	0.06
335	Electrical equipment	0.09
332	Fabricated metal products	0.1
334	Computers	0.1
333	Machinery	0.2
337	Furniture	0.3
315 and 316	Apparel	0.3
339	Miscellaneous	0.3
336	Transport: motor vehicles, aircraft, trains, boats, etc.	0.8
326	Plastic and rubber products	2.4
313 and 314	Textiles and fabrics	2.4
311 and 312	Food and beverages	2.6
321	Wood products	5.3
324	Petroleum and coal	8.9
325	Chemicals	12.4
322	Paper	17.3
327	Minerals: concrete, glass, clay, etc.	40.1
331	Metals: iron, steel, aluminum, copper, etc.	45.4

processes but can also include energy and agricultural water footprints. Two examples will illustrate the complexity of these water footprints and the importance of thinking about both the type of footprint (blue vs. green) and its location.

Paper: For many years, a sign sat above my building's copy machine with an image of a 5-gallon water jug, the tagline "Think before you ink," and the statement (in all caps) "PAPER DRINKS A LOT. 12 PAGES GULP DOWN 5 GALLONS OF WATER." Is paper really that thirsty? An analysis by the Water Footprint Network suggests that this is an underestimate: twelve sheets of paper have a (consumptive) water footprint in the range of 53–140 L, or 14–38 gallons, depending on where the wood is grown and how it is processed.[21] Importantly, however, almost all of this footprint is green water, namely, the rainwater that fed the forest or plantation where the wood was grown. The blue footprint of those twelve sheets—the blue water actually used in processing wood pulp into paper—is estimated to be about 0.3 gallons. In that sense, a better image for the sign might be rain falling on a forest, along with a small water bottle. This example also highlights the importance of plant transpiration in industrial water footprints: Any product with a significant input from plant material—whether paper or blue jeans or a T-shirt or a sugar-sweetened beverage—is likely to have an agricultural footprint that dwarfs the industrial footprint, at least if green water is included.

Shade balls: In 2015, the Los Angeles Department of Water and Power deployed 96 million black polyethylene balls to cover the surface of the Los Angeles Reservoir, with the dual goal of improving water quality (especially preventing photochemical production of bromate) and decreasing evaporative losses. A 2018 article calculated the water footprint of the shade balls, including the materials and energy required, and found that the water used to manufacture the balls was somewhere between 22 percent and 250 percent of the water saved annually by use of the balls.[22] Given that the balls have been in place for several years, the balls probably represent a net water savings, but the study does point to the importance of remembering that everything—including water-saving devices—takes water to make. However, from the perspective of California, the water footprint of the balls—probably made in a region not suffering from drought—is irrelevant. As Jay Lund, a UC Davis water specialist, said, "It's quite plausible to manufacture shading materials at a time and a place where water is relatively abundant and employ them to reduce evaporation at a time and a place when water is really valuable." The lead author of the 2018 article, on the other hand, notes that "people need to consider environmental conservation on a global scale, rather than just a local one."[23]

In addition to direct and upstream water footprints, we should also think a little about the downstream use of water over the lifetime of an object. This water use will generally show up in categories other than manufacturing (e.g., household use, water for energy), but product design can affect lifetime water use.

Computer use is perhaps the most obvious example of downstream impacts. Computer *manufacturing* has low water use (Table 17-1), but that is not true for computer *use*. The water footprint of using your computer has two primary components: the water footprint of the energy used to power the computer and the water footprint of the massive data centers that support your online use (which globally use almost 200 TWh of electricity). An analysis of US data centers found that their blue water consumption was the equivalent of 4.3 Lpcd, with about 75 percent of that associated with the energy needed to run and cool the servers; water use per computing workload has gone down over time and is about six times as high in traditional data centers as in hyperscale data centers.[24] The massive increase in the energy used for cryptocurrency mining (about 80 TWh in 2020) has driven an increase in associated water use as well. The recent Chinese crackdown on crypto mining has driven some operations to US states where the regulatory environment is welcoming and electricity rates are low, including arid states such as Texas.[25]

2.3. Pollution from Manufacturing

While municipal wastewater and agricultural runoff can certainly cause water quality problems, industry is the primary source of toxic metals and synthetic organics to the environment. Manufacturing processes of various kinds—from chemicals and pharmaceuticals to cars and steel—use large amounts of chemicals as solvents, reactants, and cleaners, and these chemicals must ultimately be disposed of. For many years, industry routinely dumped toxic chemicals into waterways and soils with little or no treatment, resulting in widespread environmental contamination. Table 17-2 highlights a few of the most infamous industrial pollution incidents of the twentieth century, but there are innumerable smaller-scale cases around the United States and the world. Note the importance of water in these cases as a vector for both environmental distribution (e.g., groundwater or surface water spreading the chemical) and human exposure (e.g., contaminated groundwater used as drinking water).

Partly in response to these cases, the United States has passed a number of laws

Table 17-2. Overview of six high-profile chemical contamination cases. The interested reader is directed to the companion website for details on each case.

Location	Company Responsible	Release Dates	Source of Contamination	Media Contaminated	Contaminants	Notable Impacts
Love Canal, Niagara Falls, NY	Hooker Chemical Company	1942–1952	Manufacture of dyes, perfumes, solvents	Soil, groundwater	Chlorinated solvents, hydrocarbons	Birth defects
Toms River, NJ	Ciba Geigy	1952–1996	Manufacture of dyes and specialty chemicals	Soil, groundwater, drinking water	Chlorinated solvents, benzene	Cancer
Hinkley, CA	Pacific Gas and Electric	1952–1966	Natural gas compression	Soil, groundwater, drinking water	Hexavalent chromium	Cancer
Bhopal, India	Subsidiary of Union Carbide	1984	Pesticide manufacture	Air	Methyl isocyanate	>3,700 killed; >500,000 injured
Minamata, Japan	Chisso	1932–1968	Chemical manufacture	Seawater, fish	Mercury	>2,000 suffering from neurologic disorders
Vietnam	US military	1961–1971	Use of Agent Orange (a defoliant mixture that was contaminated with dioxin)	Soil, food, people	Dioxin	Millions of Vietnamese (and US soldiers) exposed; health effects disputed but probably large

aimed at preventing and remediating chemical contamination (Table 17-3), although the effectiveness of these laws is sometimes reduced by **regulatory capture**, lack of political will, and scientific complexity. Many other countries have similar laws, and three international treaties—the Basel, Rotterdam, and Stockholm Conventions—address global use and trade of toxic chemicals.

Despite improvements in chemical use and disposal practices, we are still dealing with toxic industrial chemicals in at least three different contexts:

- **Legacy contaminants**: Persistent chemicals that were banned decades ago remain present in stockpiles, soil, groundwater, river sediments, and biota, and they are very expensive to clean up. **Polychlorinated biphenyls (PCBs)** are both unique and typical: unique in that the Stockholm Convention provided a long time frame for PCBs to be phased out of existing enclosed applications and typical in that many countries (including the United States) have taken little action to catalog and dispose of their massive stockpiles of PCB-containing equipment.[26]
- **Occupational exposure**: Workers are exposed to toxic chemicals used in manufacturing and other industrial processes, with an estimated 40,000 people in the

Table 17-3. Summary of US chemical laws, classified by which phase of the hazardous chemical lifecycle they address: chemical production and use, emissions, exposure, or remediation of contaminated sites. EPA, Environmental Protection Agency.

Law	Lifecycle Phase	Summary
Federal Insecticide, Fungicide, and Rodenticide Act	Production	Pesticides must be registered by EPA, including demonstration that they will not cause "unreasonable adverse effects on the environment."
Toxic Substances Control Act	Production	New chemicals must undergo toxicity review before being put into commerce; uses of both new and existing chemicals can be restricted where those uses pose unreasonable risks.
Federal Food, Drug, and Cosmetic Act	Production, exposure	The Food and Drug Administration regulates pharmaceuticals and determines allowable levels of additives and pesticide residues in food.
Emergency Planning and Community Right-to-Know Act	Emissions	Passed in response to the Bhopal disaster, the act requires facilities using certain toxic chemicals to have emergency response plans and to publicly disclose, through the Toxics Release Inventory, the amounts of these chemicals stored, used, and disposed of.
Clean Air Act	Emissions, exposure	The EPA sets National Emissions Standards for 188 Hazardous Air Pollutants and assesses exposure under the AirToxScreen program.
Occupational Safety and Health Act	Exposure	The Occupational Safety and Health Administration sets standards for chemical exposures in workplaces, although it has generally focused more on safety issues than on chemical toxicity.
Oil Pollution Act (OPA)	Emissions, remediation	Passed in response to the 1989 Exxon Valdez spill, OPA tightens oil-spill prevention and preparedness measures and makes responsible parties liable for cleanup and damages.
Safe Drinking Water Act	Emissions, exposure	The Underground Injection Control program regulates disposal wells to protect drinking-water aquifers. *Maximum Contaminant Levels* are designed to protect people from cancer and noncancer risks (Chapter 14).
Clean Water Act	Emissions, exposure	*National Pollutant Discharge Elimination System* permits for industrial facilities must include effluent limits for toxic pollutants based on EPA guidance for that type of facility. Ambient *water quality criteria* must protect humans and animals from toxic effects (Chapter 8).
Resource Conservation and Recovery Act (RCRA)	Emissions, remediation	RCRA regulates the management of hazardous wastes from generation to final disposal. RCRA's Corrective Action Program is guiding cleanup at about 4,000 sites.
Comprehensive Environmental Response, Compensation, and Liability Act (CERCLA)	Remediation	Passed in response to the Love Canal disaster, CERCLA (Superfund) requires polluters to pay for the cleanup of contaminated sites on the National Priorities List. Where no liable polluter can be found (or the polluter can't afford the cleanup), funding comes from the Superfund Trust Fund, which is funded by a tax on chemical companies.

United States[27] and 1 million around the world[28] dying each year from toxic workplace exposures.
- **Consumer exposure**: Thousands of synthetic chemicals, many of which have poorly understood toxicity profiles, are widely used in consumer products, from flame retardants in furniture to plasticizers in water bottles.

The laws in Table 17-3, and the regulations that derive from them, struggle in different ways with the complex questions of industrial pollution in the modern age:

- How do we assess the risks posed by thousands of chemicals in the face of irreducible scientific uncertainty?
- How much should we spend on better understanding risk?
- How do we balance chemical risks against other types of risk?
- Should the burden of proof be on those who claim a chemical is safe or those who claim it is not?
- If we drastically restricted the suite of existing and new chemicals that we considered safe to use, how much would we reduce innovation and economic development?
- When we are cleaning up a contaminated site, how clean is clean?
- How much should we spend on cleaning up contaminated sites versus preventing new contamination?

These are tough questions, with no simple answers. But there are also some statements that we can make with reasonable certainty:

- We have fouled our nest with toxic chemicals. This has manifested both in "sacrifice zones" of intense contamination and in the widespread distribution of persistent chemicals in soils and ecosystems around the world, producing a global signature of the ***Anthropocene Epoch***.
- These burdens are not evenly distributed. Within the United States, communities of color are more likely to be exposed to toxic water, air, and soil. Globally, rich countries often export toxicity to poorer ones.
- Prevention is cheaper than cleanup. Cleaning up contaminated sites is an extremely slow and difficult effort that can cost considerably more than the economic value originally produced at the site. The Department of Energy has environmental liabilities of at least $500 billion,[29] which is a lot even for the federal government. (The Department of Energy's total annual budget is under $50 billion.) A 2013 National Research Council report estimated a cost of $110–127 billion to clean up the 126,000 sites with groundwater contamination, but it notes that both the cost and the number of sites are almost certainly underestimates.[30]
- We have to stop playing catch-up with respect to emerging contaminants. We need to be more cautious at the upstream end of the chemical cycle, when we allow new chemicals to be put into widespread use. In the United States, the Toxic Substances Control Act is our main tool for this purpose, but we might learn from the more precautionary approach taken in other countries, including Europe's Registration, Evaluation, Authorisation and Restriction of Chemicals (REACH) regulation.

3. Water and Mining

Look around you. The objects you see form a thin thread of connection between you and distant mines, whether you are glancing at your smartphone (containing, among many others, aluminum and gallium), your jewelry (gold, silver), your LED light bulb (copper and indium, and probably powered in part by coal), your credit card (silver and tin), or your drinking glass (sand and limestone). Mining—including coal mining, ***hardrock mining*** (metals, including uranium), and ***aggregate*** mining (sand, gravel, rock)—is crucial to our modern lifestyle, but many of us are barely aware of these connections.

Depending on the location, purity, and value of the deposits they are targeting, mining operations use either subsurface mines or surface mining techniques such as strip mines or open pits. Although surface mines are cheaper and safer than deep mines, excavation and disposal of the overburden and mine tailings can rearrange the local topography, cause severe erosion and flooding problems, and require the use of massive ***tailings dams***. One form of surface mining, common in Appalachia, is mountaintop removal, in which ridgetops are dynamited to get at the coal underneath. The waste rock and debris are dumped in valley bottoms, where they smother streams, exacerbate flooding, contaminate local water sources, and lead to health effects on local communities.[31] Ironically, mountaintop removal mining in Appalachia is driven in part by air pollution regulations that have made the region's low-sulfur coal more desirable.

Although there are situations where mining's water use leads to scarcity, the greater issue is usually pollution. Common mining pollutants include the following:

- **Acidity and alkalinity**: Exposure of reduced minerals to air, a common phenomenon in mining, can lead to mineral oxidation, which can significantly reduce the pH (***acid mine drainage***). In other situations, rapid weathering of carbonate rocks can produce high-pH (alkaline) mine drainage.
- **Metals**: The combination of newly exposed rock surfaces and acidic conditions can lead to leaching of various toxic metals from rocks. One such metal, selenium, has been measured at remarkably high levels in streams draining mines in both Appalachia[32] and the remote Elk Valley in British Columbia[33]; at the latter, it is apparently responsible for fish deformities.
- **Mercury and cyanide**: The neurotoxic metal mercury has been traditionally used in gold mining (and is still used in small-scale mining, especially in Africa) because of its ability to form an amalgam with gold. It has been largely replaced with cyanide, which is also highly toxic to people and wildlife.
- **Salinity**: Mining can mobilize soluble salts, which can increase salinity in nearby soils and rivers. This is particularly likely for potash mines, since potash (KCl and other potassium salts) is found in evaporite deposits, layers of NaCl and other soluble salts that usually form when an inland sea is isolated over geologic time and loses its water to evaporation.[34]

In low- and middle-income countries, mining often takes an exploitive form. The open-pit gold and silver mine near the village of Cerro de San Pedro, Mexico, is a case in point.[35] The area had been home to small-scale mining over the centuries but was mostly dependent on subsistence agriculture and tourism. In 1996, the Canadian mining company New Gold announced plans for a large open-pit mine in Cerro de San Pedro. Despite sometimes fierce local opposition and national laws that in theory should have protected the area, the mine operated from 2007 to 2018, with annual production as high as 3 tons of gold and 43 tons of silver.[36] The mine removed the hill for which the town is named and occupied a footprint of over 370 ha, including the open-pit mine itself, waste dumps, and a cyanide leaching area that used over 30 million liters of water per day. The conflict over the mine tore the town apart and led to the assassination of the mayor. Similar situations have occurred around Latin America, as neoliberal reforms have led to privatization of land and water rights, making them available for sale to international mining companies.

Once a mining site is disturbed, pollutants can persist for many years, creating a long-term hazard that we are only starting to understand. In the United States, the Surface Mining Control and Reclamation Act of 1977 requires coal-mine operators to restore the land after the mine ceases operation, a process known as mine reclamation. But the boom-and-bust nature of mining means that there are tens of thousands of coal mines that were abandoned by their owners before these regulations were put in place. The Abandoned Mine Land Reclamation Program, funded by coal-mining fees, works to stabilize and reclaim high-priority sites, but the work is slow and expensive.

Unfortunately, there is no comparable program to address abandoned hardrock mines,[37] in part because the Mining Law of 1872—which does not address cleanup (and gives private companies royalty-free access to mineral resources on public lands)—has not been updated in 150 years. Congressional efforts to reform hardrock mining have been stymied by the economic and political power of the mining industry.[38] The number of abandoned hardrock mines in the United States is unknown and is often assumed to be somewhere around 500,000. The Government Accountability Office recently identified 140,652 specific mine features (e.g., tunnel, structure, tailings pile), of which at least 67,000 pose physical hazards (collapse, toxic gas exposure) and 22,000 pose environmental hazards (acid mine drainage, soil contamination).[39] There is no effective program to prioritize these sites. Various federal agencies (Environmental Protection Agency, US Forest Service, Bureau of Land Management, National Park Service, and the Office of Surface Mining Reclamation and Enforcement) spend about $300 million per year of taxpayer funds on stabilization and remediation (including through Superfund), although this is probably a drop in the bucket compared to what is needed. The 2015 Gold King Mine disaster brought this issue to the public's attention, but so far Congress has not passed meaningful reform.

The transition to a clean energy economy will require an acceleration in the mining of certain metals, such as cobalt and lithium, with potentially serious local environmental and social impacts.[40] Lithium—a critical component of electric vehicle batteries—is particularly concerning from a water perspective, since it is often found in arid environments where mining and processing will increase water scarcity and affect sensitive desert ecosystems. The US goal of producing more lithium domestically has led to projects such as the Thacker Pass mine in Nevada, which has attracted significant opposition because of potential impacts to water, ecosystems, and Paiute Shoshone sacred land.

Mining of aggregate (rock, gravel, sand) takes place globally at far larger volumes than coal or hardrock mining, and this activity is increasing rapidly as urban construction explodes and coastal land reclamation becomes more popular. Although aggregate mining doesn't generally produce significant toxicity, the mining of sand and gravel from riverbeds poses a serious threat to the physical, biotic, and chemical integrity of rivers.[41] In the lower Mekong River, for example, excessive sand mining has changed channel profiles, destabilized riverbanks, damaged infrastructure, and contributed to sediment starvation and land loss in the Mekong Delta.

4. Corporate Water Stewardship

High-profile multinational corporations have had their share of bad press over water issues. In an effort to improve their reputation and maintain their "social license to operate," many companies now address water stewardship as part of their corporate social responsibility portfolios. One prominent effort—founded in 2007 and now boasting almost 200 endorsing companies—is the CEO Water Mandate, part of the UN Global Compact in partnership with the Pacific Institute.[42] The CEO Water Mandate defines water stewardship as "a framework and set of practices that help businesses *manage risks*, *cut costs*, and *build trust* while promoting long-term water security for all" (italics added).

- **Manage risks**: Water-related risks for businesses are often divided into physical risks (scarcity, flooding, poor water quality), regulatory risks (inconsistent governance that may lead to sudden changes in water rights or water-treatment requirements), and reputational risks (the kind of headlines mentioned above).
- **Cut costs**: When water costs are significant to a company, managers will be motivated to use water more efficiently. This points to the importance of ensuring that companies pay the true cost of water, and—importantly—the true cost of water pollution. If we want both industry and clean water, we need to ensure that companies internalize the costs of their pollution; this requires a strong regulatory framework. Other ways that water issues can affect a company's bottom line may be less obvious. If a company's employees don't have access to safe water and sanitation at home, they will be less productive at work. When a company's supply chain ignores water risks, the costs of inputs may go up. When poor governance allows other users to pollute a river, the costs of treatment may go up. These examples suggest that companies need to be involved in water management issues well beyond their own walls.
- **Build trust**: This is perhaps the core of the issue: Can companies regain the trust of environmentally conscious consumers and local activists, or will water stewardship be seen as just another case of ***greenwashing***? The good news is that there are real opportunities for companies to reduce their water impact, and at least some companies are seizing those opportunities, whether because it makes them attractive to consumers or for other reasons. There is reason to hope that as more

people become aware of water issues, there will be more pressure on companies to both talk the talk and walk the walk.

The actual steps taken in improving water stewardship can vary, but they often start with a detailed audit of a company's water footprint and wastewater impact and setting targets to reduce them. For example, Unilever, a multinational corporation producing beauty and home care products, has reduced both water withdrawal and ***chemical oxygen demand (COD)*** emissions at its factories by about 40 percent (per unit of product) over eleven years, mostly through increased water recycling and more efficient cleaning processes. Unilever has also tried to reduce downstream water use by designing laundry soaps that require less rinsing.[43]

In August 2016, Coca-Cola announced that it had become the first Fortune 500 company to replenish all the water that it uses, an accomplishment that it touted as becoming water neutral, analogous to being carbon neutral. Let's unpack this claim on two levels: Is water neutrality a useful concept, and what is Coke actually doing?

Water is fundamentally different from carbon in ways that make water neutrality a problematic concept:

- Water is renewable. Whereas burning coal produces CO_2 that will stay in the atmosphere and affect climate for a long time, consumptive use of water does not change the global or local hydrologic cycle (with some minor exceptions).
- Water is local. Carbon offsets—planting trees in one place to make up for burning coal in another—make sense because CO_2 is globally distributed; the atmosphere doesn't care where it comes from. Water is different. As Amit Srivastava writes, "Replenishing an aquifer hundreds of miles away from the point of extraction, as Coca-Cola has often done to 'balance' their water use, has no bearing on the health of the local aquifer which Coca-Cola depletes through its bottling operations, nor the privations suffered by those who depend upon it."[44]

Looking in more detail at Coke's claims, there are two ways that it replenishes the water used in its manufacturing: wastewater treatment and "replenish projects." The former makes some sense. Discharge of appropriately treated wastewater into the same watershed from which the water was withdrawn does compensate for that water use; after all, that is the difference between withdrawal and consumption. "Replenish projects," which account for just over half of Coke's water neutrality claim, are more difficult to evaluate; they are projects that Coke funds around the world (in part through The Nature Conservancy's Water Funds) aimed at improving agricultural water management and providing water, sanitation, and hygiene access. These projects appear to be very worthwhile, and it is appropriate for a global company like Coke to help with global problems such as water, sanitation, and hygiene, but the projects do not compensate in any quantitative way for Coke's consumptive water use.

From my perspective, it would be better for Coke to avoid the meaningless goal of water neutrality and instead focus its water efforts a bit more on its own suppliers and facilities. Notably, Coke's water neutrality goal is limited to direct water use in manufacturing and does not include the water footprint of its ingredients (especially sugar), which dominate the overall footprint of a bottle of Coke.

Chapter Highlights
1. The production of electricity uses a great deal of water to cool power plants, although most of that use is nonconsumptive. Changes in how electricity is produced can have major effects on water use: A shift from **once-through** to **recirculating cooling** decreases withdrawal but increases consumption, a shift toward wind and solar tends to decrease both withdrawal and consumption, and a shift toward hydropower increases consumption due to reservoir evaporation.
2. Because of its dependence on cooling water, electricity production is vulnerable to hydrologic change, particularly drought and higher water temperatures.
3. The increased use of **biofuels** has increased the water footprint of the transportation sector, although the significance of that water footprint depends on whether it is green or blue (i.e., whether the biofuel crops were irrigated).
4. The extraction, transportation, and use of fossil fuels are significant sources of water, soil, and air pollution. Issues of greatest concern include the storage and disposal of **coal ash**, the transportation of **dilbit**, the contamination of drinking water aquifers by **fracking** operations, and the targeting of communities of color as points of least resistance for pipelines and other hazardous infrastructure.
5. **Produced water** from oil and gas extraction presents a disposal problem and, in some cases, a potential new source of water for irrigation or other uses.
6. Over the last sixty-five years, water use in US manufacturing (per unit of industrial production) has gone down by more than 90 percent, through increased water-use efficiency and a shift away from the production of water-intensive metals and chemicals.
7. In addition to direct water use, manufactured products also have upstream and downstream water footprints, but it is critical to distinguish between blue and green components of those footprints.
8. Historically, poorly regulated industrial activities have been major sources of toxic compounds to water, soil, air, fish, and people, including well-publicized tragedies such as Bhopal, Minimata, and Love Canal. Many banned **legacy contaminants** are extremely difficult to remove from soils and groundwater, so we are still dealing with their effects. Much of the burden of industrial toxins falls on low-income and minority communities.
9. The transition to a carbon-free economy will require increased mining of certain elements such as lithium and copper, but mining can be a significant source of water contamination and competition over water supply.
10. The recent corporate interest in water stewardship is encouraging, but water neutrality is not a scientifically valid goal.

Notes
1. Section 316 of the Clean Water Act requires regulation of both heat discharges and water intake structures in order to better protect aquatic life from temperature modification and from impacts associated with entrainment of organisms into the water intake.

2. Recirculating cooling can also reduce the efficiency of the power plant, resulting in greater greenhouse gas emissions.

3. Data from the Energy Information Administration.

4. McCall et al. (2016).

5. See also Henry and Pratson (2019), which applies climate change projections to analyze future cooling water issues for fifty-two power plants (all using once-through cooling), and finds that the most likely impact comes from difficulty in meeting thermal pollution regulations.

6. Bracken et al. (2015).

7. Jones et al. (2016).

8. Panat and Varanasi (2022).

9. Scherer and Pfister (2016).

10. This result holds even though, as we will see in Chapter 18, only a small fraction of US corn is irrigated.

11. Scown et al. (2011).

12. Bordonal et al. (2018).

13. https://www.epa.gov/coalash/coal-ash-basics.

14. https://www.nationalgeographic.com/environment/article/coal-other-dark-side-toxic-ash.

15. In the case of fracking, the water initially flowing from the well after fracking consists largely of flowback of the fracking water, which is then followed by produced water. Extraction of coalbed methane—natural gas trapped in coal deposits—also generates large volumes of produced water, although the water quality is often higher.

16. Scanlon et al. (2020).

17. These data are from https://www.api.org/oil-and-natural-gas/wells-to-consumer/transporting-oil-natural-gas/pipeline/where-are-the-pipelines, where you can see maps of liquid and gas pipelines.

18. https://www.stopline3.org/.

19. Marston et al. (2018).

20. Marston et al. (2020).

21. van Oel and Hoekstra (2012).

22. Haghighi et al. (2018).

23. https://www.pbs.org/newshour/science/why-96-million-plastic-shade-balls-dumped-into-the-la-reservoir-may-not-save-water.

24. Siddik et al. (2021).

25. https://www.buzzfeednews.com/article/sarahemerson/denton-texas-crypto-miner-core-scientific; https://www.bbc.com/news/world-us-canada-58414555; https://www.cnbc.com/2021/09/30/this-map-shows-the-best-us-states-to-mine-for-bitcoin.html.

26. Melymuk et al. (2022).

27. https://www.nytimes.com/2013/03/31/us/osha-emphasizes-safety-health-risks-fester.html.

28. Hämäläinen et al. (2017).

29. https://www.gao.gov/highrisk/u.s.-governments-environmental-liability.

30. National Research Council (2013).

31. Hendryx and Luo (2015).

32. Naslund et al. (2020).

33. https://thenarwhal.ca/for-decades-b-c-failed-to-address-selenium-pollution-in-the-elk-valley-now-no-one-knows-how-to-stop-it/.

34. Since the 1920s, potash mining in the Llobregat River in Spain has threatened the water supply of Barcelona. Claims by mining companies that the salinity was naturally occurring were undermined by a sharp drop in salinity when mining was disrupted during the Spanish Civil War (Gorostiza and Saurí 2019).

35. Stoltenborg and Boelens (2016).
36. https://www.newgold.com/assets/cerro-san-pedro/default.aspx.
37. Under certain limited conditions, funds from the coal-focused Abandoned Mine Land program can be used to address hardrock mines.
38. https://grist.org/energy/manchin-and-cortez-masto-kill-chances-of-reforming-outdated-hardrock-mining-law/.
39. GAO (2020a).
40. Lèbre et al. (2020).
41. Rentier and Cammeraat (2022).
42. https://ceowatermandate.org/.
43. https://www.unilever.com/news/news-search/2017/smartfoam-smart-way-to-save-time-effort-water/.
44. https://theecologist.org/2015/aug/25/never-mind-greenwash-coca-cola-can-never-be-water-neutral.

18 Agricultural Water Use: Water's Central Role in Our Food Supply

- How can agriculture become more sustainable from a water perspective while still providing enough food for a growing population?
- Which foods have the largest water footprints?
- How can we make better use of both blue and green water in growing food?
- How does agriculture pollute water and soil, and how can we minimize those impacts?

Agriculture is by far the biggest user of **blue water** globally. If we include **green water** as well (i.e., **rainfed agriculture**), food production becomes an even more dominant force in the human appropriation of the global water cycle. Agricultural activities are also major sources of water pollution.

In this chapter, we turn to the agricultural sector, with an eye to understanding its role in the global water crisis and thinking about ways to move toward a more sustainable future. We start with a general overview of the challenges facing modern agriculture and then turn to the water quantity and quality aspects of these challenges. We present the critical question—how much water will it take to feed the world in the next few decades, and where will that water come from?—and break it into three components that we discuss in some detail: what we eat, where we grow it, and how we use water to grow it. The final section of the chapter addresses the pollution problems caused by farming practices, especially salinization, nutrients, pesticides, and pathogens.

Before we begin, it may be helpful to consult Table 18-1, which provides a set of definitions that we will use throughout the chapter.

1. From the Green Revolution to Sustainable Intensification

Modern agriculture embodies a series of paradoxes:

- Very few of us (in the United States and other high-income countries) are farmers, but all of us are eaters.

Table 18-1. Definitions relevant to this chapter.

Category	Term	Definition
Land	Grazing land	Any land on which domesticated animals eat by grazing or browsing. Includes rangeland and pasture.
	Rangeland	Grazing lands, generally natural grasslands, where soils and vegetation are not actively managed.
	Pasture	Grazing lands that are managed through planting, fertilization, irrigation, drainage, pest control, or tillage.
	Cropland	Lands producing crops to be harvested for direct or indirect human use. Includes lands growing fiber crops, forage crops, and food crops.
Crops	Fiber crops	Crops grown for fiber. Mostly cotton, hemp, and jute.
	Forage crops	Grasses (e.g., timothy grass) and legumes (e.g., alfalfa) used for animal feed as either hay or silage.
	Hay	Forage crops that are harvested, dried, and stored before being fed to animals.
	Silage	Forage crops that are harvested, chopped, and fermented under moist, acidic conditions before being fed to animals.
	Forage or roughage	High-fiber foods, especially as necessary for the nutrition of ruminant animals. Includes forage crops and forage from grazing lands.
	Concentrates	Optimized, high-energy, full-nutrition animal feeds derived from soy, corn, and other ingredients.
	Annual food crops	Food crops that need to be replanted each year.
	Perennial food crops	Food crops that do not need to be replanted after harvest, including tree nuts, vineyards, fruit, cocoa, and coffee.
	Cereal crops	Grasses grown for their grain. Includes wheat, rice, corn, rye, and sorghum. Corn, rye, and sorghum can also be grown for silage, in which case the entire plant (not just the grain) is harvested, chopped, and ensilaged (fermented).
	Horticultural crops	Fruits, vegetables, tree nuts, nursery crops, and flowers.
	Nursery crops	Trees, shrubs, and herbaceous vegetation grown for sale. Includes ornamentals, Christmas trees, and fruit trees.
Animals	Ruminants	Animals that eat fibrous foods by fermenting them in their rumen, followed by regurgitation of the cud produced by initial fermentation. Includes cattle, sheep, goats, water buffalo, and yaks.
	Livestock	Farm animals, including ruminants, swine, and horses. Poultry may or may not be included.
	Poultry	Birds raised for meat or eggs, including chickens, turkeys, ducks, and geese.
	Broilers	Chickens raised for meat.
	Layers	Chickens raised for eggs.

- Food production accounts for a tiny fraction of gross domestic product (in the United States and other high-income countries), but it is arguably the most essential of all economic sectors.
- Most of us know little about how our food is grown, but agricultural practices critically affect our health and the health of animals, landscapes, and the global environment.
- Modern agriculture has been remarkably successful in increasing food production and decreasing the human effort spent on farming, but at the same time it has created massive social and environmental problems.

The topic of agricultural sustainability goes well beyond the scope of this book. But in order to understand the water problems of industrial agriculture, we do need some context, so this section provides an overview of the recent history of agriculture and the sustainability challenges faced by our current food production system.

Over the last two centuries, agriculture has undergone a seismic shift that, in Vaclav Smil's phrase, constitutes one of the "grand transitions" that define the modern world—a transition that is intimately linked to other transitions in energy, populations, economies, and the environment.[1] The agricultural transition has allowed a much smaller share of the population to produce a much greater amount of food, contributing to urbanization, declining birth rates, and the shift toward industrial and service economies. The agricultural transition has taken place over different time periods in different countries, but the period from 1960 to 1990—when modern agricultural technologies and seeds spread rapidly across the developing world—is often identified as the **Green Revolution**.

We can use global agricultural data, collected by the Food and Agriculture Organization starting in 1961, to quantify the impact of the Green Revolution (Figure 18-1). Even as population increased 2.5-fold from 1961 to 2019, food supply more than kept up, averting the **Malthusian** famines that some had predicted—indeed, leading to an extended period of historically low mortality from famine. This increase in food production was driven by a small (15 percent) increase in area devoted to crops (extensification) but, more importantly, by increased productivity on existing croplands (intensification). For example, the average global **yield** (production per unit land) for cereals increased threefold during this time period (Figure 18-1).

These increases in yields and food supply were driven by several interacting factors that together make up the modern industrial agriculture package:

- High-yielding varieties—new genetic varieties of staple crops—are more productive, have a higher edible fraction, and grow more quickly (allowing double- and triple-cropping on the same plot of land).
- Synthetic **fertilizers** increase crop productivity by providing plant nutrients, and **pesticides** reduce pest damage.
- **Irrigation** offers farmers greater control over water supply and thus provides higher yields than rainfed agriculture.
- Mechanization allows faster and more efficient cultivation and harvesting.
- **Animal feeding operations** are used to grow poultry, swine, and cattle rapidly under controlled conditions using high-energy feeds.

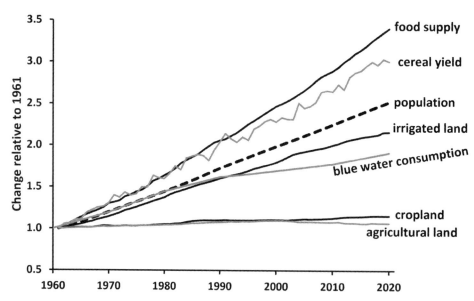

Figure 18-1. The good news. From 1961 to 2020, increases in food supply and cereal yield outpaced increases in population, even as water and land use grew more slowly. All data are global totals from the Food and Agriculture Organization (https://www.fao.org/faostat/en/#home), except blue water data from Shiklomanov/AQUASTAT (see Chapter 4). (For reference, 1961 values for each parameter are as follows: 2.5 Ecal/yr [food supply], 1.4 ton/ha [average cereal yield], 3.1 billion [population], 0.16 billion ha [irrigated cropland], 1,200 km^3/yr [agricultural blue water consumption], 1.4 billion ha [cropland], and 4.5 billion ha [agricultural land].)

- Globalization has facilitated the rapid spread of seeds, agrichemicals, and practices.
- An industrial mindset views farms as factories for food production and uses the approaches of the Industrial Revolution to optimize that production.

More recently, the first two of these factors (high-yielding varieties and pesticides) have come together in the form of genetically engineered crops that are resistant to common **herbicides** and produce their own **insecticides**. For example, almost all the corn planted in the United States is Roundup Ready (i.e., it is resistant to the herbicide glyphosate) and *Bt* (i.e., it contains a gene from the bacterium *Bacillus thuringiensis*, causing it to produce a protein that protects it from corn borers).

Despite the "good news" shown in Figure 18-1, there are also some concerning trends in the global food picture. While global yields of cereals continue to increase, the rate of increase is slowing. Yields for roots and tubers, a primary food source in much of Africa, have lagged far behind cereals, in part because of a lack of research. Climate change is already exerting a drag on global agricultural productivity, especially in warmer regions.[2] In some places, agricultural land, including irrigated land, is being lost to urban and suburban sprawl.

In addition, the increase in global calorie production shown in Figure 18-1 masks problems with the distribution of food. Progress in reducing hunger has slowed in recent years, with the number of undernourished people rising slightly from its 2014 low of 607 million to 650 million in 2019, even before the COVID-19 pandemic added another 120 million to this tally.[3] In other words, the availability of food continues to be uneven and inequitable, with lower-income populations often finding high-quality food unaffordable, especially when global food prices rise sharply, as in 2010–2012 and again in 2021–2022.

Looking to the future, food demand will continue rising due to population growth and poverty alleviation, while expansion of agricultural land is both undesirable (since it destroys habitat) and unlikely in many countries (due to rapid urbanization). Will we really be able to continue growing more and more food on the same (or decreasing) amount of land?

To make matters worse, the Green Revolution's emphasis on increasing yields, to the exclusion of other values, has had very significant environmental and social impacts, summarized briefly in Table 18-2. Agriculture is a major contributor to almost all environmental problems, with the food supply chain (mostly food production but also processing, transportation, and consumption) contributing 26 percent of greenhouse gas emissions, 32 percent of acid rain emissions, and 78 percent of nutrient (N and P) pollution.[4]

Table 18-2. The bad news: summary of environmental and social impacts associated with each of the features of the Green Revolution.

Feature	Impacts
High-yielding varieties	• More agricultural monocultures • Less crop and dietary diversity • Farmer dependence on agribusiness for seeds • Risks of genetically modified organisms
Chemical inputs	• Nutrient pollution • Pesticide contamination of food and drinking water • Energy intensity of fertilizer production • Farmer dependence on expensive inputs • Development of pesticide resistance by major pests
Irrigation	• Water scarcity • Drying rivers and lakes • Soil salinization
Mechanization	• Soil erosion • Increased energy use • Loss of jobs
Animal feeding	• Mistreatment of animals • Conversion of manure from a widely distributed resource to a highly concentrated waste • Use of human food for feeding animals
Globalization	• Consolidation and specialization • Loss of local food security • Farmer indebtedness • Diversion of habitat and cropland to fuel production
Industrial mindset	• Lack of recognition of the complexity of soil–biota interactions • Loss of natural productivity • Undervaluing of place-based agricultural adaptations

In short, then, we need to increase food production on a limited land base to meet rising demand while dramatically reducing the environmental impact of agriculture. This dual challenge has been called **_sustainable intensification_**, and various approaches to a more sustainable agriculture are starting to emerge.

2. Water and Agriculture: Overview

The remainder of the chapter will focus specifically on the role of water in current and future agriculture. This section sets the stage by providing an overview of how water is used in global agriculture and then posing the central challenge: Is there enough water to feed the world in the coming decades?

2.1. Water Use in Global Agriculture

Water is used in agriculture in a variety of ways, such as animal watering, crop processing, and equipment cleaning, but the largest volume by far is used in crop production. The crop water requirement, ET_{crop}, is the amount of water that a given crop needs to transpire in order to grow, and it is generally expressed in millimeters (or, in the United States, in acre-feet per acre, which is equivalent to feet). ET_{crop} can be estimated on a daily basis via Equation 18-1 and summed over the crop's growing period to give the total ET needed to grow one harvest of that crop, a number that can vary from about 200 to 2,000 mm. (Note that 1,000 mm is equal to 10,000 m³/ha, or 3.3 AF/acre.)

Equation 18-1. $ET_{crop} = ET_0 \times K_c$

Where

- ET_0 is the **_reference evapotranspiration_** (ET from well-watered grass), which ranges from about 1 to 10 mm/d, depending on conditions.
- K_c is the crop coefficient, which can vary from about 0.3 to 1.2, depending on the crop and stage of growth; crop coefficients are available from the Food and Agriculture Organization.[5]

The crop water requirement can be provided exclusively by soil moisture derived from rainfall (green water) or, in the case of irrigated agriculture, can be supplemented by blue water. The irrigation water requirement for a crop is equal to ET_{crop} minus whatever green water is available during the growing period.

In general, irrigation produces higher yields than rainfed agriculture, because farmers can control the availability of moisture to optimize plant growth and because irrigated areas tend to be sunnier. Irrigated land—which represents 18 percent of global cropland—makes up 24 percent of the area harvested in an average year and produces 33 percent of global crops using 17 percent of crop green water use (and, of course, 100 percent of crop blue water use). In the United States, irrigated land—about 18 percent of cropland—produces over 50 percent of crop value.[6]

Table 18-3 provides a summary of global blue and green water use by various crops. Notably, over half of global blue water use goes to just three crops—rice, wheat, and forage crops—and about two thirds of global blue water use is in Asia.

Table 18-3. Global crop water consumption, by crop and region. Data from Siebert and Döll (2010). "Forage crops" does not include cereals (e.g., corn) used for silage.

	Irrigated Land as Percentage of Total Harvested Land	Blue Water Use (km^3/yr)	Green Water Use on Irrigated Land (km^3/yr)	Green Water Use on Rainfed Land (km^3/yr)
Global total	24%	1,180	919	4,586
By Crop				
Rice	62%	307	337	297
Cotton	49%	84	46	85
Sugar crops	43%	78	74	119
Wheat	31%	208	115	535
Corn	20%	72	92	493
Forage crops	12%	102	55	604
Other	22%	329	199	2,453
By Region				
South Asia	43%	480	239	523
Asia (except South Asia)	40%	328	429	1,011
North America	15%	173	119	858
Oceania	11%	14	13	71
Africa	9%	98	27	738
South America	9%	30	40	542
Europe and Russia	7%	58	54	842

2.2. How Much Agricultural Water Do We Need?

Several trends are converging to put unprecedented stress on agricultural water availability. The supply of water for agriculture is being reduced by the depletion of groundwater supplies (Chapter 5) and by the increasing volumes of agricultural water transfers to urban and environmental uses (Chapter 11). Climate change is affecting both water supply and crop water needs. At the same time, increasing demand for food, fiber, and biofuels is translating into rapidly increasing agricultural water demand. Where will this water come from? How can we possibly feed a growing population while also providing adequate water to cities, protecting aquatic ecosystems, and facing the end of unsustainable groundwater extraction?

Equation 18-2 provides a framework that will help us understand the amount of water needed to meet the world's food (and fiber[7]) needs. It is a water-specific version of the IPAT equation, $I = P \times A \times T$: Environmental *impact* is a product of *population*, *affluence* (i.e., consumption levels), and *technology*.

Equation 18-2. $W = P \times D \times F$

Where
- W = total agricultural water required (m^3)
- P = population (number of people)

- D = average diet (calories/person)
- F = **_water footprint_** of that diet (m³/calorie)

Equation 18-2 suggests that the water needed to feed the world is a function of three factors:

- **How many people there are**: Clearly population is a key driver of agricultural water demand, and equitably reducing population growth would help reduce that demand. Population policy is outside the scope of this book.
- **What those people eat**: Diet (total calories and choice of foods) plays an important role in agricultural water demand and will be the focus of Section 3.
- **How much water it takes to grow that food**: The water footprint of a food product describes the volume of water consumed to produce a given amount of that product, and it will be the focus of much of our analysis. Since different foods have different caloric, nutritional, and economic value, the denominator of the water footprint can be expressed in various units, such as calories, to allow better comparison of apples and oranges. The **_water productivity_** of agricultural production (food produced per unit water) is simply the inverse of the water footprint.

The logic of Equation 18-2 suggests that our goal should be to minimize the water footprint of food production, and indeed this is where we will be heading in the rest of this chapter. However, we must first address two complications:

- Blue or green footprint? Unlike other sectors, agriculture can use blue and green water somewhat interchangeably. That is, we can grow food by using rainfed agriculture with essentially no blue footprint (but a large green footprint) or, alternatively, using irrigated agriculture with a large blue footprint. The two pathways have very different implications. Relying on rainfed agriculture means minimal competition with other water users (such as cities and the environment), but—given the low yields of rainfed agriculture—large land requirements. In contrast, relying on irrigation means higher yields (and thus less land dedicated to agriculture) but also depletion of surface and groundwater sources and more intense competition for water.
- Local or global footprint? As we have discussed previously, there is an ongoing debate over the appropriateness of using a volumetric blue water footprint in which all blue water is treated equally, regardless of whether that water is used in a water-scarce or water-abundant region. Given the dominance of irrigation in total blue water use—and the fact that some, but not all, of this use directly exacerbates local water scarcity and conflict—it seems important to tie agricultural blue water use to local impacts.

Our primary goal, then, is not simply to minimize the blue water footprint of food production but to minimize the blue water *impact* of food production while avoiding net increases in agricultural land. With this goal in mind, we turn to an analysis of water use in agriculture as a function of three factors: what we choose to eat, where we grow that food, and how we use water to grow that food.

3. What We Eat

The question of what we should eat—what Michael Pollan famously called "the omnivore's dilemma"[8]—has occupied humanity for millennia and spawned a great range of cultural preferences and taboos. In the modern age, globalization and industrial agriculture have dramatically changed our eating patterns, with most societies experiencing an increase in consumption of animal products and processed foods, although this trajectory is not inevitable.

Two pillars should guide our thinking about optimal diets: our own well-being and that of the planet. In considering the former, we should be looking for foods that provide sufficient calories and are culturally meaningful, personally satisfying, high in nutrients and healthy fats, low in unhealthy fats and sugars, and low in contaminants such as **polychlorinated biphenyls (PCBs)** and mercury. At the same time, we should be seeking foods that are low in environmental impact, including water use but also greenhouse gas emissions, nutrient (N and P) pollution, acid rain emissions, and habitat and biodiversity impacts. Although there are some differences between environmental metrics, animal products generally have substantially higher impact than nonanimal products.[9] Diets designed to minimize environmental impact while providing healthy nutrition tend to be heavy in fruits, vegetables, roots, grains, and seafood.[10] In this section, we look at these questions from a water perspective.

3.1. Water Footprints of Various Foods

How does our choice of diet affect the amount of water needed to grow our food? Figure 18-2 shows the average water footprints of various foods and is worth discussing in some detail.

Figure 18-2a presents total (green plus blue) footprints, expressed in both liters per gram of food product and liters per food calorie (kcal).[11] In this figure, footprint differences between foods reflect primarily the inherent water intensity of that food. For example, almond trees transpire a great deal of water relative to the mass of almonds that they produce, resulting in a high water footprint (L/g), although almonds are calorie-rich, resulting in a somewhat lower water footprint when expressed in L/kcal.

Figure 18-2b presents blue footprints, expressed only in L/kcal for simplicity. The much lower vertical scale in this figure reflects the fact that green water makes up the majority of agricultural water consumption. In this figure, differences between foods are affected by the prevalence of irrigation for that particular crop. For example, tubers (potatoes, sweet potatoes, cassava) are rarely irrigated, so they have blue footprints that are close to zero, whereas most nut orchards around the world are dependent on irrigation, leading to a high blue footprint. Rice, which is usually irrigated, has a higher blue water footprint than other cereals but still less than many other products.

The water footprint of animal products, which consists mostly of the water needed to grow animal feed, is generally higher than that of nonanimal products, because most animals, especially cattle, are inefficient at converting their feed into

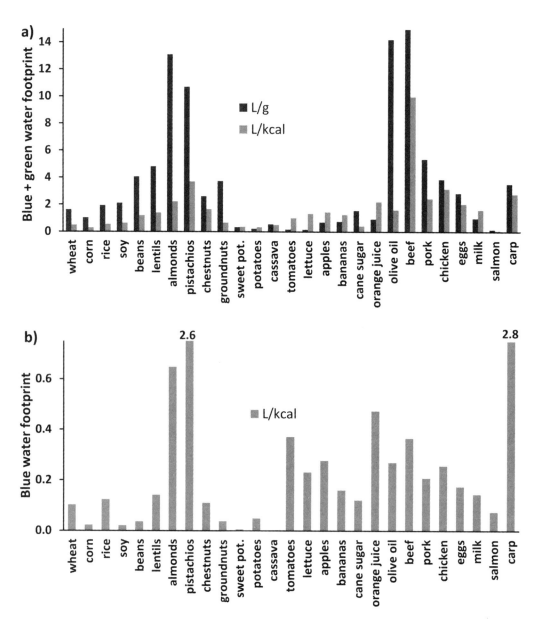

Figure 18-2. Average water footprints of various foods: (a) total; (b) blue. Note different vertical scales. Values for pistachios and carp in Figure 18-2b are off the top of the graph and are shown as numbers. Data are based on global average values from Mekonnen and Hoekstra (2011), except aquaculture, which is from Gephart et al. (2021).

biomass, an issue we will explore in more detail below. Modifying diets globally to reduce animal product consumption (excluding seafood) by 75 percent would reduce agricultural blue water consumption by 11 percent and green water consumption by 18 percent.[12]

Many tree nuts have a remarkably high blue water footprint and, from a water perspective at least, are not a good protein substitute for animal products. The recent "reference diet" published by the EAT-Lancet Commission on Food, Planet, Health—which aims to improve human health while respecting planetary boundaries—recommends increased consumption of groundnuts and tree nuts as protein substitutes for red meat[13] but has been aptly criticized for ignoring the water implications of this recommendation.[14]

Seafood is an interesting case. The water footprint of wild capture fisheries is close to zero, but there are other serious sustainability problems with those fisheries, most notably overfishing, bottom habitat destruction, and biodiversity loss. **Aquaculture**, a protein source that is rapidly increasing in importance, has a water footprint associated with production of fish feed and, in the case of species grown in freshwater ponds (such as the carp included in Figure 18-2), evaporation from those ponds, which can be very large.

Figure 18-2 reflects the water needed to *produce* a certain amount of food, but not all of that produced food is consumed. In fact, about 30–40 percent of food goes to waste (along with the water it took to grow it), whether at the farm (e.g., because it can't get to market), in the household (e.g., it spoils or is thrown away), or somewhere in between (e.g., it reaches its expiration date at the retailer).[15] One global estimate suggests that cutting food waste in half would save 12 percent of current agricultural green and blue water use.[16]

3.2. Animal Products: A Closer Look

We noted above that animal products, especially beef, have high water footprints. In order to fully understand this, we need a better understanding of animal feeding practices.

In traditional societies, meat animals were rarely fed grains that could be eaten by humans; under these conditions, animals grew slowly but were a useful mechanism for converting human-inedible resources—such as grasses, insects, and wastes—into nutritious, high-protein food. Industrial agriculture has turned the traditional model on its head, raising animals—particularly poultry and swine—in massive feeding operations where they are fed grain-based concentrates, often imported from across the country and around the world. This has dramatically accelerated the rate at which animals put on weight, shortened the time to slaughter, and driven a vast increase in production and decrease in cost. In the United States, there are approximately 145,000 animal feeding operations (AFOs),[17] defined as locations where animals are confined and fed purchased feed for at least forty-five days per year. Of these AFOs, over 21,000 qualify as **concentrated animal feeding operations (CAFOs)** based on their size or potential environmental impacts.[18]

Cattle and other ruminants (sheep and goats) are a bit harder than poultry and swine to manage in this way, since they need a certain amount of roughage (fodder with a high fiber content, as provided by forage crops and pasture) in their diet, and since they take longer to mature and are less fecund. In the United States, most dairy cattle are grown in AFOs and fed a mix of concentrates and roughage, while beef cattle are usually grazed for six to nine months before being finished on feedlots with concentrates and roughage.

The basic reason that animal products are so water intensive is that animals eat a lot of feed, and the laws of thermodynamics mean that energy and embedded water are lost in the conversion of feed into edible animal products. The efficiency of this

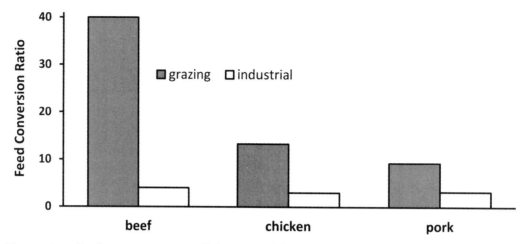

Figure 18-3. Feed conversion ratios (kilograms of feed consumed per kilogram of meat produced) for beef, chicken, and pork grown in industrial and grazing systems in North America. Data from Mekonnen and Hoekstra (2012).

conversion process is described by the *feed conversion ratio (FCR)*, defined as the dry mass of feed divided by the dry mass of animal product; higher FCRs mean less efficient feed conversion. FCRs can vary quite a bit,[19] but some average numbers for North America are shown in Figure 18-3. Beef has a higher FCR than poultry and pork, a factor that underlies its higher water (and carbon) footprint, since producing more feed requires more water.

Also notable in Figure 18-3, however, is the difference between animals that are grazed and those raised in industrial feedlots, with the latter being much more efficient. This difference is not surprising, since industrial meat production relies on concentrated high-energy feed, while grazing animals forage on low-quality grasses over large areas. The tradeoff, of course, is that producing the concentrates used in industrial meat production requires blue water and cropland.

So is grazed beef more sustainable than industrial beef? Livestock grazing is arguably an efficient way to use marginal lands where low rainfall and water scarcity limit other agricultural uses; roaming livestock capture the green water over vast areas of land and convert it into high-quality protein. Still, there are plenty of environmental and social problems associated with livestock grazing, including land degradation, soil erosion, privatization of the commons through subsidized grazing fees, and others. Allan Savory has argued that rotational grazing—in which large numbers of livestock are rotated over different parts of the landscape—is actually beneficial for the health of semiarid lands, although this approach has proven controversial.

From a water perspective, the benefits of cattle grazing are diminished when grazed cattle are fed supplemental concentrate feeds or are grazed on irrigated pasture. One analysis suggests that, in the United States, where supplemental feeds are common, grazed beef actually has a higher average water footprint—both green and blue—than industrial beef, while grazed beef from China has essentially no blue footprint (but a high green footprint).[20]

Because of differences in metabolism, FCRs for insects are generally lower than for mammals and birds, leading to growing interest in using insects—ranging from crickets to cockroaches to black soldier fly larvae—as protein sources. Insects may also be able to grow on various organic wastes that are now underused. This field is in its infancy, but there seems to be potential in growing insects both for direct human consumption and as a lower-footprint feed source for traditional food animals.

3.3. Sustainability of Blue Water Use

Up to this point, we have looked at the water footprints of different foods without getting at the question of the impact and sustainability of that water use. Understanding impact is challenging, because it requires us to understand where water is being used and how that use affects local ecosystems and **aquifers**. Here we look at two examples where researchers have been able to draw clear linkages between food consumption and local hydrologic impacts.

MEAT FROM THE DESERT

We noted above that production of meat—especially beef—is an inefficient use of blue water. Two recent studies have traced that blue water to specific watersheds.

The first study, by Richter et al. (2020), focused primarily on forage crops, most of which is used as roughage for beef cattle (with a small fraction going to dairy cattle and other ruminants). While only 18 percent of forage crop acreage in the United States as a whole is irrigated, this is not true in the arid western United States; in California, for example, 89 percent of forage crop acreage is irrigated.[21] Richter et al. show that forage crop irrigation accounts for 32 percent of blue water consumption in the water-scarce western United States, rising to 55 percent in the Colorado River basin. There is thus a direct link between industrial beef production and the ills affecting watersheds in the western United States, including flow depletion, fish declines, and conflicts over scarce water.

The second study, by Brauman et al. (2020), focused on the corn and soy used for feed and ethanol production. Overall, only 10–15 percent of corn and soy acreage in the United States is irrigated, but Brauman et al. show that reliance on irrigated feed is highly variable across ethanol and animal production facilities. For example, some ethanol producers use exclusively rainfed corn, while others use exclusively irrigated corn (and everything in between). However, there are some typical patterns for different products. Specifically, pork and ethanol are mostly produced in—and source their corn from—the Midwest, where only a small fraction of cropland is irrigated; a similar situation applies to chicken producers, who are concentrated in the humid Southeastern states. Beef production, on the other hand, is concentrated mostly in the semiarid Great Plains and sources its corn from nearby irrigated lands, where it is in direct competition with other blue water uses.

Putting these two studies together, we can conclude that most beef eaten in the United States (and exported to other countries from the United States) has serious negative impacts on watersheds in the Great Plains, the Southwest, and California. Although these impacts vary somewhat between various meat-packing plants and beef-consuming regions, the best way to get water-sustainable beef is to buy locally in wetter parts of the country from companies that are committed to avoiding grain, pasturing during the growing season, and using local forages during the winter.

Nuts from the Epicenter of Groundwater Depletion

We noted above that almonds have a very high global average blue water footprint. Understanding the local impact of this footprint is simplified by the fact that 80 percent of the world's almonds are grown in California, mostly in the San Joaquin Valley (SJV). Thus, when you eat an almond, it's a good bet that the 6 liters (!) of blue water that it took to grow that single almond came from a place where competition for water is fierce, where surface water comes at great cost to ecosystems, and where groundwater depletion has been a problem for decades. In addition, almond orchards—unlike alfalfa fields—represent a ***hardening of water demand***, in the sense that they can't be fallowed in time of drought; they need water every year in order to survive.

How did we get to the point where we are using water so unsustainably to produce massive quantities of almonds? The answers are complex, but the heart of the story is about money and power.

- **Money**: For those able to afford the up-front costs of planting an almond orchard, there are few agricultural investments that provide higher net revenue per acre, especially since labor costs are much lower than for vegetables.
- **Power**: The Wonderful Company, owned by Stewart and Lynda Resnick, controls the bulk of almond (and pistachio) production and processing in California, and it has been able to use its market and political power to control land and water rights in California and to drive up demand globally through highly successful marketing campaigns.

Are the societal financial benefits of almonds large enough to justify their continued production in the SJV? While almonds do bring in more revenue than any other California crop, their blue water use is disproportionate to their revenue; the water footprint of almonds, in cubic meters per dollar, is more than twice that of grapes and higher than that of all other products except cotton, rice, and tomatoes. The only reason that almonds are so lucrative for the Resnicks is that they are not paying the full cost of water. Indeed, estimates of operating costs for almond orchards in California suggest that farmers expect to pay only around $100 per acre-foot in pumping costs and that total operating costs for irrigation (including labor) amount to $468 per acre, or 17 percent of total operating costs.[22]

Around the world, irrigation water is often available at prices that are very low compared to the true value of the water and compared to the prices being paid by industrial and domestic users. A more reasonable pricing structure would encourage more efficient use of this vital resource. It would also allow us to better distinguish between irrigation projects that make sense and those that don't, and it would drive agriculture out of locations where higher-valued users need the water. Ultimately, charging farmers for the true value of the water they use would mean higher food prices, especially for meat. This would be a positive development for the long-term sustainability of our food system, as it would start to internalize the environmental and social costs that we now pay in other ways. At the same time, higher food prices must be met with stronger safety-net and fair-wage programs to reduce food insecurity.

4. Where We Grow It

A key factor in thinking about the geography of agricultural water use is the rapidly growing international trade of agricultural products—and of the ***virtual water*** imbedded in those products. Because of this trade, food is one of the areas where water clearly operates at a global scale; food that I eat in Connecticut can affect water scarcity in India.

From a water perspective, trade in food can potentially accomplish two distinct goals:

- Reduce local water scarcity by allowing water-scarce regions to import virtual water;
- Reduce the global amount of water used in agriculture by allowing food to be grown where its water footprint is lowest.

The amount of water that could be saved globally by virtual water trade is potentially quite large.[23] Nonetheless, there are some potentially serious drawbacks to the idea of using global food flows to solve water scarcity problems. Global agricultural trade undermines local food security, drives loss of local crop diversity and food culture, empowers large agribusiness at the expense of small farmers, perpetuates colonial relationships, and leaves the poorest around the world vulnerable to price shocks resulting from distant conflicts, such as the 2022 war in Ukraine.

Does virtual water trade, as currently practiced, accomplish both: saving water globally and alleviating water scarcity locally? Several groups have attempted to calculate virtual water flows between countries by combining global trade data with data on water footprints of agricultural products in both importing and exporting countries. These results are sometimes quite disparate,[24] but there is a general consensus that agricultural trade does save water globally. One typical study estimates global blue water savings of 119 km^3/yr (about 6 percent of total agricultural blue water use) and green water savings of 105 km^3/yr (1 percent) for the year 2008.[25] In other words, if each country grew all the food it consumed (i.e., no trade in food), the world would use about 224 km^3/yr more water to grow the same amount of food.

When it comes to alleviating local scarcity, the picture is more problematic. For example, the top three net exporters of virtual blue water identified in one study—Pakistan, India, and Egypt—are all countries with severe water scarcity.[26] Another research group found that about one third of global blue virtual water trade came from groundwater depletion, with nonrenewable groundwater, especially from Pakistan, India, and the United States, traveling around the world in the form of food.[27]

There is also a strong equity component to these questions of virtual water trade. Looking at per capita imports and exports of virtual water by country (Figure 18-4), it is clear that richer countries (left side of figure) tend to be net importers of unsustainable virtual water, while poor countries tend to be exporters; Turkmenistan, which exports large quantities of cotton at the expense of the Aral Sea, is a notable example. There are exceptions—the agricultural behemoths of the United States, Australia, and Spain are net exporters—but it seems clear that many ***low- and middle-income countries (LMICs)*** are overusing their water supplies in order to grow food for richer countries. Thus, virtual water trade—while saving water globally—also shifts water

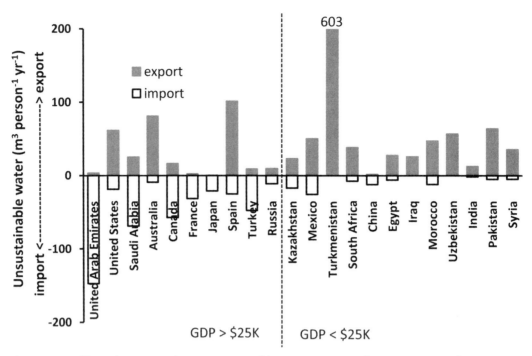

Figure 18-4. Virtual water trade in unsustainable water, expressed on a per capita basis, using 2015 data from Rosa et al. (2019), for all countries with at least 10 m³/person/yr of unsustainable import or export. Country order from left to right is from highest to lowest per-capita gross domestic product (2015). The value for Turkmenistan is off the top of the vertical scale and is shown as a number.

from poor, water-stressed countries to rich ones (in accordance with the truism "water flows uphill to money") and thus embodies some of the most fundamental inequities of global trade.

One form of virtual water flow that is particularly problematic is associated with the acquisition of agricultural land in LMICs by large multinational agricultural corporations, resulting in displacement of small farmers who had been working the land without title or legal protections. This phenomenon of ***land grabbing***—which typically involves a shift from subsistence to export-oriented agriculture—has been increasing in recent years, driven in part by higher food prices, growing demand for biofuels, and the limited availability of new agricultural land in high-income countries. A 2013 study estimated the amount of water used on grabbed land (i.e., the amount of grabbed virtual water exported from LMICs) as 308 km³/yr of green water and somewhere between 11 and 146 km³/yr of blue water,[28] although these numbers are no doubt increasing rapidly.[29]

5. How We Use Water

Having discussed what we eat and where it is grown, we now turn to the third factor determining agricultural water use: how we use water in growing food. In this section, we discuss how water is used in both irrigated and rainfed croplands and how we might grow more food while using less water.

A key concept in understanding the role of water in food production is the *yield gap*: the difference between the maximum yield attainable for a given crop in a given climate region and the actual average yield for that crop in that region. The two main factors contributing to the yield gap tend to be soil quality and water availability; the yield gap due to suboptimal water availability is called the ***water gap***. In theory, closing yield gaps could dramatically increase food production on existing croplands,[30] but research suggests that that approach would require blue water use to increase by 150 percent[31]; additional water at that scale is simply not available. So how do we best use the water we have? The answers are different for irrigated and rainfed agriculture.

5.1. Irrigation

Irrigated agriculture generally has high yields and low water gaps; on average, yields on irrigated lands fall only 6 percent below the yields that could be achieved with optimal water supply.[32] While it makes sense to close the remaining water gaps (e.g., by ensuring full water delivery to all parts of an irrigation network), the main water-management goal on irrigated land is not to increase yields but to increase water efficiency. (In certain settings, we may even be willing to sacrifice a little on yield in order to optimize water efficiency, an approach known as ***deficit irrigation***.)

We can define ***irrigation efficiency*** in two different ways (Equations 18-3 and 18-4). In both cases, the numerator is crop transpiration (the place where we want our irrigation water to end up). Eff_1 uses total withdrawal as the denominator, implying that any withdrawn water that is not transpired by crops—including return flow—is wasted, while Eff_2 uses consumptive use as the denominator and thus puts the focus on reducing nonbeneficial consumptive uses such as evaporation from irrigation ditches.

Equation 18-3. $$Eff_1 = \frac{\text{Crop transpiration}}{\text{Withdrawal}}$$

Equation 18-4. $$Eff_2 = \frac{\text{Crop transpiration}}{\text{Consumption}}$$

A focus on maximizing Eff_1 can have unintended consequences. In particular, if a farmer reduces their return flow and uses the "saved" water to increase crop production, the increased farm-level efficiency (Eff_1) may come at the expense of downstream users who had been making use of that return flow. Thus, increased "efficiency" can lead to increased depletion at the basin scale—unless appropriate governance structures are in place that recognize the value of return flows.

With definitional issues behind us, we can examine the efficiency of the three broad types of methods for applying irrigation water:

- ***Surface irrigation*** (also called gravity or flood irrigation) involves flooding entire fields or furrows with water, usually from an irrigation ditch carrying surface water; evaporation rates are typically high and efficiency is low by both definitions.
- **Sprinkler irrigation** uses sprinklers of varying levels of sophistication to apply water from the air to soil and plant surfaces. Evaporation and interception losses can be moderate, though not as high as in surface irrigation. Sprinkler irrigation often takes the form of *center pivot* systems, in which sprinklers move slowly around a

pivot to irrigate a circular field. For both sprinkler and surface irrigation, precision leveling of fields can help increase efficiency by ensuring that water reaches all plants equally.

- ***Drip irrigation*** uses surface or buried water lines with emitters that allow water to drip out in certain locations, providing water directly to the soil under individual plants. Since only the soil right around the plants is being wetted, evaporation losses are much lower and plant yields are often higher. Drip irrigation is most suitable for crops where plants are spaced far apart, such as vegetables, fruits, and pulses, but is being adapted for use with cereal crops as well. For trees that need more water than an emitter can provide, microsprinklers can be used to direct water efficiently to tree roots; drip irrigation and microsprinklers are both considered microirrigation techniques.

Drip and sprinkler irrigation—collectively called pressurized irrigation systems—are often combined with soil moisture sensors that can fine-tune water delivery schedules.

While efficiency generally increases from surface to sprinkler to drip, the exact values can vary quite a bit depending on the crop, the specific irrigation system used, and the definition of efficiency. Average values calculated from a global model (Table 18-4) show that both Eff_1 and Eff_2 are highest for drip and lowest for surface irrigation—and that the current global mix of irrigation methods is highly inefficient. The same global model suggests that an ambitious program of improved irrigation infrastructure (specifically, 100 percent drip on crops where it is appropriate, 100 percent sprinkler on all other crops except rice, which would still be surface irrigated), together with better soil moisture conservation (e.g., mulching), could decrease non-beneficial consumptive water losses by 48 percent. Using the saved water to expand irrigation could increase global food production by 26 percent.[33]

Water can also be saved by reducing ***conveyance losses***, especially where water travels hundreds of kilometers from source to field. Covering irrigation canals to prevent evaporation is prohibitively expensive, but at least two projects in India have installed solar panels atop canals, which has several benefits in addition to reduced evaporation: reductions in aquatic weed growth, improved solar efficiency (because over-water panels are cooler than over-land panels), and sparing of land habitats that would otherwise be used for solar panels. Over-canal solar is more expensive to build than regular solar, but an analysis for California showed that the co-benefits are large enough to make the projects economically sound,[34] and the Turlock Irrigation District is moving forward with a pilot project. Covering all 6,350 km of California's conveyance network with solar panels would save about 240 MCM/yr in evaporation.

The other way to reduce conveyance losses is to reduce seepage by lining canals with an impermeable layer of concrete or clay. However, since seepage is not a consumptive use, the effectiveness of this measure depends on which definition of efficiency you are using and, in particular, on whether the seepage water is being used by other farmers. A prime example of the unforeseen consequences of canal lining is the All-American Canal, which carries Colorado River water to the Imperial Irrigation District in California. The $300-million effort to line 23 miles of the canal was completed in 2010 and is "saving" over 80 MCM of water annually, most of which is

Table 18-4. Global average irrigation efficiency values for surface, sprinkler, and drip irrigation methods and for the current mix of irrigation methods. Data source: Jägermeyr et al. (2015).

	Surface	Sprinkler	Drip	Current Mix
Eff_1	24%	43%	69%	26%
Eff_2	49%	65%	79%	52%

designated for use by San Diego. However, a significant portion of this seepage had been recharging groundwater in the Mexicali Valley in Mexico, where farmers who had been using it for their own irrigation are the losers from the improved efficiency of the canal. The canal lining project has led to cross-border tensions for years and points to the importance of understanding the true fate of water that is considered "wasted."

Before we leave irrigated agriculture, we should say a word about the trend toward controlled-environment agriculture: a set of irrigated agricultural systems that includes greenhouses but increasingly also indoor farms, where plants are grown using artificial lighting in temperature- and humidity-controlled conditions (often hydroponically). The advantages of these systems include higher yields, lower water footprints, efficient use of space (including vertical space), and the potential for food production in cities for local consumption (including in winter). However, the energy and greenhouse gas emissions needed for temperature control and lighting are very high; substituting artificial lighting for sunlight is not a good strategy for sustainable agriculture.

5.2. Rainfed Agriculture

Unlike irrigated agriculture, rainfed agriculture has a large water gap; better water management could significantly improve crop yields, which would reduce the pressure on irrigated areas. Water management in rainfed agriculture needs to focus on improving soil health, particularly increasing soil **infiltration capacity** and plant-available water (the water that is held tightly enough that it is retained after a rainstorm but not so tightly that plants can't access it). Key measures to achieve these goals include reducing soil erosion through cover crops, terracing, and minimal tillage; increasing soil organic matter content through cover crops, mulching, and compost addition; and decreasing soil compaction through careful use of heavy equipment, especially in wet conditions.

In addition to soil health, farmers need to think about larger patterns of water flow on the landscape. There are significant opportunities to capture more runoff in soils, in part by emulating the **runoff farming** techniques discussed in Chapter 4. Small-scale **water harvesting** can also provide supplemental irrigation to help crops get through increasingly frequent dry periods. More widespread use of **dry farming** techniques, such as drought-tolerant crops, deeper planting, wider spacing, and dust mulching, could increase productivity in arid regions. One study suggests that implementing all these measures could shrink the water gap in rainfed croplands from 29 percent to 11 percent, resulting in an increase in global food production of 15 percent.[35]

In addition, recent research suggests that there is a subset of rainfed areas where supplemental irrigation would significantly increase yields and renewable blue water is available after accounting for *environmental flow requirements* and human water uses. These areas, estimated to be 15–25 percent of total cropland area,[36] are good candidates for irrigation expansion, which would significantly increase yields in those areas and help compensate for loss of irrigated land elsewhere. Many of these areas are in LMICs, where irrigation infrastructure may be unaffordable to local farmers; helping these farmers install efficient irrigation would improve local food security and help solve global agricultural water scarcity.

6. Pollution

This section addresses three of the main pollution problems caused by agriculture: salts, fertilizers, and manure.

6.1. Salinization

One of the biggest problems facing modern agriculture is the same problem that faced Mesopotamian civilizations (Chapter 4): soil **salinization**, sometimes referred to as secondary salinization to distinguish it from natural salinization processes.

Soil salinity is a more complex concept than water salinity, since salts can be present in soils in several different forms: dissolved in porewater, sorbed to mineral surfaces, incorporated into soil minerals, and precipitated onto soil particles. Soil salinity can be highly variable over small spatial scales and with depth in the soil profile, so assessing and managing salinization is quite data intensive.

The amount of salt in a given soil is determined by the balance between sources and sinks. Salt sources in soils consist mostly of mineral weathering, dissolution of fossil salt deposits, and sea spray; the primary salt sink is the flow of salt-containing water out of the soil to groundwater or surface water. In wet regions with lots of water moving through the soils, salts typically don't build up, as salt production in soils is balanced by export in rivers. However, in arid and semiarid regions, especially ones with flat topography, salts are less likely to be carried out by water, and saline soils are common.

While saline soils are a natural feature of some regions, salinization can also be induced by anthropogenic changes to either sources or sinks. Human salt sources to agricultural soils include applications of lime, gypsum, and fertilizer. In addition, irrigation water—even when it is quite low in salinity—brings new salts into the soil, a problem that is exacerbated when using reclaimed water or other higher-salinity sources.

Agriculture can also affect salt sinks, most commonly by changing the ways that water moves through soils. Conversion of natural ecosystems to agriculture uses (both rainfed and irrigated) can lead to higher water tables, which can make it harder for salts to be flushed out of soils; waterlogging is also bad for crops independently of salt levels. Waterlogging and salinization often lead farmers to install artificial drainage to carry saline water away from the root zone. This allows farmers to apply excess water beyond the crop water requirement to leach salts out of surface soils into deeper layers and the drainage system. However, drainage systems can cause problems in downstream ecosystems. The poster child for the problems with artificial drainage is the Kesterson National Wildlife Refuge in California, where birds suffered severe deformities due to the toxic effects of selenium. Investigations showed that the main source of selenium to Kesterson was subsurface tile drainage installed in farms in the SJV; the combined effects of irrigation and drainage had mobilized the selenium that was naturally occurring in the soils.

Besides installing drainage, another option for managing soil salinity or waterlogging problems is to plant salt-tolerant, deep-rooted cover crops that can lower the water table through transpiration. At the same time, reducing surface evaporation (e.g., through mulching) can reduce the upward transport of salts to the soil surface. Shifting to more salt-tolerant crops and higher-quality irrigation water can help, as can selective irrigation to create patches of low-salinity soils for seedling establishment.

In addition to causing soil salinization, agriculture can also increase the salinity of groundwater and surface water. In California's SJV, groundwater salinity has increased significantly over the last 100 years, with the largest increases seen in the areas with the most agriculture.[37] The areally weighted median total dissolved solids in the southern SJV has more than doubled, from 230 mg/L in 1910 to 471 mg/L in 2010, which is higher than California's water quality goal for irrigation water (450 mg/L). Geochemical modeling suggests that fertilizer (NO_3^-) and gypsum (SO_4^{-2}) are contributing to the rising salinity but that the largest factor is increased weathering of silicate minerals driven by higher groundwater recharge rates from irrigation—another example of the way that hydrology can influence water quality.

The salinity problem in the SJV is worse at shallower depths, which means that private domestic wells are more likely to be affected by salt and nitrate contamination than larger, deeper irrigation wells. A program called Central Valley Salinity Alternatives for Long-Term Sustainability is trying to address the issue through mapping, provision of alternative water supplies to homes, and improved irrigation management.[38]

6.2. Nutrient Pollution from Croplands

(For background on natural and anthropogenic N and P cycling, see the companion website.)

Humanity has vastly increased the global supply of reactive N and available P, mostly by producing agricultural fertilizers. The global increase in fertilizer application rates (Figure 18-5) has been critical in driving increased global food production. Indeed, without **Haber–Bosch** N, our existing agricultural system could feed only about half of the world's current population.[39]

However, fertilizers also have unintended environmental consequences. When we apply a fertilizer to a field, some of it is taken up by crops, but much of it escapes to the surrounding water and air. Even the nutrients removed in crops can do damage at points further down the food supply chain, whether as water or air emissions from animal operations or in the form of human waste. (Remember that typical secondary treatment doesn't effectively remove nutrients, so wastewater can be a significant source of nutrients to the environment—nutrients that were ultimately derived from fertilizer application.)

In the case of P, the main environmental impact of fertilizers is **eutrophication** of P-limited freshwater systems due to increased P runoff, although there are also

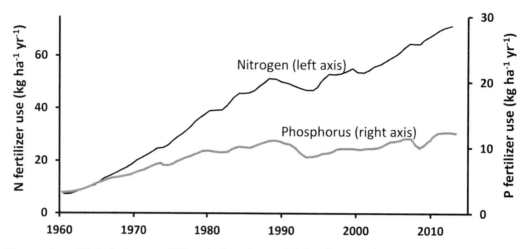

Figure 18-5. Global nitrogen (N) and phosphorus (P) fertilizer application rates (per unit of cropland). Data from Lu and Tian (2017). Note that N fertilization application rates have increased more than P, in part because N is more mobile, so it must be reapplied in large amounts every year to compensate for the N that was lost to runoff or drainage.

environmental impacts from mining of P-containing rocks, as well as concerns over the limited availability of these rocks. In the case of N, we need to worry about eutrophication of N-limited coastal and marine systems, along with a whole cascade of other N pollution issues, including nitrate contamination of drinking water.

Ideally, we would use fertilizers to improve crop growth without excessive loss to the environment. For a given area of cropland, we can define the ***nitrogen use efficiency (NUE)*** as the fraction of added N that ends up in harvested crops (Equation 18-5).

Equation 18-5. $$NUE = \frac{N \text{ harvested in crops}}{N \text{ applied to field}}$$

What management practices can be used to increase NUE? First, improvements in agronomic practices can produce higher yields and allow crops to make better use of N fertilizer. These improvements include ensuring adequate availability of water and other (non-N) nutrients, but the most important measure is probably improving soil health (e.g., organic matter content, water-holding capacity, structure) through measures such as reduced tillage, cover cropping, and more varied crop rotations.

Second, NUE can be improved by fine-tuning the amount, timing, location, and form of N fertilizer application.

- **Amount**: Given that the impacts of N fertilizer use are externalized (i.e., they aren't reflected in the price of fertilizer), many farmers overapply N as a sort of insurance policy. Governments need to use regulatory, pricing, and outreach tools to change this behavior.
- **Timing**: N should be provided when crops can best take it up and when it is least likely to be lost from the field. This varies depending on the type of crop and weather conditions, but it can often include adaptive sidedressing (applying

fertilizer to plants partway through the growing season) based on measurements of N availability and plant need.
- **Location**: Taking land out of production is the simplest way to reduce fertilizer losses. Certain areas, such as ***riparian*** strips and steep slopes, are particularly susceptible to N loss and should not be used as croplands. (Intact riparian buffers can also take up N on its way to the stream.) In addition, N applications for each part of a field should be appropriate to that location's soil N supply and plant potential; ***precision agriculture*** can help provide high-resolution spatial data to better manage small-scale variability in optimal N application rates.
- **Form**: Not all N inputs are the same. Of the three primary N sources to cropland—inorganic fertilizer, ***biological nitrogen fixation***, and manure—the first is most mobile and most likely to be lost to water through drainage or run-off.[40] Biological fixation by leguminous crops, such as soy, alfalfa, and clover, produces less mobile organic N forms; when used as a cover crop or in crop rotations, legumes can increase soil N and reduce fertilizer demand for other crops. Properly applied, manure can be a good way to close the loop and return nutrients to the soil. However, increasing agricultural specialization has created a spatial disconnect between manure sources (animal operations) and fields that need nutrients (croplands), a topic that we will address in the next section.

Managing for P is similar to managing for N, and phosphorus use efficiency (PUE) is defined analogously to NUE. One important difference between the two nutrients is that P is generally less soluble and more particle-bound than N, so measures to reduce erosion are particularly important for reducing P losses. Also, since P is more likely to persist in soils, P fertilizer application rates must be closely tied to measurements of existing available P in soils.

The Clean Water Act, with all its success in reducing point source discharges from wastewater treatment plants and industrial sources, does not have adequate tools to address agricultural nutrient runoff. Two specific provisions in the Clean Water Act protect agriculture from responsibility for overfertilization. First, the return flow exemption releases agricultural return flow from the requirement to obtain a ***National Pollutant Discharge Elimination System (NPDES)*** permit, even when that flow is conveyed in discrete channels such as ditches or tile drain outfalls. Second, the agricultural stormwater exemption provides similar relief to stormwater runoff from agricultural fields. Voluntary incentive and assistance programs have tried to pick up the slack, including the Environmental Quality Incentive Program,[41] the Conservation Stewardship Program,[42] the Conservation Reserve Program,[43] and the Conservation Technical Assistance Program,[44] all run by the US Department of Agriculture under various Farm Bill provisions.

6.3. Pollution from Animal Feeding Operations

Industrial agriculture's shift toward large confined feeding operations has created a new kind of pollution problem. Unlike traditional grazing or mixed crop–animal husbandry—where animal manure was dispersed and largely beneficial to soils—industrial animal feeding creates vast quantities of manure in concentrated areas, turning a resource into a waste that is increasingly difficult to dispose of. Farm animals in the United States produce over a billion tons of manure per year, more than an order of magnitude higher than the excreta produced by humans.

Unlike human waste, most manure from CAFOs is not effectively treated before release to the environment. Poultry CAFOs usually use dry litter systems, in which manure (often mixed with bedding) is stored in piles outdoors before being spread on fields. For swine and cattle, manure is typically flushed from barns into large open-air manure lagoons, where it is stored before being sprayed onto fields. Lagoons provide some treatment in the form of settling and anaerobic decomposition, but they sometimes leach contaminants into groundwater and can breach or overflow during large rain events, especially since manure is produced year-round but should be applied to fields only during the growing season. Nutrients and pathogens from manure can pollute surrounding areas through multiple routes: volatilization, leaching, and runoff from barns, lagoons, litter piles, and spray fields. Manure from CAFOs can cause significant harm to nearby ecosystems and communities, which are often disproportionately populated by people of color.

The Clean Water Act explicitly denotes CAFOs as point sources that require NPDES permits before discharging to the environment, providing the Environmental Protection Agency with a potentially powerful regulatory tool. However, several factors make CAFOs harder to regulate than human sewage. First, the power of the agricultural lobby and the invisibility of CAFO operations to the public make this a problem that is easy to ignore. Second, discharges from CAFOs tend to be weather and season dependent and thus episodic. Third, each animal operation is different in terms of the amount and type of manure generated and the likelihood of harm. Fourth, actually treating manure wastewater (e.g., with something like secondary treatment) would be prohibitively expensive in our current agricultural system. Finally, the application of manure to fields converts a point source into a nonpoint source that is not subject to NPDES, thanks to the agricultural stormwater and return flow exemptions. (This is the reverse of urban stormwater, where flows that are nonpoint in their origin ultimately become regulated point sources in the form of ***municipal separate storm sewer systems***.) These complications have resulted in an evolving set of regulations and court cases over the last few decades, as regulators and advocates on all sides have struggled to work out a fair and effective set of rules.

The link between manure and croplands is clearly key to sustainable management of both animals and crops. The safe application of manure to fields allows the return of nutrients and organic matter to the soils and reduces the need for fertilizer inputs. But the overapplication of manure—as a form of waste disposal rather than fertilization—can pollute soil and water with nutrients and pathogens.

One useful metric for thinking about safe manure reuse is the ***manureshed***.[45] The manureshed of a given CAFO (or other source of manure) is the geographic area surrounding the CAFO containing enough cropland to use the nutrients supplied by that CAFO's manure. Where manuresheds are small, manure recycling is efficient. Where manuresheds are large, it becomes too expensive to transport liquid manure to croplands, and CAFOs will seek other solutions, which may include problematic ones (overapplication to nutrient-saturated fields, illegal discharges) as well as appropriate ones (biogas production, separation of manure liquids and solids).

For regions with high densities of CAFOs, manuresheds may be quite large. For example, in 2012, hog and poultry CAFOs in thirty-six counties in North and South Carolina produced 24.3 million tons of manure containing 53,900 tons of P. Croplands in those thirty-six counties can safely absorb 14,300 tons of P, leaving an excess of 39,600 tons, which would need to be transported more than 300 km to find appropriate croplands.[46]

Thus, as we mentioned in the previous section, one underlying driver of both manure and fertilizer problems is the separation of crop and animal production to different regions. The result is illustrated in schematic form in Figure 18-6, which illustrates the upstream and downstream nutrient flows of a pork chop eaten in Connecticut. This hypothetical (but realistic) pork chop came from a North Carolina CAFO, which imported feed from an Iowa corn farm, which imported P fertilizer from a mine in Florida and N fertilizer from a Haber–Bosch facility in Louisiana. Our pork chop drives environmental and social damage at each step along the way, from the petrochemical industry in Louisiana to nutrient pollution in Iowa to CAFO health impacts in North Carolina to the discharge of sewage N to Connecticut's Long Island Sound. Solving these problems requires the reintegration of crop and livestock agriculture, whether through rangeland grazing (where animals fertilize the soil as they are grazing), agropastoralism (growing both crops and animals on the same land), or more closely matching manure production to the area of nearby cropland.

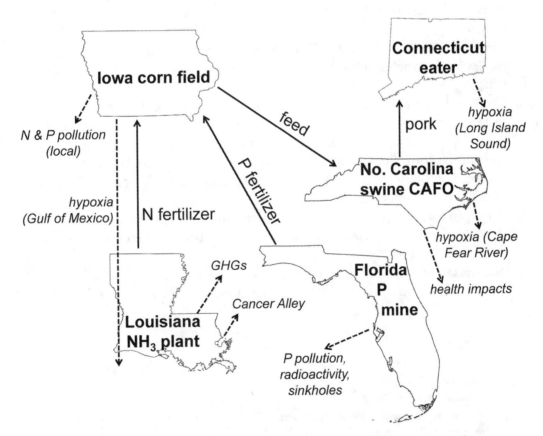

Figure 18-6. Schematic of possible agricultural flows (solid arrows) and associated environmental problems (dashed arrows) tied to pork consumption in Connecticut. CAFO, concentrated animal feeding operation; GHGs, greenhouse gases.

Chapter Highlights

1. Modern industrial agriculture has vastly increased our food supply but at high social and environmental costs. We need agriculture to undergo **sustainable intensification** to reduce its impacts on land, water, air, and biodiversity while producing enough food for a growing population.
2. The majority of water use globally goes to growing food, meaning that agriculture is at least partly responsible for groundwater depletion and impacts on aquatic ecosystems, but these linkages are often obscured by the complexity of our food systems.
3. Over the next few decades, finding enough water—and doing so sustainably and justly—will be a central challenge for global agriculture, especially given the pressure to reallocate agricultural water to meet environmental flow requirements and satisfy growing urban demand.
4. The water it takes to grow our food is determined by three factors: what we eat, where we grow it, and how we use water to grow it.
5. Our choices of diet affect our health and the health of the planet, including water scarcity and its impacts. We can make some strong generalizations about what foods have high water impact (e.g., nuts, beef), although these generalizations are tempered by local variations in footprints and impacts.
6. Trade in virtual water can save water globally and locally, but in practice it often exacerbates scarcity in LMICs by shifting water from local needs to export agriculture.
7. Large volumes of water can be saved by increasing irrigation efficiency, although we have to be careful in how we define efficiency and how we regulate what happens to the "saved" water.
8. Croplands and **concentrated animal feeding operations (CAFOs)** can pollute surface and groundwater with salts, pathogens, pesticides, and fertilizers. The Clean Water Act has weak mechanisms for regulating agricultural pollution.
9. Soil **salinization** threatens the productivity of irrigated croplands and must be managed through careful attention to soil, water, and vegetation. Large-scale drainage can transfer the problem downstream.
10. Nitrogen and phosphorus are essential plant nutrients but also significant pollutants. **Nitrogen use efficiency** can be improved by better agronomic practices and careful fertilizer use.
11. Reintegrating crops and livestock within **manuresheds** can reduce the impacts of both CAFOs and croplands by better matching nutrient sources and sinks.

Notes

1. Smil (2021).
2. Ray et al. (2019), Ortiz-Bobea et al. (2021).
3. https://www.fao.org/state-of-food-security-nutrition.
4. Poore and Nemecek (2018).
5. https://www.fao.org/3/x0490e/x0490e0b.htm.
6. Hrozencik and Aillery (2021).
7. Agricultural production of cotton and other fibers (hemp, jute) for textiles and other uses is implicitly included in Equation 18-2, as is production of other noncaloric items such as tea, coffee, tobacco, and flowers; we use the term *food* for simplicity.

8. Pollan (2006).

9. Poore and Nemecek (2018).

10. Gephart et al. (2016).

11. A food calorie (sometimes written *Calorie*) is equal to 1,000 scientific (or "small") calories, or 1 kilocalorie (kcal), where a scientific calorie is the energy needed to raise the temperature of 1 g of water by 1°C, or 4.184 joules.

12. Jalava et al. (2016); see also Davis et al. (2017a), which suggests that optimizing crop choices in the United States—growing more potatoes and groundnuts and less wheat and corn—could decrease agricultural water consumption by 5 percent while increasing calorie production by 46 percent, protein production by 34 percent, and agricultural value by more than 200 percent.

13. Willett et al. (2019).

14. Vanham et al. (2020).

15. http://www.fao.org/in-action/seeking-end-to-loss-and-waste-of-food-along-production-chain/en/. According to Food and Agriculture Organization definitions, "food waste" occurs at the level of retailers, food service providers, and consumers, and "food loss" occurs upstream in the supply chain; I use "food waste" to refer to both.

16. Jalava et al. (2016).

17. https://www.nrcs.usda.gov/wps/portal/nrcs/main/national/plantsanimals/livestock/afo/.

18. https://www.epa.gov/npdes/npdes-cafo-regulations-implementation-status-reports.

19. There are also definitional issues with FCRs, especially in terms of accounting for different life stages and interactions between meat, milk, and egg production systems.

20. Gerbens-Leenes et al. (2013).

21. USDA 2017 Census of Agriculture; https://www.nass.usda.gov/Publications/AgCensus/2017/.

22. Duncan et al. (2019).

23. Chouchane et al. (2020).

24. For example, Konar et al. (2012) estimated total blue + green virtual water trade at 672 km^3/yr (for the year 2008), whereas Hoekstra and Mekonnen (2012) estimated it at 2,038 km^3/yr (for the period 1996–2005).

25. Konar et al. (2012).

26. Chen et al. (2018).

27. The 1/3 fraction comes from combining Konar et al.'s (2012) estimate of ~78 km^3/yr of blue virtual water trade in 2008 with Dalin et al.'s (2012) estimate of 26 km^3/yr of groundwater depletion in virtual water trade in 2010.

28. Rulli et al. (2013).

29. A follow-up study (Dell'Angelo et al. 2018) defined water grabbing more narrowly to include only situations where the country whose land is being grabbed is suffering from both water scarcity and food insecurity. By this definition, approximately 4.4 km^3/yr of blue water is classified as having a moderate to high likelihood of being grabbed, with another 100 km^3/yr in deals that are under contract but not yet under production.

30. Mueller et al. (2012).

31. Davis et al. (2017b).

32. Jägermeyr et al. (2016).

33. Ibid.

34. McKuin et al. (2021).

35. Jägermeyr et al. (2016).

36. Rosa et al. (2020).

37. Hansen et al. (2018).

38. https://www.cvsalinity.org/.

39. Erisman et al. (2008).

40. Lassaletta et al. (2014) found that countries where more of the N inputs come from inorganic fertilizer tend to have a lower NUE, although the relationship was relatively weak.

41. https://www.nrcs.usda.gov/wps/portal/nrcs/main/national/programs/financial/eqip/.

42. https://www.nrcs.usda.gov/wps/portal/nrcs/main/national/programs/financial/csp/.

43. https://www.fsa.usda.gov/programs-and-services/conservation-programs/conservation-reserve-program/.

44. https://www.nrcs.usda.gov/wps/portal/nrcs/main/national/programs/technical/.

45. Spiegal et al. (2020).

46. Ibid.

Abbreviations

AI	aridity index
ALUS	aquatic life use support
ASR	aquifer storage and recovery
BCA	benefit–cost analysis
BCR	benefit/cost ratio
BCU	basin–country unit
BOD	biochemical oxygen demand
BuRec	Bureau of Reclamation
cfs	cubic feet per second
CII	commercial, industrial, and institutional
CLTS	community-led total sanitation
COD	chemical oxygen demand
CPR	common-pool resource
CSO	combined sewer overflow
CSP	concentrating solar power
DALY	disability-adjusted life year
DBP	disinfection byproduct
DO	dissolved oxygen
DPR	direct potable reuse
EFR	environmental flow requirement
EIS	Environmental Impact Statement
EJ	environmental justice
ENSO	El Niño–Southern Oscillation
EPA	Environmental Protection Agency
ESA	Endangered Species Act
ET	evapotranspiration
FAO	Food and Agriculture Organization
FDC	flow duration curve
FEMA	Federal Emergency Management Agency
FERC	Federal Energy Regulatory Commission
GHG	greenhouse gas

GRACE	Gravity Recovery and Climate Experiment
ICOLD	International Commission on Large Dams
IPCC	Intergovernmental Panel on Climate Change
IPR	indirect potable reuse
ITCZ	Intertropical Convergence Zone
IWRM	integrated water resource management
JMP	Joint Monitoring Programme
LDCs	least developed countries
LMICs	low- and middle-income countries
Lpcd	liters per capita per day
LSL	lead service line
MAF	millions of acre-feet
MAR	managed aquifer recharge
MCL	maximum contaminant level
MCLG	maximum contaminant level goal
MDGs	Millennium Development Goals
MNB	marginal net benefit
MS4	Municipal Separate Storm Sewer System
NEPA	National Environmental Policy Act
NFIP	National Flood Insurance Program
NID	National Inventory of Dams
NPDES	National Pollutant Discharge Elimination System
NPS	nonpoint source
NPV	net present value
NRW	nonrevenue water
OD	open defecation
PCB	polychlorinated biphenyl
PDF	probability density function
PDSI	Palmer Drought Severity Index
PES	payment for ecosystem services
PET	potential evapotranspiration
PFAS	per- and poly-fluoroalkyl substances
PS	point source
RBO	river basin organization
RO	reverse osmosis
SDGs	Sustainable Development Goals
SDWA	Safe Drinking Water Act
SFD	shit flow diagram
SLR	sea-level rise
SSO	sanitary sewer overflow
TDS	total dissolved solids
TMDL	total maximum daily load
TMR	terrestrial moisture recycling
TRWR	total renewable water resources
TSS	total suspended solids
TVA	Tennessee Valley Authority
UDDT	urine-diverting dry toilet
UNICEF	United Nations Children's Fund
USACE	US Army Corps of Engineers

USGS	US Geological Survey
WASH	water, sanitation, and hygiene
WCD	World Commission on Dams
WF	water footprint
WHO	World Health Organization
WQC	water quality criteria
WQS	water quality standards
WTA	withdrawal-to-availability ratio
WWTP	wastewater treatment plant

Glossary

Absolute river integrity: A water-rights doctrine claiming that a downstream *riparian* has the right to receive the natural flow into the country from upstream, undiminished in quantity or quality. Generally rejected in favor of *limited territorial sovereignty*.

Absolute sovereignty: A water-rights doctrine claiming that a country has a complete right to use the flow of any rivers within its borders, regardless of impact on downstream countries. Generally rejected in favor of *limited territorial sovereignty*.

Acid mine drainage: Acidic (low-pH) water (often with high metal concentrations) draining from a mine site, caused by the generation of acidity during the oxidation of reduced minerals.

Aerobic respiration: The conversion of organic matter to CO_2 by aerobic (oxygen-requiring) organisms, as represented in the chemical equation $C_6H_{12}O_6 + 6\ O_2 \rightarrow 6\ CO_2 + 6\ H_2O$ + energy.

Aerosol: A liquid or solid suspended in air.

Aggregate: Sand, gravel, and crushed stone used in construction.

Agroforestry: The integration of trees and shrubs into crop and animal farming systems to create environmental, economic, and social benefit.

Albedo: The reflectivity of a surface, expressed as the fraction of incoming radiation that is reflected (range = 0–1). The greater the albedo, the less radiation the surface will absorb and the cooler it will be (all other things being equal). The albedo of snow can be as high as 0.9, and the albedo of asphalt can be as low as 0.05.

Alluvial soil: Soil deposited by moving water, often highly productive for agriculture.

Anadromous fish: Fish that migrate from saltwater to freshwater for spawning.

Animal feeding operation (AFO): A facility in which animals are fed in confinement for at least forty-five days per year and forage crops are not grown on site.

Anthropocene Epoch: A proposed unit of geological time, denoting the modern era in which humanity is the dominant force affecting Earth's surface.

Aquaculture: The artificial propagation of aquatic animals or plants for consumption.

AQUASTAT: A Food and Agriculture Organization database with country-level data on water availability and use.

Aquatic life use support (ALUS): A category of *designated use* under the Clean Water Act, referring to the ability of a water body to provide habitat for a healthy community of aquatic organisms.

Aqueduct: Any conduit moving water from one location to another, often over long distances.

Aquifer: A subsurface formation through which water moves quickly, so it can serve as a water source for human use.

Aridification: A process by which an area becomes more arid (lower water availability) over time due to either natural or human causes.

Aridity index (AI): The ratio of mean annual precipitation to mean annual *evapotranspiration*. Low values of the aridity index indicate locations where little *blue water* is generated.

Assured water supply: A requirement in some western states that new residential developments secure, in advance, an adequate water supply for a defined period (often 100 years).

Atmospheric rivers: Narrow, elongated atmospheric regions with high transport of water vapor.

Avulsion: A process by which a river changes course abruptly by carving a new channel.

Beneficial use: A necessary component of a water right under *prior appropriation*. Beneficial uses generally include *irrigation*, domestic water supply, livestock, industrial use, and firefighting, and may or may not include *instream flows*.

Benefit–cost analysis (BCA): An economic analysis comparing a project's benefits to its costs, in order to help in decision making or evaluate already-completed projects.

Benefit/cost ratio (BCR): The ratio of a project's benefits to its costs, as determined by a *benefit-cost analysis*. A BCR greater than 1 indicates that the project provides enough benefits to justify its costs.

Benthic: Located in or on the bottom sediment of a water body.

Biochemical oxygen demand (BOD): A measure of the organic matter content of a water sample, expressed as the amount of O_2 that is consumed by *aerobic respiration* of that organic matter by a bacterial inoculum over five days.

Biofuel: A fuel made from biomass (e.g., ethanol from corn).

Biological nitrogen fixation (BNF): The conversion of N_2 to reactive (plant-available) nitrogen by bacteria. BNF occurs naturally but has been accelerated by the planting of legumes, whose symbiotic bacteria are nitrogen fixers.

Biomonitoring: Monitoring of the health of a group of aquatic organisms (e.g., fish, benthic macroinvertebrates) as a means of assessing the health of a water body.

Blackwater: Water containing human feces, typically from a toilet or latrine.

Blue water: Liquid water in lakes, rivers, aquifers, etc. Cf. *green water*.

Brine: Water having a salinity significantly greater than seawater.

Buy and dry: A form of water transfer in which the purchase of agricultural water rights (often by urban water providers) leads to large-scale loss of agricultural land and livelihoods within a given area.

Capacity factor (for a power plant): Actual energy generation divided by the amount of energy that would be generated if the power plant were continually running at its *installed capacity*.

Capture (groundwater): Water pumped from a well that comes not from groundwater storage but from a reduction in *groundwater discharge* or an increase in *groundwater recharge*.

Carbon dioxide equivalents (CO_2eq): The emissions of a *greenhouse gas*, converted to the equivalent amount of CO_2 using its global warming potential (a metric of its contribution to climate change).

Carcinogen: A chemical that can cause cancer.

Cenote: A water-filled sinkhole in a *karst* landscape, especially as used by the Maya for water supply and religious ritual.

Channelization: A suite of practices designed to convert a natural, dynamic river into a stable transportation channel with improved capacity to carry floodwaters.

Chemical oxygen demand (COD): A measure of the organic matter content of a water sample, expressed as the amount of O_2 that is consumed by oxidation of that organic matter using a chemical oxidant.

Climate forcer: A factor external to the climate system that drives changes in Earth's climate. Includes *greenhouse gases*, orbital forcing, solar output, and volcanic eruptions, but not internal *climate variability*.

Climate variability: Variation in climate that is internal to the climate system, rather than forced by an external factor. *El Niño Southern Oscillation (ENSO)* is a prominent example.

Cloud seeding: Techniques that aim to modify clouds to increase precipitation.

Coal ash: The material that remains after burning coal. High in toxic compounds.

Combined sewer: A sewer system that carries both household sewage and *stormwater*.

Common-pool resource (CPR): A resource characterized by *rivalry* and non-*excludability*. Susceptible to the *tragedy of the commons*.

Common property: A property rights system, especially for a *common-pool resource*, in which a defined group of people collectively share benefits and responsibilities, following a set of formal or informal rules.

Concentrated animal feeding operation (CAFO): An *animal feeding operation* that has at least 1,000 beef cattle (or equivalent) or otherwise poses a significant threat to water quality.

Concentrating solar power (CSP): A form of renewable energy in which mirrors concentrate solar radiation onto a receiver, where a fluid is heated and either used directly for thermal energy or converted to electrical energy.

Confined aquifer: An *aquifer* overlain by a confining layer of low hydraulic conductivity (e.g., clay), through which water moves slowly. Water in a confined aquifer is often at higher-than-atmospheric pressure. Cf. *unconfined aquifer*.

Conservationism: An early-twentieth-century environmental ethic that saw nature as a resource to be exploited for the benefit of humanity but was concerned about the destruction of that resource by improper management, especially overharvesting, erosion, and pollution, and called for more sustainable management to ensure "the greatest good for the greatest number *for the longest time*." Cf. *preservationism*.

Consumptive use (of water): A use that removes water from the local hydrologic cycle, generally through *evapotranspiration*, pollution, or *interbasin transfer*.

Conveyance losses: The loss of water during transportation from one location to another (e.g., evaporation and seepage from aqueducts).

Cost recovery: An approach for setting water prices based on the cost of providing that water (including maintenance).

Cultural burn: A fire set by Indigenous people in accordance with cultural practices, often with the goal of encouraging the regeneration of certain culturally important species.

Daylighting: The process of restoring a stream that has been routed underground, typically in an urban environment, by bringing it back to the surface.

Debris flow: A type of flood-related landslide in which a slurry of water, soil, and rock moves rapidly down a steep slope.

Deficit irrigation: An *irrigation* approach in which the farmer applies less water than the crop needs, usually resulting in a lower *yield* but also a lower *water footprint*.

Desertification: The conversion of vegetated semiarid areas into sparsely vegetated deserts.

Designated uses: The uses that a given water body should be able to support under the CWA. Typical designated uses include habitat (referred to as *aquatic life use support, ALUS*), recreation, finfish and shellfish harvesting, aesthetics, municipal water supply, industrial water supply, agricultural water supply, and navigation.

Dilbit: Diluted bitumen. Bitumen—a viscous, often semisolid petroleum product derived from tar sands—is diluted with lighter petroleum products for transport through pipelines.

Disability-adjusted life year (DALY): A measure of the impact of a disease on years of healthy life.

Discount rate: In *discounting*, the degree to which the future is considered less valuable than the present. If the discount rate is 3 percent, then next year's benefits are valued at 3 percent less than their nominal value.

Discounting: The practice of treating future monetary flows (both costs and benefits) as less valuable than flows in the present.

Disinfection byproducts (DBPs): A group of toxic chemicals produced during drinking water disinfection.

Dissolved oxygen (DO): The concentration of O_2 in water.

Dose–response curve: A tool for assessing a substance's toxicity. Typically, lab animals are exposed to different doses of the substance and a biological response is plotted as a function of dose.

Drip irrigation: An *irrigation* technology that uses surface or buried water lines with emitters that allow water to drip out in certain locations, providing water directly to the soil under individual plants.

Dry farming: Farming techniques for growing crops in *drylands* without irrigation.

Drylands: Areas with an *aridity index* less than 0.65 and therefore little to no generation of *runoff*.

Ecosystem services: The value provided by healthy ecosystems to human society.

El Niño Southern Oscillation (ENSO): A mode of *climate variability* centered in the equatorial Pacific Ocean with *teleconnections* that affect weather around the world.

Endocrine-disrupting chemicals: A group of synthetic chemicals that interfere with the body's hormone system and may cause toxic effects at low doses.

Environmental flow requirement (EFR): The amount and timing of water necessary to support environmental values in rivers, wetlands, and other water bodies.

Environmental pathogens: Pathogens that are widely distributed in soil and water.

Eutrophication: An increase in the inputs of nutrients or organic matter to a water body, leading to symptoms such as algal blooms, *hypoxia*, and loss of submerged vegetation.

Evapotranspiration (ET): The combined processes of evaporation (the physical–chemical conversion of water from liquid to gas) and transpiration (the conversion of water from liquid to gas by plants).

Exceedance probability: The probability that an event of a given magnitude will be equaled or exceeded. Usually expressed on an annual basis, in units of "per year." The inverse of the *recurrence interval*.

Excludability (economics): A characteristic of a resource in which potential users can easily be excluded from using that resource.

Exotic river: A dryland river whose flow is generated in a wetter environment upstream.

Exposure: The degree to which people and property are in the path of a natural hazard.

Externality: A cost or benefit that is borne not by a party carrying out a given action but by other parties or the public at large.

Extreme event attribution: The science of determining to what extent a particular extreme event (e.g., drought) was made more likely by climate change.

Falkenmark indicator: A water scarcity indicator based on the amount of renewable water flow available per person in a given region.

Fecal sludge: Human excreta stored in cesspits or other containers. May contain toilet paper, other waste streams, and water but is not liquid enough to flow through a pipe.

Feed conversion ratio (FCR): The ratio of the feed consumed by an animal to the biomass produced by the animal. A higher FCR means less efficient conversion of feed to biomass.

Feedback (climate): A process that leads to a change in climate due to the secondary response of the climate system to the changes caused by an external *climate forcer*.

Feedback loop: A set of causal relationships that either amplify (positive feedback loop) or diminish (negative feedback loop) the effects of an external perturbation to a system.

Fertilizers: Substances added to croplands to provide plant nutrients, most commonly nitrogen (N), phosphorus (P), and potassium (K).

Filter feeder: An organism that eats by filtering organic matter out of large volumes of water.

Fish passage: Any structure or process designed to help fish get around a dam or other barrier. May have low effectiveness.

Flashy (hydrology): Responding quickly and strongly to storm events, resulting in high peak flows.

Floatable: A piece of trash that floats on the water surface, such as a plastic bag.

Flood frequency analysis: A hydrologic method for predicting the distribution of floods of different sizes.

Floodplain: The area adjacent to a river that is normally prone to *fluvial* flooding.

Floodway: An area designated to receive flood waters during a large flood event.

Flow duration curve: A graph of flows (vertical axis) against their *exceedance probabilities* (horizontal axis). Generally constructed for a given river based on historical flow data.

Flow regime: The particular flow characteristics (including magnitude, timing, duration, frequency, rate of change, and predictability of different flows) that are typical of a given river or other water body.

Fluvial: Associated with a river.

Fog: Condensed water vapor forming at or near Earth's surface.

Fog catcher: A high-surface-area device designed to capture *fog* for use.

Fossil aquifer: An *aquifer* that was formed under different climatic conditions and is no longer being recharged at any significant rate.

Fossil fuel: A form of organic matter, generated over geologic time, that can be used to generate energy through combustion. Includes coal, oil, and natural gas.

Fracking (hydraulic fracturing): The process of artificially creating fractures in a fine-grained rock (e.g., shale) to increase the rock's permeability so that oil or gas can be extracted. Involves the injection of a mixture of water, sand, and chemicals under high pressure into the formation.

Fugitive resource: A resource, such as groundwater, that moves freely between properties.

Gaining stream: A stream that receives water from *groundwater discharge*.

General stream adjudications: The process of resolving all the water rights claims on a river system.

Global greening: The net global increase in vegetation cover over the last several decades.

Gravity Recovery and Climate Experiment (GRACE): A pair of satellites used to measure changes in the distribution of mass (mostly water) on Earth.

Gray infrastructure: Large-scale, centralized, engineered water infrastructure requiring high inputs of materials and energy. Includes aqueducts, levees, dams, and water and wastewater treatment plants. Cf. *green infrastructure*.

Graywater: Used water that is only lightly contaminated. Includes water from sinks, showers, and washing machines.

Great Acceleration: The simultaneous, near-exponential increase in many measures of the human impact on the planet during the modern era.

Green infrastructure (GI): Water infrastructure that uses or mimics natural ecosystems and cycles and requires low inputs of external materials and energy. Includes **water harvesting**, ecosystem restoration, decentralized stormwater retention, constructed wetlands, and watershed protection. Also referred to as "nature-based solutions." Cf. **gray infrastructure**.

Green Revolution: The period (roughly 1960–1990) when modern agricultural technologies and seeds spread rapidly across the developing world.

Green water: Water adsorbed to soil particles in the unsaturated portion of the soil. Can be used by plants for transpiration (green water flow) but not moved elsewhere by people.

Greenhouse gas (GHG): A gas that traps Earth's outgoing infrared radiation, including carbon dioxide (CO_2), **methane (CH_4)**, water, and nitrous oxide (N_2O). Anthropogenic increases in greenhouse gas concentrations in the atmosphere are the leading cause of global warming and climate change.

Greenwashing: The false or unsubstantiated claim that a particular practice or company is environmentally friendly.

Groundwater: Liquid water found beneath the ground surface, present in the pore space of unconsolidated materials or the fractures in rocks.

Groundwater discharge: The outward flow of groundwater from an aquifer to a stream, spring, oasis, or transpiring vegetation.

Groundwater mining or depletion: The decline in the amount of groundwater stored in a given aquifer, caused by extraction rates that exceed recharge plus capture.

Groundwater recharge: The flow of water into an aquifer, generally from percolation from above.

Haber–Bosch process: The industrial process by which N_2 gas (from air) is converted to reactive (plant-available) nitrogen in the form of NH_4^+.

Hard path: An approach to water management characterized by the use of **gray infrastructure**, a desire to control nature, and top-down decision-making. Cf. **soft path**.

Hardening of water demand: A phenomenon in which the amount of water needed in a given region becomes less capable of being reduced during times of scarcity. Often driven by planting of crops that can't be fallowed, such as trees.

Hardrock mining: Mining of metals (as opposed to coal or **aggregate**).

Hazard: Something that can potentially cause harm. The physical component of risk. Some measure of the physical magnitude of a natural disaster, such as flood height or drought severity.

Head: The change in water level between a hydropower system's high point (e.g., the elevation of the reservoir) and its low point (e.g., the elevation of the river at the base of the dam).

Helminth: A parasitic worm.

Herbicide: A *pesticide* used to kill plants (weeds).

Hungry water: Water that is erosive (i.e., it tends to erode the river's bed and banks), usually because the amount of sediment that it carries has been artificially lowered (e.g., due to sediment trapping by dams).

Hydraulic gradient: The slope of the water table (in an **unconfined aquifer**) or of the pressure surface (in a **confined aquifer**). The hydraulic gradient determines the direction and (together with the hydraulic conductivity) rate of groundwater movement.

Hydraulic societies: A term used by Karl Wittfogel to describe despotic civilizations where control over water was used to solidify the power of an elite ruling class.

Hydroelectric power: The use of water to generate electricity.

Hydrograph: A graph of flow over time in a river.

Hydro-illogical cycle: A feature of human psychology in which concern builds to panic during drought periods, but apathy follows as soon as it rains.

Hydromechanical power: The use of water to generate mechanical power, such as that used in a gristmill.

Hypoxia: A condition characterized by low dissolved O_2 concentrations in a water body.

Impaired water body: A water body that is not supporting its **designated uses** under the Clean Water Act.

Impervious surfaces: Areas that have very low capacity to infiltrate water, such as roads and buildings.

Incised stream: A river that is significantly lower than—and thus disconnected from—its **floodplain**.

Indicator bacteria: Bacterial groups used to monitor the possible presence of pathogens in water sources. These bacteria are generally not harmful in themselves but co-occur with fecal-derived pathogens.

Infiltration capacity: The ability of a particular soil to allow water at the surface (e.g., from rain) to enter the soil. Infiltration capacity is higher for coarse-textured soils with high organic matter.

Insecticide: A **pesticide** used to kill insects.

Installed capacity: The amount of power that a given hydropower facility or set of facilities can generate when operating at maximum capacity. Units: megawatts (MW).

Institutional resilience: A set of institutional factors (e.g., robust treaties and river basin organizations) that give a transboundary basin resilience to changes that might otherwise cause conflict.

Instream flow right: A water right (usually in **prior appropriation** systems) to keep water in a river to support environmental and recreational values.

Instream water use: A human use of water that does not remove the water from the river. Generally includes hydropower, navigation, fishing, recreation, waste disposal, and environmental uses.

Interbasin transfer: Transportation of water from one **river basin** to another, usually through an **aqueduct**.

Interception: The process by which vegetation captures precipitation on leaves, causing it to evaporate without reaching the soil. A part of the hydrologic cycle.

Intermittency (of water supply): A condition in which household taps only have water flowing through them periodically. Leads to poor water quality.

Intertropical Convergence Zone (ITCZ): An area near the equator where high temperatures cause air to rise, leading to high precipitation and winds converging from both north and south.

Iron triangle: The exchange of power, money, and influence between an executive-branch agency, a legislative committee, and special interest groups. Can lead to policy choices that favor special interests over the public good.

Irrigation: The application of *blue water* to agricultural fields to grow crops.

Irrigation efficiency: The amount of irrigation water that is transpired by crops, divided by either the amount of water withdrawn or the amount of water consumed.

Karst: A type of easily dissolved rock that supports rapid *groundwater recharge* and flow. Karst landscapes are characterized by sinkholes, underground caverns, productive springs, and few rivers.

Keystone species: A species that has a large influence on ecosystem structure and function, usually despite small numbers.

Lacustrine: Related to lakes.

Land grabbing: The acquisition of agricultural land in developing countries by large multinational agricultural corporations, resulting in the displacement of small farmers.

Land subsidence: A decline in the elevation of the land surface, often caused by the collapse of subsurface formations due to *groundwater* depletion.

Lead service lines (LSLs): Lead-containing water pipes running from a water main to a building.

Legacy city: A city that has experienced population loss and economic decline.

Legacy contaminant: A contaminant that is no longer being used but is still present in the environment.

Lentic: Characteristic of still-water environments.

Levee: A vertical structure, often made of earth, that is designed to prevent a flooding river from reaching certain areas. Often runs along a riverbank.

Levee arms race: A phenomenon in which the construction (or expansion) of a levee by one community increases the flood risk to another community, prompting further levee construction or expansion.

Levee effect: A phenomenon in which construction of a levee increases economic development in the area protected by the levee, increasing the damages when the levee fails.

Limited territorial sovereignty: The international water rights doctrine that claims that all *riparians* have limited rights of use.

Listed species: A threatened or endangered species under the Endangered Species Act.

Lock: A structure designed to move vessels from one water level to another in a river or canal used for navigation.

Lockage: A transit through a navigation *lock*.

Losing stream: A stream that recharges groundwater.

Lotic: Characteristic of flowing water environments.

Low- and middle-income countries (LMICs): Countries classified by the World Bank as low income (per capita income of $1,085 or less in 2021), lower middle income ($1,086–$4,255), or upper middle income ($4,256–$13,205).

Malthusianism: The perspective that population growth will outstrip food production, resulting in famine. Named for Robert Malthus.

Managed aquifer recharge (MAR): Purposeful *groundwater recharge* (often through infiltration basins) with the goal of increasing groundwater storage and avoiding the harmful effects of *groundwater* depletion.

Managed retreat: An approach to flooding and sea level rise that involves moving people and structures out of flood zones in purposeful, managed ways.

Manureshed: The geographic area around a manure source that contains enough active cropland to productively use the nutrients supplied by that manure.

Marginal benefit: The benefit from consuming one additional unit of a good.

Maximum Contaminant Level (MCL): A regulatory standard under the Safe Drinking Water Act specifying the maximum level of a contaminant that is allowed in drinking water. Determined by balancing public health with cost and feasibility.

Maximum Contaminant Level Goal (MCLG): A nonregulatory standard under the Safe Drinking Water Act specifying the level of a contaminant in drinking water that is likely to result in no harm.

Megadrought: A drought lasting for a long time period.

Membrane bioreactor: A wastewater treatment technology that combines biological treatment (oxidation of organic matter) with *membrane filtration*. Available in a variety of configurations and sizes.

Membrane filtration: Use of semipermeable membranes to remove materials of different sizes (from particles to ions) from water.

Methane (CH_4): A *greenhouse gas* with a high global warming potential. The main component of natural gas. An end-product of anaerobic respiration (respiration in the absence of O_2).

Millennium Development Goals (MDGs): A set of eight goals for *sustainable development* that guided global development efforts during the 2000–2015 time period.

Monsoon: A seasonal change in wind direction, usually associated with a seasonal change in precipitation.

Moral hazard: A situation in which people do not bear the consequences of their decisions, leading them to act in ways that increase risk.

Multiple-barrier approach: An approach to water quality protection, especially for drinking water, that involves multiple layers of protection, from watershed management to water treatment.

Municipal separate storm sewer system (MS4): A sewer system carrying (at least in principle) only *stormwater*. Also used to refer to the entity (e.g., municipality) managing the sewer system.

National Pollutant Discharge Elimination System (NPDES): Under the Clean Water Act, a system of permits for all *point sources*.

Net present value (NPV): The total value produced by a project, *discounted* to the present.

Networked sanitation: Sanitation systems that involve collection of sewage from multiple households, generally through sewers.

Nitrogen use efficiency (NUE): For an agricultural field, the ratio of the N removed in crops to the N added to the field (e.g., as *fertilizer*).

Nonconsumptive water use: Water that is withdrawn from the environment for human use but is returned to the environment (in liquid form) rather than undergoing *consumptive use*.

Nonnative invasive species: A species that has recently appeared in a given area (nonnative) that is expanding rapidly at the expense of other species (invasive).

Nonpoint source: A source of pollution that is distributed across the landscape, making it hard to measure and control. Cf. *point source*.

Nonrevenue water (NRW): Water delivered by a utility for which it does not receive payment. Comprised of real losses, apparent losses, and authorized unbilled uses.

Offstream dams: Dams not located on a stream.

Offstream water use: A human use of water that removes water from a river, *aquifer*, or other water body.

Once-through cooling: A cooling system for a power plant in which water is removed from the environment, used for cooling, and returned directly to the environment. Has higher *withdrawal* and lower *consumption* than *recirculating cooling*.

Onsite sanitation: A sanitation solution that involves storing human waste where it is generated, such as a *pit latrine*. Cf. *networked sanitation*.

Opportunity cost: The forgone value of water in other uses when it is tapped for a particular use.

Organic contaminants: Carbon-based chemicals, usually toxic, that are either synthetic or petroleum derived.

Paleoclimatology: The science of studying past climates.

Palmer Drought Severity Index (PDSI): A measure of drought based on deviation from normal temperature and precipitation, meant to be roughly comparable across locations. PDSI values are categorized as follows: −4 or less, extreme drought; −3.99 to −3, severe drought; −2.99 to −2, moderate drought; −1.99 to −1, mild drought; −0.99 to −0.5, incipient dry spell; −0.49 to 0.49, near normal; 0.5 to 0.99, incipient wet spell; 1 to 1.99, slightly wet; 2 to 2.99, moderately wet; 3 to 3.99, very wet; 4 or more, extremely wet.

Paper water right: A water right that is unlikely to yield any actual water for use, often because it is a very junior water right in an overallocated basin or because a tribe does not have the resources to build infrastructure to use the water right.

Pareto-efficient: A situation in which an outcome for one user (or decision criterion) is as good as it can be, given the outcome for other users. A situation is Pareto-efficient (Pareto-optimal) if any change that improves the outcome for one user (or decision criterion) will provide a worse outcome for another user. A situation is Pareto-inefficient (Pareto-inferior) if there is a change that will benefit at least one user and will not hurt anyone.

Pareto-optimal: See *Pareto-efficient*.

Payment for ecosystem services (PES): Payment made with the goal of preserving *ecosystem services*. Payments generally flow from those benefiting from the ecosystem services to those who can take action (or inaction) to preserve them.

Pesticide: A chemical used to kill pests of various kinds. Includes *herbicides* and *insecticides*.

PFAS: Per- and poly-fluorinated alkyl substances. A group of chemicals widely used in surfactants, firefighters, and other applications, known as "forever chemicals" because of their persistence in the environment.

Phreatophyte: A plant, often deep-rooted, that uses groundwater as a water supply.

Pit latrine: A pit used for human excreta, often with some type of structure over it.

Plant water-use efficiency: The ratio of a plant's transpiration rate to its photosynthetic rate.

Playa: A dry lakebed.

Pluvial: Associated with rainfall.

Point source: A discrete source of pollution, generally a pipe discharging from an industrial facility, *WWTP*, or *MS4*. Cf. *nonpoint source*.

Polychlorinated biphenyls (PCBs): A group of chemicals widely used in electrical equipment, now restricted under the Stockholm Convention because of their persistence, bioaccumulation, and toxicity.

Porosity: The fraction of a soil or other formation that is made up of pores (as opposed to solids).

Potential evapotranspiration (PET): The amount of water that would move to the atmosphere through *evapotranspiration* if water availability were not limiting.

Precautionary principle: The idea that new chemicals should be assumed unsafe until proven safe.

Precipitationshed: The area whose evapotranspiration contributes to precipitation in a given location.

Precision agriculture: Agricultural practices that rely heavily on high-resolution spatial data to determine optimal rates of planting and chemical application.

Prescribed burn: The purposeful use of fire, especially to reduce forest fuel loads and prevent catastrophic fire.

Preservationism: An early-twentieth-century environmental ethic that posited a moral duty to protect nature for its own sake, not just for the benefits it provided to humans, and argued for the preservation of wilderness as a place for nature to thrive. Cf. *conservationism*.

Primary treatment: An initial stage of wastewater treatment, usually consisting primarily of screening (to remove debris) and settling (to remove solids).

Prior appropriation: A water-rights doctrine common in the American West, in which senior water rights are obtained by putting water to *beneficial use* and are lost with nonuse.

Private good: A good that is *rival* and *excludable* and is thus efficiently allocated by markets.

Probability density function (PDF): A graph, often roughly bell-shaped, representing the probability of events of different magnitudes.

Produced water: Water that is extracted from the subsurface during oil and gas production. May contain high salinity, toxic metals, and radioactive elements.

Public good: A good that is non-*rival* and non-*excludable*. Undervalued by markets.

Public trust doctrine: The idea that rivers and other water bodies are fundamentally owned by the public, so private use of them should not ruin their value.

Pulse flow: A form of environmental flow in which a large pulse of water is released into a river, generally with the goal of simulating natural flood events.

Pumped-storage hydropower: A hydropower facility in which water is pumped uphill when external energy is available and released to generate hydropower when energy is needed. Functions as a very large battery.

Qanat: A shallow sloping tunnel that taps groundwater in its upper end and directs that groundwater to the surface by gravity.

Rainfall intensity–duration–frequency (IDF) curves: Curves showing rainfall amounts for events of different durations and *recurrence intervals*.

Rainfed agriculture: Agriculture that relies only on green water from rainfall.

Recirculating cooling: A cooling system for a power plant in which cooling water withdrawn from the environment is used multiple times within the facility. Has lower *withdrawal* and higher *consumption* than *once-through cooling*.

Recurrence interval: The inverse of the annual exceedance probability. Expresses event likelihood in units of years (e.g., the hundred-year flood), but it should not be understood as expressing the frequency with which this event will occur. Sometimes called return period.

Redlining: The process of designating neighborhoods dominated by people of color as "hazardous" and unfavorable for home loans. Areas designated as red or yellow in maps created by the federal Home Owners' Loan Corporation in the 1930s saw systematic underinvestment for decades.

Reference evapotranspiration (ET_o): The amount of evapotranspiration from a well-watered reference crop of short grass.

Regulated riparianism: A water rights doctrine in which water rights are based primarily on *riparian* principles but are enshrined in government-issued water use permits.

Regulatory capture: A situation in which government agencies that are meant to regulate a certain industry are controlled (directly or indirectly) by that industry, so the industry remains underregulated.

Regulatory takings: A situation in which government regulation reduces the value of private property to the point where the government must compensate the property owner to avoid violating the Fifth Amendment ("nor shall private property be taken for public use, without just compensation").

Renewable resource: A resource that can be used without being used up; a resource that is renewed by natural cycles.

Repetitive loss property: A National Flood Insurance Program–insured structure for which at least two payouts ($1,000 minimum) were made within a ten-year period.

Reserved water rights: Water rights for tribes and federal lands that supersede water rights under state law. See also *Winters doctrine*.

Reservoir-induced seismicity: The phenomenon by which the lubricating and pressure effects of a reservoir can cause an earthquake.

Reservoir sedimentation: The process by which a reservoir loses storage space for water due to sediment trapping.

Return flow: The portion of water *withdrawal* that is not *consumed* but rather flows back into the environment.

Reverse osmosis (RO): A desalination method in which pressurized water is forced through salt-rejecting membranes (against the osmotic gradient).

Rights of nature: The idea that natural features (e.g., mountains, rivers, fish species) should have legal personhood or other form of legal standing.

Riparian: River bordering. Also used as a noun to mean a country or individual whose land abuts a given river.

Riparian doctrine: See *riparianism*.

Riparianism: A water rights doctrine where water rights originate from ownership of *riparian* land and where water is shared based on reasonable use.

Rivalry (economics): A characteristic of a resource in which one person's use of the resource precludes another person from using it.

River basin: The land area draining to a river. The river's *watershed*.

River-basin organization (RBO): An institutional structure, typically composed of members from various government and nongovernment agencies, designed to collaboratively manage water issues within a *river basin*.

Roughness: A characteristic of a land surface, expressing its geometric variation and relating to its ability to efficiently transfer heat and moisture to the atmosphere.

Rule of capture: A water-rights doctrine under which a landowner can pump groundwater from their land without limit. Also known as "law of the largest pump."

Run-of-river: Operated in a way that does not substantially alter the flow of water in a river. Used of a reservoir to indicate that outflow from the reservoir equals inflow to the reservoir. Used of a hydropower project to indicate that power generation is proportional to whatever flow is in the river. Cf. *storage dam or reservoir*.

Runoff (R): Generally, the flow of liquid water from an area. For more on the multiple meanings of *runoff*, see the companion website.

Runoff farming: A set of practices for capturing *stormwater* for *irrigation*, especially in desert settings.

Runoff ratio: The ratio of *runoff* to precipitation.

Safe yield: The amount of water that can reliably be obtained from a given water source. For an aquifer, often defined as equal to *groundwater recharge* plus *capture*.

Salinization: An increase in the salinity of a soil or water body.

Saltwater intrusion: The movement of saltwater (usually, though not always, from the ocean) into an *aquifer*.

Sanitary Revolution: The realization, during the nineteenth century, that disease can be spread by contamination of drinking water with sewage and thus can be prevented by the provision of clean water and effective sanitation.

Schistosomiasis: A *water-based* disease caused by *helminths* in the genus *Schistosoma*.

Sea-level rise (SLR): The increase in mean sea level at a given location. Affected by ocean warming; loss of ice from mountain glaciers, Greenland, and Antarctica; local vertical land movement; and water management activities that change the distribution of freshwater on Earth.

Secondary treatment: A type of wastewater treatment focused mostly on removal of organic matter and pathogens through biological treatment.

Sediment passage: Operation of a dam to allow incoming or accumulated sediment to pass downstream.

Setback levee: A *levee* that is moved farther away from a river to give the river more room to flood.

Settling basin: A section of a water system with low velocities, allowing sediments to settle out.

Shit flow diagram (SFD): A graphic for quantitatively illustrating safe and unsafe pathways for fecal material, usually in a city.

Siltation: A phenomenon in which canals or aqueducts fill with silt, obstructing the flow of water.

Soft path: An approach to water management characterized by the use of *green infrastructure*, the protection and restoration of natural ecosystems, and decentralized or collaborative decision making. Cf. *hard path*.

Stationarity: The idea that water availability, or climate more broadly, is highly variable, but the parameters of that variability (e.g., the *probability density function*) are not changing systematically over time.

Steam turbine: A turbine that is turned by steam (water vapor).

Stepwell: A type of well found in South Asia that has many levels of steps (often beautifully designed).

Storage dam or reservoir: A *dam* or *reservoir* that is used to store water for later release. Outflows from the reservoir at any point in time do not generally equal inflows to the reservoir. When inflows are greater than outflows, reservoir storage is increasing. When inflows are less than outflows, reservoir storage is decreasing. Cf. *run-of-river*.

Storm surge: In a coastal region, the increase in water level associated with a storm, especially one with winds that blow water toward the shore.

Stormwater: The water flowing from the landscape in response to a rain event, especially water flowing off *impervious surfaces* and into an *MS4* or *combined sewer*.

Stormwater utility: A local or regional agency that manages a stormwater system and charges customers (i.e., homeowners) a fee for this service.

Structural adjustment program: Economic reform programs required by the World Bank and International Monetary Fund as a condition of loans to countries experiencing economic crises. These reforms emphasize privatization and the treatment of natural resources (including water) as economic goods.

Surface irrigation: Irrigation techniques that involve diverting the flow of a canal or ditch into a field. Surface irrigation generally inundates furrows or the entire field.

Sustainable development: The dual goal of improving people's lives, especially in low-income countries (development), without destroying the ecological life-support systems needed by future generations (sustainability).

Sustainable Development Goals (SDGs): A set of 17 goals for *sustainable development*, agreed to by the international community in 2015, to be achieved by 2030. Preceded by the *Millennium Development Goals*.

Sustainable intensification: The dual goal of increasing production on existing agricultural land (intensification) while reducing the environmental impact of agriculture (sustainability).

Tailings dam or pond: A dam or reservoir for storing mine tailings (the waste material remaining after valuable ores are extracted).

Technology-based effluent limit: Under the Clean Water Act, a limit on the amount of a chemical that is allowed to be discharged, where the value of that limit is derived from the capacity of wastewater treatment technology to remove that chemical.

Teleconnections: Complex causal relationships between oceanic and atmospheric conditions in different parts of Earth.

Terminal lake: A lake that has no surface water outflow.

Terracing: An agricultural practice in which sloped land is converted to a series of level terraces in order to increase arable land, decrease erosion, and increase water retention.

Terrestrial moisture recycling (TMR): The phenomenon in which evapotranspiration from one location on land contributes to precipitation in another location on land.

Tertiary treatment: Any wastewater treatment beyond secondary; most commonly involves nutrient or pathogen removal.

Threshold toxicant: A toxic substance that is not a *carcinogen* and is therefore assumed to have a threshold exposure below which it is safe.

Total dissolved solids (TDS): The mass of material dissolved in a water sample, usually expressed in mg/L; salinity.

Total maximum daily load (TMDL): The maximum amount of a pollutant that a water body can receive while still meeting *water quality criteria*. Also, a planning document designed to achieve that load, required for certain *impaired water bodies* under the Clean Water Act.

Total renewable water resources (TRWR): The annual renewable water flows within a geographic area. For river basins, this is identical to *runoff*. For countries, external inflows and outflows must be accounted for.

Total suspended solids (TSS): The mass of material suspended in a water sample, usually expressed in mg/L.

Tragedy of the commons: The overuse of a *common-pool resource* by individuals each acting in their own short-term self-interest.

Transaction costs: The costs (in money, water, time, or other resources) of a given water transfer.

Transpiration: The loss of water from plants in the form of water vapor.

Tropical cyclone: A rotating storm that originated in the tropics. Strong tropical cyclones are called hurricanes.

Turbidity: Cloudiness, as of a water sample. Samples with high turbidity generally have high *total suspended solids*.

UN Watercourses Convention: An international treaty establishing principles for shared use of international basins.

Unconfined aquifer: An *aquifer* with no confining layer above it. The water at the top of an unconfined aquifer (the water table) is at atmospheric pressure.

Urban heat island effect: The phenomenon in which urban areas are warmer than surrounding areas.

Urban runoff: See *stormwater*.

Urban stream syndrome: The consistent pattern of ecological degradation of urban streams.

Urine-diverting dry toilet (UDDT): A toilet designed to separate urine from feces that does not use water for flushing or anal cleansing.

Utilitarianism: An ethic that sees the maximization of utility as a primary goal. "The greatest good for the greatest number."

Virtual water: The water footprint of a product, especially one that is traded around the world. For example, moving a kilogram of wheat from one country to another also represents movement of the virtual water that it took to grow that wheat.

Vulnerability: The social component of risk. The degree to which *exposure* to a given *hazard* will cause harm. Determined by the ways that individuals and societies prepare themselves for, and respond to, natural disasters.

Wadi: A channel that was carved by water but is dry most of the time. Also called an arroyo, wash, or dry creek.

Water 3.0: In David Sedlak's usage, centralized water-supply systems with drinking water treatment and wastewater treatment.

Water-based diseases: Diseases caused by pathogens whose life cycles require an intermediate water-based host. Can be prevented by adequate sanitation.

Water–energy nexus: An umbrella term for the various ways that the water and energy sectors intersect.

Water footprint: The water used (directly and indirectly) to produce a given product, or the water used (directly and indirectly) by a sector, individual, or country.

Water gap: The yield gap associated with insufficient (or excessive) soil water.

Water harvesting: Various approaches, generally small in scale, for capturing water for human use.

Water productivity: The amount of food produced per unit water. The inverse of the food's *water footprint*. Similar to *water-use efficiency*.

Water quality criteria (WQCs): Under the Clean Water Act, the levels of various water quality parameters that, if achieved, will protect the *designated uses* of a water body.

Water-related diseases: Infectious diseases where water can play a role in transmission. Includes *water-based diseases*, *waterborne diseases*, and *water-washed diseases*, as well as some *environmental pathogens* and insect-borne diseases.

Water tower: A mountain range that receives high levels of precipitation and stores that precipitation in the form of snow and ice, gradually releasing meltwater to downstream rivers and communities.

Water-use efficiency: The amount of a product produced per unit of water. The inverse of the water footprint.

Water-washed diseases: Eye and skin infections, such as trachoma, body lice, and scabies, whose transmission can be prevented by adequate washing.

Water year: In the United States, the period from October 1 of one year to September 30 of the following year.

Water yield: The flow of water from a given area, usually expressed in depth units. Area-normalized flow.

Waterborne diseases: Diseases transmitted when feces from an infected person contaminates the household or environment, ultimately leading a susceptible individual to ingest an infectious dose of the pathogen.

Watershed: The land area draining to a specific point on a river or other water body.

Weir: A low dam designed for water to flow over its crest.

Winans rights: The rights of tribes to exercise their traditional practices, including hunting and fishing, on lands that the tribes gave up under coercion as part of the treaties establishing reservations.

Winters doctrine: The legal principle that tribes have *reserved water rights* to sufficient water to fulfill the purposes of their reservations.

Withdrawal: Water that is removed from the environment for human use. Withdrawal = consumption + return flow.

Withdrawal-to-availability indicator (WTA): A water scarcity indicator based on the ratio of water withdrawal to renewable water resources.

Yield (agriculture): The amount of crop produced per unit land.

About the Author

Shimon Anisfeld is Senior Lecturer II and Research Scientist at the Yale School of the Environment, where he teaches courses in water management, coastal ecology, organic pollutants, and physical science for environmental managers. His research and writing interests include both freshwater and coastal issues, which he sees as central to the challenges facing the world in the coming decades.

References

Abbott, B. W., K. Bishop, J. P. Zarnetske, C. Minaudo, F. S. Chapin, S. Krause, D. M. Hannah, L. Conner, et al. 2019. "Human Domination of the Global Water Cycle Absent from Depictions and Perceptions." *Nature Geoscience* 12(7): 533+.

Acreman, M., A. Smith, L. Charters, D. Tickner, J. Opperman, S. Acreman, F. Edwards, P. Sayers, et al. 2021. "Evidence for the Effectiveness of Nature-Based Solutions to Water Issues in Africa." *Environmental Research Letters* 16(6): 063007.

Adegoke, A. A., I. D. Amoah, T. A. Stenström, M. E. Verbyla, and J. R. Mihelcic. 2018. "Epidemiological Evidence and Health Risks Associated with Agricultural Reuse of Partially Treated and Untreated Wastewater: A Review." *Frontiers in Public Health* 6: 337.

Alam, S., M. Gebremichael, Z. Ban, B. R. Scanlon, G. Senay, and D. P. Lettenmaier. 2021. "Post-Drought Groundwater Storage Recovery in California's Central Valley." *Water Resources Research* 57(10): e2021WR030352.

Alexander, J. S., R. C. Wilson, and W. R. Green. 2012. *A Brief History and Summary of the Effects of River Engineering and Dams on the Mississippi River System and Delta*. USGS. Circular 1375.

Allaire, M., T. Mackay, S. Zheng, and U. Lall. 2019. "Detecting Community Response to Water Quality Violations Using Bottled Water Sales." *Proceedings of the National Academy of Sciences* 116(42): 20917–22.

Allaire, M., H. Wu, and U. Lall. 2018. "National Trends in Drinking Water Quality Violations." *Proceedings of the National Academy of Sciences* 115(9): 2078–83.

Almeida, R. M., Q. Shi, J. M. Gomes-Selman, X. Wu, Y. Xue, H. Angarita, N. Barros, B. R. Forsberg, et al. 2019. "Reducing Greenhouse Gas Emissions of Amazon Hydropower with Strategic Dam Planning." *Nature Communications* 10(1): 4281.

American Rivers, Friends of the Earth, and Trout Unlimited. 1999. *Dam Removal Success Stories: Restoring Rivers through Selective Removal of Dams That Don't Make Sense*. ISBN 0-913890-96-0.

An, L. L., J. D. Wang, J. P. Huang, Y. Pokhrel, R. Hugonnet, Y. Wada, D. Caceres, H. M. Schmied, et al. 2021. "Divergent Causes of Terrestrial Water Storage Decline between Drylands and Humid Regions Globally." *Geophysical Research Letters* 48: e2021GL095035.

Anastasiou, E., and P. D. Mitchell. 2015. "Human Intestinal Parasites and Dysentery in Africa and the Middle East prior to 1500." In *Sanitation, Latrines, and Intestinal Parasites in Past Populations*. P. D. Mitchell, ed. Burlington, VT: Ashgate Publishing Company.

Annandale, G. 2013. *Quenching the Thirst: Sustainable Water Supply and Climate Change.* North Charleston, SC: CreateSpace Independent Publishing Platform.

Ansar, A., B. Flyvbjerg, A. Budzier, and D. Lunn. 2014. "Should We Build More Large Dams? The Actual Costs of Hydropower Megaproject Development." *Energy Policy* 69: 43–56.

Arellano-Gonzalez, J., A. AghaKouchak, M. C. Levy, Y. Qin, J. Burney, S. J. Davis, and F. C. Moore. 2021. "The Adaptive Benefits of Agricultural Water Markets in California." *Environmental Research Letters* 16(4): 044036.

ASCE (American Society of Civil Engineers). 2021a. 2021 Infrastructure Report Card: Dams. https://infrastructurereportcard.org/

ASCE (American Society of Civil Engineers). 2021b. 2021 Infrastructure Report Card: Drinking Water. https://infrastructurereportcard.org/

ASCE (American Society of Civil Engineers). 2021c. 2021 Infrastructure Report Card: Levees. https://infrastructurereportcard.org/

ASCE (American Society of Civil Engineers). 2021d. 2021 Infrastructure Report Card: Wastewater. https://infrastructurereportcard.org/

Association of State Floodplain Managers. 2020. Flood Mapping for the Nation: A Cost Analysis for Completing and Maintaining the Nation's NFIP Flood Map Inventory. Madison, WI: Author. https://floodsciencecenter.org/products/flood-mapping-for-the-nation/

Averyt, K., J. Meldrum, P. Caldwell, G. Sun, S. McNulty, A. Huber-Lee, and N. Madden. 2013. "Sectoral Contributions to Surface Water Stress in the Coterminous United States." *Environmental Research Letters* 8(3): 9.

Baek, S. H., and J. M. Lora. 2021. "Counterbalancing Influences of Aerosols and Greenhouse Gases on Atmospheric Rivers." *Nature Climate Change* 11: 958–65.

Bain, R., R. Cronk, J. Wright, H. Yang, T. Slaymaker, and J. Bartram. 2014. "Fecal Contamination of Drinking-Water in Low- and Middle-Income Countries: A Systematic Review and Meta-Analysis." *PLoS Med* 11(5): e1001644.

Bair, E., T. Stillinger, K. Rittger, and M. Skiles. 2021. "COVID-19 Lockdowns Show Reduced Pollution on Snow and Ice in the Indus River Basin." *Proceedings of the National Academy of Sciences* 118(18): e2101174118.

Balazs, C., R. Morello-Frosch, A. Hubbard, and I. Ray. 2011. "Social Disparities in Nitrate-Contaminated Drinking Water in California's San Joaquin Valley." *Environmental Health Perspectives* 119(9): 1272–8.

Barbier, E. B. 2003. "Upstream Dams and Downstream Water Allocation: The Case of the Hadejia-Jama'are Floodplain, Northern Nigeria." *Water Resources Research* 39(11): 9.

Barskey, A. E., G. Derado, and C. Edens. 2022. "Rising Incidence of Legionnaires' Disease and Associated Epidemiologic Patterns, United States, 1992–2018." *Emerging Infectious Diseases* 28(3): 527–38.

Beach, T., S. Luzzadder-Beach, S. Krause, T. Guderjan, F. Valdez, J. C. Fernandez-Diaz, S. Eshleman, and C. Doyle. 2019. "Ancient Maya Wetland Fields Revealed under Tropical Forest Canopy from Laser Scanning and Multiproxy Evidence." *Proceedings of the National Academy of Sciences* 116(43): 21469–77.

Behnke, R. H., and M. Mortimore, Eds. 2016. *The End of Desertification? Disputing Environmental Change in the Drylands.* New York: Springer.

Belletti, B., C. G. de Leaniz, J. Jones, S. Bizzi, L. Borger, G. Segura, A. Castelletti, W. van de Bund, et al. 2020. "More Than One Million Barriers Fragment Europe's Rivers." *Nature* 588(7838): 436+.

Benson, D., A. K. Gain, and C. Giupponi. 2020. "Moving beyond Water Centricity? Conceptualizing Integrated Water Resources Management for Implementing Sustainable Development Goals." *Sustainability Science* 15(2): 671–81.

Berg, N., and A. Hall. 2017. "Anthropogenic Warming Impacts on California Snowpack during Drought." *Geophysical Research Letters* 44(5): 2511–8.

Bernauer, T., and T. Böhmelt. 2020. "International Conflict and Cooperation over Freshwater Resources." *Nature Sustainability* 3(5): 350–6.

Bibby, K., and J. Peccia. 2013. "Identification of Viral Pathogen Diversity in Sewage Sludge by Metagenome Analysis." *Environmental Science & Technology* 47(4): 1945–51.

Biswas, A. K. 2004. "Integrated Water Resources Management: A Reassessment: A Water Forum Contribution." *Water International* 29(2): 248–56.

Boelens, R., J. Vos, and T. Perreault. 2018. Introduction: The Multiple Challenges and Layers of Water Justice Struggles. In *Water Justice*. R. Boelens, J. Vos, and T. Perreault, eds. Cambridge, UK: Cambridge University Press.

Bordonal, R. d. O., J. L. N. Carvalho, R. Lal, E. B. de Figueiredo, B. G. de Oliveira, and N. La Scala. 2018. "Sustainability of Sugarcane Production in Brazil. A Review." *Agronomy for Sustainable Development* 38(2): 13.

Borgomeo, E., A. Jägerskog, E. Zaveri, J. Russ, A. Khan, and R. Damania. 2021. *Ebb and Flow Volume 2. Water in the Shadow of Conflict in the Middle East and North Africa*. Washington, DC: World Bank Group.

Boulay, A. M., J. Bare, L. Benini, M. Berger, M. J. Lathuilliere, A. Manzardo, M. Margni, M. Motoshita, et al. 2018. "The WULCA Consensus Characterization Model for Water Scarcity Footprints: Assessing Impacts of Water Consumption Based on Available Water Remaining (AWARE)." *International Journal of Life Cycle Assessment* 23(2): 368–78.

Bovee, K. D., B. L. Lamb, J. M. Bartholow, C. B. Stalnaker, J. Taylor, and J. Henriksen. 1998. *Stream Habitat Analysis Using the Instream Flow Incremental Methodology*. US Geological Survey Report USGS/BRD-1998-0004.

BP (British Petroleum). 2022. *BP Statistical Review of World Energy 2022*. 71st edition. https://www.bp.com/content/dam/bp/business-sites/en/global/corporate/pdfs/energy-economics/statistical-review/bp-stats-review-2022-full-report.pdf

Bracken, N., J. Macknick, A. Tovar-Hastings, P. Komor, M. Gerritsen, and S. Mehta. 2015. *Concentrating Solar Power and Water Issues in the U.S. Southwest*. Joint Institute for Strategic Energy Analysis Technical Report NREL/TP-6A50-61376.

Brauman, K. A., A. L. Goodkind, T. Kim, R. E. O. Pelton, J. Schmitt, and T. M. Smith. 2020. "Unique Water Scarcity Footprints and Water Risks in US Meat and Ethanol Supply Chains Identified via Subnational Commodity Flows." *Environmental Research Letters* 15(10): 105018.

Brauman, K. A., B. D. Richter, S. Postel, M. Malsy, and M. Florke. 2016. "Water Depletion: An Improved Metric for Incorporating Seasonal and Dry-Year Water Scarcity into Water Risk Assessments." *Elementa-Science of the Anthropocene* 4: 12.

Briscoe, J. 2015. "Water Security in a Changing World." *Daedalus* 144(3): 27–34.

Broad, R., and J. Cavanagh. 2021. *The Water Defenders: How Ordinary People Saved a Country from Corporate Greed*. Boston: Beacon Press.

Brouwer, R., S. Akter, L. Brander, and E. Haque. 2007. "Socioeconomic Vulnerability and Adaptation to Environmental Risk: A Case Study of Climate Change and Flooding in Bangladesh." *Risk Analysis* 27(2): 313–26.

Brown, K. P. 2017. "Water, Water Everywhere (or, Seeing Is Believing): The Visibility of Water Supply and the Public Will for Conservation." *Nature + Culture* 12(3): 219–45.

Bruijnzeel, L. A., M. Mulligan, and F. N. Scatena. 2011. "Hydrometeorology of Tropical Montane Cloud Forests: Emerging Patterns." *Hydrological Processes* 25(3): 465–98.

Bruno, E. M., and K. Jessoe. 2021. Using Price Elasticities of Water Demand to Inform Policy. In *Annual Review of Resource Economics*, Vol 13. G. C. Rausser and D. Zilberman. 13: 427–41.

Bureau of Reclamation. 2018. *Colorado River Basin Ten Tribes Partnership Tribal Water Study Report.* https://www.usbr.gov/lc/region/programs/crbstudy/tws/finalreport.html

Butzer, K. W. 2012. "Collapse, Environment, and Society." *Proceedings of the National Academy of Sciences* 109(10): 3632–9.

California Energy Commission (CEC). 2005. California's Water–Energy Relationship, Final Staff Report CEC-700-2005-011-SF.

Campbell, W. 2020. Western Kentucky University Stormwater Utility Survey 2020. *SEAS Faculty Publications* Paper 3. https://digitalcommons.wku.edu/seas_faculty_pubs/3

Cao, S., L. Chen, D. Shankman, C. Wang, X. Wang, and H. Zhang. 2011. "Excessive Reliance on Afforestation in China's Arid and Semi-Arid Regions: Lessons in Ecological Restoration." *Earth-Science Reviews* 104(4): 240–5.

Capodaglio, A. G. 2021. "Fit-for-Purpose Urban Wastewater Reuse: Analysis of Issues and Available Technologies for Sustainable Multiple Barrier Approaches." *Critical Reviews in Environmental Science and Technology* 51(15): 1619–66.

Carey, M. P., B. L. Sanderson, K. A. Barnas, and J. D. Olden. 2012. "Native Invaders: Challenges for Science, Management, Policy, and Society." *Frontiers in Ecology and the Environment* 10(7): 373–81.

Ceballos, G., P. R. Ehrlich, A. D. Barnosky, A. García, R. M. Pringle, and T. M. Palmer. 2015. "Accelerated Modern Human–Induced Species Losses: Entering the Sixth Mass Extinction." *Science Advances* 1(5): e1400253.

CH2M Hill Engineers, Inc. 2018. *City of New Haven Combined Sewer Overflow Long-Term Control Plan Update.* Final Report for Greater New Haven Water Pollution Control Authority. https://gnhwpca.com/combined-sewer-overflow-documents/

Chen, B., M. Y. Hanb, K. Peng, S. L. Zhou, L. Shao, X. F. Wu, W. D. Wei, S. Y. Liu, et al. 2018. "Global Land–Water Nexus: Agricultural Land and Freshwater Use Embodied in Worldwide Supply Chains." *Science of the Total Environment* 613: 931–43.

Chinnasamy, C. V., M. Arabi, S. Sharvelle, T. Warziniack, C. D. Furth, and A. Dozier. 2021. "Characterization of Municipal Water Uses in the Contiguous United States." *Water Resources Research* 57(6): e2020WR028627.

Chouchane, H., M. S. Krol, and A. Y. Hoekstra. 2020. "Changing Global Cropping Patterns to Minimize National Blue Water Scarcity." *Hydrology and Earth System Sciences* 24(6): 3015–31.

Clair, R. P., R. Rastogi, S. Lee, R. A. Clawson, E. R. Blatchley, and C. Erdmann. 2019. "A Dialectical and Dialogical Approach to Health Policies and Programs: The Case of Open Defecation in India." *Health Communication* 34(11): 1231–41.

Climate Central. 2019. *Ocean at the Door: New Homes and the Rising Sea.* Research brief by Climate Central. https://sealevel.climatecentral.org/news/ocean-at-the-door-new-homes-and-the-rising-sea/

Collier, S., L. Deng, E. Adam, K. Benedict, E. Beshearse, A. Blackstock, B. Bruce, G. Derado, et al. 2021. "Estimate of Burden and Direct Healthcare Cost of Infectious Waterborne Disease in the United States." *Emerging Infectious Disease Journal* 27(1): 140–9.

Colton, R. D. 2020. "The Affordability of Water and Wastewater Service in Twelve U.S. Cities: A Social, Business and Environmental Concern." *The Guardian.* https://www.theguardian.com/environment/2020/jun/23/full-report-read-in-depth-water-poverty-investigation

Congressional Budget Office. 2018. Public Spending on Transportation and Water Infrastructure, 1956 to 2017. https://www.cbo.gov/publication/54539

Congressional Research Service (CRS). 2016. Clean Water Act: A Summary of the Law. Report RL30030. https://crsreports.congress.gov/product/pdf/RL/RL30030

Cook, B. I., R. L. Miller, and R. Seager. 2009. "Amplification of the North American 'Dust Bowl' Drought through Human-Induced Land Degradation." *Proceedings of the National Academy of Sciences* 106(13): 4997–5001.

Cornejo, P. K., M. V. E. Santana, D. R. Hokanson, J. R. Mihelcic, and Q. Zhang. 2014. "Carbon Footprint of Water Reuse and Desalination: A Review of Greenhouse Gas Emissions and Estimation Tools." *Journal of Water Reuse and Desalination* 4(4): 238–52.

Corringham, T. W., F. M. Ralph, A. Gershunov, D. R. Cayan, and C. A. Talbot. 2019. "Atmospheric Rivers Drive Flood Damages in the Western United States." *Science Advances* 5(12): eaax4631.

Costanza, R., R. de Groot, L. Braat, I. Kubiszewski, L. Fioramonti, P. Sutton, S. Farber, and M. Grasso. 2017. "Twenty Years of Ecosystem Services: How Far Have We Come and How Far Do We Still Need to Go?" *Ecosystem Services* 28: 1–16.

Costanza, R., R. de Groot, P. Sutton, S. van der Ploeg, S. J. Anderson, I. Kubiszewski, S. Farber, and R. K. Turner. 2014. "Changes in the Global Value of Ecosystem Services." *Global Environmental Change: Human and Policy Dimensions* 26: 152–8.

Couto, T. B. A., and J. D. Olden. 2018. "Global Proliferation of Small Hydropower Plants—Science and Policy." *Frontiers in Ecology and the Environment* 16(2): 91–100.

Cox, M., G. Arnold, and S. V. Tomas. 2010. "A Review of Design Principles for Community-Based Natural Resource Management." *Ecology and Society* 15(4): 19.

Crawford, S. E., M. Brinkmann, J. D. Ouellet, F. Lehmkuhl, K. Reicherter, J. Schwarzbauer, P. Bellanova, P. Letmathe, et al. 2022. "Remobilization of Pollutants during Extreme Flood Events Poses Severe Risks to Human and Environmental Health." *Journal of Hazardous Materials* 421: 126691.

Crozier, L. G., M. M. McClure, T. Beechie, S. J. Bograd, D. A. Boughton, M. Carr, T. D. Cooney, J. B. Dunham, et al. 2019. "Climate Vulnerability Assessment for Pacific Salmon and Steelhead in the California Current Large Marine Ecosystem." *PLoS One* 14(7): e0217711.

Dahlke, H. E., A. Brown, S. Orloff, D. H. Putnam, and T. O'Geen. 2018. "Managed Winter Flooding of Alfalfa Recharges Groundwater with Minimal Crop Damage." *California Agriculture* 72(1): 65–75.

Dai, A. G. 2016. "Historical and Future Changes in Streamflow and Continental Runoff: A Review." In *Terrestrial Water Cycle and Climate Change: Natural and Human-Induced Impacts*, ed. Q. Tang and T. Oki, 221: 17–37. Washington, DC: American Geophysical Union.

Dalin, C., M. Konar, N. Hanasaki, A. Rinaldo, and I. Rodriguez-Iturbe. 2012. "Evolution of the Global Virtual Water Trade Network." *Proceedings of the National Academy of Sciences of the United States of America* 109(16): 5989–94.

Dalin, C., Y. Wada, T. Kastner, and M. J. Puma. 2017. "Groundwater Depletion Embedded in International Food Trade." *Nature* 543(7647): 700–4.

Danert, K., and G. Hutton. 2020. "Shining the Spotlight on Household Investments for Water, Sanitation and Hygiene (WASH): Let Us Talk about HI and the Three 'T's." *Journal of Water, Sanitation and Hygiene for Development* 10(1): 1–4.

Davis, K. F., M. C. Rulli, F. Garrassino, D. Chiarelli, A. Seveso, and P. D'Odorico. 2017a. "Water Limits to Closing Yield Gaps." *Advances in Water Resources* 99: 67–75.

Davis, K. F., M. C. Rulli, A. Seveso, and P. D'Odorico. 2017b. "Increased Food Production and Reduced Water Use through Optimized Crop Distribution." *Nature Geoscience* 10(12): 919–24.

Davis, M. 2001. *Late Victorian Holocausts: El Niño Famines and the Making of the Third World*. New York: Verso.

Dawson, T. E. 1998. "Fog in the California Redwood Forest: Ecosystem Inputs and Use by Plants." *Oecologia* 117(4): 476–85.

de Bruin, S. P., S. Schmeier, R. van Beek, and M. Gulpen. 2023. "Projecting Conflict Risk in Transboundary River Basins by 2050 Following Different Ambition Scenarios." *International Journal of Water Resources Development* 40(1): 7–32.

De Stefano, L., J. D. Petersen-Perlman, E. A. Sproles, J. Eynard, and A. T. Wolf. 2017. "Assessment of Transboundary River Basins for Potential Hydro-Political Tensions." *Global Environmental Change* 45: 35–46.

Debaere, P., and T. Li. 2020. "The Effects of Water Markets: Evidence from the Rio Grande." *Advances in Water Resources* 145: 103700.

Deemer, B. R., J. A. Harrison, S. Li, J. J. Beaulieu, T. DelSontro, N. Barros, J. F. Bezerra-Neto, S. M. Powers, et al. 2016. "Greenhouse Gas Emissions from Reservoir Water Surfaces: A New Global Synthesis." *BioScience* 66: 949–64.

Dell'Angelo, J., M. C. Rulli, and P. D'Odorico. 2018. "The Global Water Grabbing Syndrome." *Ecological Economics* 143: 276–85.

Dery, J. L., C. M. Rock, R. R. Goldstein, C. Onumajuru, N. Brassill, S. Zozaya, and M. R. Suri. 2019. "Understanding Grower Perceptions and Attitudes on the Use of Nontraditional Water Sources, Including Reclaimed or Recycled Water, in the Semi-Arid Southwest United States." *Environmental Research* 170: 500–9.

DeSimone, L., P. McMahon, and M. Rosen. 2014. *Water Quality in Principal Aquifers of the United States, 1991–2010*. USGS Circular 1360.

DeSimone, L. A., P. A. Hamilton, and R. J. Gilliom. 2009. *Quality of Water from Domestic Wells in Principal Aquifers of the United States, 1991–2004: Overview of Major Findings*. US Geological Survey Circular 1332.

Destouni, G., F. Jaramillo, and C. Prieto. 2013. "Hydroclimatic Shifts Driven by Human Water Use for Food and Energy Production." *Nature Climate Change* 3(3): 213–17.

Dickerson-Lange, S. E., J. A. Vano, R. Gersonde, and J. D. Lundquist. 2021. "Ranking Forest Effects on Snow Storage: A Decision Tool for Forest Management." *Water Resources Research* 57(10): e2020WR027926.

Dieperink, C. 2011. "International Water Negotiations under Asymmetry, Lessons from the Rhine Chlorides Dispute Settlement (1931–2004)." *International Environmental Agreements: Politics, Law and Economics* 11(2): 139–57.

Dieter, C. A., M. A. Maupin, R. R. Caldwell, M. A. Harris, T. I. Ivahnenko, J. K. Lovelace, N. L. Barber, and K. S. Linsey. 2018. *Estimated Use of Water in the United States in 2015*. US Geological Survey Circular 1441.

Diffenbaugh, N. S., F. V. Davenport, and M. Burke. 2021. "Historical Warming Has Increased U.S. Crop Insurance Losses." *Environmental Research Letters* 16(8): 084025.

DigDeep and US Water Alliance. 2019. *Closing the Water Access Gap in the United States: A National Action Plan*. https://www.digdeep.org/close-the-water-gap

Dillehay, T. D., H. H. Eling, and J. Rossen. 2005. "Preceramic Irrigation Canals in the Peruvian Andes." *Proceedings of the National Academy of Sciences of the United States of America* 102(47): 17241–44.

Dimitrova, A., A. Gershunov, M. C. Levy, and T. Benmarhnia. 2023. "Uncovering Social and Environmental Factors that Increase the Burden of Climate-Sensitive Diarrheal Infections on Children." *Proceedings of the National Academy of Sciences* 120(3): e2119409120.

Dinar, S., and A. Dinar. 2017. *International Water Scarcity and Variability: Managing Resource Use across Political Boundaries*. Oakland: University of California Press.

Döll, P., H. Mueller Schmied, C. Schuh, F. T. Portmann, and A. Eicker. 2014. "Global-Scale Assessment of Groundwater Depletion and Related Groundwater Abstractions: Combining Hydrological Modeling with Information from Well Observations and GRACE Satellites." *Water Resources Research* 50(7): 5698–720.

Doyle, M. W., and L. A. Patterson. 2019. "Federal Decentralization and Adaptive Management of Water Resources: Reservoir Reallocation by the U.S. Army Corps of Engineers." *JAWRA Journal of the American Water Resources Association* 55(5): 1248–67.

Drechsel, P., M. Qadir, and J. Baumann. 2022a. "Water Reuse to Free Up Freshwater for Higher-Value Use and Increase Climate Resilience and Water Productivity." *Irrigation and Drainage*. 2022: 1–10.

Drechsel, P., M. Qadir, and D. Galibourg. 2022b. "The WHO Guidelines for Safe Wastewater Use in Agriculture: A Review of Implementation Challenges and Possible Solutions in the Global South." *Water* 14(6): 864.

Duflo, E., and R. Pande. 2007. "Dams." *Quarterly Journal of Economics* 122(2): 601–46.

Duncan, R. A., P. E. Gordon, B. A. Holtz, and D. Stewart. 2019. *Sample Costs to Establish an Orchard and Produce Almonds: San Joaquin Valley North, Micro-Sprinkler*. University of California Agriculture and Natural Resources Cooperative Extension, Extension Agricultural Issues Center, UC Davis Department of Agricultural and Resources Economics. https://coststudyfiles.ucdavis.edu/uploads/cs_public/79/86/79863d8a-8f8b-4379-91c9-0335e20a2dd2/2019almondssjvnorth.pdf

ECONorthwest and Martin & Nicholson Environmental Consultants. 2019. *Assessment of Potential Costs of Declining Water Levels in Great Salt Lake*. Report prepared for Great Salt Lake Advisory Council. https://documents.deq.utah.gov/water-quality/standards-technical-services/great-salt-lake-advisory-council/activities/DWQ-2019-012913.pdf

Edwards, M. B. P., and P. J. T. Roberts. 2006. "Managing Forests for Water: The South African Experience." *The International Forestry Review* 8(1): 65–71.

Eke, J., A. Yusuf, A. Giwa, and A. Sodiq. 2020. "The Global Status of Desalination: An Assessment of Current Desalination Technologies, Plants and Capacity." *Desalination* 495: 114633.

Elliott, J. R., P. L. Brown, and K. Loughran. 2020. "Racial Inequities in the Federal Buyout of Flood-Prone Homes: A Nationwide Assessment of Environmental Adaptation." *Socius* 6: 2378023120905439.

Ellis, E. C., A. H. W. Beusen, and K. K. Goldewijk. 2020. "Anthropogenic Biomes: 10,000 BCE to 2015 CE." *Land* 9(5): 129.

Ellis, E. C., N. Gauthier, K. Klein Goldewijk, R. Bliege Bird, N. Boivin, S. Díaz, D. Q. Fuller, J. L. Gill, et al. 2021. "People Have Shaped Most of Terrestrial Nature for at Least 12,000 Years." *Proceedings of the National Academy of Sciences* 118(17): e2023483118.

Environmental Integrity Project. 2022. *The Clean Water Act at 50: Promises Half Kept at the Half-Century Mark*. https://environmentalintegrity.org/reports/the-clean-water-act-at-50/

EPA. 2007. *National Management Measures to Control Nonpoint Source Pollution from Hydromodification*. EPA 841-B-07-002.

EPA. 2012. *Recreational Water Quality Criteria*. EPA 820-F-12-058.

EPA. 2019. *Strategies to Achieve Full Lead Service Line Replacement*. EPA 810-R-19-003.

Erisman, J. W., M. A. Sutton, J. Galloway, Z. Klimont, and W. Winiwarter. 2008. "How a Century of Ammonia Synthesis Changed the World." *Nature Geoscience* 1(10): 636–9.

Espinoza, V., D. E. Waliser, B. Guan, D. A. Lavers, and F. M. Ralph. 2018. "Global Analysis of Climate Change Projection Effects on Atmospheric Rivers." *Geophysical Research Letters* 45(9): 4299–308.

Evans, S., C. Campbell, and O. V. Naidenko. 2019. "Cumulative Risk Analysis of Carcinogenic Contaminants in United States Drinking Water." *Heliyon* 5: e02314.

Evenari, M., L. Shanan, N. Tadmor, and Y. Aharoni. 1961. "Ancient Agriculture in the Negev." *Science* 133(3457): 979–96.

Exley, J. L. R., B. Liseka, O. Cumming, and J. H. J. Ensink. 2015. "The Sanitation Ladder: What Constitutes an Improved Form of Sanitation?" *Environmental Science & Technology* 49(2): 1086–94.

Faunt, C. C., M. Sneed, J. Traum, and J. T. Brandt. 2016. "Water Availability and Land Subsidence in the Central Valley, California, USA." *Hydrogeology Journal* 24(3): 675–84.

Ferguson, G., J. C. McIntosh, O. Warr, B. Sherwood Lollar, C. J. Ballentine, J. S. Famiglietti, J.-H. Kim, J. R. Michalski, et al. 2021. "Crustal Groundwater Volumes Greater than Previously Thought." *Geophysical Research Letters* 48(16): e2021GL093549.

Filoso, S., M. O. Bezerra, K. C. B. Weiss, and M. A. Palmer. 2017. "Impacts of Forest Restoration on Water Yield: A Systematic Review." *PLoS One* 12(8): e0183210.

Flammer, P. G., H. Ryan, S. G. Preston, S. Warren, R. Přichystalová, R. Weiss, V. Palmowski, S. Boschert, et al. 2020. "Epidemiological Insights from a Large-Scale Investigation of Intestinal Helminths in Medieval Europe." *PLOS Neglected Tropical Diseases* 14(8): e0008600.

Flörke, M., C. Schneider, and R. I. McDonald. 2018. "Water Competition between Cities and Agriculture Driven by Climate Change and Urban Growth." *Nature Sustainability* 1(1): 51–8.

Foley, M. M., J. R. Bellmore, J. E. O'Connor, J. J. Duda, A. E. East, G. E. Grant, C. W. Anderson, J. A. Bountry, et al. 2017. "Dam Removal: Listening In." *Water Resources Research* 53(7): 5229–46.

Friedrich, K., K. Ikeda, S. A. Tessendorf, J. R. French, R. M. Rauber, B. Geerts, L. Xue, R. M. Rasmussen, et al. 2020. "Quantifying Snowfall from Orographic Cloud Seeding." *Proceedings of the National Academy of Sciences* 117(10): 5190–5.

Furlong, C. 2016. *SFD Promotion Initiative: Lima, Peru*. Final Report. https://sfd.susana.org/about/worldwide-projects/city/52-lima

Furlong, K., N. Petter Gleditsch, and H. Hegre. 2006. "Geographic Opportunity and Neomalthusian Willingness: Boundaries, Shared Rivers, and Conflict." *International Interactions* 32(1): 79–108.

GAO. 2013. *Clean Water Act: Changes Needed If Key EPA Program Is to Help Fulfill the Nation's Water Quality Goals*. GAO-14-80.

GAO. 2019. *Superfund: EPA Should Take Additional Actions to Manage Risks from Climate Change*. GAO-20-73.

GAO. 2020a. *Abandoned Hardrock Mines: Information on Number of Mines, Expenditures, and Factors That Limit Efforts to Address Hazards*. GAO-20-238.

GAO. 2020b. *Drinking Water: EPA Could Use Available Data to Better Identify Neighborhoods at Risk of Lead Exposure*. GAO-21-78.

Garrick, D., L. De Stefano, W. Yu, I. Jorgensen, E. O'Donnell, L. Turley, I. Aguilar-Barajas, X. P. Dai, et al. 2019. "Rural Water for Thirsty Cities: A Systematic Review of Water Reallocation from Rural to Urban Regions." *Environmental Research Letters* 14(4): 14.

Garrick, D. E., and L. De Stefano. 2016. "Adaptive Capacity in Federal Rivers: Coordination Challenges and Institutional Responses." *Current Opinion in Environmental Sustainability* 21: 78–85.

Ge, S. M., M. A. Liu, N. Lu, J. W. Godt, and G. Luo. 2009. "Did the Zipingpu Reservoir trigger the 2008 Wenchuan earthquake?" *Geophysical Research Letters* 36: 5.

Gebrehiwot, S. G., D. Ellison, W. Bewket, Y. Seleshi, B.-I. Inogwabini, and K. Bishop. 2019. "The Nile Basin Waters and the West African Rainforest: Rethinking the Boundaries." *WIREs Water* 6(1): e1317.

Gephart, J. A., K. F. Davis, K. A. Emery, A. M. Leach, J. N. Galloway, and M. L. Pace. 2016. "The Environmental Cost of Subsistence: Optimizing Diets to Minimize Footprints." *Science of the Total Environment* 553: 120–7.

Gephart, J. A., P. J. G. Henriksson, R. W. R. Parker, A. Shepon, K. D. Gorospe, K. Bergman, G. Eshel, C. D. Golden, et al. 2021. "Environmental Performance of Blue Foods." *Nature* 597(7876): 360–5.

Gerbens-Leenes, P. W., M. M. Mekonnen, and A. Y. Hoekstra. 2013. "The Water Footprint of Poultry, Pork and Beef: A Comparative Study in Different Countries and Production Systems." *Water Resources and Industry* 1–2: 25–36.

Giannini, A., and A. Kaplan. 2019. "The Role of Aerosols and Greenhouse Gases in Sahel Drought and Recovery." *Climatic Change* 152(3–4): 449–66.

Gibson, J. M., M. Fisher, A. Clonch, J. M. MacDonald, and P. J. Cook. 2020. "Children Drinking Private Well Water Have Higher Blood Lead than Those with City Water." *Proceedings of the National Academy of Sciences* 117: 16898–907.

Giordano, M., A. Drieschova, J. A. Duncan, Y. Sayama, L. De Stefano, and A. T. Wolf. 2014. "A Review of the Evolution and State of Transboundary Freshwater Treaties." *International Environmental Agreements: Politics, Law and Economics* 14(3): 245–64.

Giordano, M., and T. Shah. 2014. "From IWRM Back to Integrated Water Resources Management." *International Journal of Water Resources Development* 30(3): 364–76.

Gleick, P., and C. Iceland. 2018. *Water, Security, and Conflict*. World Resources Institute and Pacific Institute issue brief. https://www.wri.org/research/water-security-and-conflict

Gleick, P. H. 2003. "Global Freshwater Resources: Soft-Path Solutions for the 21st Century." *Science* 302(5650): 1524–8.

Gleick, P. H. 2014. "Water, Drought, Climate Change, and Conflict in Syria." *Weather, Climate, and Society* 6(3): 331–40.

Gleick, P. H., and H. Cooley. 2021. "Freshwater Scarcity." *Annual Review of Environment and Resources*. 46: 319–48.

Global Water Intelligence and International Desalination Association. 2022. *IDA Desalination and Reuse Handbook 2022–2023*. Oxford, UK: Media Analytics Limited.

Gorostiza, S., and D. Saurí. 2019. "Naturalizing Pollution: A Critical Social Science View on the Link between Potash Mining and Salinization in the Llobregat River Basin, Northeast Spain." *Philosophical Transactions of the Royal Society B: Biological Sciences* 374(1764): 20180006.

Goyal, K., and A. Kumar. 2021. "Development of Water Reuse: A Global Review with the Focus on India." *Water Science and Technology* 84: 3172–90.

Grafton, R. Q., and J. Horne. 2014. "Water Markets in the Murray–Darling Basin." *Agricultural Water Management* 145: 61–71.

Grafton, R. Q., G. D. Libecap, E. C. Edwards, R. J. O'Brien, and C. Landry. 2012. "Comparative Assessment of Water Markets: Insights from the Murray–Darling Basin of Australia and the Western USA." *Water Policy* 14(2): 175–93.

Greeley, J. 2017. "Water in Native American Spirituality: Liquid Life-Blood of the Earth and Life of the Community." *Green Humanities* 2: 156–79.

Grey, D., and C. W. Sadoff. 2007. "Sink or Swim? Water Security for Growth and Development." *Water Policy* 9(6): 545–71.

Griffin, R. C. 2016. *Water Resource Economics: The Analysis of Scarcity, Policies, and Projects*. Cambridge, MA: MIT Press.

Grubert, E. A. 2016. "Water Consumption from Hydroelectricity in the United States." *Advances in Water Resources* 96: 88–94.

Guiteras, R., J. Levinsohn, and A. M. Mobarak. 2015. "Encouraging Sanitation Investment in the Developing World: A Cluster-Randomized Trial." *Science* 348: 903–6.

Gulev, S. K., P. W. Thorne, J. Ahn, F. J. Dentener, C. M. Domingues, S. Gerland, D. Gong, D. S. Kaufman, H. C. Nnamchi, J. Quaas, J. A. Rivera, S. Sathyendranath, S. L. Smith, B. Trewin, K. von Schuckmann, and R. S. Vose. 2021. "Chapter 2. Changing State of the Climate System." Pp. 287–422 in *Climate Change 2021: The Physical Science Basis. Contribution of Working Group I to the Sixth Assessment Report of the Intergovernmental Panel on Climate Change*. V. Masson-Delmotte, P. Zhai, A. Pirani, S. L. Connors, C. Péan, S. Berger, N. Caud, Y. Chen, L. Goldfarb, M. I. Gomis, M. Huang, K. Leitzell, E. Lonnoy, J. B. R. Matthews, T. K. Maycock, T. Waterfield, O. Yelekçi, R. Yu, and B. Zhou, eds. Cambridge, MA: Cambridge University Press.

Guo, L. L., T. W. Li, D. L. Chen, J. G. Liu, B. He, and Y. F. Zhang. 2021. "Links between Global Terrestrial Water Storage and Large-Scale Modes of Climatic Variability." *Journal of Hydrology* 598: 126419.

Haag, W. R., and J. D. Williams. 2014. "Biodiversity on the Brink: An Assessment of Conservation Strategies for North American Freshwater Mussels." *Hydrobiologia* 735(1): 45–60.

Haghighi, E., K. Madani, and A. Y. Hoekstra. 2018. "The Water Footprint of Water Conservation Using Shade Balls in California." *Nature Sustainability* 1(7): 358–60.

Hämäläinen, P., J. Takala, and T. B. Kiat. 2017. *Global Estimates of Occupational Accidents and Work-Related Illnesses 2017*. Singapore: Workplace Safety and Health Institute.

Hanemann, M., and M. Young. 2020. "Water Rights Reform and Water Marketing: Australia vs the US West." *Oxford Review of Economic Policy* 36(1): 108–31.

Hansen, J., B. Jurgens, and M. S. Fram. 2018. "Quantifying Anthropogenic Contributions to Century-Scale Groundwater Salinity Changes, San Joaquin Valley, California, USA." *Science of the Total Environment* 642: 125–36.

Hashizume, M., Y. Wagatsuma, A. S. G. Faruque, T. Hayashi, P. R. Hunter, B. Armstrong, and D. A. Sack. 2008. "Factors Determining Vulnerability to Diarrhoea during and after Severe Floods in Bangladesh." *Journal of Water and Health* 6(3): 323–32.

Heine, R. A., and N. Pinter. 2012. "Levee Effects upon Flood Levels: An Empirical Assessment." *Hydrological Processes* 26(21): 3225–40.

Heistermann, M. 2017. "HESS Opinions: A Planetary Boundary on Freshwater Use Is Misleading." *Hydrology and Earth System Sciences* 21(7): 3455–61.

Helms, S. W. 1981. *Jawa: Lost City of the Black Desert*. Ithaca, NY: Cornell University Press.

Hendryx, M., and J. H. Luo. 2015. "An Examination of the Effects of Mountaintop Removal Coal Mining on Respiratory Symptoms and COPD Using Propensity Scores." *International Journal of Environmental Health Research* 25(3): 265–76.

Henry, C. L., and L. F. Pratson. 2019. "Differentiating the Effects of Climate Change-Induced Temperature and Streamflow Changes on the Vulnerability of Once-Through Thermoelectric Power Plants." *Environmental Science & Technology* 53(7): 3969–76.

Hertwich, E. G. 2013. "Addressing Biogenic Greenhouse Gas Emissions from Hydropower in LCA." *Environmental Science & Technology* 47(17): 9604–11.

Hock, R., G. Rasul, C. Adler, B. Cáceres, S. Gruber, Y. Hirabayashi, M. Jackson, A. Kääb, S. Kang, S. Kutuzov, Al. Milner, U. Molau, S. Morin, B. Orlove, and H. Steltzer. 2019. "Chapter 2. High Mountain Areas." Pp. 131–202 in *IPCC Special Report on the Ocean and Cryosphere in a Changing Climate*. D. C. R. H.-O. Pörtner, V. Masson-Delmotte, P. Zhai, M. Tignor, E. Poloczanska, K. Mintenbeck, A. Alegría, M. Nicolai, A. Okem, J. Petzold, B. Rama, and N. M. Weyer, eds. Cambridge, UK: Cambridge University Press.

Hoekstra, A. Y. 2017. "Water Footprint Assessment: Evolvement of a New Research Field." *Water Resources Management* 31(10): 3061–81.

Hoekstra, A. Y., and M. M. Mekonnen. 2012. "The Water Footprint of Humanity." *Proceedings of the National Academy of Sciences of the United States of America* 109(9): 3232–7.

Hoffmann, S., U. Feldmann, P. M. Bach, C. Binz, M. Farrelly, N. Frantzeskaki, H. Hiessl, J. Inauen, et al. 2020. "A Research Agenda for the Future of Urban Water Management: Exploring the Potential of Nongrid, Small-Grid, and Hybrid Solutions." *Environmental Science & Technology* 54(9): 5312–22.

Hrozencik, R. A., and M. Aillery. 2021. *Trends in U.S. Irrigated Agriculture: Increasing Resilience Under Water Supply Scarcity*. US Department of Agriculture, Economic Research Service, Economic Information Bulletin 229.

Huang, J., Y. Li, C. Fu, F. Chen, Q. Fu, A. Dai, M. Shinoda, Z. Ma, et al. 2017. "Dryland Climate Change: Recent Progress and Challenges." *Reviews of Geophysics* 55(3): 719–78.

Hughes, R. M. 2015. "Recreational Fisheries in the USA: Economics, Management Strategies, and Ecological Threats." *Fisheries Science* 81(1): 1–9.

Human Rights Watch. 1995. *The Three Gorges Dam in China: Forced Resettlement, Suppression of Dissent and Labor Rights Concerns*. https://www.hrw.org/legacy/summaries/s.china952.html

Human Rights Watch. 2019. *Basra Is Thirsty: Iraq's Failure to Manage the Water Crisis*. https://www.hrw.org/report/2019/07/22/basra-thirsty/iraqs-failure-manage-water-crisis

Hummel, M. A., M. S. Berry, and M. T. Stacey. 2018. "Sea Level Rise Impacts on Wastewater Treatment Systems along the US Coasts." *Earth's Future* 6(4): 622–33.

Hutton, G. 2013. "Global Costs and Benefits of Reaching Universal Coverage of Sanitation and Drinking-Water Supply." *Journal of Water and Health* 11(1): 1–12.

Hutton, G., and M. Varughese. 2016. *The Costs of Meeting the 2030 Sustainable Development Goal Targets on Drinking Water, Sanitation, and Hygiene.* Water and Sanitation Program and World Bank. https://www.worldbank.org/en/topic/water/publication/the-costs-of-meeting-the-2030-sustainable-development-goal-targets-on-drinking-water-sanitation-and-hygiene

Hutton, G., and M. Varughese. 2020. *Global and Regional Costs of Achieving Universal Access to Sanitation to Meet SDG Target 6.2.* New York: UNICEF.

Interstate Council on Water Policy. 2020. *Interstate Water Solutions: Lessons from the Past and Recommendations for the Future: A Look toward 2050.* https://icwp.org/wp-content/uploads/2020/09/Interstate-Water-Solutions_Lessons-Recommedations_Full-Report_FINAL_9_29_20.pdf

Interstate Council on Water Policy. 2021. "State Water Plan Review." https://icwp.org/wp-content/uploads/2021/06/2021-ICWP-State-Water-Plan-Report.pdf

Irvine, K. N., L. H. C. Chua, and H. S. Eikass. 2014. "The Four National Taps of Singapore: A Holistic Approach to Water Resources Management from Drainage to Drinking Water." *Journal of Water Management Modeling* C375.

Izhitskiy, A. S., P. O. Zavialov, P. V. Sapozhnikov, G. B. Kirillin, H. P. Grossart, O. Y. Kalinina, A. K. Zalota, I. V. Goncharenko, et al. 2016. "Present State of the Aral Sea: Diverging Physical and Biological Characteristics of the Residual Basins." *Scientific Reports* 6(1): 23906.

Jägermeyr, J., D. Gerten, J. Heinke, S. Schaphoff, M. Kummu, and W. Lucht. 2015. "Water Savings Potentials of Irrigation Systems: Global Simulation of Processes and Linkages." *Hydrology and Earth System Sciences* 19(7): 3073–91.

Jägermeyr, J., D. Gerten, S. Schaphoff, J. Heinke, W. Lucht, and J. Rockström. 2016. "Integrated Crop Water Management Might Sustainably Halve the Global Food Gap." *Environmental Research Letters* 11(2): 14.

Jain, M., R. Fishman, P. Mondal, G. L. Galford, N. Bhattarai, S. Naeem, U. Lall, S. Balwinder, et al. 2021. "Groundwater Depletion Will Reduce Cropping Intensity in India." *Science Advances* 7(9): eabd2849.

Jalava, M., J. H. A. Guillaume, M. Kummu, M. Porkka, S. Siebert, and O. Varis. 2016. "Diet Change and Food Loss Reduction: What Is Their Combined Impact on Global Water Use and Scarcity?" *Earth's Future* 4(3): 62–78.

James, T., A. Evans, E. Madly, and C. Kelly. 2014. *The Economic Importance of the Colorado River to the Basin Region.* L William Seidman Research Institute, W. P. Carey School of Business, Arizona State University. https://businessforwater.org/wp-content/uploads/2016/12/PTF-Final-121814.pdf

Jaramillo, F., and G. Destouni. 2015. "Comment on 'Planetary Boundaries: Guiding Human Development on a Changing Planet.'" *Science* 348(6240): 1217.

Jasechko, S., D. Perrone, H. Seybold, Y. Fan, and J. W. Kirchner. 2020. "Groundwater Level Observations in 250,000 Coastal US Wells Reveal Scope of Potential Seawater Intrusion." *Nature Communications* 11(1): 3229.

Javidi, A., and G. Pierce. 2018. "U.S. Households' Perception of Drinking Water as Unsafe and Its Consequences: Examining Alternative Choices to the Tap." *Water Resources Research* 54(9): 6100–13.

Jenkins, M. 2017. *The Impact of Corruption on Access to Safe Water and Sanitation for People Living in Poverty.* Bergen, Norway: U4 Anti-Corruption Resource Centre. https://www.u4.no/publications/the-impact-of-corruption-on-access-to-safe-water-and-sanitation-for-people-living-in-poverty

Jimenez, B., P. Drechsel, D. Kone, A. Bahri, L. Raschid-Sally, and M. Qadir. 2010. "Wastewater, Sludge and Excreta Use in Developing Countries: An Overview." In *Wastewater Irrigation and Health: Assessing and Mitigating Risk in Low-Income Countries*. P. Drechsel, L. Raschid-Sally, M. Redwood, A. Bahri and C. A. Scott, eds. London: Earthscan: 3–28.

Johnstone, J. A., and T. E. Dawson. 2010. "Climatic Context and Ecological Implications of Summer Fog Decline in the Coast Redwood Region." *Proceedings of the National Academy of Sciences of the United States of America* 107(10): 4533–8.

Jones, E., M. Qadir, M. T. H. van Vliet, V. Smakhtin, and S. M. Kang. 2019. "The State of Desalination and Brine Production: A Global Outlook." *Science of the Total Environment* 657: 1343–56.

Jones, R. K., A. Baras, A. A. Saeeri, A. A. Qahtani, A. O. A. Amoudi, Y. A. Shaya, M. Alodan, and S. A. Al-Hsaien. 2016. "Optimized Cleaning Cost and Schedule Based on Observed Soiling Conditions for Photovoltaic Plants in Central Saudi Arabia." *IEEE Journal of Photovoltaics* 6(3): 730–8.

Jongman, B., P. J. Ward, and J. C. J. H. Aerts. 2012. "Global Exposure to River and Coastal Flooding: Long Term Trends and Changes." *Global Environmental Change* 22(4): 823–35.

Jongman, B., H. C. Winsemius, J. Aerts, E. C. de Perez, M. K. van Aalst, W. Kron, and P. J. Ward. 2015. "Declining Vulnerability to River Floods and the Global Benefits of Adaptation." *Proceedings of the National Academy of Sciences of the United States of America* 112(18): E2271–80.

Jordan, C. E., and E. Fairfax. 2022. "Beaver: The North American Freshwater Climate Action Plan." *WIREs Water* 9(4): e1592.

Karger, D. N., M. Kessler, M. Lehnert, and W. Jetz. 2021. "Limited Protection and Ongoing Loss of Tropical Cloud Forest Biodiversity and Ecosystems Worldwide." *Nature Ecology & Evolution* 5: 854–62.

Katz, D. (2021). "Desalination and Hydrodiplomacy: Refreshening Transboundary Water Negotiations or Adding Salt to the Wounds?" *Environmental Science & Policy* 116: 171–80.

Keiser, D. A., and J. S. Shapiro. 2019. "Consequences of the Clean Water Act and the Demand for Water Quality." *Quarterly Journal of Economics* 134(1): 349–96.

Kelley, C. P., S. Mohtadi, M. A. Cane, R. Seager, and Y. Kushnir. 2015. "Climate Change in the Fertile Crescent and Implications of the Recent Syrian Drought." *Proceedings of the National Academy of Sciences* 112: 3241–6.

Keys, P. W., M. Porkka, L. Wang-Erlandsson, I. Fetzer, T. Gleeson, and L. J. Gordon. 2019. "Invisible Water Security: Moisture Recycling and Water Resilience." *Water Security* 8: 100046.

Keys, P. W., L. Wang-Erlandsson, and L. J. Gordon. 2018. "Megacity Precipitationsheds Reveal Tele-Connected Water Security Challenges." *PLoS One* 13(3): e0194311.

Kimmerer, R. W. 2013. *Braiding Sweetgrass: Indigenous Wisdom, Scientific Knowledge, and the Teachings of Plants*. Minneapolis, MN: Milkweed Editions.

Kirchhoff, C. J., and L. Dilling. 2016. "The Role of U.S. States in Facilitating Effective Water Governance under Stress and Change." *Water Resources Research* 52(4): 2951–64.

Kishimoto, S., E. Lobina, and O. Petitjean (Eds.). 2015. *Our Public Water Future: The Global Experience with Remunicipalisation*. Amsterdam: Transnational Institute, Public Services International Research Unit, Multinationals Observatory, Municipal Services Project, and the European Federation of Public Service Unions.

Klare, M. T. 2001. "The New Geography of Conflict." *Foreign Affairs* 80(3): 49+.

Klein, C. A., and S. B. Zellmer. 2014. *Mississippi River Tragedies: A Century of Unnatural Disaster*. New York: New York University Press.

Kohli, A., and K. Frenken. 2015. *Evaporation From Artificial Lakes and Reservoirs*. FAO AQUASTAT Report. Rome: Food and Agriculture Organization.

Konar, M., C. Dalin, N. Hanasaki, A. Rinaldo, and I. Rodriguez-Iturbe. 2012. "Temporal Dynamics of Blue and Green Virtual Water Trade Networks." *Water Resources Research* 48(7): W07509.

Kondolf, G. M., R. J. P. Schmitt, P. Carling, S. Darby, M. Arias, S. Bizzi, A. Castelletti, T. A. Cochrane, et al. 2018. "Changing Sediment Budget of the Mekong: Cumulative Threats and Management Strategies for a Large River Basin." *Science of the Total Environment* 625: 114–34.

Konisky, D. M., C. Reenock, and S. Conley. 2021. "Environmental Injustice in Clean Water Act Enforcement: Racial and Income Disparities in Inspection Time." *Environmental Research Letters* 16(8): 084020.

Kumpel, E., and K. L. Nelson. 2016. "Intermittent Water Supply: Prevalence, Practice, and Microbial Water Quality." *Environmental Science & Technology* 50(2): 542–53.

Kundzewicz, Z. W., M. Szwed, and I. Pinskwar. 2019. "Climate Variability and Floods: A Global Review." *Water* 11(7): 24.

Lagarde, F., C. Beausoleil, S. M. Belcher, L. P. Belzunces, C. Emond, M. Guerbet, and C. Rousselle. 2015. "Non-Monotonic Dose-Response Relationships and Endocrine Disruptors: A Qualitative Method of Assessment." *Environmental Health* 14(1): 13.

Lansing, J. S., and T. A. de Vet. 2012. "The Functional Role of Balinese Water Temples: A Response to Critics." *Human Ecology* 40(3): 453–67.

Lansing, J. S., S. Thurner, N. N. Chung, A. Coudurier-Curveur, Ç. Karakaş, K. A. Fesenmyer, and L. Y. Chew. 2017. "Adaptive Self-Organization of Bali's Ancient Rice Terraces." *Proceedings of the National Academy of Sciences* 114(25): 6504–9.

Lark, T. J., N. P. Hendricks, A. Smith, N. Pates, S. A. Spawn-Lee, M. Bougie, E. G. Booth, C. J. Kucharik, et al. 2022. "Environmental Outcomes of the US Renewable Fuel Standard." *Proceedings of the National Academy of Sciences* 119(9): e2101084119.

Larsen, T. A., S. Hoffmann, C. Luthi, B. Truffer, and M. Maurer. 2016. "Emerging Solutions to the Water Challenges of an Urbanizing World." *Science* 352(6288): 928–33.

Lassaletta, L., G. Billen, B. Grizzetti, J. Anglade, and J. Garnier. 2014. "50 Year Trends in Nitrogen Use Efficiency of World Cropping Systems: The Relationship between Yield and Nitrogen Input to Cropland." *Environmental Research Letters* 9(10): 9.

Lawson, M. L. 2009. *Dammed Indians Revisited: The Continuing History of the Pick–Sloan Plan and the Missouri River Sioux*. Pierre, SD: South Dakota State Historical Society Press.

Lebel, L., and P. Garden. 2008. Deliberation, Negotiation and Scale in the Governance of Water Resources in the Mekong Region. In *Adaptive and Integrated Water Management: Coping with Complexity and Uncertainty*. C. Pahl-Wostl, P. Kabat, and J. Möltgen. ed., Berlin, Heidelberg: Springer: 205–225.

Lèbre, É., M. Stringer, K. Svobodova, J. R. Owen, D. Kemp, C. Côte, A. Arratia-Solar, and R. K. Valenta. 2020. "The Social and Environmental Complexities of Extracting Energy Transition Metals." *Nature Communications* 11(1): 4823.

Ledec, G., and J. D. Quintero. 2003. *Good Dams and Bad Dams: Environmental Criteria for Site Selection of Hydroelectric Projects*. The World Bank. https://documents1.worldbank.org/curated/en/224701468332373651/pdf/303600NWP0Good000010Box18600PUBLIC0.pdf

Lee, J. A. 2021. "Turning Participation into Power: A Water Justice Case Study." *Environmental Law Reporter*. https://www.eli.org/sites/default/files/files-pdf/ParticipationIntoPower.pdf

Leopold, A. 1949. *A Sand County Almanac: And Sketches Here and There*. New York: Oxford University Press.

Levintal, E., M. L. Kniffin, Y. Ganot, N. Marwaha, N. P. Murphy, and H. E. Dahlke. 2023. "Agricultural Managed Aquifer Recharge (Ag-MAR): A Method for Sustainable Groundwater Management: A Review." *Critical Reviews in Environmental Science and Technology* 53(3): 291–314.

Levy, Z. F., B. C. Jurgens, K. R. Burow, S. A. Voss, K. E. Faulkner, J. A. Arroyo-Lopez, and M. S. Fram. 2021. "Critical Aquifer Overdraft Accelerates Degradation of Groundwater Quality in California's Central Valley during Drought." *Geophysical Research Letters* 48(17): e2021GL094398.

Lichatowich, J. 2013. *Salmon, People, and Place: A Biologist's Search for Salmon Recovery*. Corvallis: Oregon State University Press.

Liu, N., G. R. Dobbs, P. V. Caldwell, C. F. Miniat, G. Sun, K. Duan, S. A. C. Nelson, P. V. Bolstad, et al. 2022. "Inter-Basin Transfers Extend the Benefits of Water from Forests to Population Centers across the Conterminous U.S." *Water Resources Research* 58(5): e2021WR031537.

Lo, M. H., and J. S. Famiglietti. 2013. "Irrigation in California's Central Valley Strengthens the Southwestern U.S. Water Cycle." *Geophysical Research Letters* 40(2): 301–6.

Logan, T. M., D. Guikema, and J. D. Bricker. 2018. "Hard-Adaptive Measures Can Increase Vulnerability to Storm Surge and Tsunami Hazards over Time." *Nature Sustainability* 1(9): 526–30.

London, J., A. Fencl, S. Watterson, J. Jarin, A. Aranda, A. King, C. Pannu, P. Seaton, et al. 2018. *The Struggle for Water Justice in California's San Joaquin Valley: A Focus on Disadvantaged Unincorporated Communities.* UC Davis Center for Regional Change. https://regionalchange.ucdavis.edu/sites/g/files/dgvnsk986/files/inline-files/The%20Struggle%20for%20Water%20Justice%20FULL%20REPORT.pdf

Lopes, A. F., J. L. Macdonald, P. Quinteiro, L. Arroja, C. Carvalho-Santos, M. A. Cunha-e-Sá, and A. C. Dias. 2019. "Surface vs. Groundwater: The Effect of Forest Cover on the Costs of Drinking Water." *Water Resources and Economics* 28: 100123.

Lu, C. Q., and H. Q. Tian. 2017. "Global Nitrogen and Phosphorus Fertilizer Use for Agriculture Production in the Past Half Century: Shifted Hot Spots and Nutrient Imbalance." *Earth System Science Data* 9(1): 181–92.

Luijendijk, E., T. Gleeson, and N. Moosdorf. 2020. "Fresh Groundwater Discharge Insignificant for the World's Oceans but Important for Coastal Ecosystems." *Nature Communications* 11(1): 1260.

Lusardi, R. A., and P. B. Moyle. 2017. "Two-Way Trap and Haul as a Conservation Strategy for Anadromous Salmonids." *Fisheries* 42(9): 478–87.

MacDonald Gibson, J., N. DeFelice, D. Sebastian, and H. Leker. 2014. "Racial Disparities in Access to Community Water Supply Service in Wake County, North Carolina." *American Journal of Public Health* 104(12): e45–e45.

MacDonnell, L. J. 2015. "Rethinking the Use of General Stream Adjudications." *Wyoming Law Review* 15(2): 347–81.

Macfarlane, D. 2020. *Fixing Niagara Falls: Environment, Energy, and Engineers at the World's Most Famous Waterfall.* Vancouver: UBC Press.

Mach, K. J., C. M. Kraan, M. Hino, A. R. Siders, E. M. Johnston, and C. B. Field. 2019. "Managed Retreat through Voluntary Buyouts of Flood-Prone Properties." *Science Advances* 5(10): eaax8995.

Mack, E. A., and S. Wrase. 2017. "A Burgeoning Crisis? A Nationwide Assessment of the Geography of Water Affordability in the United States." *PLoS One* 12(1): e0169488.

Ma'Mun, S. R., A. Loch, and M. D. Young. 2020. "Robust Irrigation System Institutions: A Global Comparison." *Global Environmental Change-Human and Policy Dimensions* 64: 15.

Mankin, J. S., R. Seager, J. E. Smerdon, B. I. Cook, and A. P. Williams. 2019. "Mid-Latitude Freshwater Availability Reduced by Projected Vegetation Responses to Climate Change." *Nature Geoscience* 12, no. 12: 983–8.

Mankin, J. S., D. Viviroli, D. Singh, A. Y. Hoekstra, and N. S. Diffenbaugh. 2015. "The Potential for Snow to Supply Human Water Demand in the Present and Future." *Environmental Research Letters* 10(11): 114016.

Mapulanga, A. M., and H. Naito. 2019. "Effect of Deforestation on Access to Clean Drinking Water." *Proceedings of the National Academy of Sciences* 116(17): 8249–54.

Marston, L., Y. Ao, M. Konar, M. M. Mekonnen, and A. Y. Hoekstra. 2018. "High-Resolution Water Footprints of Production of the United States." *Water Resources Research* 54(3): 2288–316.

Marston, L., and X. M. Cai. 2016. "An Overview of Water Reallocation and the Barriers to Its Implementation." *Wiley Interdisciplinary Reviews-Water* 3(5): 658–77.

Marston, L. T., G. Lamsal, Z. H. Ancona, P. Caldwell, B. D. Richter, B. L. Ruddell, R. R. Rushforth, and K. F. Davis. 2020. "Reducing Water Scarcity by Improving Water Productivity in the United States." *Environmental Research Letters* 15(9): 094033.

McCall, J., J. Macknick, and D. Hillman. 2016. *Water-Related Power Plant Curtailments: An Overview of Incidents and Contributing Factors*. Washington, DC: National Renewable Energy Laboratory.

McCracken, M., and A. T. Wolf. 2019. "Updating the Register of International River Basins of the World." *International Journal of Water Resources Development* 35(5): 732–77.

McDonald, R. I., K. F. Weber, J. Padowski, T. Boucher, and D. Shemie. 2016. "Estimating Watershed Degradation over the Last Century and Its Impact on Water-Treatment Costs for the World's Large Cities." *Proceedings of the National Academy of Sciences* 113(32): 9117–22.

McDonald, R. I., K. Weber, J. Padowski, M. Flörke, C. Schneider, P. A. Green, T. Gleeson, S. Eckman, et al. 2014. "Water on an Urban Planet: Urbanization and the Reach of Urban Water Infrastructure." *Global Environmental Change* 27: 96–105.

McKenna, M. L., S. McAtee, P. E. Bryan, R. Jeun, T. Ward, J. Kraus, M. E. Bottazzi, P. J. Hotez, et al. 2017. "Human Intestinal Parasite Burden and Poor Sanitation in Rural Alabama." *American Journal of Tropical Medicine and Hygiene* 97(5): 1623–8.

McKuin, B., A. Zumkehr, J. Ta, R. Bales, J. H. Viers, T. Pathak, and J. E. Campbell. 2021. "Energy and Water Co-Benefits from Covering Canals with Solar Panels." *Nature Sustainability* 4(7): 609–17.

McMahon, A. 2015. Waste Management in Early Urban Southern Mesopotamia. In *Sanitation, Latrines, and Intestinal Parasites in Past Populations*. P. D. Mitchell, ed. Burlington, VT: Ashgate Publishing Company.

McMahon, T. A., G. G. S. Pegram, R. M. Vogel, and M. C. Peel. 2007. "Review of Gould–Dincer Reservoir Storage–Yield–Reliability Estimates." *Advances in Water Resources* 30(9): 1873–82.

Meehan, K., J. R. Jurjevich, N. M. J. W. Chun, and J. Sherrill. 2020. "Geographies of Insecure Water Access and the Housing–Water Nexus in US Cities." *Proceedings of the National Academy of Sciences* 117(46): 28700–7.

Mekonnen, M. M., and A. Y. Hoekstra. 2011. "The Green, Blue and Grey Water Footprint of Crops and Derived Crop Products." *Hydrology and Earth System Sciences* 15(5): 1577–600.

Mekonnen, M. M., and A. Y. Hoekstra. 2012. "A Global Assessment of the Water Footprint of Farm Animal Products." *Ecosystems* 15(3): 401–15.

Meldrum, J., S. Nettles-Anderson, G. Heath, and J. Macknick. 2013. "Life Cycle Water Use for Electricity Generation: A Review and Harmonization of Literature Estimates." *Environmental Research Letters* 8(1): 18.

Melymuk, L., J. Blumenthal, O. Sáňka, A. Shu-Yin, V. Singla, K. Šebková, K. Pullen Fedinick, and M. L. Diamond. 2022. "Persistent Problem: Global Challenges to Managing PCBs." *Environmental Science & Technology* 56: 9029–40.

Mickley, M. 2018. *Updated and Extended Survey of U.S. Municipal Desalination Plants*. US Bureau of Reclamation, Desalination and Water Purification Research and Development Program Report no. 207.

Milly, P. C. D., J. Betancourt, M. Falkenmark, R. M. Hirsch, Z. W. Kundzewicz, D. P. Lettenmaier, and R. J. Stouffer. 2008. "Climate Change—Stationarity Is Dead: Whither Water Management?" *Science* 319(5863): 573–4.

Milly, P. C. D., J. Betancourt, M. Falkenmark, R. M. Hirsch, Z. W. Kundzewicz, D. P. Lettenmaier, R. J. Stouffer, M. D. Dettinger, et al. 2015. "On Critiques of "Stationarity Is Dead: Whither Water Management?" *Water Resources Research* 51(9): 7785–7789.

Milly, P. C. D., and K. A. Dunne. 2020. "Colorado River Flow Dwindles as Warming-Driven Loss of Reflective Snow Energizes Evaporation." *Science* 367: 1252–5.

Mithen, S. 2012. *Thirst: Water and Power in the Ancient World*. Cambridge, MA: Harvard University Press.

Modesto, V., M. Ilarri, A. T. Souza, M. Lopes-Lima, K. Douda, M. Clavero, and R. Sousa. 2018. "Fish and Mussels: Importance of Fish for Freshwater Mussel Conservation." *Fish and Fisheries* 19(2): 244–59.

Molle, F. 2009. "River-Basin Planning and Management: The Social Life of a Concept." *Geoforum* 40(3): 484–94.

Morris-Iveson, L., E. Granillo, and S. Grundin. 2021. *Water Under Fire Volume 3: Attacks on Water and Sanitation Services in Armed Conflict and the Impacts on Children.* New York: UNICEF.

Mosse, D. 2003. *The Rule of Water: Statecraft, Ecology, and Collective Action in South India.* New York: Oxford University Press.

Mueller, N. D., J. S. Gerber, M. Johnston, D. K. Ray, N. Ramankutty, and J. A. Foley. 2012. "Closing Yield Gaps through Nutrient and Water Management." *Nature* 490(7419): 254–7.

Mulligan, M., A. van Soesbergen, and L. Sáenz. 2020. "GOODD, a Global Dataset of More Than 38,000 Georeferenced Dams." *Scientific Data* 7(1): 31.

Mullin, M. 2020. "The Effects of Drinking Water Service Fragmentation on Drought-Related Water Security." *Science* 368(6488): 274–7.

Murray, C. J. L., A. Y. Aravkin, P. Zheng, C. Abbafati, K. M. Abbas, M. Abbasi-Kangevari, F. Abd-Allah, A. Abdelalim, et al. 2020. "Global Burden of 87 Risk Factors in 204 Countries and Territories, 1990–2019: A Systematic Analysis for the Global Burden of Disease Study 2019." *The Lancet* 396(10258): 1223–49.

Nancey Green, L., and H. Lee. 2019. "Sustainable and Resilient Urban Water Systems: The Role of Decentralization and Planning." *Sustainability* 11(3): 918.

Naslund, L. C., J. R. Gerson, A. C. Brooks, D. M. Walters, and E. S. Bernhardt (2020). "Contaminant Subsidies to Riparian Food Webs in Appalachian Streams Impacted by Mountaintop Removal Coal Mining." *Environmental Science & Technology* 54(7): 3951–9.

National Academies of Sciences, Engineering, and Medicine. 2015. *Funding and Managing the U.S. Inland Waterways System: What Policy Makers Need to Know.* Washington, DC: The National Academies Press.

National Intelligence Council. 2020. *Memorandum: Water Insecurity Threatening Global Economic Growth, Political Stability.* Report NIC-2021-02489. https://www.dni.gov/files/images/globalTrends/GT2040/NIC_2021-02489_Future_of_Water_18nov21_UNSOURCED.pdf

National Intelligence Council. 2021. "Climate Change and International Responses Increasing Challenges to US National Security through 2040." Report NIC-NIE-2021-10030-A. https://www.dni.gov/files/ODNI/documents/assessments/NIE_Climate_Change_and_National_Security.pdf

National Research Council. 2013. *Alternatives for Managing the Nation's Complex Contaminated Groundwater Sites.* Washington, DC: The National Academies Press.

Nelson, K. S., and J. Camp. 2020. "Quantifying the Benefits of Home Buyouts for Mitigating Flood Damages." *Anthropocene* 31: 100246.

Nover, D. M., M. S. Dogan, R. Ragatz, L. Booth, J. Medellin-Azuara, J. R. Lund, and J. H. Viers. 2019. "Does More Storage Give California More Water?" *Journal of the American Water Resources Association* 55(3): 759–71.

Null, S. E., A. Farshid, G. Goodrum, C. A. Gray, S. Lohani, C. N. Morrisett, L. Prudencio, and R. Sor. 2021. "A Meta-Analysis of Environmental Tradeoffs of Hydropower Dams in the Sekong, Sesan, and Srepok (3S) Rivers of the Lower Mekong Basin." *Water* 13(1): 63.

Nylen, N. G., C. Pannu, and M. Kiparsky. 2018. *Learning from California's Experience with Small Water System Consolidations: A Workshop Synthesis.* Berkeley, CA: Center for Law, Energy & the Environment, UC Berkeley School of Law.

Öberg, G., G. S. Metson, Y. Kuwayama, and S. A. Conrad. 2020. "Conventional Sewer Systems Are Too Time-Consuming, Costly and Inflexible to Meet the Challenges of the 21st Century." *Sustainability* 12(16): 17.

Ochoa-Tocachi, B. F., J. D. Bardales, J. Antiporta, K. Perez, L. Acosta, F. Mao, Z. Zulkafli, J. Gil-Rios, et al. 2019. "Potential Contributions of Pre-Inca Infiltration Infrastructure to Andean Water Security." *Nature Sustainability* 2(7): 584–93.

Ocko, I. B., and S. P. Hamburg. 2019. "Climate Impacts of Hydropower: Enormous Differences among Facilities and over Time." *Environmental Science & Technology* 53: 14070–82.

Office of the Director of National Intelligence. 2012. "Global Water Security Intelligence Community Assessment." Report ICA 2012-08. https://www.dni.gov/files/documents/Special%20Report_ICA%20Global%20Water%20Security.pdf

Ojha, C., S. Werth, and M. Shirzaei. 2019. "Groundwater Loss and Aquifer System Compaction in San Joaquin Valley during 2012–2015 Drought." *Journal of Geophysical Research: Solid Earth* 124(3): 3127–43.

Omernik, J. M., and R. G. Bailey. 1997. "Distinguishing between Watersheds and Ecoregions." *JAWRA Journal of the American Water Resources Association* 33(5): 935–49.

Ortiz-Bobea, A., T. R. Ault, C. M. Carrillo, R. G. Chambers, and D. B. Lobell. 2021. "Anthropogenic Climate Change Has Slowed Global Agricultural Productivity Growth." *Nature Climate Change* 11(4): 306–12.

Ostrom, E. 1990. *Governing the Commons: The Evolution of Institutions for Collective Action*. Cambridge, UK: Cambridge University Press.

Ostrom, E., J. Burger, C. B. Field, R. B. Norgaard, and D. Policansky. 1999. "Sustainability—Revisiting the Commons: Local Lessons, Global Challenges." *Science* 284(5412): 278–82.

Oswald, W. E., G. C. Hunter, M. R. Kramer, E. Leontsini, L. Cabrera, A. G. Lescano, and R. H. Gilman. 2014. "Provision of Private, Piped Water and Sewerage Connections and Directly Observed Handwashing of Mothers in a Peri-Urban Community of Lima, Peru." *Tropical Medicine & International Health* 19(4): 388–97.

Paltan, H., D. Waliser, W. H. Lim, B. Guan, D. Yamazaki, R. Pant, and S. Dadson. 2017. "Global Floods and Water Availability Driven by Atmospheric Rivers." *Geophysical Research Letters* 44(20): 10,387–95.

Panat, S., and K. K. Varanasi. 2022. "Electrostatic Dust Removal Using Adsorbed Moisture-Assisted Charge Induction for Sustainable Operation of Solar Panels." *Science Advances* 8(10): eabm0078.

Pataki, D. E., M. M. Carreiro, J. Cherrier, N. E. Grulke, V. Jennings, S. Pincetl, R. V. Pouyat, T. H. Whitlow, et al. 2011. "Coupling Biogeochemical Cycles in Urban Environments: Ecosystem Services, Green Solutions, and Misconceptions." *Frontiers in Ecology and the Environment* 9(1): 27–36.

Peccia, J., and P. Westerhoff. 2015. "We Should Expect More out of Our Sewage Sludge." *Environmental Science & Technology* 49(14): 8271–6.

Peña-Arancibia, J. L., L. A. Bruijnzeel, M. Mulligan, and A. van Dijk. 2019. "Forests as 'Sponges' and 'Pumps': Assessing the Impact of Deforestation on Dry-Season Flows across the Tropics." *Journal of Hydrology* 574: 946–63.

Perera, D., S. Williams, and V. Smakhtin. 2023. "Present and Future Losses of Storage in Large Reservoirs Due to Sedimentation: A Country-Wise Global Assessment." *Sustainability* 15(1): 219.

Perrone, D., and S. Jasechko. 2017. "Dry Groundwater Wells in the Western United States." *Environmental Research Letters* 12(10): 104002.

Perrone, D., and S. Jasechko. 2019. "Deeper Well Drilling an Unsustainable Stopgap to Groundwater Depletion." *Nature Sustainability* 2(8): 773–82.

Peterson, K., E. Apadula, D. Salvesen, M. Hino, R. Kihslinger, and T. K. BenDor. 2020. "A Review of Funding Mechanisms for US Floodplain Buyouts." *Sustainability* 12(23): 10112.

Peterson, T. J., M. Saft, M. C. Peel, and A. John. 2021. "Watersheds May Not Recover from Drought." *Science* 372(6543): 745–9.

Pincetl, S., E. Porse, and D. Cheng. 2016. "Fragmented Flows: Water Supply in Los Angeles County." *Environmental Management* 58(2): 208–22.

Pluth, T. B., D. A. Brose, D. W. Gallagher, and J. Wasik. 2021. "Long-Term Trends Show Improvements in Water Quality in the Chicago Metropolitan Region with Investment in Wastewater Infrastructure, Deep Tunnels, and Reservoirs." *Water Resources Research* 57(6): e2020WR028422.

Poff, N. L., J. D. Allan, M. B. Bain, J. R. Karr, K. L. Prestegaard, B. D. Richter, R. E. Sparks, and J. C. Stromberg. 1997. "The Natural Flow Regime: A Paradigm for River Conservation and Restoration." *BioScience* 47: 769–84.

Poff, N. L., B. D. Richter, A. H. Arthington, S. E. Bunn, R. J. Naiman, E. Kendy, M. Acreman, C. Apse, et al. 2010. "The Ecological Limits of Hydrologic Alteration (ELOHA): A New Framework for Developing Regional Environmental Flow Standards." *Freshwater Biology* 55(1): 147–70.

Pohl, B., A. Kramer, W. Hull, S. Blumstein, I. Abdullaev, J. Kazbekov, T. Reznikova, E. Strikeleva, et al. 2017. *Rethinking Water in Central Asia: The Costs of Inaction and Benefits of Water Cooperation*. adelphi and CAREC. https://carececo.org/Rethinking%20Water%20in%20Central%20Asia.pdf

Pollan, M. 2006. *The Omnivore's Dilemma: A Natural History of Four Meals*. New York: Penguin Group.

Pollock, M. M., T. J. Beechie, J. M. Wheaton, C. E. Jordan, N. Bouwes, N. Weber, and C. Volk. 2014. "Using Beaver Dams to Restore Incised Stream Ecosystems." *BioScience* 64(4): 279–90.

Poore, J., and T. Nemecek. 2018. "Reducing Food's Environmental Impacts through Producers and Consumers." *Science* 360(6392): 987–992.

Popelka, S. J., and L. C. Smith. 2020. "Rivers as Political Borders: A New Subnational Geospatial Dataset." *Water Policy* 22(3): 293–312.

Porse, E., K. B. Mika, A. Escriva-Bou, E. D. Fournier, K. T. Sanders, E. Spang, J. Stokes-Draut, F. Federico, et al. 2020. "Energy Use for Urban Water Management by Utilities and Households in Los Angeles." *Environmental Research Communications* 2(1): 015003.

President's Water Resources Policy Commission. 1951. "A Water Policy for the American People." *Journal (American Water Works Association)* 43(2): 91–112.

Proctor, C. R., J. Lee, D. Yu, A. D. Shah, and A. J. Whelton. 2020. "Wildfire Caused Widespread Drinking Water Distribution Network Contamination." *AWWA Water Science* 2(4): e1183.

Prüss-Ustün, A., J. Wolf, J. Bartram, T. Clasen, O. Cumming, M. C. Freeman, B. Gordon, P. R. Hunter, et al. 2019. "Burden of Disease from Inadequate Water, Sanitation and Hygiene for Selected Adverse Health Outcomes: An Updated Analysis with a Focus on Low- and Middle-Income Countries." *International Journal of Hygiene and Environmental Health* 222(5): 765–77.

Puttock, A., H. A. Graham, J. Ashe, D. J. Luscombe, and R. E. Brazier. 2021. "Beaver Dams Attenuate Flow: A Multi-Site Study." *Hydrol Process* 35(2): e14017.

Qiang, Y. 2019. "Disparities of Population Exposed to Flood Hazards in the United States." *Journal of Environmental Management* 232: 295–304.

Quansah, J. K., C. L. Escalante, A. P.-H. Kunadu, F. K. Saalia, and J. Chen. 2020. "Pre- and Post-Harvest Practices of Urban Leafy Green Vegetable Farmers in Accra, Ghana and Their Association with Microbial Quality of Vegetables Produced." *Agriculture* 10(1): 18.

Rabaey, K., T. Vandekerckhove, A. V. de Walle, and D. L. Sedlak. 2020. "The Third Route: Using Extreme Decentralization to Create Resilient Urban Water Systems." *Water Research* 185: 116276.

Raucher, R. S., S. J. Rubin, D. Crawford-Brown, and M. M. Lawson. 2011. "Benefit–Cost Analysis for Drinking Water Standards: Efficiency, Equity, and Affordability Considerations in Small Communities." *Journal of Benefit–Cost Analysis* 2(1): 1–24.

Ray, D. K., P. C. West, M. Clark, J. S. Gerber, A. V. Prishchepov, and S. Chatterjee. 2019. "Climate Change Has Likely Already Affected Global Food Production." *PLoS One* 14(5): e0217148.

Reich, K., N. Berg, D. B. Walton, M. Schwartz, F. Sun, X. Huang, and A. Hall. 2018. *Climate Change in the Sierra Nevada: California's Water Future*. Los Angeles: UCLA Center for Climate Science.

Reij, C., and D. Garrity. 2016. "Scaling Up Farmer-Managed Natural Regeneration in Africa to Restore Degraded Landscapes." *Biotropica* 48(6): 834–43.

Rentier, E. S., and L. H. Cammeraat. 2022. "The Environmental Impacts of River Sand Mining." *Science of the Total Environment* 838: 155877.

Reynolds, T. S. 1984. "Medieval Roots of the Industrial Revolution." *Scientific American* 251(1): 122–30.

Richter, B. D., D. Bartak, P. Caldwell, K. F. Davis, P. Debaere, A. Y. Hoekstra, T. S. Li, L. Marston, et al. 2020. "Water Scarcity and Fish Imperilment Driven by Beef Production." *Nature Sustainability* 3(4): 319–28.

Richter, B. D., M. M. Davis, C. Apse, and C. Konrad. 2012. "A Presumptive Standard for Environmental Flow Protection." *River Research and Applications* 28(8): 1312–21.

Richter, B. D., S. Postel, C. Revenga, T. Scudder, B. Lehner, A. Churchill, and M. Chow. 2010. "Lost in Development's Shadow: The Downstream Human Consequences of Dams." *Water Alternatives* 3: 14-42.

Ripple, W. J., C. Wolf, M. K. Phillips, R. L. Beschta, J. A. Vucetich, J. B. Kauffman, B. E. Law, A. J. Wirsing, et al. 2022. "Rewilding the American West." *BioScience* 72: 931–5.

Ritchie, M., and B. Angelbeck. 2020. "'Coyote Broke the Dams': Power, Reciprocity, and Conflict in Fish Weir Narratives and Implications for Traditional and Contemporary Fisheries." *Ethnohistory* 67(2): 191–220.

Rockström, J., J. Gupta, D. Qin, S. J. Lade, J. F. Abrams, L. S. Andersen, D. I. Armstrong McKay, X. Bai, et al. 2023. "Safe and Just Earth System Boundaries." *Nature*. 619: 102–11.

Rodell, M., H. K. Beaudoing, T. S. L'Ecuyer, W. S. Olson, J. S. Famiglietti, P. R. Houser, R. Adler, M. G. Bosilovich, et al. 2015. "The Observed State of the Water Cycle in the Early Twenty-First Century." *Journal of Climate* 28(21): 8289–318.

Root, J. C., and D. K. Jones. 2022. "Elevation–Area–Capacity Relationships of Lake Powell in 2018 and Estimated Loss of Storage Capacity since 1963." US Department of the Interior and US Geological Survey Scientific Investigations Report 2022–5017.

Rosa, L., D. D. Chiarelli, M. C. Rulli, J. Dell'Angelo, and P. D'Odorico. 2020. "Global Agricultural Economic Water Scarcity." *Science Advances* 6(18): 10.

Rosa, L., D. D. Chiarelli, C. Tu, M. C. Rulli, and P. D'Odorico. 2019. "Global Unsustainable Virtual Water Flows in Agricultural Trade." *Environmental Research Letters* 14(11): 114001.

Roth, M. B., and A. Tal. 2022. "The Ecological Tradeoffs of Desalination in Land-Constrained Countries Seeking to Mitigate Climate Change." *Desalination* 529: 115607.

Rulli, M. C., A. Saviori, and P. D'Odorico. 2013. "Global Land and Water Grabbing." *Proceedings of the National Academy of Sciences* 110(3): 892–7.

Sampaio, G., M. H. Shimizu, C. A. Guimarães-Júnior, F. Alexandre, M. Guatura, M. Cardoso, T. F. Domingues, A. Rammig, et al. 2021. "CO_2 Physiological Effect Can Cause Rainfall Decrease as Strong as Large-Scale Deforestation in the Amazon." *Biogeosciences* 18(8): 2511–25.

Sanchez, L., E. C. Edwards, and B. Leonard. 2020. "The Economics of Indigenous Water Claim Settlements in the American West." *Environmental Research Letters* 15(9): 14.

Sanchez, L., B. Leonard, and E. C. Edwards. 2022. *The Long-Term Outcomes of Recognizing Indigenous Property Rights to Water*. Minneapolis: Federal Reserve Bank of Minneapolis, Center for Indian Country Development, Working Paper Series 2022-01.

Sanders, K. T., and M. E. Webber. 2012. "Evaluating the Energy Consumed for Water Use in the United States." *Environmental Research Letters* 7(3): 034034.

Sanderson, B. L., K. A. Barnas, and A. M. W. Rub. 2009. "Nonindigenous Species of the Pacific Northwest: An Overlooked Risk to Endangered Salmon?" *BioScience* 59(3): 245–56.

Saravanan, V. S., G. T. McDonald, and P. P. Mollinga. 2009. "Critical Review of Integrated Water Resources Management: Moving beyond Polarised Discourse." *Natural Resources Forum* 33(1): 76–86.

Savchenko, O. M., M. Kecinski, T. Li, and K. D. Messer. 2019. "Reclaimed Water and Food Production: Cautionary Tales from Consumer Research." *Environmental Research* 170: 320–31.

Scamardo, J. E., S. Marshall, and E. Wohl. 2022. "Estimating Widespread Beaver Dam Loss: Habitat Decline and Surface Storage Loss at a Regional Scale." *Ecosphere* 13(3): e3962.

Scanlon, B. R., C. C. Faunt, L. Longuevergne, R. C. Reedy, W. M. Alley, V. L. McGuire, and P. B. McMahon. 2012. "Groundwater Depletion and Sustainability of Irrigation in the US High Plains and Central Valley." *Proceedings of the National Academy of Sciences of the United States of America* 109(24): 9320–5.

Scanlon, B. R., R. C. Reedy, P. Xu, M. Engle, J. P. Nicot, D. Yoxtheimer, Q. Yang, and S. Ikonnikova. 2020. "Can We Beneficially Reuse Produced Water from Oil and Gas Extraction in the U.S.?" *Science of the Total Environment* 717: 137085.

Schaider, L. A., L. Swetschinski, C. Campbell, and R. A. Rudel. 2019. "Environmental Justice and Drinking Water Quality: Are There Socioeconomic Disparities in Nitrate Levels in U.S. Drinking Water?" *Environmental Health* 18(1): 3.

Scherer, L., and S. Pfister. 2016. "Global Water Footprint Assessment of Hydropower." *Renewable Energy* 99: 711–20.

Schmitt, R. J. P., S. Bizzi, A. Castelletti, and G. M. Kondolf. 2018. "Improved Trade-Offs of Hydropower and Sand Connectivity by Strategic Dam Planning in the Mekong." *Nature Sustainability* 1(2): 96–104.

Schneiders, R. K. 1997. "Flooding the Missouri Valley: The Politics of Dam Site Selection and Design." *Great Plains Quarterly* 1954.

Schwabe, K., M. Nemati, R. Amin, Q. Tran, and D. Jassby. 2020a. "Unintended Consequences of Water Conservation on the Use of Treated Municipal Wastewater." *Nature Sustainability* 3(8): 628–35.

Schwabe, K., M. Nemati, C. Landry, and G. Zimmerman (2020b). "Water Markets in the Western United States: Trends and Opportunities." *Water* 12(1): 15.

Scown, C. D., A. Horvath, and T. E. McKone. 2011. "Water Footprint of U.S. Transportation Fuels." *Environmental Science & Technology* 45(7): 2541–53.

Scudder, T. 2005. *The Future of Large Dams: Dealing with Social, Environmental, Institutional and Political Costs*. London: Earthscan.

Seager, R., M. Hoerling, S. Schubert, H. L. Wang, B. Lyon, A. Kumar, J. Nakamura, and N. Henderson. 2015. "Causes of the 2011-14 California Drought." *Journal of Climate* 28(18): 6997–7024.

Sedlak, D. 2014. *Water 4.0: The Past, Present, and Future of the World's Most Vital Resources*. New Haven, CT: Yale University Press.

Shiklomanov, I. A. 2003. World Water Use and Water Availability. In *World Water Resources at the Beginning of the 21st Century*. I. A. Shiklomanov and J. C. Rodda, eds. Cambridge, UK: Cambridge University Press.

Shlezinger, M., Y. Amitai, A. Akriv, H. Gabay, M. Shechter, and M. Leventer-Roberts. 2018. "Association between Exposure to Desalinated Sea Water and Ischemic Heart Disease, Diabetes Mellitus and Colorectal Cancer; A Population-Based Study in Israel." *Environmental Research* 166: 620–7.

Shumilova, O., K. Tockner, A. Sukhodolov, V. Khilchevskyi, L. De Meester, S. Stepanenko, G. Trokhymenko, J. A. Hernández-Agüero, et al. 2023. "Impact of the Russia–Ukraine Armed Conflict on Water Resources and Water Infrastructure." *Nature Sustainability* 6: 578–86.

Siddik, M. A. B., A. Shehabi, and L. Marston. 2021. "The Environmental Footprint of Data Centers in the United States." *Environmental Research Letters* 16(6): 064017.

Siebert, S., and P. Döll. 2010. "Quantifying Blue and Green Virtual Water Contents in Global Crop Production as Well as Potential Production Losses without Irrigation." *Journal of Hydrology* 384(3–4): 198–217.

Simkin, R. D., K. C. Seto, R. I. McDonald, and W. Jetz. 2022. "Biodiversity Impacts and Conservation Implications of Urban Land Expansion Projected to 2050." *Proceedings of the National Academy of Sciences* 119(12): e2117297119.

Simonit, S., J. P. Connors, J. Yoo, A. Kinzig, and C. Perrings. 2015. "The Impact of Forest Thinning on the Reliability of Water Supply in Central Arizona." *PLoS One* 10(4): e0121596.

Smil, V. 2010. *Energy Transitions: Global and National Perspectives*. Santa Barbara, CA: Praeger.

Smil, V. 2021. *Grand Transitions: How the Modern World Was Made*. New York: Oxford University Press.

Smith, C. 2013. *City Water, City Life: Water and the Infrastructure of Ideas in Urbanizing Philadelphia, Boston, and Chicago*. Chicago: University of Chicago Press.

Spiegal, S., P. J. A. Kleinman, D. M. Endale, R. B. Bryant, C. Dell, S. Goslee, R. J. Meinen, K. C. Flynn, et al. 2020. "Manuresheds: Advancing Nutrient Recycling in US Agriculture." *Agricultural Systems* 182: 102813.

Stafford, L., D. Shemie, T. Kroeger, T. Baker, and C. Apse. 2019. *Greater Cape Town Water Fund Business Case: Assessing the Return on Investment for Ecological Infrastructure Restoration*. Arlington, VA: The Nature Conservancy.

Stanford, B., E. Zavaleta, and A. Millard-Ball. 2018. "Where and Why Does Restoration Happen? Ecological and Sociopolitical Influences on Stream Restoration in Coastal California." *Biological Conservation* 221: 219–27.

Steffen, W., K. Richardson, J. Rockström, S. E. Cornell, I. Fetzer, E. M. Bennett, R. Biggs, S. R. Carpenter, et al. 2015. "Planetary Boundaries: Guiding Human Development on a Changing Planet." *Science* 347(6223).

Stoltenborg, D., and R. Boelens. 2016. "Disputes over Land and Water Rights in Gold Mining: The Case of Cerro de San Pedro, Mexico." *Water International* 41(3): 447–67.

Stone, C. D. 1972. "Should Trees Have Standing? Towards Legal Rights for Natural Objects." *Southern California Law Review* 45: 450–501.

Storlazzi, C. D., S. B. Gingerich, A. v. Dongeren, O. M. Cheriton, P. W. Swarzenski, E. Quataert, C. I. Voss, D. W. Field, et al. 2018. "Most Atolls Will Be Uninhabitable by the Mid-21st Century Because of Sea-Level Rise Exacerbating Wave-Driven Flooding." *Science Advances* 4(4): eaap9741.

Strategic Foresight Group. 2014. *Water and Violence: Crisis of Survival in the Middle East*. https://www.files.ethz.ch/isn/188318/63948150123-web.pdf

Subbaraman, R., L. Nolan, K. Sawant, S. Shitole, T. Shitole, M. Nanarkar, A. Patil-Deshmukh, and D. E. Bloom. 2015. "Multidimensional Measurement of Household Water Poverty in a Mumbai Slum: Looking beyond Water Quality." *PLoS One* 10(7): e0133241.

Sun, G., P. V. Caldwell, and S. G. McNulty. 2015. "Modelling the Potential Role of Forest Thinning in Maintaining Water Supplies under a Changing Climate across the Conterminous United States." *Hydrological Processes* 29(24): 5016–30.

Suri, M. R., J. L. Dery, J. Pérodin, N. Brassill, X. He, S. Ammons, M. E. Gerdes, C. Rock, et al. 2019. "U.S. Farmers' Opinions on the Use of Nontraditional Water Sources for Agricultural Activities." *Environmental Research* 172: 345–57.

Swain, D. L., D. Singh, D. Touma, and N. S. Diffenbaugh. 2020. "Attributing Extreme Events to Climate Change: A New Frontier in a Warming World." *One Earth* 2(6): 522–7.

Taber, B. 2012. *Recreation in the Colorado River Basin: Is America's Playground under Threat?: The 2012 State of the Rockies Report Card*. Colorado Springs: Colorado College.

Tal, A. 2016. "Rethinking the Sustainability of Israel's Irrigation Practices in the Drylands." *Water Research* 90: 387–94.

Tanana, H., J. Garcia, A. Olaya, C. Colwyn, H. Larsen, R. Williams, and J. King. 2021. *Universal Access to Clean Water for Tribes in the Colorado River Basin*. Water & Tribes Initiative, Colorado River Basin.

Tanoue, M., Y. Hirabayashi, and H. Ikeuchi. 2016. "Global-Scale River Flood Vulnerability in the Last 50 Years." *Scientific Reports* 6: 36021.

Tao, S., H. Zhang, Y. Feng, J. Zhu, Q. Cai, X. Xiong, S. Ma, L. Fang, et al. 2020. "Changes in China's Water Resources in the Early 21st Century." *Frontiers in Ecology and the Environment* 18(4): 188–93.

Tellman, B., J. A. Sullivan, C. Kuhn, A. J. Kettner, C. S. Doyle, G. R. Brakenridge, T. A. Erickson, and D. A. Slayback. 2021. "Satellite Imaging Reveals Increased Proportion of Population Exposed to Floods." *Nature* 596(7870): 80–6.

Tennant, D. L. 1976. "Instream Flow Regimens for Fish, Wildlife, Recreation and Related Environmental Resources." *Fisheries* 1(4): 6–10.

Teodoro, M. P., M. Haider, and D. Switzer. 2018. "US Environmental Policy Implementation on Tribal Lands: Trust, Neglect, and Justice." *Policy Studies Journal* 46(1): 37–59.

Thebo, A. L., P. Drechsel, E. F. Lambin, and K. L. Nelson. 2017. "A Global, Spatially-Explicit Assessment of Irrigated Croplands Influenced by Urban Wastewater Flows." *Environmental Research Letters* 12(7): 074008.

Thompson, J. R., and G. Polet. 2000. "Hydrology and Land Use in a Sahelian Floodplain Wetland." *Wetlands* 20(4): 639–59.

Tickner, D., J. J. Opperman, R. Abell, M. Acreman, A. H. Arthington, S. E. Bunn, S. J. Cooke, J. Dalton, et al. 2020. "Bending the Curve of Global Freshwater Biodiversity Loss: An Emergency Recovery Plan." *Bioscience* 70(4): 330–42.

Tosca, M. G., J. T. Randerson, C. S. Zender, M. G. Flanner, and P. J. Rasch. 2010. "Do Biomass Burning Aerosols Intensify Drought in Equatorial Asia during El Niño?" *Atmospheric Chemistry and Physics* 10(8): 3515–28.

Tow, E. W., A. L. Hartman, A. Jaworowski, I. Zucker, S. Kum, M. AzadiAghdam, E. R. Blatchley, A. Achilli, et al. 2021. "Modeling the Energy Consumption of Potable Water Reuse Schemes." *Water Research X* 13: 100126.

Treuer, D. 2021, May. "Return the National Parks to the Tribes." *The Atlantic*.

Turner, S. W. D., J. S. Rice, K. D. Nelson, C. R. Vernon, R. McManamay, K. Dickson, and L. Marston. 2021. "Comparison of Potential Drinking Water Source Contamination across One Hundred US Cities." *Nature Communications* 12(1): 7254.

Twain, M. 1883. *Life on the Mississippi*. Boston: James R. Osgood and Company.

Twichell, J. H., K. K. Mulvaney, N. H. Merrill, and J. J. Bousquin. 2022. "Geographies of Dirty Water: Landscape-Scale Inequities in Coastal Access in Rhode Island." *Frontiers in Marine Science* 8: 760684.

UNESCO-IHP and UNEP. 2016. "Transboundary Aquifers and Groundwater Systems of Small Island Developing States: Status and Trends." Nairobi: UNEP.

Uría-Martínez, R., M. M. Johnson, and R. Shan. 2021. *US Hydropower Market Report*. Oak Ridge, TN: US Department of Energy Office of Scientific and Technical Information.

van Oel, P. R., and A. Y. Hoekstra. 2012. "Towards Quantification of the Water Footprint of Paper: A First Estimate of Its Consumptive Component." *Water Resources Management* 26(3): 733–49.

Vanham, D., M. M. Mekonnen, and A. Y. Hoekstra. 2020. "Treenuts and Groundnuts in the EAT-Lancet Reference Diet: Concerns Regarding Sustainable Water Use." *Global Food Security-Agriculture Policy Economics and Environment* 24: 7.

Vedachalam, S., and R. Dobkin. 2021. *H2Affordability: How Water Bill Assistance Programs Miss the Mark*. Washington, DC: Environmental Policy Innovation Center.

Venkataramanan, V., J. Crocker, A. Karon, and J. Bartram. 2018. "Community-Led Total Sanitation: A Mixed-Methods Systematic Review of Evidence and Its Quality." *Environmental Health Perspectives* 126(2): 026001.

Vincent, L., L. Michel, C. Catherine, and R. Pauline. 2014. "The Energy Cost of Water Independence: The Case of Singapore." *Water Science and Technology* 70(5): 787–94.

Wada, Y., M. Florke, N. Hanasaki, S. Eisner, G. Fischer, S. Tramberend, Y. Satoh, M. T. H. van Vliet, et al. 2016. "Modeling Global Water Use for the 21st Century: The Water Futures and Solutions (WFaS) Initiative and Its Approaches." *Geoscientific Model Development* 9(1): 175–222.

Wada, Y., D. Wisser, and M. F. P. Bierkens. 2014. "Global Modeling of Withdrawal, Allocation and Consumptive Use of Surface Water and Groundwater Resources." *Earth System Dynamics* 5(1): 15–40.

Waehler, T. A., and E. S. Dietrichs. 2017. "The Vanishing Aral Sea: Health Consequences of an Environmental Disaster." *Tidsskrift for Den Norske Laegeforening* 137(18): 1443–5.

Walter, R. C., and D. J. Merritts. 2008. "Natural Streams and the Legacy of Water-Powered Mills." *Science* 319(5861): 299–304.

Wang-Erlandsson, L., I. Fetzer, P. W. Keys, R. J. van der Ent, H. H. G. Savenije, and L. J. Gordon. 2018. "Remote Land Use Impacts on River Flows through Atmospheric Teleconnections." *Hydrology and Earth System Sciences* 22(8): 4311–28.

Warner, J., N. Mirumachi, R. L. Farnum, M. Grandi, F. Menga, and M. Zeitoun. 2017. "Transboundary 'Hydro-Hegemony': 10 Years Later." *Wiley Interdisciplinary Reviews-Water* 4(6): 13.

Weinkle, J., C. Landsea, D. Collins, R. Musulin, R. P. Crompton, P. J. Klotzbach, and R. Pielke. 2018. "Normalized Hurricane Damage in the Continental United States 1900–2017." *Nature Sustainability* 1(12): 808–13.

Weiss, H. (2017). Megadrought, Collapse, and Causality. In *Megadrought and Collapse: From Early Agriculture to Angkor*. H. Weiss, ed. New York: Oxford University Press.

WHO. 2019. *National Systems to Support Drinking-Water, Sanitation and Hygiene: Global Status Report 2019*. UN-Water Global Analysis and Assessment of Sanitation and Drinking-Water (GLAAS) 2019 Report.

WHO/UNICEF. 2021. *Progress on Household Drinking Water, Sanitation and Hygiene 2000–2020: Five Years into the SDGs*. Geneva: World Health Organization (WHO) and the United Nations Children's Fund (UNICEF).

Wilber, C. D. 1881. *The Great Valleys and Prairies of Nebraska and the Northwest*. Omaha, NE: Daily Republican.

Willett, W., J. Rockström, B. Loken, M. Springmann, T. Lang, S. Vermeulen, T. Garnett, D. Tilman, et al. 2019. "Food in the Anthropocene: The EAT-*Lancet* Commission on Healthy Diets from Sustainable Food Systems." *The Lancet* 393(10170): 447–92.

Williams, A. P., B. I. Cook, and J. E. Smerdon. 2022. "Rapid Intensification of the Emerging Southwestern North American Megadrought in 2020–2021." *Nature Climate Change* 12: 232–4.

Williams, J. 2018. "Diversification or Loading Order? Divergent Water–Energy Politics and the Contradictions of Desalination in Southern California." *Water Alternatives: An Interdisciplinary Journal on Water Politics and Development* 11(3): 847–65.

Wing, O. E. J., W. Lehman, P. D. Bates, C. C. Sampson, N. Quinn, A. M. Smith, J. C. Neal, J. R. Porter, et al. 2022. "Inequitable Patterns of US Flood Risk in the Anthropocene." *Nature Climate Change* 12: 156–62.

Winsemius, H. C., B. Jongman, T. I. E. Veldkamp, S. Hallegatte, M. Bangalore, and P. J. Ward. 2018. "Disaster Risk, Climate Change, and Poverty: Assessing the Global Exposure of Poor People to Floods and Droughts." *Environment and Development Economics* 23(3): 328–48.

Wisser, D., S. Frolking, S. Hagen, and M. F. P. Bierkens. 2013. "Beyond Peak Reservoir Storage? A Global Estimate of Declining Water Storage Capacity in Large Reservoirs." *Water Resources Research* 49(9): 5732–9.

Wolf, A. T. 2007. "Shared Waters: Conflict and Cooperation." *Annual Review of Environment and Resources* 32: 241–69.

Womble, P., and W. M. Hanemann. 2020a. "Legal Change and Water Market Transaction Costs in Colorado." *Water Resources Research* 56(4): 24.

Womble, P., and W. M. Hanemann. 2020b. "Water Markets, Water Courts, and Transaction Costs in Colorado." *Water Resources Research* 56(4): 28.

Wood, W. W., and D. W. Hyndman. 2017. "Groundwater Depletion: A Significant Unreported Source of Atmospheric Carbon Dioxide." *Earth's Future* 5(11): 1133–5.

Wright, D. B., C. D. Bosma, and T. Lopez-Cantu. 2019. "U.S. Hydrologic Design Standards Insufficient due to Large Increases in Frequency of Rainfall Extremes." *Geophysical Research Letters* 46(14): 8144–53.

Xu, Z., L. Zhang, L. Zhao, B. Li, B. Bhatia, C. Wang, K. L. Wilke, Y. Song, et al. 2020. "Ultrahigh-Efficiency Desalination via a Thermally-Localized Multistage Solar Still." *Energy and Environmental Science* 13: 830.

Yan, X. C., V. Thieu, and J. Garnier. 2021. "Long-Term Evolution of Greenhouse Gas Emissions from Global Reservoirs." *Frontiers in Environmental Science* 9: 705477.

Yao, F., B. Livneh, B. Rajagopalan, J. Wang, J.-F. Crétaux, Y. Wada, and M. Berge-Nguyen. 2023. "Satellites Reveal Widespread Decline in Global Lake Water Storage." *Science* 380(6646): 743–9.

Yin, J. B., P. Gentine, S. Zhou, S. C. Sullivan, R. Wang, Y. Zhang, and S. L. Guo. 2018. "Large Increase in Global Storm Runoff Extremes Driven by Climate and Anthropogenic Changes." *Nature Communications* 9: 10.

Yogananth, N., and T. Bhatnagar. 2018. "Prevalence of Open Defecation among Households with Toilets and Associated Factors in Rural South India: An Analytical Cross-Sectional Study." *Transactions of the Royal Society of Tropical Medicine and Hygiene* 112(7): 349–60.

Young, O. R. 2011. "Effectiveness of International Environmental Regimes: Existing Knowledge, Cutting-Edge Themes, and Research Strategies." *Proceedings of the National Academy of Sciences* 108(50): 19853–60.

Zaveri, E., J. Russ, A. Khan, R. Damania, E. Borgomeo, and A. Jägerskog. 2021. *Ebb and Flow* Volume 1. *Water, Migration, and Development.* https://doi.org/10.1596/978-1-4648-1745-8

Zeitoun, M., and J. Warner. 2006. "Hydro-Hegemony–A Framework for Analysis of Trans-Boundary Water Conflicts." *Water Policy* 8(5): 435–60.

Zhang, H., I. Jarić, D. L. Roberts, Y. He, H. Du, J. Wu, C. Wang, and Q. Wei. 2020. "Extinction of One of the World's Largest Freshwater Fishes: Lessons for Conserving the Endangered Yangtze Fauna." *Science of the Total Environment* 710: 136242.

Zhang, W., G. Villarini, G. A. Vecchi, and J. A. Smith. 2018. "Urbanization Exacerbated the Rainfall and Flooding Caused by Hurricane Harvey in Houston." *Nature* 563(7731): 384+.

Index

Note: page numbers followed by b, f, or t refer to boxes, figures, or tables, respectively.

absolute river integrity doctrine, 223b
absolute sovereignty doctrine, 223b
access, 11, 253, 265–66, 299–300. *See also* availability; drinking water
acid mine drainage, 45, 330
Active Management Areas (AMAs), 195
adaptive management, 211
affordability of water, 283–84, 301
agriculture and food production
    animal products, 339, 345–49
    controlled-environment systems, 355
    crop water requirement and global crop water consumption, 342, 343t
    definitions, 338t
    diet and water footprints, 344–47, 346f
    distribution, inequitable, 340–41
    flooding and, 137
    food waste, 347
    Green Revolution, 337–42, 340f, 341t
    irrigation, 59–61, 66, 67, 188–89, 190b, 339, 342, 353–56
    pollution from, 356–61, 358f, 361f
    premodern, 58–64
    rainfed, 23, 355–56
    runoff farming techniques, 61–62, 355
    sustainability and, 349–50
    sustainable intensification, 342
    total water requirement, 343–44
    as use, 76
    virtual water trade and, 351–52, 352f
    wastewater reuse and, 239–41
    yield gap, 352–53
agroforestry, 53–54
allocation. *See also* reallocation
    common-pool resources and tragedy of the commons, 186–87
    economic efficiency, 184–86, 185f
    goals for, 183–84
    groundwater, 194–95
    interstate, 196–97
    prior appropriation doctrine, 191–94, 205–6, 211
    privatization, government intervention, and communal management, 187–89, 190b
    regulated riparianism, 191
    reserved rights, 195–96
    riparian doctrine, 190–91
almonds, 350
animal products, 339, 345–49
aquatic life use support (ALUS), 148
aqueducts, 64–66, 178–79
aquifers. *See also* groundwater
    confined and unconfined, 89, 91, 260
    depletion of, 88–94, 90f, 93t
    discharge, recharge, and capture, 89–91
    fossil, 91, 260
    fracking and, 321
    indirect potable reuse (IPR), 241
    marginal net benefits, 184
    as resource, 6
    safe yield of, 92
    as stocks, 21
    storage, artificial, 246–47, 249t
aquifer storage and recovery (ASR), 247
Aral Sea, 85–86

aridification, 39, 44
aridity index (AI), 24, 25f
atmospheric rivers, 122b
availability (supply). *See also* scarcity; scarcity solutions
   about, 6
   deforestation and, 46–47, 47f
   drought, 32–39
   forest management and, 50–52
   global stocks and flows, 21–23, 22t, 23f
   premodern urban, 64–65
   reliability and, 276–77, 276f
   spatial variation, 24, 24f, 26–27b
   temporal variation, 27–32, 42
   treatment regulations, 260–61

Bali, 190b
basin–country units (BCUs), 221
beaver dams, 133b
benefit–cost analysis (BCA), 12, 131–33, 154–55, 174, 310
benefit sharing, 227
biochemical oxygen demand (BOD), 143t, 146b
biofuels, 318–19, 320b
biomonitoring, 153
bioswales, 290t
bitumen, 322–23
blackwater, 286
blue water vs. green water, 5, 71, 74, 343t, 344
bofedales (peatlands), 245
boron toxicity, 238
bottled water, 270
brine, 238
British thermal units (BTUs), 108b
Bureau of Reclamation (BuRec), 67–68, 110, 165, 213
buyouts, 136–37

California
   Central Valley, 93–94, 94f, 247, 266–67, 350, 357
   Groundwater Replenishment System, Orange County, 241, 242f, 247
   planning process, 210–12
canals, 101–3, 103b, 354–55
capacity factor, 108b
carbon dioxide (CO_2). *See* greenhouse gases (GHGs) and emissions
cenotes, 62
CEO Water Mandate, 332
channelization, 104–5, 104t, 129
chemical toxicants, 144
cisterns, 63

Clean Water Act (CWA)
   about, 144–45
   agricultural runoff and, 359
   CAFOs and, 360
   chemical pressures, 145–48
   effectiveness, 153–55, 154f
   nonchemical stressors, 151–53
   secondary treatment and, 262
   stormwater and, 291
   structure of, 145f
   water quality goals, 148–51
   wetlands and, 152b
climate change
   about, 41–42
   flooding and, 125
   as global issue, 6b
   hydropower and, 113
   indirect effects, 45–46
   observation, 43–45
   salmon and, 115
   sea-level rise (SLR), 45–46, 45f, 122, 262
   theory and prediction, 42–43
climate forcers, 38
climate variability, 33–34, 125
cloud seeding, 246
coal, 319f, 320–21, 331
Coca-Cola, 333–34
Colorado, 210–12
Colorado doctrine, 192–93
Colorado River, 87b, 89f, 108b, 205, 250
combined sewers and CSOs, 262–64, 263f, 269
commercial, industrial, and institutional (CII) water use, 277–78
common-pool resources (CPRs), 187–88, 189t
common-property management, 188–89, 189t, 190b, 217
community of nature, 14
concentrated animal feeding operations (CAFOs), 347, 360–61
concentrating solar power (CSP), 318, 319f
conflict
   antigovernment protests, 228
   casualty of war, water as, 219b
   crisis of, 11
   desalination and, 237b
   drivers of, 224–26, 225t
   event data and hydro-hegemony, 223–24
   farmer–herder, 228
   international law and, 222–23
   national security and, 218–19
   resilience, institutional, 225–27
   scale and, 217
   transboundary waters, extent of, 220–22

water protectors, 228–29
"water wars" argument, 218
weapon of war, water as, 220b
Western US, 228–29
Connecticut, 210–13
conservationism, 13–14, 69
consumptive uses, 71, 72, 74–76, 75f
Contaminant Candidate Lists (CCLs), 259
cooling systems, 317
cyanide, 330

dams. *See also* hydropower; reservoirs
  antidam movement and WCD report, 165–66
  aqueducts and, 178–79
  beaver dams, 133b
  better dams, building, 174–76
  boom era, 164–65, 164f, 168–69, 169f
  categories of, 159–60
  components and terminology, 160–61, 161f
  ecological impacts, 170–71
  elevation–storage curve, 162f
  fish and, 115
  flood-control, 130
  as gray infrastructure, 66
  history of hydropower and, 107–8
  Hoover, 67–68, 108b
  lifespan, 173–74
  Mekong Basin, 174–75, 175f
  multi-purpose management, 161–63
  navigation and, 105
  number, purposes, and storage capacity data, 166–70, 167t, 168f, 169f
  Oroville, 160–61, 161f
  population impacts, 171–72, 172b
  removal, 176–78
  reservoir storage–yield relationship, 163–64, 163f
  Roman, 64
  safety and failures, 172–74
  sediment-passage and fish-passage management, 176
  Stairway of Water, Mississippi River, 105
  as supply strategy, 249t
  tailings, 330
  Tiga and Chowalla Gorge dams, Nigeria, 88b
debris flows, 123
deforestation, 46–49
demand. *See* history of water use and management; scarcity; use
depletion. *See also* scarcity
  groundwater, 88–94, 90f, 93t, 184–86, 350
  surface-water, 85–88
desalination, 83, 236–38, 237b, 249t
desertification, 52–54, 53b

development, global, 7, 9, 9t, 70–71. *See also* low- and middle-income countries
diamond-water paradox, 7, 8b
difference principle, 13
dilbit (diluted bitumen), 322–23
direct potable reuse (DPR), 241, 243, 248
disability-adjusted life years (DALYs), 296–97
discounting, 131, 184
disease. *See* health
disinfection byproducts (DBPs), 260–61
dissolved oxygen (DO), 143t
drinking water. *See also* health, human
  access to, 11, 253, 265–67
  bottled, 270
  disinfection byproducts (DBPs), 260–61
  distribution system and lead contamination, 261–62
  LMICs and JMP ladder for, 298–99, 298t
  standards and treatment, 258–61
drought, 32–34, 36–39, 39f, 236–37, 277
drylands, 24

ecological limits of hydrologic alteration (ELOHA), 204t
economic water scarcity, 81–82
ecosystem services, 10. *See also* payment for ecosystem services
effluent limits, technology-based, 145
electricity-related use, 316–18, 319f. *See also* hydropower
El Niño–Southern Oscillation (ENSO), 34
Emergency Recovery Plan for Freshwater Biodiversity, 10
Endangered Species Act (ESA), 114, 155–57, 156b, 201, 323
endocrine-disrupting chemicals, 144
energy. *See also* hydropower
  electricity and transportation, water for, 316–19, 319f, 320f
  modern infrastructure and, 67
  pollution from production of, 319–23
  for wastewater reuse, 243
energy intensity, 247–49, 248f
environmental flow requirements (EFRs), 81, 203–5, 204t, 356
Environmental Impact Statements (EISs), 155
environmental justice. *See* justice and equity
Environmental Protection Agency (EPA), 146, 148–51, 241, 259–61
Equitable Access Score-Card, 300–301
equitable apportionment doctrine, 197b
equity. *See* justice and equity
Erie Canal, 102–3, 103b

ethics, 12–15
evapotranspiration (ET)
    climate change and, 42
    as consumptive use, 72
    crop water requirement (ET_{crop}), 342
    deforestation and, 47
    flows and, 22
    TMR and, 49b
event data approach, 223–24
exceedance probabilities, 28, 31f
externalities, 16, 134, 186, 262, 280
extreme event attribution, 43

Falkenmark indicator, 79, 80t
Federal Emergency Management Agency (FEMA), 135–36
Federal Energy Regulatory Commission (FERC), 110
federal reserved flows, 206
Federal river basins, 213
feedback loops, 34, 34f, 35b, 42
feedbacks, 34
feed conversion ratios (FCRs), 348–49, 348f
fertilizers, 357–59, 358f, 361
fish, 176, 209, 347
fishing, 113–16, 196
FloodFactor model, 136
flooding and flood management
    agricultural, 137
    beaver dams, 133b
    building in floodplains, 133–35
    buyouts and managed retreat, 136–37
    crisis of, 11
    definition of flood, 120
    exposure, 126–27, 127f
    flood maps and national insurance, 135–36
    frequency analysis, 29–30, 31f
    hazard, 124–25
    impacts and trends, 123–24, 123t, 124f
    levee effect, 130–31, 131f
    recurrence intervals, 30
    sea-level rise and, 122, 262
    structural flood control, 127–33
    types of floods, 121–23
    vulnerability, 127
floodplains, 121, 128, 133–35, 289
floodways, 129
flow duration curves (FDCs), 28–29, 29f
flow regimes, 86–88
flows, 22–23, 23f, 80
fluvial geomorphology, 105
fog catchers, 245
food and biofuels, 320b

food production. *See* agriculture and food production
forest management, 50–52
fossil fuels, 41, 67, 319f, 320–23, 331
fracking, 317, 321–22
fugitive resource, 6, 223b

gender in water management, 70
general stream adjudications, 196
global water crisis, 10–12, 14, 17
governance systems, 16, 210–14, 267–70, 268f
Gravity Recovery and Climate Experiment (GRACE) and GRACE-FO, 92–93
gray infrastructure, 10, 15, 66
graywater, 285–86
Great Acceleration, 1–4, 3t
Great Drought (1870s), 36–37
Great Green Wall of Africa project, 53–54
Great Salt Lake, 86
greenhouse gases (GHGs) and emissions
    carbon sequestration, 290–91
    climate change and, 41, 43
    as climate forcers, 38
    CO_2 equivalent, 111
    desalination and, 237–38
    groundwater depletion and, 92
    from hydropower, 111–12
    methane, 111–12, 321
green infrastructure, 15, 264, 289–92, 290t
greening, global, 46
Green Revolution, 337–42, 340f, 341t
green water vs. blue water, 5, 71, 74, 343t, 344
groundwater. *See also* aquifers
    allocation of, 194–95
    ancient use of, 62
    contaminated, 266–67, 321
    depletion of, 88–94, 90f, 93t, 184–86, 350
    recharge, 23, 246–47
    as resource, 5–6
    as stock, 21
    transboundary, 221–22
groundwater banks, 202
Groundwater Replenishment System (GWRS), Orange County, CA, 241, 242f, 247

Hadejia-Nguru floodplain wetlands, Nigeria, 88b
hand-washing, 306–7
hard path, 10–11, 65–67, 69t, 104, 127–33. *See also* dams
hatcheries, 115
Haudenosaunee Confederacy, 103b
hazard, 36–37, 124–25
hazard, moral, 135, 262, 280

Hazard Mitigation Grant Program (HMGP), 136–37
health, human
   chemical toxicants and, 144
   crisis of, 11
   infectious and water-related disease, 66, 142, 254–57, 255f, 265, 295–97, 296t
   LMICs and disease burden, 295–97, 296t
   US water system and, 265
   wastewater reuse and, 242
history of water use and management
   global development agenda, 70–71
   modern hard path and soft path, 65–70
   modern quantitative analysis, 71–76
   premodern agriculture, 58–64
   premodern urban water supply, 64–65
hungry water, 105, 170
hurricanes, 122, 125
hydraulic gradient, 45, 89–91, 90f
hydraulic societies, 60b
hydrographs, 27–28, 28f
hydro-hegemony framework, 224
hydro-illogical cycle, 277
hydrologic cycle, 6, 16, 23f, 41–43
hydromodification, 151
hydropower, 106–13, 108b, 109f, 116–17, 160, 318, 319f. *See also* dams
hygiene, 306–7

impervious surfaces, 54, 121, 243, 264, 288
India, 62, 244, 306b
indicator bacteria, 143t, 148–49, 151
Indigenous peoples
   adjudication of rights, 207–8
   dams, displacement by, 171, 172b
   fishing rights, 196
   Haudenosaunee Confederacy and Erie Canal, 103b
   Navajo Water Project, 266
   Pacific Northwest salmon and, 114–15
   pipeline protests, 322–23
   privatization of reservation lands, 188
   water access, 266
   WaterBack, 208–10
   water rights, 195–96
indirect potable reuse (IPR), 241, 243, 248
Industrial Revolution, 107, 113
industrial water use
   chemical contamination cases, 326, 327t
   for electricity, 316–18, 319f
   for energy production, 319–23
   for manufacturing, 323–29, 325t
   mining, 330–32
   stewardship, corporate, 332–34
   for transportation, 318–19, 320b, 320f
   US chemical laws, 328t
   withdrawal over time, 324f
information, 17
infrastructure. *See also* hard path
   failings in, 15
   gray, 10, 15, 66
   green, 15, 264, 289–92, 290t
   maintenance of, 15
   modern vs. premodern, 67
Infrastructure Investment and Jobs Act (IIJA), 281
installed capacity, 108b
instream flow incremental methodology (IFIM), 204t
instream uses
   about, 5
   fishing, 113–16
   hydropower, 106–13, 109f
   navigation, 101–6, 106t
   offstream vs., 71
   recreation, 116–17
Integrated Urban Water Management (One Water), 275
Integrated water resource management (IWRM), 70–71
intensity–duration–frequency (IDF) curves, 30–32, 32f, 134
interbasin transfers, 27b, 72, 86, 178–79
interception, 47
Intertropical Convergence Zone (ITCZ), 24, 29b
invasive species, nonnative, 115–16
iron triangle, 132
irrigation
   controlled-environment systems, 355
   conveyance losses, 354–55
   deficit, 353
   as gray infrastructure, 66
   irrigation efficiency, 353
   large-river and small-river, premodern, 59–61
   local, communal systems, 188–89, 190b
   types of, 353–54
   western US history of, 67
   yields and, 339, 342
issue linkage, 227

joint conservation pool, 162
Joint Monitoring Programme (JMP), 298–300, 303–4, 304t
justice and equity. *See also* low- and middle-income countries
   access to clean water, 265–67
   affordability of water, 283–84, 301

allocation and, 183, 191, 193
distributive justice, 13
environmental justice (EJ) movement, 13
Equitable Access Score-Card, 300–301
floods and, 126–27, 132–33, 137
food availability, 340–41
global development and, 71
Just Transition, 4
participatory justice, 13
pipeline protests, 322–23
scarcity vs. injustice, 82
UN Watercourses Convention and, 222
Just Transition, 1–4

Kaldor–Hicks efficiency, 12

lakes, shrinking, 85–86
land subsidence, 92, 94
land-use change (LUC)
    about, 46
    coastal storm hazard and, 125
    deforestation, 46–49
    desertification, 52–54, 53b
    managing forests for water supply, 50–52
lead service lines (LSLs), 261–62
legacy cities, 269–70
Leopold, Aldo, 17, 156b
levee effect, 130–31, 131f
levees, 128–29, 137
limited territorial sovereignty doctrine, 223b
linear flows, 270–71
lithium, 331–32
locks, 102, 102f
low- and middle-income countries (LMICs)
    development and, 7
    disease burden, 295–97, 296t
    drinking water, 253, 298–301, 298t, 300f
    economic burden, 297
    hard path and, 70
    hydropower and, 108–10
    hygiene, 307–8
    mining and, 330–31
    sanitation, 301–7, 302f, 304t, 305f, 306b
    virtual water trade and, 351–52
    WASH funding, 310–13, 310f
    WASH goals, 297–98, 298t
    WASH solutions, 308–9
    waterborne diseases and, 66
Low Impact Hydropower Institute (LIHI) certification, 112

managed aquifer recharge (MAR), 247
managed retreat, 136–37

manufacturing, 323–29, 325t, 327t
manureshed, 360
marginal benefits, 7, 8b
marginal net benefits, 184, 185f
Maximum Contaminant Level (MCL), 259
Maximum Contaminant Level Goal (MCLG), 259
mean monthly flow (MMF) and mean annual flow (MAF), 204t
megadroughts, 38–39, 39f, 61b
membrane filtration, 147b, 236, 236t
menstrual hygiene, 307
mercury, 330, 345
Mesopotamia, ancient, 60, 255, 356
methane (CH_4), 111–12, 321
Millennium Development Goals (MDGs), 7, 297–98, 298t
mining, 330–32
monsoon systems, 29b
Muir, John, 14, 165
multicriteria decision making, 174–75
multiple-barrier approach, 260
municipal separate storm sewer system (MS_4), 288
mussels, freshwater, 156b

National Environmental Policy Act (NEPA), 155
National Flood Insurance Program (NFIP), 135–36
National Pollutant Discharge Elimination System (NPDES), 145–47, 146b, 291–92, 359, 360
national security, 218–19
Native Americans. *See* Indigenous peoples
natural disasters, 15, 36–37
natural gas, 318, 319f
navigation, 101–6, 106t, 161
needs, 227
net present value (NPV), 131
New Zealand, 209
Niagara Falls, 116–17
nitrogen, 154, 357–59, 358f
nonconsumptive use, 5, 262
nonrevenue water (NRW), 277–78, 279
nonuse value, 132
Norte Chico, 60–61

official development assistance (ODA), 312
offstream uses, 5, 71, 162
oil, 321
One Water (Integrated Urban Water Management), 275
Ostrom, Elinor, 188

Palmer Drought Severity Index (PDSI), 33, 39f
paper manufacturing, 325
Pareto efficiency, 12, 175
pathogens, environmental, 254, 255f
payment for ecosystem services (PES), 50, 51, 245, 260
peatlands, 245
Peru, 244–45, 305–7, 307f
phosphorus, 112, 151, 154, 357–59, 358f
phreatophytes, 47
Pick–Sloan Program, 172b
Pinchot, Gifford, 14
pipelines, oil and gas, 322–23
pit latrines, 302–3, 302f
planetary boundary framework, 81
plant water-use efficiency, 46
pollution. *See also* water quality
    from animal feeding operations, 359–61, 361f
    as consumptive use, 72
    dams and, 177
    from energy production, 319–23
    industrial incidents, 326, 327t
    from manufacturing, 326–29
    from mining, 330–32
    nutrient, from fertilizers, 357–59, 358f
    point and nonpoint sources, 142, 145–48, 291
    prevention, urban, 290t
    salinization, 46, 60, 356–57
    toxic flooding, 122–23
    water quality parameters, 142–44, 143t
polychlorinated biphenyls (PCBs), 327, 345
population- attributable fraction (PAF), 296, 296t
potential evapotranspiration (PET), 24, 42
Powell, John Wesley, 53b, 196
precipitation (P), 22, 42, 48–49, 246
precipitationsheds, 49b
preservationism, 13–14
pricing of water, 16–17, 280–85, 301
prior appropriation doctrine, 191–94, 205–6, 211
private goods, 186
privatization, 187–88
probability density function (PDF), 30, 31f, 43, 44f, 55
probable maximum flood (PMF), 30
produced water, 321–22
public goods, 16, 186–87
public trust doctrine, 206
pulse flow, 205

qanats, 62, 63f

rain barrels, 290t
rain gardens, 290t

Ramsar Convention on Wetlands, 152b
rate structures, 284–85
reallocation
    ag-to-urban negotiation, 202–3
    environmental flow requirements (EFRs), 203–5, 204t
    Indigenous rights and WaterBack, 207–10
    legal tools for EFRs, 205–7
    water markets, 201–2
recreation, water-based, 116–17, 162
recurrence intervals, 30
regulated riparianism, 191
reliability, 275–77, 276f
reserved water rights, 195–96
reservoirs
    ecological impacts, 170–71
    GHG emissions from, 111–12
    Lake Mead, 162–63, 162f
    multi-purpose management, 161–63
    premodern, 63–64
    sedimentation, 169, 176
    seismicity, reservoir-induced, 173, 177
    storage–yield relationship, 163–64, 163f
resilience, 55, 225–27, 249–50
retention ponds, 290t
return flow, 71
revealed-preference methods, 132
reverse osmosis (RO), 236, 237, 243
rights of nature approach, 207, 209
riparian doctrine, 190–91
risk, 36–37, 130, 275–76, 332
river-basin organizations (RBOs), 213–14, 226
river-basin planning, 68
river basins, 24, 26–27b, 219–21
rivers. *See also* flooding and flood management
    channelization, 104–5, 104t, 129
    depletion and altered flow regimes, 86–88, 87b, 89f
    erosion flooding and avulsion, 121
    fishing, 113–16
    large-river and small-river irrigation, premodern, 59–61
    navigation and, 101–6, 106t
Romans, 64–65, 255–56
Roosevelt, Theodore, 68
roughness, 47, 47f
runoff, 22, 42, 54, 291
runoff farming, 61–62
runoff ratio, 27b, 42
rural areas
    allocation and, 195
    dams and, 172
    federal programs for, 281

rural areas (*continued*)
   flood exposure, 126
   in India, 244, 306b, 311
   rural-to-urban transfers, 83–84, 201, 203, 260
   septic tanks and drainfields, 264
   small utilities in, 269
   wells, 269

Safe Drinking Water Act (SDWA), 259, 260
salinity from mining, 330
salinization, 46, 60, 356–57
salmon, 114–16, 209
saltwater intrusion, 45–46, 45f, 86, 92
San Francisco, 286
sanitary sewer overflow (SSO), 264
sanitation. *See* wastewater and sanitation
scarcity. *See also* allocation; depletion
   assessment of, 82–85, 84f
   as conflict driver, 224
   crisis of, 11
   hard-path assumptions, 65
   indicators for, 79–82
   injustice and, 82
   management of, 94–95, 95t
   reliability and, 276–77, 276f
   virtual water trade and, 83b
scarcity solutions
   about, 235
   aquifer storage, 246–47
   desalination, 236–38, 237b
   energy intensity, 247–49, 248f
   summary, 249–51, 249t
   wastewater reuse, 239–43, 239f
   water harvesting (WH), 243–46
sea-level rise (SLR), 45–46, 45f, 122, 262
securitization, 219
seismicity, reservoir-induced, 173
self-interest, enlightened, 14
septic tanks, 264
setback levees, 137
settling basins, 64–65
sewers. *See* wastewater and sanitation
shade ball manufacturing, 325–26
shit flow diagram (SFD), 305–7, 307f
side payments strategy, 226–27
Singapore, 250b
Smith, Adam, 7
Snow, John, 256–57
soft path, 14–15, 69t
solar energy, 318, 319f, 354
source protection, 260
state coordination and planning, 210–13
stationarity, 30, 41, 210–11

steam turbines, 317
stepwells, 62
stewardship, corporate, 332–34
stocks, 21, 22t
storage of water, 63–64, 276–77, 276f
storm surges, 122
stormwater, 54, 262–64, 263f, 269, 288–92
streams, 93, 288–91
strict equality, 13
substitutability, 6
sufficiency principle, 13
supply. *See* availability
surface water depletion, 85–88
sustainability
   agriculture and, 349–50
   allocation and, 183, 191, 193
   conservationism vs. preservationism and, 13–14
   defined, 13
   hydropower and, 112–13
   Just Transition and, 4
   of meat and nut production, 349–50
   White Commission on, 69
sustainability boundary approach, 204t
sustainable development framework, 7
Sustainable Development Goals (SDGs), 7, 9, 9t, 80, 297–308, 298t, 304t, 312–13

tailings dams, 330
tailings ponds, 160
tar sands, 322
teleconnections, 6b, 33
Tennessee Valley Authority (TVA), 68, 321
terminal lakes, 85
terrestrial moisture recycling (TMR), 49, 49b
threshold toxicants, 144
total maximum daily load (TMDL), 150, 292
total renewable water resources (TRWR), 24, 79, 81, 83
tragedy of the commons, 187, 236
transaction costs, 7
transboundary conflict and cooperation. *See* conflict
transpiration, 46
transportation, 318–19, 320b, 320f
tree plantations, 52
turbidity, 48, 143t
Twain, Mark, 104

Unilever, 333
UN TWAP Groundwater, 221–22
UN Watercourses Convention, 222–23, 226
urban stream syndrome, 288
urban water

conservation, 279–80, 281–82
flooding, 121
history, 255–58
pricing, 280–85
reliability, 275–77, 276f
runoff, 54
rural-to-urban transfers, 83–84, 201, 203, 260
stormwater management, 288–92
stream restoration and green infrastructure, 288–91, 290t
urbanization, 54, 274
use, amount and types of, 277–79
wastewater reuse, 285–88, 285t, 287f
urban water depletion, 83–85
US Army Corps of Engineers (USACE), 104, 110, 128–29, 151, 213, 323
use. *See also* history of water use and management; industrial water use; instream uses
  about, 5
  blue water vs. green water, 5, 71, 74, 343t, 344
  consumptive, 71, 72, 74–76, 75f
  data, global and US, 74–76
  demand, measurement of, 6–8
  demand elasticity, 282
  nonconsumptive, 5, 262
  outdoor, CII, and NRW, 277–78
  per capita, 76
  water-use efficiency, 76
use value, 132
utilitarianism, 12
utility shortage, 81

values, 12–15, 212
vapor capture, 245–46
virtual water, 6b, 72–73, 83b, 351–52, 352f

WASH (water, sanitation, and hygiene). *See* low- and middle-income countries
wastewater and sanitation. *See also* health, human
  combined sewers, CSOs, and SSOs, 262–64, 263f
  community-led total sanitation (CLTS), 306b
  ecological sanitation, 286
  evaluation of US system, 265–71
  history of, 255–58
  JMP sanitation ladder, 303–5, 304t
  linear flows, 270–71
  open defecation (OD), 304t, 306b
  pit latrines, 302–3, 302f
  reuse, 239–43, 239f, 249t, 285–88, 285t, 287f
  Sanitary Revolution, 257, 257f, 271
  septic tanks and drainfields, 264
  urban sanitation in US, 262

urine-diverting dry toilets (UDDTs), 286, 303
WASH and, 301–7, 305f, 306b
wastewater treatment plants (WWTPs), 146–47b, 262
water
  centrality of, 4–10, 16
  as economic good, 70, 71
  ecosystems and, 9–10
  as human right, 4–5
  human spirit and, 4
  as renewable resource, 6
  as resource, 5–8, 65
WaterBack, 208–10
water banking, 202
water budgets, 94f
water crisis, 6b, 10–12, 14, 17
water–energy nexus, 9, 247, 316
water ethic, 12–15
water footprints (WFs), 72–74, 73f, 83b, 318–19, 325t, 344–47, 346f
water gap, 353
water harvesting (WH), 243–46, 249t, 355
Water Infrastructure Finance and Innovation Act, 281
Water Infrastructure Improvements for the Nation Act, 281
water management history. *See* history of water use and management
water markets, 201–2
water neutrality, 333–34
water poverty, 81–82
water productivity, 76
water protectors, 228–29
water quality. *See also* Clean Water Act; pollution
  about, 141
  deforestation and, 48
  Endangered Species Act (ESA), 155–57, 156b
  forest management and, 50–51
  groundwater depletion and, 94
  National Environmental Policy Act (NEPA), 155
  parameters, 142–44, 143t
water quality criteria (WQC), 148–54
Water Resources Council, 214
watersheds, 24, 26–27b, 46–48, 289
water towers, 24, 44, 245
water-use efficiency, 76
water yield, 27b, 51–52
wells, 66, 269
wetlands, 62–63, 152b, 290t
White, Gilbert, 68–70, 134
White Commission, 68–69
Wild and Scenic Rivers Act, 206–7
wildfires, 51

Winans rights, 196, 208–9
Winters doctrine, 195–96, 206, 207–8
withdrawal. *See also* depletion
   consumption vs., 71, 333
   for cooling systems, 317
   EFRs and, 205
   global data, 74–76, 75f, 77f, 93t
   irrigation efficiency, 353
   for manufacturing, 323, 324f
   for transportation, 318, 319f, 320f
   Wyoming v. Colorado and, 197b
withdrawal-to-availability (WTA) indicator, 80, 80t, 82–84, 84f
World Commission on Dams (WCD), 165–66

zombie registrations, 211